Anthropocene – Humanities and Social Sciences

This series is the first collection to address the subject of education within the Anthropocene. The inspiration behind the series is a dialogue between Earth system science (ESS) and human and social sciences (which, when their object of study is the socio-political implications of our entering the Anthropocene are now officially called the environmental humanities). The series offers perspectives from a variety of disciplines. The raison d'être of this new collection is to draw together high quality studies dealing with the gravity of today's environmental situation whilst also viewing the concept of the Anthropocene from a critical perspective by expanding on work engaged in in human and social sciences over recent years.

The main objectives of this series are to:

- bring together studies from Earth system science and the environmental humanities in the same collection, from an educational angle
- engage with the question of the future and how to prepare for it in a time of uncertainty

This series welcomes topics including, but not limited to:

- pedagogy
- didactics in the Anthropocene
- health education
- education for governance and democracy
- addressing the challenges of contemporary societies and ecological changes
- transforming agricultural production methods
- preparing for the future

The collection would make a valuable contribution to the academic world and research, which is currently seeking new editorial spaces for us to publish scientific works on the implications of entering into a new geological epoch.

Works published in this series include monographs, edited volumes, and handbooks and textbooks.

The series is intended for researcher working on the Anthropocene with all its socio-political implications and also students in different disciplines following courses on environmental issues. It may, however, also be of interest to all social stakeholders interested in environmental questions, who are becoming increasingly numerous including, but not limited to, policy makers, activists, entrepreneurs.

We aim to make a first decision within 1 month of submission.

In case of a positive first decision the work will be provisionally contracted: the final decision about publication will depend upon the result of the anonymous peer review of the complete manuscript. We aim to have the complete work peer-reviewed within 3 months of submission.

Proposals should include:

- A short synopsis of the work or the introduction chapter
- The proposed Table of Contents
- CV of the lead author(s)
- List of courses for possible course adoption

The series discourages the submission of manuscripts that are below 75,000 words in length

For more information, please contact the Publishing Editor: Floor.Oosting@springer.com

Christoph Antweiler

Anthropology in the Anthropocene

An Earthed Theory for Our Extended Present

Springer

Christoph Antweiler
Institut für Orient- und Asienwissenschaften
Universität Bonn
Bonn, Nordrhein-Westfalen, Germany

ISSN 2731-6343 ISSN 2731-6351 (electronic)
Anthropocene – Humanities and Social Sciences
ISBN 978-3-031-74590-4 ISBN 978-3-031-74591-1 (eBook)
https://doi.org/10.1007/978-3-031-74591-1

The text has been heavily revised and updated compared to the original German edition.

English translation of the original German edition published by WBG Academic, 2022.

The original submitted manuscript has been translated into English. The translation was done using artificial intelligence. A subsequent revision was performed by the author(s) to further refine the work and to ensure that the translation is appropriate concerning content and scientific correctness. It may, however, read stylistically different from a conventional translation.

This Springer imprint is published by the registered company Springer Nature Switzerland AG
The registered company address is: Gewerbestrasse 11, 6330 Cham, Switzerland

Preface: Amy, a Geoanthropologist in the Future Wonders

Anthropocene is the name given to the fact that humans are currently shaping the surface of the Earth so strongly that this will be recognizable in the distant geological future. No place in the world is completely untouched by humans. The influence of humans is no longer limited to local interventions in nature. Human interventions have radically changed the geosphere. Natural scientists have come to the conclusion that since the middle of the twentieth century at the latest, human activity has been shaping changes to the Earth's surface in a way that is unprecedented in the planet's history.

In the history of mankind, the exploitation of nature has reached completely new heights with the utilization of fossil fuels. For about 200 years, human influences on the Earth's surface have been so strong that they can be considered a natural force in their own right. Human activities affect every aspect of nature on the Earth's surface, they are global in scale, and they are indelible—these are the central empirical findings. This means that humans have the potential to inadvertently create everything from instabilities up to catastrophic changes in the entire system of the geosphere, this is the apprehension.

The word Anthropocene denotes this break in the recent history of the Earth, or more accurately the geosphere—more precisely the suffix "-cene," that is, the "new, unusual" (ancient Greek καινός, *kainos*), which is produced by humans (ἄνθρωπος, ánthrōpos). In the established geological chronology, we have been living for almost 12,000 years in the epoch of the Holocene—the most recent, post-glacial, and climatically stable section of the Quaternary period. The results of human activities are permanently deposited in the sediment. Concrete will be a completely normal type of rock in the geology of the future. Due to the extent of human interventions in the Earth's crust (geosphere) and their suddenness in geological terms, this new phase of history should be given the rank of a *geological* epoch, the Anthropocene—that is the main idea.

Unlike other geological periods, which last many millions of years, this epoch of the Anthropocene has so far only lasted the extremely short duration of a human life. From a geological point of view, the Anthropocene is not simply the "epoch of man" or the "human age." It is rather the age whose current *rock strata* are characterized by residues of recent human activity or, since it is still ongoing, will be in the future. The Anthropocene is thus composed of two things: on the one hand, a thing for which there are clear geological evidence, and on the other, an idea, a concept. Let us illustrate the significance of the geological findings and the conceptual earthquake they have triggered by means of a journey into the future.

Amy is a geoanthropologist logist living in the future, in the geological age of the Quintary. She specialized in paleontology during her studies and chose anthropology as an obvious minor. She has just started her current job as a cultural paleontologist. At the moment, Amy is in the process of evaluating data on sediments from recent Earth history. They date back to the late Quaternary period, just over ten million years before her time, more precisely to the 2000s, according to the Christian calendar. Amy knows that in the layers of this age, she and her colleagues from around the world will always find the index fossils that mark the "Anthropocene," the phase in which humanity became a real geofactor: plastic pieces, concrete remnants, and artificial radionuclides, like the stable lead-207 at the end of the decay series of radioactive uranium-235. All these had become more decisive in the shaping of the Earth's surface than earthquakes, volcanic eruptions, and tsunamis.

In the cultural-paleontological archives, which contain remnants from the cultures of that time, Amy and her colleagues study not only human fossils but also electronic documents and, very rarely, preserved written pieces, which bear testimony to the discussions from this distant past. And sometimes Amy can only wonder. At the time, many scientists, including geologists and paleontologists, had actually argued against such an epoch called the "Anthropocene" being formally introduced into stratigraphy and thus being equated with the previous geological eras.

Amy thinks: Okay, back then, the effects of human activity had not been around for so long. Between the years 2000 and around 2020 of the Christian calendar, there were mainly scientific controversies about how long human geo-influence had been so strong. Had this started 50 years earlier, 500 years, or 5000 years before, or even longer, perhaps it had begun with agriculture or the first cities? As scientists think in very long time frames, many of her colleagues at the time considered this to be a temporary event in the history of humankind. Some politicians even denied the findings.

Back in the Quaternary, humanity had just managed to get its act together. After long negotiations, globally accepted agreements were actually reached, which limited human impact on the environment. The Anthropocene ended around 2100 and has since been the shortest period in the geological timetable. Owing to lectures on the history of the subject, Amy knows that the first attempts to measure the geological influence of individual societies and even individual persons were made as early as 2020. This so-called biographical Anthropocene method was then introduced with the aid of a politician named Donald Trump. However, as historians specializing on the subject know, it took a long time for the method to become established as a standard procedure among Earth historians. The Anthropocene concept as well as the sciences of geoanthropology, cultural geology, and cultural paleontology had taken a long time to be accepted. But Amy also tells herself: Now, in the post-Anthropocene era, and at a considerable distance in time, it is of course easy to clearly recognize the geohistorical threshold of that time, indeed this break in human history. Now, in the post-Anthropocene, indeed in the Quintary, it is easier to see that the Anthropocene was more than just a vulgar scientific crisis topos of the early twenty-first century, according to the Christian calendar of that time.

Bonn, Germany Christoph Antweiler

About the Author and Acknowledgments

I was born at a level of 308 ppm CO_2, and I wrote the book between the levels of 414 ppm and 421 ppm. According to the currently dominant geological version, the phase of the Anthropocene roughly corresponds to my life data so far. Science is done by humans, and that implies imprints and biases. Therefore I hereby disclose some personal background information. I feel a direct connection to the topic mainly due to the fact that I am an anthropologist and have a strong quasi-English penchant for natural history. Since my childhood in the Lower Rhine and in Franconia, I have been interested in nature and landscapes. As a ten-year-old, I was very interested in art. I was particularly fascinated by the detailed paintings of Albrecht Dürer, especially the landscape and nature paintings as well as the large portraits. I would spend hours "searching" through pictures such as "The Great Piece of Turf" from 1503, which shows a meadow from a frog's perspective, so that it looks like a whole forest. I was also captivated by maps, atlases, and space images of the Earth from an early age. At flea markets, I still look out for old wall maps from schools and globes. In his book on the Rhine, Balmes reflects my attitudes to nature well:

> Ever since I went swimming in the Rhine and Moselle as a child, I have thought again and again, in this blue floating state, about the mud, the silt, and the minerals dissolved in the river. If you spend your whole life by the Rhine, I wondered, will you then have soaked up the whole river from the mountains through the water? As a riverside dweller, do you carry the geology of the river in your own body? Did the minerals become the building material of your own body? (Balmes 2021: 263)

In my parents' house, there were thousands of books across all disciplines. My mother was an art librarian and my father was a pharmacologist and toxicologist with an interest in the humanities. He was involved in research on lung diseases and was involved in controversies about smog and SO_2. The human-environment issue fascinated me, especially since the findings of the *Club of Rome* (Meadows et al. 1972), but above all through the book *The Ecological Context* by John McHale, which I read shortly before my high school graduation (McHale 1970). McHale placed humans in the context of the biosphere—and he did so as a sociologist. The book was also fascinating because of its many figures, tables, and graphs, even though the volume was published by Suhrkamp, then rather known for theory-heavy books with endless text deserts.

My interest in the intertwining of natural history and cultural history led me to initially complete a diploma program in geology and paleontology, while also attending cultural anthropology seminars on the side; I later studied cultural anthropology. During my geology studies from 1975 onwards, I dealt a lot with evolution within the context of paleontology and in my minor subject of biology. In my cultural anthropology studies, this eventually developed into an interest in long-term cultural change (cultural evolution, social evolution). Here I was mainly interested in approaches that criticized the teleological stage models and progress theories of classical evolutionism as—despite their name—non-Darwinian (pioneering: Greenwood 1984). More recent models based on generalized Darwinian principles focus more on mechanisms of microevolution rather than directed processes of macroevolution and are known as "truly evolutionary models" (*Darwinian models*). I received my doctorate in Cologne, with an anthropological thesis on the criticism of social evolutionism and on the mechanisms of transgenerational cultural change (Antweiler 1988).

My regional interest is particularly in people and societies in Asia. After having gone on many trips to Asia during my studies, I now have a special fondness for tropical living environments. During my cultural anthropology studies, I mainly focused on South Asia, especially India and Nepal, and later on Southeast Asia. The switch to Southeast Asia was ultimately characterized by romanticizing ideas. I still have images of misty, steaming rice fields in my head, even though I have seen often enough that a concrete house typically stands in the middle of a rice field in Southeast Asia today. As my wife, who is a mathematician, has been working on the culture of the sari for years, I have not only been to Southeast Asia almost every year for the past ten years, but also to India for about a month each year. In Southeast Asian studies, where I work today, my department at the IOA in Bonn deals with issues of political ecology in Southeast Asia. Hence my interest in the question of what significance the Anthropocene has in southern and eastern Asia and how it is experienced and discussed there.

I would like to thank people from various contexts and disciplines for their helpful comments and reactions to drafts of this work. Some gave me detailed tips on content, others provided small but important hints on colleagues and literature, and there were those who asserted that "Anthropocene" is just a cheap buzzword. First of all, I would like to thank my wife Maria and my sons Dario and Craig for the many debates we have had about nature and culture at home and on many trips, and also for our constant mutual "fact checks." My thanks go to the anthropologists Dorle Dacklé, Daniel Grana-Behrens, Chris Hann, Michaela Haug, Olaf Kaltmeier, Karl-Heinz Kohl, Anna Kalina Krämer, Werner Krauß, Karoline Noack, Jan Oberg, Anette Rein, Thomas Reuter, Nikolaus Schareika, Judith Schlehe, Michael Schönhuth, Peter Schröder, Guido Sprenger, and Svenja Völkel. I would like to thank the anthropologist and media scientist Erhard Schüttpelz for the exchange. In geology, my thanks go to my friends and fellow geology students, Ulrich Milatz and Günter Wahlefeld.

As for the historical sciences, I received advice from the environmental historians Bernd Herrmann, Christof Mauch, and Fabienne Will as well as the historian of science Jürgen Renn and the world historian Patrick Manning. I would like to thank the archaeologists Daniel Gronenborn, Tobias Kienlin, Leonie Koch, Tilman Lenssen-Erz, Christian Mader, Christopher Pare, and Andreas Zimmermann. My thanks also go to the chemist and philosopher Jens Soentgen and the psychiatrist Peter Mehne. Advice from the primatologists Andrea Höing and Volker Sommer was also helpful. I was likewise supported by the sociologist of knowledge Anna-Katharina Hornidge and the sociologist Andreas Folkers. Furthermore, I received tips from the philosophers Matthias Herrgen and Christian Thies. As for literary studies, my thanks go to Hans Bergthaller, Eva Horn, Rudy Simek, and Herbert Uerlings. Among the Southeast Asian scholars in my department in Bonn, I would like to thank Berthold Damshäuser, Jessica Riffel, Kristina Großmann, and Frank Seemann. Many thanks to Kathryn Abaño for significantly improving the translation. I have received a lot of very different and also contradictory suggestions, which is no wonder given this young and multifaceted topic. Therefore, it is all the more important to say here, as always: All rights to the certainly numerous gaps and other deficiencies remain solely with the author.

Contents

List of Figures

List of Tables

Chapter 1
Humans Make Earth History—New Earth and New Anthropology

Abstract This chapter deals with the existential significance of the Anthropocene for the future of humanity. We have only been living in the Anthropocene for the length of a human life. If *geologists*, with their extremely long-term perspective, are already proclaiming a new phase of the Earth's history, then we should be alarmed. A meaningful scientific approach that also leads to realistic solutions combines natural sciences and a whole range of human sciences. Of central importance are a significantly time-extended cultural anthropology and an anthropologically informed geology. In short: We need geoanthropology!

> *Are we good ancestors?*
> Jonas Salk (1992: 16)

> *We live in a world where the fate of every human being is linked to that of many others and also to that of the Earth, which we think we own, but which we only inhabit.*
> Jedediah Purdy (2019: 7)

> *We must not focus our attention*
> *solely on the climate.*
> Johan Rockström (2021: 102)

The word "Anthropocene" is a meme that has reached many scientific fields as well as the arts and the mass media, but much less so the general public. The word is filled with very different content, which often has little to do with the conceptual geological core of the Anthropocene concept.

There have already been many crises in this world, including global ones. However, the current generation of children in this world will be the first global cohort to become aware that the world they will inherit will be significantly less habitable for human and non-human life than that of their parents. Let me give you an economic example. Well-known economic players in relation to man-made climate change are oil companies, energy companies and states. But the insurance industry is also heavily affected: Insurers need climate data to be able to assess the risks they want to insure or not. This also applies to insurers of insurers. It is therefore no wonder that *Munich Re*, the world's largest reinsurance company, has the

C. Antweiler, *Anthropology in the Anthropocene*, Anthropocene – Humanities and Social Sciences, https://doi.org/10.1007/978-3-031-74591-1_1

latest data and global maps of natural risks compiled. But this also applies to local environments. In Florida, for example, there used to be a hurricane or storm surge every 12 years. However, as this now happens almost every year, nobody can or wants to insure against it any more. There will soon be entire areas where residents will no longer be able to insure their homes against damage caused by hurricanes and flooding, even if they want to. Accordingly, investors will also look for places where they can deploy their capital with less risk (Bhattacharyya and Santos 2023).

Until the modern era, natural disasters and resource scarcity were considered problems whose causes lay outside society. With the Anthropocene, we know that elementary threats can no longer be attributed exclusively to external causes. They can therefore no longer be viewed in purely scientific terms and addressed as problems to be solved technologically (Niewöhner 2013: 43–45). The signal words 'worldwide' and 'global' invite blurred use, which is particularly problematic in the Anthropocene debate. In view of spatially distributed environmental effects in the Anthropocene, we should distinguish between two types of global anthropogenic change (Smil 2019: 13–14, 207): (a) impacts of anthropogenic phenomena distributed over the entire surface of the planet—such as climate gases, radio-nuclides and increasingly plastic particles—and (b) ubiquitous human-induced environmental change, i.e. environmental degradation, which can be found wherever humans live on a massive and permanent scale—such as photochemical smog, soil erosion and dead zones in coastal waters).

The term "Anthropocene" refers to something that is much more comprehensive in space and time than climate change. The Anthropocene refers to the geohistorical phase in which humanity has become a strong or even dominant geological factor in changing the geosphere of our planet. This involves much more than just the global warming of the atmosphere. Everything that starts with "Anthropo-" sounds enticing at first, except perhaps when followed by "-phagy." The word "Anthropocene" is seductive and invites media to exaggerations and scientists to engage in academic trend surfing. The word is often used to simply express concerns about the state of the Earth or to criticize the consequences of technological development (Malhi 2017: 93, Uekötter 2021). I believe that the Anthropocene as a problem is too important, and as a word too rich in content, for it to be used as a shorthand for talking about globalization, global environmental problems, sustainability issues, or human-made climate change, or to be misused for purely rhetorical maneuvers.

1.1 The "Age of Man"—Welcome to the Anthropocene?

One species transforms the planet.
Andrew Knoll (2021: 195)

We as a species have proven to be good historians,
but bad futurologists.
Robert Macfarlane (2016: 97)

The Anthropocene is also known as the "Age of Man," the "Age of Mankind," or the "Human Age" (Monastersky 2015). Why? The Anthropocene is not only characterized by an abrupt climate change in geological terms, but also by a dramatically high consumption of natural resources, water, and fertilizers, an acidification of the seas, a drastic decline in the diversity of living things and a sealing of large areas by asphalt (Fig. 1.1). In addition, there is soil loss due to erosion and sealing, the loss of large parts of peatlands, the rapid transformation of deltas, and the sharp increase in the amount of human objects, especially plastic waste, in the oceans. The most important feature of our planet is the existence of life. All living beings survive by extracting resources from the environment. But only humans continue to extract from the environment after their needs have been met. Fundamental assessments of environmental dynamics should therefore pay attention to human-induced changes in the biomass and proportions of life on Earth (Smil 2021: 236, Headrick 2021: 3).

The human footprint is omnipresent, and it will be at least partly geologically permanent due to the sometimes extremely long residence times of plastics, for example (Table 1.1). Pieces of asphalt, plastic particles, and radioactive substances will become permanent components of rock strata: concrete and asphalt will become the rocks of the future. The diversity of human-made objects already exceeds today's biodiversity. The human technosphere is growing in breadth, but also vertically. In Pudong, the dynamic district of Shanghai, more skyscrapers over 100 meters tall were erected in 2011 than in the rest of the world combined. Humans

Fig. 1.1 Land sealing on Tempelhofer Feld in Berlin. (Photo: author)

Table 1.1 Diversity of indicators of the new Earth dynamics in the Anthropocene

Indicator	Current state, change in comparison period	Source
Humans: Number	8,2 billion (1960s: 3 billion, 1900: around 1.5 billion, 1 AD: 200–400 million, 10,000 BC: 1–10 million)	Smil (2019: 307–331, 2021: 25, 69)
Humans: Carbon in biomass	25 Mt. humans/129 Mt. domesticated mammals/5 Mt. wild land mammals (in 2000), compared to 10 Mt. humans/10 Mt. wild land mammals/35 Mt. domesticated mammals (in 1900)	Christian (2018: 311)
People: Increase	Approx. 80 million/year = approx. population of Germany	Smil (2021: 25–69)
People: Mass (*anthropomass*)	0.3 Gt, together with domesticated animals, 97% of the total mass of terrestrial mammals	Smil (2013, 2019), Bar-On et al. (2018)
Diversity of terrestrial large mammals	1/3 humans, 2/3 domestic land mammals, around 3% remaining land mammals (vs. 350 pre-human species)	Barnosky (2008)
Livestock	>2 billion cattle and water buffalo, 2–4 times the live weight of humans, >1 billion pigs	Thomas et al. (2020: 83), Marks (2020: 231)
Domestic chickens (Gallus gallus)	23.7 billion, biomass 2.5 × wild birds, 1/10 live weight of humanity, since the mid-twentieth century anthropogenic morphospecies (Gallus Gallus domesticus)	Reichholf (2011), Bennett et al. (2018), Thomas et al. (2020: 98–99)
Energy use	23–75 gigajoules per head/year, Holocene: 3–10, 90% consumed since 1950	Christian (2018: 349) Syvitski et al. (2020)
Energy consumption	572 EJ/year (2014), 13.7 million t oil/year (2014), 6.1 million t oil/year (1973), <100 EJ/year (1850)	Morris (2015), Smil (2021)
Carbon dioxide (CO_2)	CO_2 levels around 400 ppm, higher than ever in the last 800,000 years or at any time in the quaternary	National Research Council (1986), Summerhayes (2020)
Nitrogen fertilizer	Emission of reactive nitrogen higher than from natural sources, 70% from USA, India, China	Marks (2020: 227)
Sea level rise	20 cm rise in the last 100 years, 3 mm/year since 2000	Summerhayes (2020)
Technosphere: human-made objects (*anthropogenic mass*)	30 trillion tons (tT), = approx. 4000 tons/person buildings and infrastructure 1100 gigatons = 50 kg/m ^{2}, 5 times the anthropomass	Zalasiewicz et al. (2017c: 19)
Technosphere: Plastic	8.3 gigatons = more than the total mass of all land and aquatic animals (4 Gt)	Elhacham et al. (2020: 2–3), Zalasiewicz et al. (2018a, b)

(continued)

Table 1.1 (continued)

Indicator	Current state, change in comparison period	Source
Technosphere: potential technofossils	Technofossil diversity > biodiversity	Elhacham et al. (2020)
Transported rocks, sediments due to resource exploitation, buildings	Since the beginning of the Anthropocene, i.e., since 2015, 30 times as much as in the 70 years before, equivalent to 3–7 times the sediment transported by rivers into the sea, 21 T/person	Hooke (2000: 844–845), Bridge (2009), Price et al. (2011); Cooper et al. (2018)
Dams	45,000 (2007), ½ of all rivers with water structures	Duflo and Pande (2007)
Artificial minerals, synthetic crystalline components	208 recognised new minerals, 193,000 (vs. 5000 natural) mineral-like components, more than in the entire history of the earth	Hazen et al. (2017), according to Thomas et al. (2020: 2, 56)
Human landscape shaping	95% of the land area (excluding Antarctica), 5% in remote areas, 1700: 5%	Ellis and Ramankutty (2008)
Anthrome	40% of the terrestrial land area used, especially agriculture and pasture land	Ellis and Ramankutty (2008), Ellis (2018), LALE (2020) Ceballos et al. (2020)
Agricultural land	20% of the planet's land mass used	Marks (2020: 231)
Deforestation	Half of all deforestation in human history between 1945 and 1995	Williams (2006: 395–496)
Biodiversity, species transport, neobiota	Extinction rate 10,000% of the normal rate, <1% of species, transport fast and over long distances	IPBES 2019, Almond et al. 2020
Extinction of land mammals	Background: 1.8 E/MSY (extinct species per million mammal species/year), from 1500: 14.0 E/MSY, from 1900: 28.0 E/MSY	IUCN, Wignall (2019: 19–20), Ceballos et al. (2020), Hannah (2021: 22–26)
Individuals per species	60% reduction of mammals, birds, fish, and reptiles since 1970, smaller number of individuals due to reduced habitats	Thomas et al. (2020: 173)
Diversity: Domesticates	¾ of food sources from 12 plant and 5 animal species	Thomas et al. (2020: 83)
Geographical shift of biota	Northward shift of plankton distribution, 200 km/decade for the past 50 years	Thomas et al. (2020: 83)
Primary production consumption / destruction	30% of above-ground plant net production, equivalent to 373 EJ/year.	Zalasiewicz et al. (2020a, b)

have not only drilled thousands of meters into the ground, but have also recently erected buildings over 1 kilometer high (Graham 2018: 371).

In this current phase of our lifetime, the mass of things produced by humans (*anthropogenic mass*), which doubles approximately every 20 years, has begun to exceed the mass of living matter (*biomass*) (Elhacham et al. 2020: 1). Since the beginning of the Anthropocene, strictly speaking in the nearly 70-year period from 1950 to 2015, humans have transported about 30 times as much rock or sediment as in the 70 years before. By the year 2000, a total of 21 tons of rock and soil per Earth inhabitant had been moved. This corresponds to seven times the amount of natural sediment transported by rivers into the sea and is two orders of magnitude higher than the amount of magma ejected by volcanoes worldwide during this time (Cooper et al. 2018; Maslin 2021: 48). The technosphere weighs five orders of magnitude more than the mass of humans (anthropomass).

Since 1950, a total of 99 percent of all cement (Waters and Zalasiewicz 2018) and 99 percent of all synthetic plastic in the world have been produced. On average, each person in the world today produces more than their body weight in anthropogenic things, objects, animals, etc., each week. The entire technosphere of the Earth currently has a mass of 30 trillion tons. This can only be conceivable if one realizes that, when evenly distributed, this corresponds to a load of 50 kilograms on *every* square *meter* of the Earth's surface (Zalasiewicz et al. 2017a, b, c: 12, 19). This is five *orders of magnitude* more than the biomass of humanity.

The Anthropocene can be seen as a dangerous phase in which life based on fossil fuels disrupts the geosphere, on which human life depends, in unpredictable ways. Given the spatially global effects and, above all, the permanence, massive environmental damage has been interpreted as a form of *distributed violence*. The exploitative treatment of nature is an integral part of today's economy. It is a violence that is currently being generated but is unfolding with some delay: a slow violence, as elucidated by the literary critic Rob Nixon with the outsourced damage to the environment, just as the costs of the imperial lifestyle had been outsourced in the form of human physical suffering in societies of the Global North ("slow violence", Nixon 2011, 2016). Seen in geological temporality, the change is sudden, but from a human perspective, it appears incremental and gradual, like the wear and tear of a bicycle tire. Thus, for the perpetrators, the crisis initially remains almost invisible. Similar to structural violence, it is either not seen as violence or it is deliberately ignored. This violence is then unevenly distributed over time to the people of future generations, and even to people in the distant future.

1.1.1 Deep Past and Deep Futures

> *We now live more than ever in a human-created, but nonetheless unintentional, "anthroposphere" ... That is a "Great Departure" from past historical patterns.*
> Robert Marks (2020: 241)

In view of the Anthropocene, the 'communitarian credibility' that progress, limitless consumption and supposedly unlimited growth have long brought has increasingly been lost (Niewöhner 2013: 44). The topic of the Anthropocene concerns human existence—physically as well as psychologically. It concerns the future of humanity. The number of people will increase significantly, and according to the "global footprint," humanity is already consuming more than 1.5 times as many resources as can be replaced by natural processes. Technology has developed faster than societies, and it now surrounds each and every one of us. As Marc Augé said, we are exhausting ourselves in the consumption of devices. Many people today have the feeling that they are being sucked into a future instead of being determined by the past, a future about which little societal thought has been given, but which is particularly frightening in the Anthropocene and often leads to not even wanting to imagine them::

> The great paradox of our era is: We no longer dare to imagine the future, although the progress of science gives us access to the infinitely large as well as the infinitely small (Augé 2017: 17).

Possible futures can be envisioned by means of changes in the atmosphere. Earth scientists and historians see two simplified future scenarios: collapse or sustainability (Costanza et al. 2007), or rather "Hothouse Earth" vs. "Stabilized Earth" (Steffen et al. 2018: 8252, 8258). We could be heading towards an overheated atmosphere if the countermeasures remain so small that certain thresholds are soon exceeded. The atmosphere and the entire geosphere would then enter an irreversible path out of Holocene stability. Extreme pessimists are no longer giving current institutions and measures a chance, and they have either given up because it is simply too late, or they are demanding, for example, the abolition of capitalism, an eco-authoritarian policies imposed by strong nations, a centralised global political power or an additional 'fourth estate' (*future branch*) that safeguards the interests of future generations..

However, we could also cooperate so intensively and reduce inequality that we would be able to manage the Earth system through fundamental societal changes, to such an extent that the climate system, while experiencing permanent anthropogenic changes in structure and function, could still be stabilized overall. We would live in a "moderated Anthropocene" instead of the uncontrolled Anthropocene in which we now live (Thomas et al. 2020: 171, 188). In this scenario, the Earth would indeed become warmer, sea levels would rise, the diversity of life would decrease, *and* significant protective measures would be required, but the world would at least remain habitable. At first glance, this may seem naive. Or it could be an attitude that relies too much on the truly ingenious abilities of people and cultures to adapt to new circumstances.

From a macro-historical perspective, however, this optimistic scenario appears to be quite realistic. The actual global extent of anthropogenic effects is recent (about 70 years), as is the global orientation towards growth as a cultural value (about 200 years). Even today, not all people and communities live a consumptive and resource-intensive lifestyle. However, the majority of these people live in the Global South and are unintentionally resource-poor (Brand and Wissen 2017). Nevertheless, a less imperial lifestyle does exist, albeit to a lesser extent, as a freely chosen way of life. The well-founded expectation that global population growth will decline in the long term (Wilson 2016: 210) offers hope, especially for the preservation of biodiversity, even though population increase has not historically been the main driver of the Anthropocene in a comprehensive sense (Malm 2016: 255–257, 268). Furthermore, the initial signs of unintended *degrowth* in the Global North will become apparent when economic developments are partially flattened to sigmoidal curves instead of the dizzying exponential increases in human-made effects.

However, a less imperial way of life does exist, albeit to a lesser extent, as a freely chosen form of existence. History and cultural anthropology have documented many cultures practicing a way of life that is *more* adapted to the actual local environmental conditions. However, they should not be romanticized as "primitive peoples" or idealized as "eco-saints," as is often the case in popular literature. From an anthropological point of view, it is necessary to distinguish between statements asserting that indigenous communities had and partially still have have a resource-efficient way of life (which is true *on average*) and the widespread statement among some environmentalists and in popular cultural anthropology that they do so out of an ecological or environmental ethic (which is generally *not* true. People in prehistory also intervened courageously in the environment for their own benefit according to their technical possibilities. A conscious ecosensitivity can hardly be detected. Neanderthals were already consciously setting fires and hunting elephants 125,000 years ago (Gronenborn 2024: 50–1; cf. Krech 1999, Milton 1996: 31, 109–114). What many anthropologically documented communities do or did have, however, is at best a guiding cosmological and cosmogonic concept of *embedding* their way of life in the local environment and a concrete knowledge of adaptation options to environmental threats (Ehrlich and Ehrlich 2022).

From a long-term historical perspective, modern civilization represents an anomaly in this respect, a "new anormal" (Ehrlich and Ehrlich 2022). This is demonstrated by studies in global history and economic history. Not only is the idea of growth young, but the concept of "the Economy" as an economic totality, according to whose standard governments are evaluated, only emerged in the 1930s. The idea of unlimited growth, which implies that one could push boundaries at will or somehow circumvent the laws of entropy—the idea of an endless cornucopia, so to speak—only emerged in the 1970s and the 1980s within the context of financial capitalism. But one can learn from mistakes. Whole countries such as Bhutan, Cuba, and Costa Rica are now trying to establish a society that would only consume as much as the Earth can provide.

Science and technology play a dual role: On the one hand, their innovations were the accelerants that allowed colonialism and capitalism to bring about the destruction of the geosphere, but they were also the creators of Cassandra who warned us early on (Renn 2021: 3). I would like to emphasize that we now have a lot of fact-based knowledge about the Anthropocene. There is now a robust body of knowledge on the subject (Horn 2020). In addition, people and cultures actually have an enormous capacity to adapt or even reinvent themselves. Based on current knowledge, the most effective measures for the reduction of CO_2 emissions as a core problem of the Anthropocene are probably the de-carbonization of the construction industry, since concrete, sand, and cement account for more than 30 percent of CO_2 emissions, and a switch to plant-based nutrition. This implies societal transformations rather than just individual behavioral changes, and it is also the societal transformations that could (probably) efficiently mitigate not only climate change, but Anthropocene effects as a whole because they generally reduce the flows of material and energy that are damaging to the geosphere (Barnosky et al. 2012; Thomas et al. 2020: 177–180). However, these are all orientations that have so far hardly shaped education and politics worldwide, neither in the Global South nor in the Global North. This shortcoming is manifested in schools as well as in museums (Isager et al. 2021).

In the Anthropocene, humans appear to be an Earth system-relevant factor for environmental change. Humans are a central driver (i.e. a global agent, a force), as are solar radiation, volcanism, and natural selection (Crutzen and Stoermer 2000; Steffen et al. 2011a, b). Human activity changes geophysical trends and global geochemical material cycles For example, the use of fertilisers alters the phosphorus and nitrogen cycles. This also applies to flora and fauna. From a paleobiological perspective, this means that “human activity massively changes the composition of communities and thus, in the long term, even the fossil record of the future” (Schwägerl and Leinfelder 2014: 235). This can hardly be overstated: the planet is at a crossroads, because altered exchange processes and ecosystems simultaneously entail the irreversible transformation of the basic living conditions for all living beings on the planet, including humans:

> The significance of the Anthropocene lies not in its discovery of the first traces of our species, but in the magnitude, significance and future longevity of the planetary system's transformation (Thomas et al. 2020: 19).

The Anthropocene is rarely seen in a positive light, but rather as an empirically substantiated warning about the global consequences of a market-focused economy and the unchecked growth of desires and consumption. Plastic particles can now be found in the depths of oceanic deep-sea trenches. The global mass of produced plastic alone is already greater than the total weight of all land and water animals (Zalasiewicz et al. 2015; Elhacham et al. 2020). From an evolutionary perspective, the astonishing thing about the Anthropocene is that a single species has managed to change both the biosphere and the geosphere extremely quickly. As “productivores” who produce our own food, we are more successful than all carnivores and omnivores combined. The Anthropocene appears to be a catastrophic consequence of the outsize success of *Homo sapiens*: we are overachievers in the planetary context (Oeser 1987: 52, Eriksen 2016: 17–18). This applies even physically: together with the animals they have domesticated, humans make up 97 percent of the total mass of terrestrial mammals (Fig. 1.2). During the writing of this book, the sheer mass of human-made materials had begun to surpass the total mass of living beings on this planet.

1.1.2 Earth System Science

These findings do not primarily come from field-based geology, but from biology, ecology, and especially from Earth System Science. The interdisciplinary field of *Earth System Science* (*ESS*) emerged between 1974 and 1980 (Bretherton 1985;

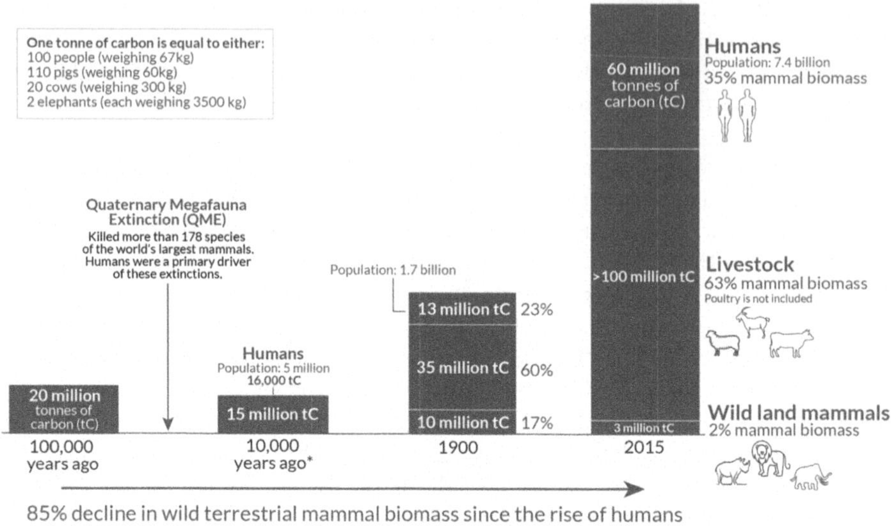

Fig. 1.2 Changing Distribution of the World's Land Mammals. (https://en.wikipedia.org/wiki/Extinction#/media/File:Decline-of-the-worlds-wild-mammals.png; CC_BY)

Heymann and Dalmedico 2019). The Earth system is understood here as a self-regulating system of the geosphere that is formed by the interaction of non-human physical, chemical, and biological forces, as well as human forces, including their interactions and repercussions within the spheres on planet Earth (Steffen et al. 2004: 298). The system extends from the upper limit of the atmosphere to the uppermost layer of the rock strata. The *ESS* mainly works with large amounts of data and global datasets and promotes large-scale research (*big science*). Earth system scientists use the data to develop global system models in order to establish current trends and simulate future developments (Cook et al. 2015: 2–3, 9–10; Roscoe 2016; Hirsbrunner 2021).

Geologists, on the other hand, work with a hammer and a magnifying glass and deal with past states and processes from a very long time ago. They do this not with experiments, and hardly ever with simulation, but by means of rock layers, local fossil findings, reconstruction, and deduction. Geology is a multi-layered science in many respects (Zalasiewicz 2017), which is also a rich sociological topic (Clark and Szerszynski 2021). Geology is a science that—like cultural anthropology and ecology—works primarily with field-based rather than laboratory-based evidence (Will 2021: 20–21, 143–201). It is important to note the methodological differences between Earth System Science and geology as they are consequential and are often overlooked by cultural scientists. For this reason, I will address the geoscientific time concepts and methods of stratigraphy that are relevant to the Anthropocene in further chapters.

In any case, it is true that in the Anthropocene, humans are the ones writing natural history. Humans leave behind geological signatures in the sediments. In stratigraphic terms, the traces left by humans are comparable to the extraterrestrial mineral iridium, which can be found in rocks all over the world as a result of an asteroid impact in Mexico 66 million years ago (Glaubrecht 2019: 693). The proclamation of a geological epoch, whose godfather was the recently deceased meteorologist and atmospheric chemist Paul Josef Crutzen (1933–2021), reflects this profound transformation, this historical turning point or even break, and is conceptually clearer than the term globalization or global change. We are living in a period of strong human growth, increasing globalization, and, above all, increasing connectedness (Krogh 2020). Both the economy and the environment are experiencing an "overheating" and a rapidly intensifying increase in the exploration and extraction of natural resources (Eriksen 2014: 140, 2016). Unlike previous waves of globalization, the phenomenon of the Anthropocene actually affects the entire surface of the Earth.

Consequently, we need a scientific approach that would link Earth history, life history, and human history with each other. Instead of just talking about powerful epochs, systems, or cultures—such as early empires, colonialism, and capitalism—we need a term that would emphasize the relationship to the species *Homo sapiens*. It is about "… not only about the Earth, but also about humans, their society, and culture" (Folkers 2020: 591). This is also probably one reason for the enthusiasm with which the Anthropocene was initially received in the social and cultural sciences (see Sect. 2.1). Within the humanities and cultural sciences, this insight was most succinctly formulated by Dipesh Chakrabarty, an Indian historian with a postcolonial orientation and a global sphere of influence. In his facetiously titled work "Climate of History," an essay on anthropogenic climate change, he says that humans have gone from being a biological *agent* to a geological *power*, which requires an integrated "geohistory". The genuinely modern distinction between human history and the history of the earth would thus become obsolete (Chakrabarty 2009). Chakrabarty has continuously developed this idea further, which has also led to fundamental discussions about the consequences of the Anthropocene in the humanities (Emmett and Lekan 2016; McAfee 2016; Dube et al. 2020; Domańska 2020; Chakrabarty 2018, 2020a; Will 2021: 202–229).

The term "Anthropocene," which is new in the history of science, initially concerned the geosciences, where the expected long-term lithological traces are referred to as a new "era" or "epoch." However, it is also relevant for the historical sciences, such as when global historians speak of the "age of the Anthropocene," which has (so far) occurred within the span of a single human lifetime (Brooke 2014: 529–530, McNeill 2015, 2016, McNeill and Engelke 2016, Headrick 2021: 353). The Anthropocene illustrates an older insight that one cannot "save the environment" because "the environment" as such does not exist (Lewontin 2000). However one evaluates the idea, Anthropocene thinking has demonstrated to natural scientists that humans are not just external and "disruptive natural processes"; it has also shown humanities scholars, social scientists, and historians that they can play an important role in addressing the current environmental crisis (Morrison 2015: 76, Magni 2019).

> The overlap of the two forms of study, fusing observations of processes active today and deductions from ancient strata, is a good deal of what gives the Anthropocene concept its veracity and power (Thomas et al. 2020: 20, see Thomas 2022).

With Amy, we saw the current world situation through the eyes of the future geo-anthropologist. The thought experiment highlighted many of the burning questions around the Anthropocene. When did the geological phase of human dominance begin? What does it mean for our image of humanity and our image of nature when we "geologize" human history? With the popularization of the idea of the Anthropocene through the natural sciences, we have come to speak of the "Age of Man" and the "Human Age." But why is a human-determined phase being introduced into the geological stratigraphy—and thus into natural history—instead of establishing it merely in cultural history?

Is the Anthropocene perhaps just a young and particularly profound, globalized phase of cultural history? We document our material culture worldwide in topographic maps. Will we need geological maps in the future for the documentation of human influence? We can keep track of global environmental change quantitatively, and we can in fact do so across the Earth, for example, through geodesy and remote sensing from space. Our civilization is constantly in the process of historically documenting its own epoch, and so it will also leave behind *historical* archives of its effects on the geosphere (Finney 2014; Visconti 2014: 384). However, why do we need geological archives and the introduction of a *geological* epoch? Is the Gregorian calendar not enough?

Is the Anthropocene a form of "Eco-Enlightenment" or "Eco-Modernism"? That is, a form of Eurocentric environmental history that seeks to distinguish modern society or its own western subject from premodern societies, just with different words? From the perspective of the humanities and cultural sciences, the idea of the Anthropocene indicates the "return of man at the moment of his supposedly final farewell" (Bajohr 2019: 64). This farewell seemed to be sealed, at the latest, with Foucault's dictum on the end of humanity (Foucault 1974: 462). The idea of the Anthropocene now emphasizes the planetary significance of humans at a time when scientific representatives of posthumanist currents want to abolish humanity (Orr et al. 2015: 162). Foucault's dictum now proves to be premature:

> Man does not simply disappear like a face in the sand, but remains eerily present as flotsam washed up on the beach. In this sense, the question is not only 'what is the Anthropocene?' but 'what will it have been?' (Folkers 2020: 592)

How should we now rethink the relationship between humans and technology, between nature and culture? Genuinely anthropological and philosophical questions arise in an intensified form in view of the Anthropocene (Noller 2023: 8, cf. Morcillo 2022). It is these provocative questions that the humanities and cultural sciences must respond to if they aspire to be a human science, an anthropology. The inclusion of cultural anthropology (also known as social anthropology, sociocultural anthropology, and, increasingly, simply *anthropology* in Anglo-American scholarship) and other cultural sciences is far more than their mere incorporation into an already broad range of Earth sciences (Krauß 2015a: 61). It is also more than the simple addition of a local perspective to the global view. A side effect of this reciprocal opening up could be the insight that a number of research approaches in the humanities, cultural studies, and, above all, history are not only somehow similar to the questions and methods in some natural sciences, but are closely related to them. In Sect. 6.6, I will be discussing the relationship of geology as natural history with the approaches and methods employed in the historical sciences, which is particularly relevant to the topic of the Anthropocene.

1.1.3 Anthropologization of Geology

Geology is traditionally a natural science discipline that operates on time scales beyond the human imagination, and so far it has hardly had to take social conditions into account. This is at least true for classical geology, but not for applied environmental geology (e.g. Hilberg 2022). Taken together with the Anthropocene, however, geology is fundamentally discovering humans, which is interesting in terms of the sociology of science (Görg 2016: 9). The following example may demonstrate what I see as the productive uncertainty that the idea of the Anthropocene entails for field geology. In the legend of geological maps, one often finds the unit "artificial deposits". These are, for example, soils and debris that have been altered by humans, or what used to be layers of garbage, now covered and sealed in by concrete. From a geological perspective, however, not all artificial deposits are considered "geological." The buildings standing on a mapped surface do not appear on these maps even though they are made of "geological" materials, such as rocks, limestone, and sand (Thomas et al. 2020: 58 Since the production of maps notoriously combines factual information with interpretations, such questions arise in an intensified form for geography and geology in the Anthropocene.

These human traces can only be found (partially) on topographical maps or road maps. From the perspective of the Anthropocene, all larger objects and materials produced by humans should appear on geological maps, as they could form a consolidated Anthropocene layer in the future. Concrete or asphalt surfaces should not only appear on cultural or anthropogeographical maps, but also on physical-geographical maps... and they should even be represented on geological ones. Similar uncertainties give rise to questions about temporality, such as the relationships between our life cycles and the larger cycles of the terrain around us (Bjornerud 2018; Irvine 2020: 10).

Geology is not the sole multi-layered science, nor is cultural anthropology the only science that is oriented towards diversity and is internally disparate. No, the other sciences are as well. The journey into the future, to the time of geoanthropologist Amy, should also make my position clear that a productive use of the Anthropocene concept can only be achieved if humanities scholars and cultural scientists open up to the particular way of thinking that geologists have; on the other hand, geologists and geosystem scientists must also understand the particular approaches used in the cultural sciences, such as in cultural anthropology, as well as those employed by historians. In addition, there are clear differences within the geosciences between the laboratory-based approach of Earth system scientists on the one hand, and the field-based approach of geologists on the other. Hence, this book by a cultural anthropologist who is equally fascinated by geology and Earth history is intended to contribute to this.

One example that is central to the topic of the Anthropocene, and where mutual understanding is crucial, is related to periodization. Geologists construct their stratigraphies (at least ideally) based on globally *synchronous* time periods. The geological ages of the Devonian or the Holocene, for example, have been set with the use of certain time markers as simultaneously beginning and ending *worldwide*. In contrast, prehistorians and early historians, particularly archaeologists, historians as well as cultural and social scientists, speak of "ages," just like geologists, but the boundaries between these ages are set at different times (*diachronic*). The Stone Age, human history, or the Modern Age, for example, began as periods in different regions of the Earth at very different times.

1.2 *In a Nutshell*—Questions and Argumentation of the Book

The Anthropocene is real.
Julia Thomas et al. (2020: 196)

Since the year 2000, geoscientists have been discussing whether we have been living in a new geological era, one that started in the mid-twentieth century: the Anthropocene, the "Age of Humans." On the one hand, this book deals with the Anthropocene as a phenomenon, and on the other hand, as the first major and global discourse of the twenty-first century. The *case* consists in the geological findings involving the worldwide and completely new imprint of humans on the geosphere. The *discourse* consists of a variety of currents, especially in cultural studies and civil society, which have controversial interpretations of these findings and are searching for countermeasures. When I speak of the "Anthropocene" in this book without the use of any further adjectives, I am referring to the (postulated) epoch of human impact on the geosphere.

This book, penned by a cultural anthropologist as well as a geologist, is proof that geology is a special, historical natural science and therefore relevant for a historically informed anthropology. We need not only a geologically informed cultural anthropology, but conversely also an anthropologically trained geology. A comprehensive geology must include material human culture as a fundamental geological phenomenon. With respect to cultural anthropology, I will be discussing the challenges that the Anthropocene, a macro theme in both temporal and spatial terms, poses for cultural anthropology as traditionally micro-oriented social science. More generally said, this book will show that cultural anthropology and geology can mutually enrich each other, empirically and theoretically, through the Anthropocene.

Taken together, they can contribute to an understanding of the pressing existential questions facing humanity as a result of the Anthropocene. This book is a heavily revised and updated English version of my now out-of-print German book "Anthropologie im Anthropozän" (Antweiler 2022a).

1.2.1 Anthropocene as a Phenomenon and an Idea

The Anthropocene has "two lives," as once asserted by environmental historian and geographer Jason W. Moore. One is the scientific concept and the subject of an earth science debate ("Geological Anthropocene"). The other life is that of an idea that has spread from the natural sciences into the social sciences and into the public sphere and raises fundamental questions about the relationship between humans and the non-human world ("Popular Anthropocene"; Moore 2016: 80, Moore 2021a: 36). It could be that talk of the Anthropocene provides us with a vocabulary that allows us to describe the current developments in world culture more precisely than others, which becomes clear when one takes the facts in the latest World Development Report seriously (Conceição et al. 2020).

Here, I will examine the Anthropocene both in terms of the facts and findings about our planet, and as a mode of societal world perception and self-description. A broad anthropological view of the Anthropocene raises three fundamental questions: (1) How can humanity as a special active force of the Earth system be explained and understood? (2) What does the diagnosis of the Anthropocene mean for conceptions of man, of humanity, and thus for the orientation of anthropology as a broad human science? (3) What clues does scientific knowledge about people and cultures give us so that we can develop alternatives to the current trend towards a largely uninhabitable geosphere?

This book will be discussing where the blind spots lie in the deeply interdisciplinary discussion about the factual changes. I will also question the narratives about human history and humanity's future, which are motivated by the idea of the Anthropocene and are currently dominating the discussion. What are the most convincing criticisms of Anthropocene terminology, and what are the arguments in favor of alternative terms such as Capitalocene or Urbanocene? At the same time, I will critically ask whether "Anthropocene" is less a fact-based diagnosis than a depoliticizing program, a neoliberal or neocolonialist ideology. One of the open questions is whether comprehensive anthropogenic changes to the Earth are problematized less in some societies and cultures, or thought of through concepts, foils, or narratives other than the Anthropocene.

I will ask what contribution cultural anthropology can make to the research and the societal problematization of the topic, and what an intensive engagement with the Anthropocene could mean for the future of our field. This implies further questions. What are the possible empirical, methodical-analytical, interpretative, and ethical-political contributions? How can cultural anthropology contribute to "anthropologizing" the subject of the Anthropocene, which is geoscientifically founded and motivated, but also increasingly reflected in cultural studies? I will explore the potential of the experiential methodology used in cultural anthropology for a phenomenon that is usually only discussed in a global or planetary context. Contemporary cultural anthropology is locally oriented despite all the conducted migration studies, network research, and multi-local (sic!) field research. The topic of the Anthropocene is thus a real challenge here. However, the local focus of the discipline may explain cultural anthropology's limited influence on public debates about the Anthropocene:

> Unfortunately, ethnography tends to have very little to contribute to public debate on global processes of climate change, environmental degradation and rising inequalities. The logic of such processes is rarely grasped at the local level. Ethnography is inevitably confined to a given space, but the processes of the Anthropocene are not (Hornborg 2020b: 1).

Conversely, I will be discussing the implications that the Anthropocene thesis has for the subject of my discipline and in molding it into *anthropology*. What challenges does the Anthropocene—a macro-scale topic in space and time—pose for micro-oriented anthropological research that is traditionally focused on field methodology and interested in local and contemporary living environments? Can the notorious problematization of the nature/culture dichotomy in Anthropocene debates stimulate a renewed discussion of the anthropological concept of culture? We will also reflect on whether cultural anthropology should make a normative contribution to this scientific but also highly politicized topic. This could be related to the appreciation of cultural diversity or the social cohesion of humanity. Finally, we will discuss which aspects of the Anthropocene should be included in a future-oriented curriculum of general cultural anthropology.

What can the humanities, cultural studies, and social sciences contribute to this? What significance do questions from these disciplines have for the discussion of the Anthropocene? What expert knowledge can fields outside the natural sciences contribute? What potential for anthropological syntheses lies in the concept of the Anthropocene? From an anthropological perspective, what are the blind spots in the existing theorization of this truly interdisciplinary topic? What can cultural anthropology, as a science that traditionally takes on a micro perspective, contribute to the Anthropocene in terms of empirical findings, and what new subject areas need to be researched?

I will also critically ask why anthropological contributions on the subject notoriously argue against a dualism of culture and nature, but then still cling to what I believe is an ultimately idealistic monism under the banner of "flat ontology," which maintains the dualism, albeit subcutaneously. In a nutshell, the question is: How can the genuinely geological category be "socialized," and how can sociology and cultural anthropology be geologized (Delanty and Mota 2017; Clark and Szerszynski 2021: 38–41,46–49; Leggewie and Hanusch 2020; Hanusch et al. 2021)? What could an interdisciplinary synthesis of a "geoanthropology"(Barker and Barker 1988; Renn 2020: 375–376; Antweiler 2022a, b) look like?

From the perspective of human experience, and therefore necessarily anthropocentric, the question is: Under what conditions could we as humans—and which humans in which living conditions—feel "welcome" in the Anthropocene? After all, in German, the term "human age" is not always translated as the "age of man," but as a "humane age" (Gesing et al. 2019: 8). This book offers a critical orientation on these questions based on the current state of discussion in various disciplines. The topic of the Anthropocene demands genuine attentiveness rather than good neighborly relations. My main perspective concerns a contemporary-oriented cultural anthropology. However, one guiding thesis is that in order to understand the Anthropocene, we need a human science of the whole person and the whole of humanity—Anthropology:

> The Anthropocene has produced an existential crisis in proposing a new, and hitherto unknown, figure of, the human as a planetary force. No consideration of what it means to be human cannot ignore this vision of the *anthropos* as the sum total of human impacts on the planet. Yet this singular figure does not erase the many other *anthropos* operating at multiple scales of time and space, contributing differently to the Anthropocene trajectory, and suffering its effects unequally (Thomas et al. 2020: 196–197).

1.2.2 Focus: Anthropology and Geology in the Multidisciplinary Field

The topic of the Anthropocene can only be explored through the collaboration of many disciplines. It is appropriate for the Anthropocene—indeed it requires it—that various disciplines and positions are brought together in all their diversity; that is, if they are to seriously consider the following motto from a recent collaboration between a historian of ideas, an Earth system scientist, and a paleobiologist:

> Evidence is crucial, whether it comes from rocks, artifacts, or archives, but the categories and concepts that organize the evidence are not inherent in the evidence itself (Thomas et al. 2020: xiii).

However, as this book deals with topics that cut across disciplinary boundaries, it is deliberately not organized by discipline, but is rather thematically structured. The price of doing it in this manner is that some repetitions are necessary from time to time. Thus, the problem of scales appears several times. Despite the relevance of interdisciplinary approaches, disciplines are still needed for their particular contributions. Here, two subjects are given more consideration than others. On the one hand, the principles of geology, including those of paleontology and evolutionary biology, play a continuous role because the Anthropocene is primarily a geological category, and changes in humans and other living beings represent a central aspect. Secondly, cultural anthropology is used because it is—if understood within a broad anthropology– the broadest cultural and social science as well as the broadest human science.

A personal reason for my focus on these two disciplines lies in my interests—but also limited competencies—as an author. I am a cultural anthropologist, but in my "first life" I studied geology and paleontology. With respect to the literature used, this means that I will be discussing a great deal of original literature from the fields of cultural anthropology and cultural studies as well as from geology and Earth sciences, while I drew on overview publications for the discussions on historical and prehistoric or archaeological topics.

The Anthropocene can only be understood through the combination of many disciplines from the natural sciences as well as the humanities and cultural sciences, namely through "great interdisciplinarity." As an author with limited knowledge, I must nevertheless restrain myself. In addition to cultural anthropology and geology, this book particularly considers other humanities, cultural and social sciences, especially sociology, history, and archaeology. Contributions from economics, political science, technosciences and engineering, on the other hand, are only considered through their basic statements, whereby I refer to fundamental literature. I believe that we must first understand the phenomenon of the Anthropocene, even if it requires an urgent response. This means that in this book, I am primarily concerned with the definition and diagnosis of the problem of the Anthropocene and less with political or technological solutions, for which I will also refer to important research literature.

I will touch on exciting non-scientific fields in which the idea of the Anthropocene has triggered many new developments, although only to a limited extent. The reason is my lack of competence, and it is explicitly not my opinion that works of visual art or literature on the topic of the Anthropocene are unimportant. The opposite is the case, as works of fictional literature may be of utmost importance for creating intelligent reactions to the Anthropocene. They (Read and Alexander 2019: 115). They can create radical but perhaps realistic futures, such as in Ursula Le Guin's classic anarchist and anthropologically informed science fiction novel "The Dispossessed" (Le Guin 1947). Novels can change perceptions, such as of disasters, through new narratives as in Max Frisch's "Man in the Holocene" (Frisch 1979). Geological events have been and continue to be a central theme in literature, arts and design (Bostic and Howey 2017; Ray 2024). More recent literature, music, visual media, and performing arts responding to the Anthropocene are the subject of intensive research and overviews. This is particularly true for the broad field of ecocriticism (*or environmental criticism*) and the potential of global environmental issues for literature and art (for example, Dürbeck and Stobbe 2015; Falb 2015; Gibson et al. 2015; Bayer and Seel 2016; Anselm and Hoiß 2017; Clark 2019; Comos and Rosenthal 2019; Schär 2021). These discourses, which are primarily external to science, have a significant impact on the Anthropocene, both in public and in the sciences (Matejovski 2016: 5). Debates about the Anthropocene as a societal phenomenon have become highly differentiated since 2014. In approaching these fields, I will focus on the analysis of metaphors, narratives, large numbers, and the role of visual representations.

1.2.3 Line of Reasoning and Text Form

This chapter deals with the existential significance of the Anthropocene for the future of humanity. We have only been living in the Anthropocene for the length of a human life. If *geologists*, with their extremely long-term perspective, are already proclaiming a new phase of the Earth's history, then we should be alarmed. A meaningful scientific approach that also leads to realistic solutions combines natural sciences and a whole range of human sciences. Of central importance are a significantly time-extended cultural anthropology and an anthropologically informed geology. In short: We need geoanthropology!

Chapter 2 presents the fascinating discovery of the Anthropocene in the year 2000 and follows the diverse paths of its rapid popularization. I must explain that the Anthropocene is not simply old wine in new bottles, and that the Anthropocene is much more than human-made climate change: it concerns large parts of the geosphere. The chapter discusses a variety of proposals for the beginning of the Anthropocene in human history, which date back to many thousands of years from today. These historical and archaeological debates about periodization contain a hitherto untapped social and cultural-theoretical potential for research on cultural change and the question of the main factors of long-term cultural revolution.

Chapter 3 is dedicated to the strong metaphors and mostly alarming narratives about the Anthropocene. The ideas concerning the Anthropocene are related to conceptions of the Earth, globe, and planet, which were strongly influenced by space travel. At the center are images of the sick planet on the one hand, and immeasurably large numbers, alarming curves, and drastic diagrams on the other. There is a wide panorama of powerful narratives that simultaneously, and often implicitly, carry political ideals and moral concepts. The chapter shows how the Anthropocene is linked with certain images of humanity in the discourse, and that these metaphors represent a social fact that influences societal change.

Chapter 4 asks about the strengths and weaknesses of the idea of the Anthropocene. While the geological-stratigraphic criticisms were discussed in the first chapter, this chapter deals with criticisms relating to the social, ethical and political relevance of the hypothesis. A clear benefit is that the natural sciences are now actually working together with the humanities and cultural studies on a very concrete topic. A problem with the term Anthropocene is its increasing diffusiveness. Important criticisms suggest that the idea of the Anthropocene overestimates the power of humans and is anthropocentric. More fundamental criticisms see the idea as Eurocentric and criticize its blindness to the diversity of the perpetrators and also the inequality of the victims. Critics object to its depoliticizing tendency or simply consider the Anthropocene to be a distraction from the capitalist causes of the world crisis. Accordingly, a multitude of alternative terms are being discussed. If we look at the scientific landscape, we have now arrived in the "Neologismocene." My own criticism is that many contributions wrongly reduce the Anthropocene to anthropogenic climate change or misuse the word for the rhetorical enhancement of known concepts or other topics.

Chapter 5 shows that cultural anthropology has always had a professional proximity to the subject of nature and has dealt with the environment in various ways. The causes of the Anthropocene may not be perceived by us as individuals, but the effects will be experienced locally and differently in the long term. The concepts and goals of countermeasures are also local. This shows that the local access and the holistic approach of cultural anthropology opens a fruitful window to understanding the Anthropocene. In this section of the book, I criticize currently dominant approaches to the Anthropocene in cultural anthropology because they avoid scientific findings, obscure the specific geological deep time, and generally provide more artistic contributions. The chapter presents very different anthropological approaches, and I also attempt to do justice to directions and positions that I do not represent based on my scientific assumptions (see appendix), such as the *more-than-human* approaches or posthumanism.

Chapter 6 asks what the Anthropocene means for a new science of man. The idea of the Anthropocene clearly goes far beyond the widespread concept of "anthropogenic," which is elucidated by the clarification of the difference between geological and historical or archaeological periodization concepts. Only by connecting the materially open concept of culture in cultural anthropology with the concept of planetarity will the fundamental material entanglement of man and nature become clear. For a causal explanation of transgenerational change and especially the dynamics of the Anthropocene, I connect the idea of the Anthropocene with the concept of niche construction from evolutionary ecology. This reveals a threefold heritage that is crucial for understanding the Anthropocene: the *genetic* heritage, the *cultural* heritage, and the anthropogenic *geospheric* heritage.

Chapter 7 brings together what the culturalization of the planet and the geologization of culture mean for societies and for us as individuals. In view of the diverse debates, only building blocks can be provided here, which will not be enough to produce a conceptually inhabitable building. Hence, the controversial discussion about scales is combined with concepts of world citizenship in the Anthropocene. I have then formulated the guiding question as a cosmopolitan-informed question of world ecology: *How can intensively networked cultures peacefully coexist on a planet that is heavily influenced by humans but limited in size, without all having to become indistinguishable?* Using Southeast Asia as an example, I finally show what a "provincialization" and a localization of the Anthropocene could look like on site.

The final chapter summarizes the main empirical findings and theoretical insights and ends with theses that hopefully invite discussion. An extensive glossary is provided to facilitate an understanding of the terminology, which has been built from various disciplines. As a supplement to the extensive bibliography, I provide a media guide to help readers find their way through the seemingly endless jungle of literature and websites.

Due to the breadth of the subject matter, the texts I have used take very different forms. This is also manifested in the still-existing—sometimes even widening—gap between the natural sciences on the one and the humanities, cultural studies, and social sciences on the other hand. Especially when the intention is to bridge gaps, it is important not to underestimate differences in professional routines, especially in the way concrete practice constructs arguments and claims evidence. I have opted for a compromise between a more scientific writing style (hardly any direct quotes) and one that is more frequently used in the humanities or cultural studies (long footnotes). The text presents direct quotes where the exact wording matters and where other authors have presented a thought or context in a concise way. However, I refrain from footnotes because they constantly require the reader to jump down to the bottom of the page and also notoriously invite digressions, which is dangerous for a topic with a wide range of aspects. The price for this is that not every ramification of an argument is explained, and where I refer to further literature, a small cascade of up to five sources is often cited. This study takes into account scientific literature published till August 2024.

1.3 Planetary Space and Deep Time—A Geo-Sociocultural Mega-Macroepoch

> *The change in the scale of human life*
> *is the decisive development of our epoch.*
> Marc Augé (2017: 46)

> *Our brain is against thinking*
> *ahead in geological timeframes.*
> Robert Macfarlane (2016: 97)

The Anthropocene thesis asserts a geologically *abrupt* phase change, from a time when humans only acted as agents of local or regional environmental changes, to their role as the *main* agents of geophysical change. With the new phase, humans have the potential to dominate the entire living world through increasing civilizational interventions in the Earth system. This thesis, or rather diagnosis, is based on statements about geophysical processes over very large geographical areas and also over very long periods of time, covering the whole of the Earth's surface and thousands of years, up to about 12,000 years before present (BP). Since societies and cultures are at the center, we can describe the Anthropocene as a geosociocultural mega- and macro-epoch (Werber 2014; Bajohr 2019: 64). The thesis raises the question of whether it is possible to see humanity as a truly global community and to act accordingly.

1.3.1 Earth's History Leaves the Slow Motion *Mode*

The Anthropocene is about an unprecedented dimensional shift that is leading humanity into a planetary age. The central causes of the Anthropocene are economic growth and consumerist orientation in the context of industrialization and globalization. The categories with which we characterize capitalism and globalization, which also dominate the political debates on climate change, are not sufficient for an understanding of climate change. Climate change has a lot to do with capitalism, but it is not reducible to it: the politics of climate change is more than the politics of capitalism (Chakrabarty 2009: 212, 2017). Contrary to criticism of the Anthropocene concept that is critical of colonialism, which is acknowledged in Chap. 4, Chakrabarty states that a "hermetic of suspicion" against nationally and globally dominant power formations is not sufficient to deal with climate change. For an analysis of the Anthropocene as a planetary dynamic with far-reaching geo-aspects, it is therefore not enough to use imperial rule, capitalism and globalization as a framework for analysis (Chakrabarty 2021: 68–92). The capabilities of humans as a species, the "agency" of the Earth system and non-human time scales would otherwise be lost from view. This insight is remarkable and courageous, as it comes from one of the most important minds in the research field of subaltern studies, which is critical of capitalism. To be able to explain, for example, the rapid extinction of large mammals, we *also* need to use the categories of the living and the species (Martin and Wright 1967; Chakrabarty 2018: 32–34, 2021: 35).

The world's energy consumption in 1990 was 16 times higher than in 1900. In the twentieth century, the era of oil, humanity consumed more energy than in the approximately 11,000 years between the Neolithic and industrial revolutions. Human-made material culture is extraordinarily energy-hungry. With the onset of urbanization, consumption rose from a few hundred watts to the current average of more than 3000 watts per day. Through devices and infrastructure, a person consumes on average 30 times our natural energy needs worldwide (West 2017: 241). Ecologically speaking, it is crucial that only by way of industrialization were people allowed to economically break away from the limitations of the system referred to by Fernand Braudel as the "old biological regime" (*ancien régime*, Braudel 1981: 70–72). Until then, the available energy and nutrients and the amount of food that could be gathered or grown were essentially limited.

From the perspective of energy theory, humans were until then "mammals like any other," and there were only a few million of them (West 2017: 221). The world population was around 5 million in the early Holocene, around 200 million at the turn of the era, around 500 million around 1650, around 1 billion in 1804, 2 billion in 1927, 3 billion in 1960, 6 billion in 1999—and it is expected to reach almost 8 billion by the time this book is published. In the old regime, there was a widespread awareness of the connection with the landscape. From that time on, mineral energy sources were used, especially coal and oil, that is, fossil energy that had been formed over long geological periods (Malm 2016; Morris 2015: 127–179).

Previous production systems relied on renewable forms of energy such as the sun, land, wind, water, and biomass. All of these are limited in terms of their materiality. Accordingly, economic activity was restricted, for example, by the amount of available land, climate-related growth phases in agriculture, or seasonal monsoon winds in shipping. Coal, on the other hand, as an energy source for the steam engine, was only available as a raw material in some places, but its use was independent of location (Schmelzer and Vetter 2019:49). A similar periodization was used with an eco-feminist perspective by Carolyn Merchant, when she set the first "death of nature" at the transition of the organic relationship with nature into the mechanistic relationship with the world in the seventeenth century, and the "second death of nature" at the onset of the Anthropocene, with the development of the steam engine by James Watt in 1784 and the many subsequent innovations that it made possible (Merchant 2020a: xvi, 2, 26–44, 2020b).

Taken together, this means that the organic-based economy has largely been abandoned by fossil-fueled societies. Environmental historian Robert Marks speaks of the "great departure" (Marks 2020: 20–34, 175–177, 226–229) with the abandonment of the limitations and rhythms of the old regime by the end of the nineteenth century. To highlight the significance of this historical break, the term he chose echoes well-known formulations. The first is the "great transformation," which came with the beginning of industrialization around 1800 (Polanyi 1944), and the other is the "great divergence," referring to the divergent development of Asian empires, especially China, and Northwestern Europe that came with coal-based industrial growth from around 1750 (Pomeranz 2000).

Coal, and soon after oil, facilitated a constantly available and increasable flow of concentrated energy. In addition, fertilizer (guano) began to be used, thus providing an external source of nutrients. All of this enabled an enormous increase in production, the spread of industrial wage labor, and the expansion of capitalist agriculture. For the first time, fossil energy sources allowed for a life and an economy that were independent of time and space, a key feature of the modern age. However, this is only the short-term perspective, because in deep time, the industrial revolution was an energetically fundamental system change. Human economic activity underwent a dramatic transition from a solar-powered *open* system to a closed system designed for fossil fuels. It was therefore a change into a system in which entropy, according to the second law of thermodynamics, continues to grow. An external, limited, but reliable and constant energy source was replaced by an internal, temporarily reliable, fluctuating energy source that was, moreover, dependent on market forces (West 2017: 244).

Until the late nineteenth century, however, leaving the old environmental regime hardly had any influence on the atmosphere. In sediments, there is also no clear, sharp, and, above all, globally synchronous time marker, which is why the beginning of the Anthropocene cannot be established from a geological point of view. From an Earth history perspective, the Earth entered a distinct new phase in its approximately 4.5 billion-year history in the middle of the twentieth century. This rapid transformation is unprecedented in the history of the geosphere and is likely to cause problems for the continued existence of humanity in the near future. To put it in extreme terms, human activities are not only significantly shaping the geosphere, but they have substantially altered the *functioning* of the Earth system. It is important to note that the boundaries between the major eras in geology are usually marked by mass extinctions, that is, by synchronous and worldwide extinction events in which a large number of plant and animal species become extinct across a wide range of systematic units.

From a historical perspective, the reality of the Anthropocene has been diagnosed with varying degrees of enthusiasm: through alarming words (e.g.,

Chakrabarty 2009; Morton 2013: 7), as a historical finding (e.g., McNeill 2000; McNeill and Engelke 2013; Marks 2020), or also more optimistically (e.g., Ackerman 2014). Anthropogenic climate change, due to its global scale and suddenness, is currently the dramatic focus of mass media attention, a paradigmatic example of the effects of human activity on the Earth. Some go so far—in my opinion, too far—as to call the Anthropocene a "slogan for the age of climate change" (Purdy 2015: 2).

1.3.2 Anthropocene Is Much more Comprehensive than Climate Change

The Anthropocene, as I will emphasize repeatedly, is by no means the same as "climate change," "global warming," "environmental problems," or similar terms used for changes in the world. While climate change is suitable for headlines, the term Anthropocene opens up fundamental questions about the relationship between humans and the material environment or the natural world. The idea of the Anthropocene has implications for fundamental conceptions and categories, such as nature, culture, and society; it questions these by means of a concrete problem and "destabilizes" the concepts. Thus, the Anthropocene expands and *multiplies* the question raised by climate change, as to whether we are exceeding the limits of the Earth (Clark and Szerszynski 2021: 3).

The Anthropocene represents a geological concept for a phenomenon that includes many Earth-wide phenomena and places them in the context of a long-term geological perspective ("deep time"), which I will discuss in detail in Sect. 6.6. The deep time of the Earth is orders of magnitude longer than what religions or common understanding can assume. Only from this time perspective can the current and, geologically speaking, *very recent and abrupt* transformation of the planet be accurately characterized and made *visible* through science. This also applies to social institutions that have been established to distinguish a society or a population of humans from an area of nature thought to be external to it, such as an "environment ministry." In addition—and for some, even above all else –, the Anthropocene as a diagnosis of the world situation has very practical implications; it raises political questions and, ultimately, value decisions. It is becoming increasingly clear that the window of opportunity to halt or reverse key indicators of the environmental crisis is rapidly shrinking (Pálsson et al. 2013: 3). Even though the Anthropocene represents a global problem and perhaps involves the most pressing ethical problem of our time, political questions are by no means only global or geopolitical questions. The Anthropocene raises questions about the possibilities of democratizing nature relations and about bringing forms of action, from civil disobedience up to practical theological questions, sharply to the forefront (Vogt 2016; Bertelmann and Heidel 2018; Heichele 2020; Hetzel et al. 2020).

The anthropogenic contribution to climate change also played a role in the popularization of the concept by Paul Crutzen and the limnologist and diatom researcher Eugene Filmore Stoermer (1934–2012). They proclaimed the concept in two very short essays (Crutzen and Stoermer 2000; Crutzen 2002; Leinfelder 2012). Climate warming due to carbon dioxide and methane emissions was seen as the primary signal of the Anthropocene (Rudiak-Gould 2015: 48). Methane is mainly produced when using chemical fertilizers in cattle breeding and wet rice cultivation. The change brought about by the Anthropocene affects an extensive range of biotic and geophysical relationships and processes, such as the nitrogen cycle, the loss of biodiversity, and the chemical changes in the soil. In short, not only the atmosphere, but large parts of the geosphere are increasingly being influenced by humans. Human-induced change in the geosphere can therefore not be reduced to anthropogenic climate effects, which is often forgotten in public and scientific debates.

The parallel between human-induced climate change and the Anthropocene also applies to the enormous rift between understanding the urgency of globally coordinated action and strongly worded discourses on the one hand, and lacking or ineffective action on the other (Tsing 2005a, b; Krauß 2015a: 60, 64; Pálsson et al. 2013: 7). This is already true for anthropogenic climate change, but the Anthropocene is not simply a new collective term for the sum of all geological transgressions, for the "environmental problem" as such, nor is it necessarily for something that, *per se*, is to be avoided (Schwägerl and Leinfelder 2014: 235).

1.3.3 We Are Consuming the Earth

> *Planetary thinking means taking the Earth seriously*
> *as a planet, from the Earth's core to interstellar space, from the nanosecond to deep time, from elementary particles to the Earth's mass.*
> Frederic Hanusch et al. (2021: 24)

The shaping of the Earth's surface by humans has long been an established topic in geomorphology, especially the anthropogenically accelerated processes of "quasi-natural surface formation" (Busche et al. 2021: Chap. 23, 256–265). There are extensive monographs on the geomorphic effects of humans (Rathjens 1979; Goudie and Viles 2016; Busche et al. 2021), but the Anthropocene as a topic is still hardly present in the academic canon. A recent German-language textbook on geomorphology extensively discusses human-influenced landscapes under the heading of socio-ecological geomorphology or historical geomorphology (Dikau et al. 2019: 4, 7, 10, 98, 417–435), but without strongly emphasizing the Anthropocene concept. This also applies to general geography, and not only in this country. Recent English-language college textbooks on human geography and also physical geography only touch on the topic of the Anthropocene at best (Knox and Marston 2016: 144–152) or do not mention it at all (for example, Rubenstein 2019).

The Anthropocene is about global human-induced change. With respect to both space and time, it is about transformation processes that largely lie beyond our sensory experience. In addition, the phenomena are scientifically difficult to access, which subjects them to assumptions and uncertainties so that a lot of speculation is inevitable. Critics see the use of the term as a sign of increasing human overestimation and hubris, and question whether the influence is actually as spatially extensive and lasting as often claimed, or rather sporadic and perhaps short-term from a geological perspective (Krämer 2016: 24–44). In contrast, the dialectical aspect of the diagnosis of the Anthropocene must be emphasized. At its core, the Anthropocene is characterized by the paradox between profound human interventions in the geosphere and at the same time an increasing loss of control over social relations of nature (Fladvad 2021: 145). "Although mankind has 'subdued' the Earth, it is increasingly unable to control its natural conditions" (Görg 2016: 13).

The diagnosis of the Anthropocene is currently challenging because it is about much more than just climate change, namely the transformation of soils and seas along with their living world, and thus assumptions about the state and dynamics of the world in general (Sayre 2012). Accordingly, very broad terms such as "planet" and "humanity" are quickly used in debates, often in an alarmist tone and with disease metaphors. This leads to a diagnostic problem framing of the discussion, which increases public attention, but is also accompanied by exaggerated simplifications (Matejovski 2016: 7).

An almost unavoidable, though fundamental exaggeration arises simply from the fact that the Earth or the planet is repeatedly referred to as the "human-made Earth" or even the "human Earth" (Schwägerl and Leinfelder 2014: 234). Enormous quantities are emphasized, such as how humans move almost 30 times more material through construction activities and agriculture than was the case in the last 500 million years without their intervention. All these are now very well-documented, as is the fact that half of the permanently available freshwater is now used by humans (for example, Zalasiewicz et al. 2018a, b).

Exaggerations can be found in the mass media, but also among proponents and critics of the Anthropocene, such as in phrases and titles like "New World" (Zalasiewicz et al. 2010), "Big World, Small Planet" (Rockström and Klum 2016a, b), or "Earth Transformed" (Ruddiman 2013). They can be found in many critical voices, for example, in "Damaged Earth" (Haraway 2016: 2; Tsing et al. 2017), "Defiant Earth" (Hamilton 2017), when we talk about "Humans versus Earth" (Subramanian 2019) or "The Ecological Rift. Capitalism's War on the Earth" (Foster et al. 2010), and when sociologist John Urry refers to us humans as "excessively consuming the planet" (Urry 2010).

In their epoch-making paper in *Science*, Crutzen and Stoermer formulated more cautiously "… that the Earth – *as we know it today* – may cease to exist in the future" (Crutzen and Stoermer 2000, own emphasis). Therefore, for the sake of simplicity, the term "Earth" is used throughout this book to refer to the geosphere, that is, the Earth's shell, which includes the various spheres of air, soil, water, ice, and rocks. I emphasize these exaggerations because critics have objected, with some justification, that the geologically short-term human influence on the Earth is notoriously exaggerated by proponents of the Anthropocene in view of the slow geological processes (Visconti 2014: 389).

Since the Anthropocene concerns the geosphere allusions to a "burning planet" (Hornborg 2017b), a "destroyed planet" seem exaggerated. This holds also for Anna Lowenhaupt Tsing's among cultural scientists widely circulated talk of an earth in the "ruins" of capitalism (Tsing 2015: 6, 205–213; Gan et al. 2017; cf. González-Ruibal 2018), A goal of "cooperating with the planet" (Folke et al. 2021: 835) to create a world worth living in also seems grossly exaggerated. AThere are also exaggerations in other important new approaches, such as the call for a *planetary turn* in the Earth and life sciences, which is now presented with a comparative planetary perspective. This, in my opinion, is also true with regard to the formation of a "planetary social thought." It is an exaggeration when the sociologist Szerszynski explicitly says that humans are changing the Earth *as a whole*.

Emphasizing the dynamics and variability of the geosphere at all levels is important. However, I find it problematic when the planet is portrayed almost as a person, for which many thought-provoking examples can be found in fiction (see Heise 2008, Hanusch et al. 2021: 16–24). It does become problematic, however, when such questions are posed in scientific analyses as "...what enables the Earth to do new things" or "...how has the Earth discovered to do things differently over time," or when speaking of the "self-organizing tendencies of the Earth" or "planetary anamnesis" (Clark and Szerszynski 2021: 4, 30, 54, 84, 92 see Szerszynski 2019). This is similar to Manuel DeLanda's talk of "non-organic life" and his statement that the Earth has its own probes for exploration (Delanda 1992: 161), or Eduardo Kohn's famous talk among anthropologists of the "forest that thinks" (Kohn 2013). These are stimulating metaphors, but as analytical terms they are, as I will argue, hardly suitable. Such flowery formulations are probably one of the reasons why some geoscientists and biologists avoid dealing with cultural science contributions on the Anthropocene.

The diagnostic framing and the exaggeration when speaking of "the world" or "the planet" can also be found in some jokes and cartoons on the subject, such as the joke about the encounter of two planets, which exists in various versions: One planet says to the other planet, "You look bad. What's wrong with you?" The other replies, "Yes, I have Homo sapiens." The first one says, "Don't worry. It's nothing. I had it once too. It will pass on its own." If you take a closer look and think less anthropocentrically but more geocentrically, it suggests that the Anthropocene is "just" a "skin disease." It requires a dermatologist, but not a surgeon or a specialist in internal medicine. However, chronic skin diseases are not only annoying, they can become dangerous, too. These are all exaggerations from a scientific point of view, but from an anthropological and psychological point of view, they are themselves a part of social reality that should be taken seriously.

1.3.4 Anthromes Instead of Biomes

Humans are so effective, especially with the use of technology, that there is not a single ecosystem on our planet today without an anthropogenic component. Something like an "untouched tropical rainforest" no longer exists on this planet. Currently, more than 75 percent of the planet's habitable surface area can be considered as human-shaped habitats, as anthropogenic biomes, whereas in 1700, only about five percent were used. But even then, less than half of the land regions were without detectable human influence. More than half of the earth's surface has only become agricultural areas over the last hundred years, many only after the Second World War. Thus, the usual ways of defining "natural" habitats and ecozones are no longer suitable, and new hybrid habitats are emerging. In order to highlight the human–environment interaction as the main factor, the actual conditions of use, and the degree of transformation, biogeographers have introduced a specific term for human-made biomes: Anthromes (*or anthropogenic biome*, Ellis and Ramankutty 2008: 445; Fuentes and Baynes-Rock 2017; Ellis et al. 2010, 2020).

Anthromes are large areas that have been culturally shaped (anthropogenic) and significantly transformed by humans (Fig. 1.3). These spaces, also referred to as "Anthropozones" in German, provide an alternative to the traditional classification of landscape belts and geo-ecological zones (Glaser et al. 2012: 162). Ellis and Ramankutty have identified a mosaic of 19 anthromes worldwide. Among the anthromes, a distinction is made between "used" landscapes, such as through

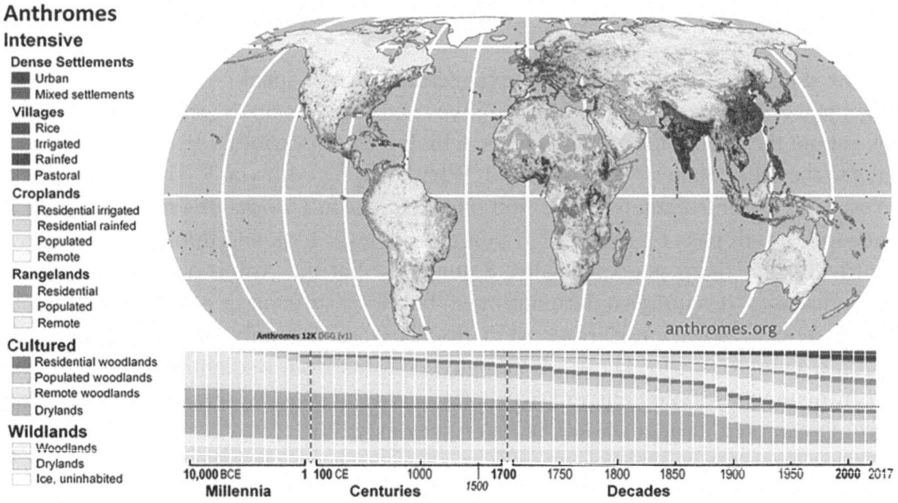

Fig. 1.3 Historical development of Anthropogenic Biomes (Anthromes). (https://commons.wikimedia.org/wiki/File:Anthromes_map_and_timeline_(10,000_BCE_to_2017_CE).png)

intensive agriculture or in cities, and only two "semi-natural" landscapes, such as inhabited forests and natural areas ("*wildlands*"). These are areas whose symbiotic communities depend not only on the inanimate and animate environment, but also on the human population and their land use.

The concept of anthromes makes it clear that the world now no longer consists of natural ecosystems in which humans intervene, but of large landscapes determined by commercialization, into which more or less altered natural systems are integrated. The few near-natural areas largely coincide with the distribution of indigenous groups, as the latest maps show (LALE, Laboratory for Anthropogenic Landscape Ecology 2020). However, it should be noted that the scales and recognizability of human influences play a role here. Just as some landscapes that appear "pristine" to the untrained eye were altered by humans in the past, such as the *Terra Preta* soils in the Amazon lowlands (WinklerPrins 2014), conversely, even large parts of the USA still show few signs of a "humanized" or "domesticated landscape," such as in Yosemite National Park (Vale 1998; Denevan 1992; Ziegler 2019: 272).

Whether we feel welcome in an Anthropocene "nature" that has now been transformed by humans on a large and planetary scale remains questionable to say the least. The topic of the Anthropocene is central to current debates in politics and economics, but also in science. These debates are about the ecology of the world, global solidarity, global governance, global inequality, and biodiversity (Graeber and Wengrow 2021). However, the new theses on the "Age of Man" also give new interpretation to old fundamental questions about the relationship between societies and material environments. What is nature, what is culture, what should we understand with the word "environment"? Should we regard the Anthropocene as the "Age of Man" or rather as the "Age of Humanity" (Merchant 2020a, b: 13, 22)?

Triggered by scientific findings, fundamental questions about the position of humans in nature emerge in a new light. The Anthropocene is a phenomenon first diagnosed by natural scientists and thus opens up new perspectives on fundamental—indeed anthropological—questions about humanity and human existence. In addition to philosophy and politics, areas such as the humanities, cultural studies, and social sciences are particularly challenged by the new scientific findings and interpretations.

Two main lines of questioning arise here, both of which I will pursue in this book. How should the social sciences react to the term and the matter (Adloff and Neckel 2020: 11–12)? On the one hand, it is necessary to ask which actors in science, politics, and the media have taken up the term—or just the word—and refer to it positively or critically. On the other hand, the much more difficult question arises as to whether and to what extent the Anthropocene is suitable as a conceptual framework for social science analysis, in order to advance understanding and explanatory theory formation as well as guide empirical research.

1.3.5 The Anthropocene Working Group (AWG)

The central geological institution for this topic is the Anthropocene Working Group (AWG), a multidisciplinary expert committee of the Subcommission on Quaternary Stratigraphy (SQS) within the International Union of Geological Sciences (IUGS), which has been in existence since 2009. The AWG currently consists of almost 40 scientists from 14 countries who work on a voluntary basis. This was prompted by a study by the Geological Society of London, which was initially empirically quite general, but which argued that there was that even on the basis of a "conservative" approach an 'unmistakable biostratigraphic signal' for the lasting effects of human civilisation in our time (Zalasiewicz et al. 2008a, b). The task of the AWG was now to explore whether the Anthropocene can be geologically substantiated as an interval. The aim was to examine dating proposals for the Anthropocene, to develop a plausible periodisation and also to carry out our own stratigraphic investigations. Under the leadership of palaeobiologist Jan Zalasiewicz, stratigraphic geologists make up the majority of the AGW. However, experts from earth system science, ecology, archaeology, geography, oceanography, soil science and even philosophy, history and international law are also represented (SQS 2009; Waters et al. 2014; Zalasiewicz et al. 2021a, b: 2, 2023). The humanities scholars and social scientists in the group represent a completely new cause and process orientation for geologists compared to their traditional orientation, which is more focussed on effects. The AWG is therefore not only important as the most important institutionalisation of the question, but also represents an absolute novelty for stratigraphy in its disciplinary diversity (Lundershausen 2018; Will 2021: 33, 107–113, 216; Zalasiewicz et al. 2023:316 f.).

If the *Anthropocene Working Group* currently favors the beginning of the Anthropocene to be set in 1950, then this would mark a *geological* epoch that began only *one human generation* ago. It is important to realize here the much longer timescape of the *longue durée* of human history compared to this short period and the even longer "deep time" of planetary history (Gould 1987). This includes the assumption or insight that time is not *only* socially constructed (Gell 1992: 241), an important point that I will discuss in Chap. 6.6. The history of the Earth began about 4.54 billion years ago. *Homo sapiens* emerged about 200,000 years ago and slowly became an important factor. Humans have had a demonstrable influence on the world for about 50,000 years and only really made a major impact about 10,000 years ago with the first permanent settlements. In the middle of the twentieth century, they suddenly became a force in geohistory that significantly shaped the planet, or more precisely, shaped the biosphere and the geosphere. An empirically proven break was the worldwide trend change, the transition from the use of bricks to concrete in buildings since about 1950 and the sudden worldwide increase in the utilization of asphalt as road surface from 1960 onwards (Elhacham et al. 2020: 3).

This planetary transformation, which, for example, drives the atmosphere out of the range of minor fluctuations of the Holocene (only by about one degree from the mean value of 14 degrees Celsius), therefore occurred *within the lifespan of a human being*. The Anthropocene, however, is more than just this feared "hot age," it signals the overall effect of human intervention on the geosphere. Similarly, it refers to the change in the geosphere, more precisely an Earth-wide surface phenomenon, and on the local level a phenomenon that affects individual humans, collectives, and entire societies and is simultaneously influenced or produced by them. This means that the ideas, innovations, political systems, and social formations, as well as those who resist this destructive change, are all part of the Anthropocene (Thomas et al. 2020: vii):

> (...) to come to grips with the Anthropocene, one needs to span the enormity of geological time and its processes, and also delve into the complexities and sheer quirkiness of human behaviour and institutions on more intimate timescales – hence our plea for a multidisciplinary understanding (Thomas et al. 2020: ix).

The editors of a comprehensive encyclopedia on the Anthropocene argue in their foreword that the word "Anthropocene" is suitable as a red warning flag to humanity, but that in view of the catastrophic human impact on the environment, it would be rather irrelevant whether geologists and stratigraphers identify the Anthropocene as a formally legitimized geological period. This is rather a consequence of the human urge to pigeonhole the world around us in order to derive meaning from it (DellaSala et al. 2018b: 7). However, this overlooks the fact that geologists usually only identify time intervals based on *worldwide, synchronous, and extreme* changes, such as in the case of mass extinctions. In this respect, a geological period boundary has a completely different relevance than a purely historical threshold. Such a boundary can also trigger much broader debates, for which the hefty 2280-page encyclopedia (DellaSala and Goldstein 2018) is itself a sign.

A blatant conceptual contradiction, which Burghard Müller has summarised under the motto 'Anthropocene my arie' ("Von wegen Anthropozän"), is challenging in terms of social theory and cultural studies. The Anthropocene appears paradoxically as a gigantic dwarf. On the one hand, the Anthropocene is supposed to have begun as little time ago as possible in order to clearly show how rapidly we humans are rapidly remodelling the world. On the other hand, it should nevertheless be given the status of a great and temporally deep geological age, "... by virtue of appointment, participate in the dignity of the great, deep, unavailable time" (Müller 2021b: 12). The core message is probably to motivate us to achieve a rapid social transformation. In view of the geological and socio-theoretical challenges and, above all, in view of the de facto rapid global environmental change, I conclude: we need a geoanthropology.

1.4 Deep Time, Periodization Concepts and Contested Geo-Chronology

> *There is nothing without us. We are in everything.*
> Andreas Meier (2011: 49)

> *Stone would call you transient, sporadic.*
> Jeffrey Jerome Cohen (2015: 30)

> *Nothing has the right scale.*
> Bruno Latour (2017b: 109)

The Earth's history found in rocks and fossils is challenging our imaginative power. For laypeople, geology is a provocation. But even "geophiles" and professional geologists ultimately remain speechless in the face of sheer infinite duration, the slowest movements, and inhumanly large spaces. Geological temporality is very difficult for humans to grasp because it lies outside of our normal mesocosmic horizon of experience (Bjornerud 2018; Irvine 2020: 106–128). The Anthropocene as a matter and a problem leads to the encounter of incompatible scales (Cohen 2015: 20). Similar problems arise here, as that in understanding the mechanisms of organic evolution, which, alongside religious reservations, is one of the reasons for the rejection of modern evolutionary theory (Kampourakis 2021: 42–69). In order to name the challenge posed by phenomena and interdependencies, the term "deep time" was introduced by journalist John McPhee in the 1980s (McPhee 1981: 21, 77). Deep time sums up the concept of the long-term dynamics that go beyond our human experience and are difficult to grasp even with numerical data, and the term has been fruitfully used ever since (Lyle 2016; Rossi 1984; Schulz 2020; Gamble 2021).

1.4.1 A Natural Reactor in Gabon—A Window to the Entanglement of Time and Space

The Anthropocene connects geological realities with *different* local cultural and political realities. An example of this is the exploitation of raw materials. Gabrielle Hecht, an environmental historian, used the example of the African country Gabon to analyze the entanglements of many time dimensions and different spaces in the Anthropocene, and in doing so shed light on spatial networks and political contexts (Hecht 2018). Gabon is one of the most resource-rich countries in Africa. The country has extensive oil reserves off its coast. Its main exports are crude oil and petroleum products, which account for about 82 percent of its export revenues. The Central African country was a French colony from 1839 to 1960. Even in postcolonial times, Gabon remained dependent on France. Right at the beginning of independence, in 1961, France began extracting uranium-bearing rock from the Precambrian era. It was the time of the Cold War and the nuclear arms race. The uranium in the Oklo open-pit mine was used for French nuclear weapons and energy production in nuclear power plants. In Gabon, the extraction of raw materials provided income to the young "developing country," especially to the local elites.

Various time dimensions and different spaces as well as political questions come together here on the basis of a material geological good, the uranium-bearing rock. Although the rock was spatially located in Africa, its use arose from a nexus of external political motives of France in the global context of nuclear armament, and internal economic interests of Gabonese elites. The rock had been formed in the prehistoric age, in the Precambrian, over a process spanning many millions of years, but its exploitation took place in just a few years in the present. Unlike globalization, this is not just about spatial contraction.

The extraction of the raw material produced waste and emissions in Gabon, but not in France. Thus, a situation triggered by global political circumstances influenced the local environmental situation in Gabon. Then, in 1972, French raw material experts discovered that the uranium content was much lower than expected. It turned out that this open-pit mine was a so-called "natural reactor." In the distant geological past, a nuclear chain reaction had started in the naturally occurring uranium concentration together with water. To date, the remains of a total of 15 such "natural reactors" have been found in the basin of Franceville. The reactors of Oklo have now been completely or largely exhausted.

Due to the low uranium content, the extraction of this rock formed in deep time ended abruptly. The open-pit and underground mines were flooded, so only the smallest of the known reactors in Bangombé could be preserved for further scientific studies. France sought other sources of fissionable uranium, thereby transferring the environmental problems brought about by uranium extraction to other parts of the world; meanwhile, Gabon embarked on a dependent path toward an economy that was largely based on raw materials. Different local and power-asymmetrical realities are intertwined here. A circumstance in the early geological past, namely water contamination, was found to be the cause of a current problem.

Hecht's interpretation of the case does not simply demonstrate the interweaving of the specific with the general. It is also not just about the entanglement of the local with the political, social, and cultural, as in studies on cultural globalization or on the location- and culture-specific localization of global influences (cultural globalisation, localisation). Hecht's approach establishes a connection to geological deep time as well as to the spatially global and the nuclear economy (Masco 2020). It is about capitalist-oriented relationships, although it is not just about material goods and economy, but also about categories: for example, in the wake of the transformation of colonized landscapes through cartography, surveying, descriptions, and forms of travel in the tropics (the so-called *worlding*, Spivak 1985; Tsing 2010a, b: 49).

Thus, the categorical handling and monitoring of waste and emissions in Hecht's analysis, which was inspired by waste studies, prove to be a key epistemological technique for Anthropocene problem contexts (Hecht 2018: 111). Owing to human activity, the current situation—in Gabon as well as in France and the later alternative uranium mining sites—has been shown to be localized to the ground and temporally connected with deep time, but without an elimination of the planetary conditions.

Hecht used the empirical object of the uranium-bearing rock as an "interscalar object" to show the relationships between different spaces and times (*theory of interscalarity*, Hecht 2018: 109, 122, 131). The respective *Scales* are historical in their emergence and decay. Using the perspective of historian Fabienne Will, Hecht's case analysis employed a narrative interpretation to localize and contextualize a specific material good as evidence of dynamics in the Anthropocene (Will 2021: 27). It thus became possible for the spatial as well as the temporal multidimensionality of the Anthropocene to be addressed methodologically in empirical case studies. The "interscalar object" here is uranium-bearing rock. But plastic residues in the oceans, technofossils, or even the broiler chicken, which tend to serve as "secondary markers" in geoscientific analyses, can also be used for such analyses (Will 2021: 170–179, for example, Swanson 2018).

Objects can be used as interscalar instruments for Anthropocene narratives, whether it be in history or in cultural anthropology. Methodologically, this appears to be a fruitful approach that could be exemplary and could be combined with *Material Studies,* the archaeology of *Entanglements*, and the history of technology. The example also shows how important it is to investigate the Anthropocene in non-European spaces, a point that I will address in Chap. 7. Raw material extraction and also connections through nuclearity are prime examples of this, as Hecht demonstrated in a more extensive study (Hecht 2018). Collaboration with archaeology can also have practical benefits, for example, when knowledge about networks of water supply in early cities provides solutions for today's problems in arid habitats (Scarborough and Isendahl 2020).

1.4.2 Geology as Natural History and the Geological Grounding of Culture

If we were to address a letter, not to a person at a place of residence, but to the geological location of humanity, the correct address would have to read: "Earth, Anthropocene epoch, Quaternary period, in the era of the Cenozoic and the Phanerozoic eons" (Thomas et al. 2020: viii). Herein lies the geological statement of the Anthropocene. The data that Anthropocene proponents repeatedly refer to are the figures put forth by the geologists. The idea of the Anthropocene began with the argument of earth scientists that there are good *geological* reasons to see the beginning of industrialization and the year 1800 as the epochal break because this is a "very well established" boundary (Crutzen 2000; Crutzen and Stoermer 2000). With this, however, they would also be referring to the *historical* upheaval that occurred from the end of Old Europe up to the beginning of modernity.

It is noteworthy, however, that the protagonists only do this by means of geological and paleoclimatological measurement series (as overviews, see Bradley 2015; Summerhayes 2020), without referring to the justifications of such a periodization in the historical sciences, media studies, and historical sociology, as in the works of Wehler, Kosseleck, Luhmann, Beck, Kittler, or Habermas (Werber 2014: 242). These authors, however, set the epochal break from pre-modernity to modernity quite differently. In view of the supposed static nature of the Earth, seeking connections between the *slow motion* of the Earth's history and the rapid movement of humanity, its current overheating, initially seems absurd. In fact, within the gradual geological change, humankind represents a force of rapid *geological* transformation. The rocks of the future will serve as the litmus test of whether we as humanity are in the Anthropocene. A geological view of human life and human cultures can reveal phenomena that would otherwise not come into view. This is particularly true of our relationship to the material environment. What is the relationship of our short and fast biographical life cycle to the slow cycles of development of the landscape in which we move (Thomas and Zalasiewicz 2020)? The Anthropocene topic is focused on the temporal break caused by human interaction with the geosphere and the biosphere, which is primarily manifested in the consequences of resource extraction.

Geology is a historically oriented science. It is concerned with natural history and its deep-time dimensions, which only became apparent in the second half of the eighteenth century. More decisive than the discovery of the extensive nature of Earth and life history (*deep time*), however, was its interpretation as being historical. Of central importance here were the collection and systematic classification of finds which began during the Enlightenment (Rudwick 2014: 79–102). Since its inception, geology has been a science that not only emphasizes the ongoing change in supposedly static things, but also the vitality of its subject. James Hutton (1726–1796), the "father of geology," wanted to see geology as the study of living things (Hutton 1788: 216). Irvine points out that Hutton's talk of the geological object as an "organized body"—as opposed to a machine—was not just meant metaphorically. Hutton saw that the systems of circulation between plant, animal, and geological life are intertwined and mutually dependent. This is also evident in the illustrations in Hutton's works. There we see not only stacks of layers, but also the fertile soil on which agriculture depends, and on top of it all, human life, as a kind of rider. We depend on the deep past, and we intervene in it. Both are most evident in the construction industry, and research can be conducted on specific objects such as clinker bricks. But the surface view alone does not show the whole picture of human culture (Irvine 2020: 34, 37–38, 42–43).

Hutton hierarchized the systems and thought of them in terms of their usefulness to humans, as the "owners of their world." He considered the Earth to be perfectly organized for humans, thus expressing what is essentially an early representative of the so-called "anthropic principle." Despite all the anthropocentrism, there is already an approach used in early geology to soften the sharp distinction between living and non-living things, which the sciences have emphasized so strongly and in surveillance mode (*geontopolitics,* Povinelli 2016). This coincides with the early approaches used in cultural anthropology, which saw social systems as being embedded in historically evolved ecological systems, thus taming anthropocentrism. This is the case, for example, with the concepts put forth by Edward Evan Evans-Pritchard (Evans-Pritchard 1940: 94) and in ecological anthropology, especially in early cultural ecology, which emphasized the adaptation of culture to the environment (see Sect. 5.3).

Irvine sees Hutton's concepts as complementary to the anthropological view of time. Evans-Pritchard described the embedding of social time into the meteorological structure in short time horizons, and he emphasized that it would mitigate the influence of the environment through structural time. Hutton has provided us with a complement to this—deep-time embeddedness. It shows that the present life on the surface is to be understood as being part of a larger geosystem. However, taking Hutton and Evans-Pritchard's ideas together also means asking what kind of break the Anthropocene, with its short-term exploitation of resources, entails in view of the immense periods of the Earth's history. In short: We humans live on the surface of the Earth now, but we will not always remain on the surface (Irvine 2020: 37).

Even if one is critical of a formalization of the Anthropocene, it is clear that the Anthropocene debate has brought to light the strong anthropogenic impacts on the geosphere. It is probable that these impacts extend back, at least regionally, almost to the beginning of the Holocene ("early Anthropocene"). Therefore, it seems appropriate that archaeologists, and especially prehistorians and early historians, should devote themselves to *all* human-made changes in both the abiotic and biotic environments, via material culture in its truest sense (Schüttpelz 2016). Even if the Anthropocene hardly appears in the leading textbooks (Renfrew and Bahn 2020), archaeologists are trying to make the multi-level selection model along with the Anthropocene concept useful for concrete empirical research (Braje and Erlandson 2013, for examples, see McCorriston and Field 2020). An example of an application on a meso-scale is provided by Felix Riede in a study on the Late Paleolithic in Southern Scandinavia, where he identified a weakly developed long-period niche in the domestication of reindeer (*Rangifer*) (Riede 2011, 2019; see Riede et al. 2014, 2019).

Irvine makes clear why human culture and its history are grounded in geological history, and why a geological time perspective can elucidate the presentism in the way nature is currently being handled. Only a deep-time expansion of the time horizon can show how abstracted, how disembedded, this extractive and present benefit-focused interaction with the environment that characterizes the Anthropocene is:

> The inhumanity of geology (...) is not an inevitable by-product of the disparity of scale between human biography and deep time. Rather it is a product of rupture: the abstraction of humanity from the time depth of the resources and the ecosystems upon which it depends. Life is hacked out from the ground where it resides (Irvine 2020: 189).

A deep-time expanded perspective of time is what is known from geology, paleontology and evolutionary biology as "deep history" (Rudwick 2014). This also goes in spatial terms beyond a *longue durée* perspective and today mainly uses aDNA (*ancient DNA*) to clarify prehistorical processes. In relation to humans, it is the view that the entire past of the anatomically modern *Homo sapiens*—contrary to the idea of a "prehistory"—is to be considered historical. Accordingly, the early increase in the size of the cerebrum relative to the total weight (encephalization) and the late and, in the geological time frame, sudden colonization of the planet are central themes. Such a perspective focuses on the *supposedly* flat part of the exponential curve that represents the effects of human activity ("hockey stick-curves"), which is often not investigated further (Stiner et al. 2011: 246). The resulting division into prehistory vs. history shows that this part of the history of human impact was not so flat after all (Schmidt and Mrozowski 2013). A consideration of the relevant scales also opens up a view of reverse processes, for example, the decline of cloven-hoofed animals in the Neolithic due to the collapse of many hunter-gatherer economies (Stiner et al. 2011: 246, 250–253, Gamble 2013: 2–6).

If we extend the time horizon further into the past, the question arises as to whether human history can be integrated not only into the history of the planet, the Earth's history, but also into cosmic history, starting from the *Big Bang* . This constitutes the manifesto of *Big History* (Spier 2010; Shryock and Smail 2011; Christian 2011, 2018, 2020; Benjamin et al. 2020; Fagan 2020). Together with geology, *Big History* opens up new perspectives, such as the following: A tiny part of the universe, which has only existed for a very short time in terms of deep time, is beginning to learn collectively and cumulatively, to create networks globally, to be the first species in nearly four billion years of life to change the biosphere globally, to understand itself (to some extent)... and to question itself (Christian et al. 2014: 244–247, 2018: 134–137, 179, 297; Wood 2020; Gamble 2021). Humans have attempted to develop a comprehensive understanding of space and time, a motive that underlies all modern origin narratives.

Homo sapiens, as a single biological species, exerts a global influence on the biosphere and geosphere. In the approximately 4.5 billion years of the Earth's history, this has only been achieved once by a group of organisms: the cyanobacteria (Christian et al. 2014: 244; Dartnell 2018: 195–197). These oceanic blue-green algae significantly changed the atmosphere over two billion years ago: permanently, globally, and synchronously. They gained their energy from photosynthesis and produced oxygen as a metabolic product, which built up the atmosphere. Anaerobic life was almost wiped out, and about 2500 new mineral forms were created (Cohen 2015: 35). Compared to the impact caused by the blue-green algae, human influence on the Earth system is still limited. The cyanobacteria needed several millions of years to have this effect, while the Anthropocene, according to the currently dominant view, only began about 70 years ago. Thus, *Big History* also provides humorous insights. With respect to the Haber-Bosch process, the highly significant innovation for the Anthropocene of reducing atmospheric nitrogen to ammonium for the production of artificial fertilisers:that prokaryotes had already solved the problem millions of years ago, but Haber and Bosch were the first multicellular organisms to be capable of extracting atmospheric nitrogen (Christian 2018: 297).

In order to formulate a response to the Anthropocene, humans cannot wait for their biological adaptation because it is too slow. We must manage this through cultural change. A deep-time *Big History* perspective can thus contribute to a less anthropocentric as well as a more human-centered classification of humans. Big History follows extremely long periods of time and—often over-generalising the basic factors—addresses a wide variety of temporalities (Wood 2020), but it can also contribute to "small big stories". *Little Big Histories* are detailed stories of small spaces, such as cities or rivers, across time (Quaedackers 2020). For example, a river in Spain can be studied from the first local traces of humans from the Lower Paleolithic over 200,000 years ago, to the end of industrialization in the 1980s within a standardised analytical framework (García-Moreno et al. 2020).

1.4.3 Strong Histories, Histories, and Counter-Histories—Lévi-Strauss Revisited

Any attempt to link natural and human history faces the problem of chronology, especially the very different chronologies of various disciplines (Oeser 1987: 52). Individual data only gain significance if they can be related to other data and can refer to opposites. However, chronologies always involve two things: (a) the temporal sequence of events and (b) the temporal distance from today. In the broad spectrum that spans from cosmic history to the biographical history of individual people, however, the type of data and the time dimensions are very different because the "events" last for such different lengths of time. This ranges from billions of years to hours or even seconds. Lévi-Strauss, who is sometimes seen as an "anti-time man" (Gell 1992: 42), made important remarks on the subject of periodization. In "The Savage Mind," he argued that history is never just a continuous or aperiodic series, but falls into "history areas" with their own classes of data:

> History is a discontinuous whole consisting of historical areas, each of which is defined by its own frequency and a differential coding of before and after (Lévi-Strauss 1973: 299).

According to Lévi-Strauss, the French count and private historian Henri Bolainvilliers (1658–1722) was one of the first to address the various fields of historical research and discontinuous chronologies. Bolainvilliers wrote both biographies and texts on the history of his country. Lévi-Strauss also spoke of a discontinuous *whole,* and it follows that the specializations in the sciences that deal with change, from cosmology to geology to individual history, are only a result of methodical abstraction, but not separate realities. Such an approach would allow for transitions between historical approaches to be established, but at the same time it would permit them to be analytically distinguished.

Lévi-Strauss made this clear using the example of prehistory, which is relevant for the Anthropocene if one considers early anthropogenic changes to the geosphere (early Anthropocene, "long Anthropocene"). He said that the "code" of prehistory is not the preparation of the codes we use for modern and recent history, and that each code refers to a system of meaning that is in essence theoretically applicable to the totality of human history. In terms of the deep-time concepts of history, this would argue against incorporating human history into a unified scheme à la *Big History* (Christian 2018). Nevertheless, Lévi-Strauss created a restriction that is relevant to the Anthropocene:

> The events that are significant for one code are not for another. The most famous episodes of recent and contemporary history lose their relevance when they are encoded in the system of prehistory, *except* perhaps (but we know nothing about this) for some *massive* aspects of demographic development considered *on a global scale*, for the invention of the steam engine, electricity, and nuclear energy (Lévi-Strauss 1973: 299–300, emphasis mine).

Lévi-Strauss's considerations can be extended to natural history without every event in human history needing to be traced back to evolutionary change (Oeser 1987: 52–55). In one of his ever-stimulating dichotomous distinctions, Levi-Strauss vividly differentiated between "weak history" and "strong history" in terms of explanatory depth (Lévi-Strauss 1973: 300–301). A weak history is thus a history of short periods, which is detailed in time and rich in content but is not comprehensible in itself. In a strong history, on the other hand, the specifics of temporal change become less important, which inevitably makes it more schematic or less rich in content. The strong history is strong, however, because it can explain a lot. In Levi-Strauss's words, the choice is between histories that explain more and teach less, and histories that teach more but explain less. The greater the explanatory power, the poorer its content: what the historian gains in information, he loses in understanding, and *vice versa*.

There is a structural similarity here to the distinction between the proximate and ultimate explanations of synchronous phenomena in evolutionary research (Uller and Laland 2019). Therefore, the desideratum would be to connect these distinctions of historical chronology with the four levels of explanation (phylogenetic, functional, ontogenetic, actual genetic) distinguished in the philosophy of biology and which can ultimately be traced back to Aristotle. Such a comprehensive historical approach has consequences for the depth and detail of the explanations, which simultaneously overlap and encapsulate, but do not result in a total history. According to Lévi-Strauss, every history thus has an indefinite number of possible "counter-histories" (Lévi-Strauss 1973: 301). Each complements the other, and a hierarchy emerges because the most evident counter-story in terms of content is based on the next higher or next lower time dimension (Oeser 1987: 55). Cultural history would then not be subsumed under natural history, but rather *included* in natural history.

> The recourse to natural history for the explanation of human events is therefore, in this sense, *neither a non-binding analogy nor an impermissible reduction*, but the introduction of a more general, i.e., stronger, explanatory principle (Oeser 1987: 53, emphasis mine).

Lévi-Strauss does not go as far as Oeser, namely, to say that the history of the evolution of life according to his criteria is the strongest of all histories. From the perspective of research on general evolution and evolutionary epistemology, the history of humanity is not simply a replacement for natural history. Just as the development of the universe does not end with the history of the Earth, biotic evolution does not end with the development of man, which does not end with the emergence of culture. With the evolution of human consciousness, a new form of evolution began, but the other stories did not end (Oeser 1987: 55). All these stories do not end with the Anthropocene.

Prominent critics see the fusion of human history and non-human change as a product of post-natural ontology. From a critical perspective, thinking in Anthropocene categories is characterized by assumptions that are post-naturalistic and post-social, but above all post-political. Humans domesticate, technologize, and capitalize nature: they "humanize" the spheres previously seen as "natural" environments (Lövbrand et al. 2015: 213; Rival 2020, 2021). Consequently, these critics recommend a radicalization of post-natural thinking in the sense of a fundamental critique of natural science, an emphasis on social diversity and difference, and an explicit reintroduction of the political into Anthropocene thinking (Swyngedouw 2007; Lövbrand et al. 2015: 212–216, 2020, similar to Lorimer's "symbiopolitics" Lorimer 2012, 2020).

1.4.4 Misunderstandings—What Will the Anthropocene Have Been?

Geological epochs are dated when they have passed.
Guido Visconti (2014: 382)

The working practice of the AWG has shown that a humanities approach to geological concepts, and especially to stratigraphic thinking, is essential (Will 2021: 216). In some critical works, which will be discussed in detail in Chap. 4, there are misconceptions about geology and paleontology. Danowski & Viveiros do Castro, for example, write: "The Anthropocene could only make way for another geological epoch after our disappearance from the surface" (Danowski and de Castro 2019: 11). This is not the case because, firstly, geological periods are only determined according to their beginning, not their end. Secondly, the impact caused by humans *could* indeed become less significant in the future than they currently are due to the measures introduced today, as I optimistically assumed at the beginning of this book using the example of the cultural paleontologist Amy. There are various proposals with respect to this, for example, the concept of *degrowth* or the conservation of biodiversity (Wilson 2016: 210; Lange et al. 2020). This would certainly be compatible with a humanity that produces fewer anthropogenic effects and yet continues to exist. The end of humanity, "a world without us" (Weisman 2008), is not a *prerequisite* for the formal end of the Anthropocene as a phenomenon. From a geological perspective, the only prerequisite would be for the *fossil record* indicating human influence to no longer be quantitatively *dominant*.

One prerequisite to an Anthropocene epoch is for traces of human activity to remain permanently preserved, regardless of the assumption as to whether humanity will still exist in the distant future and thus be able to *geologically* document the anthropogenic impacts on the geosphere. Critics of the Anthropocene thesis argue that major geological events—for instance, volcanic disasters like those that occurred in historical times in Indonesia (Tambora, Krakatau, Toba), earthquakes, or an asteroid impact—could erase all traces of human activity. Indeed, the history of the Earth will continue geologically, and in the distant future, only traces of pervasive *and* long-term disturbances inflicted by humans will remain (Visconti 2014: 382, 384–387). Given the limited depth of intervention in the earth and the uncertainties of preserving human traces, the proclamation of the Anthropocene is perhaps premature (Smil 2020: 280–284).

All these speculative considerations provoked by the real dynamics of the current geosphere, as well as the questions with respect to dating, have consequences for the classical assumptions in the human sciences. With their fixation on anthropogenic effects, the proponents of the Anthropocene would be ignoring the main geological drivers of Earth's history. Visconti also argues that the concept of the Anthropocene overlooks the fact that natural events in the Earth system could end human development (Visconti 2014: 382). However, this does not necessarily mean that no traces would be left. Another pessimistic variant is possible, which figures in current narratives about the future. At some point, humanity will have abolished itself through anthropogenic effects and will then only be history. Instead of Amy, a *non-human* geologist would have to reconstruct humans from technofossil remains. She would probably have to do this, not so much from the remains of outstanding monuments of human pride (*World Cultural Heritage*), but from plastic waste, nuclear waste, and chicken bones. Traces of the metal components of Covid-19 masks could serve as a geological marker that is currently emerging before our eyes, to be used for periodization by a non-human observer in the future.

In the face of the impending catastrophe, we may need a modified or new approach to knowledge, a new *episteme* (Egner and Zeil 2019, 18–19, 28–30). Instead of the power of humans and their emancipation from nature, which as an idea dominated the nineteenth century, it is an awareness of the great crisis of humanity that is at center stage in the twenty-first century (Jordheim and Wigen 2018). In a "state of litigation" on the edge of the Earth's history, we find ourselves in a "trial" about who or what is responsible for the crisis (Sloterdijk 2017: 26–29). Instead of Foucault's analytics of finitude, we would have to deal with an epistemology of "after finitude" in the spirit of Quentin Meillassoux (Folkers 2020: 591). This perceptual break also seems to be one of the attractors of ever new disaster stories, not only in the sciences, but also in visual media (Horn 2014). Humans would only be an object of the *Earth's history* and no longer the subject of *history*.

1.4.5 The Principles of Stratigraphic Periodization and "Anthropocene Layers"

> The Anthropocene Epoch is dead;
> long live the Anthropocene.
> Erle C. Ellis (2024: 3)

In relation to the Anthropocene, geologists, paleontologists, and stratigraphers are concerned with clear physical traces (*signals*, *markers*) of anthropogenic effects, especially in sediments. For Earth scientists, the colorful table of geological time is a kind of sacred scripture of stratigraphy and a sanctuary of the Earth sciences in general. In this respect, the *Geological Time Scale* (GTS) and the *International Chronostratigraphic Chart* (*ICC*) are comparable to the periodic table of chemists and the family tree of biologists (Glaubrecht 2019: 697). Temporal classification and categorization are specific goals of these disciplines (Von Engelhardt and Zimmermann 1982; Will 2021: 113–126). This is important because it differs from the thematization of human influences on the climate and on the other geophenomena of other Earth sciences, such as geography and Earth System Science. In putting forward proposals for the formalization of a geological time period, concrete evidence must be presented, and the way in which geological evidence is formed in geology is rather special, one that differs from other sciences (for details, see Will 2021: 143–201).

Within stratigraphic geology, a distinction must be made between geochronology and chronostratigraphy (Salvador 1994, short version Murphy and Salvador 2000). Geochronology is the area of geological stratigraphy in which geological events and time are categorized abstractly (immaterially) and hierarchically. An epoch represents a medium-length phase of ten million years in the planet's history. This phase is longer than an age, which comprises several million years, but less comprehensive than a period or even an era, with its several hundred million years, and an aeon, which consists of at least half a billion years (Davies 2016: 2). Central to both approaches are the methods of relative and absolute dating, for example, through isotope measurement.

Together, the units form the *International Chronostratigraphic Chart*, which in turn forms the basis of the official *Geological Time Scale, GTS*. All units have an isochronous base, which represents a conceptual surface of *identical time across the entire globe*. The proposal here is to introduce an "Anthropocene epoch." This time surface is then "correlated" in field geology (with varying degrees of precision) with concrete rocks, fossils, or traces (Zalasiewicz et al. 2021b: 2). This brings us to chronostratigraphy, which is materially and field-geologically oriented. It constitutes the area of stratigraphy in which time is applied to geological sequences (mostly rock). Accordingly, there is a separate chronostratigraphic terminology that is parallel to the geochronological nomenclature. The corresponding proposal for the Anthropocene is the introduction of an "Anthropocene series," which would hierarchically be located one level above the "stage" as the smallest unit that could be used on a global scale.

It is important that this Anthropocene series includes all geological events and deposits (rocks or perhaps ice cores) of the period, regardless of whether they are of human origin or not. The "Anthropocene layers" in this chronostratigraphic framework would consist of *all* the layers deposited within the precisely defined time interval, including those that are (a) human-made, such as the sealed soil under cities; (b) partially "natural" but within anthropogenic contexts, such as lake deposits behind large dams, or natural sediment accumulations containing minimal anthropogenic traces, such as microplastics (Bancone et al. 2020) or artificial radionuclides; and also (c) completely "natural" sediments or other rocks with few or no such human traces (Zalasiewicz et al. 2010, 2021a, b: 4).

Following this distinction, geological boundaries would be underpinned by *Global Boundary Stratotype Sections and Points* (GSSP) or by *Global Standard Stratigraphic Ages* (GSSA). GSSPs (also known as *Golden Spikes*) are chronostratigraphic horizons within a stratigraphic sequence that are connected to other stratigraphic sequences and serve as a *global* reference point (Waters et al. 2018). Thus, while GSSAs simply represent geochronological horizons, GSSPs require the detection of physical markers in a specific sequence of sediment layers, such as ubiquitous plastic in ice deposits or fly ash in lake sediments (Rose 2015a, b). The most important candidate is radionuclides from nuclear tests, which are recognised as the basis for the onset of the Anthropocene (Zalasiewicz et al. 2015). However, there are also a number of other markers. In the case of recognition, a corresponding 'footprint' is placed at the lower end of a layer in a so-called type locality. The marking is crucial here in order to clearly determine the boundaries of a global chronostratigraphic unit. The recognition of a GSSP requires globally synchronous (or isochronous) changes, different from historical or archaeological boundaries (Fig. 1.4):

> The synchronicity and precision of definition of both epoch and series (by GSSP) is essential to geoscientists, as the boundary then acts as a time reference surface, around which (commonly complex and diachronous) events and processes in different parts of the world can be located and ordered in time and space, so as to construct a meaningful Earth history (Zalasiewicz et al. 2021a, b: 4).

After years of preparatory work and systematic comparison of many candidates for a reference point, Crawford Lake in Ontario, Canada, was officially proposed by the AWG as a 'golden spike' in 2023 (Waters et al. 2023a, b). This is a small lake above a collapsed karst cave, which represents a natural archive of environmental change due to its low level of water movement (a meromictic lake). As critics emphasize, this excludes diachronic changes (time-transgressive) and local or regional, even 'continent-wide' variations (Ruddiman 2018). A documentation of the gradual accumulation of specific human impacts is not possible with this approach, nor is it the intention (Edgeworth et al. 2015). However, so far there are only very few *spikes* that have actually been documented stratigraphically. An exception is a two-centimeter-thick layer of organic substances in a lake sediment (Waters et al. 2014, 2016).

Within geology, (a) the extreme brevity of the postulated Anthropocene as a geological interval, (b) the novelty of geological signals resulting from human activity

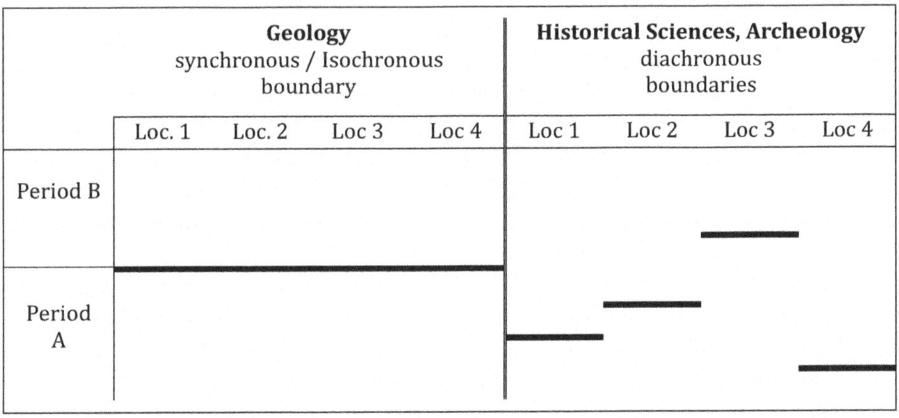

Fig. 1.4 Synchronous periodization in geology and diachronic periodization in historical sciences (A = Anthropocene, Pre-A = before the Anthropocene; Loc. = locality). (Author's work)

(technofossils, such as plastic), and (c) the linking of geological consequences with social causes and also political issues have attracted criticism (Klein 2015; Finney and Edwards 2016; Foster 2018; Finney 2018; Rival 2021). The few spikes so far have prompted critics to compare it to the Gregorian calendar. The latter has indeed recorded human environmental impacts, but also numerous geological events such as volcanic eruptions, and so, as Finney flatly states, it has replaced the geological time scale with good reason (Finney 2018: 217). From this perspective, a *geo*chronology of human influences on the geosphere appears anachronistic.

However, this criticism fails to recognize that the period is only considered to be so short *up until the present time* and could therefore still be extended; nonetheless, it marks the unique geophysical change in scope, rate, and form through formalization and thus represents an important insight in geology (Head 2019). Moreover, the most recent units of the geological time scale all have the shortest duration within their ranks: the Cenozoic era with 66 million years, the Quaternary period with 2.6 million, and the Holocene epoch with 11,700 years and its stages with 3465–4250 years in duration (Zalasiewicz et al. 2021a, b: 6, 2023). As is usual in geological stratigraphy, the comparisons that feature variations in the geosphere in time intervals can only become possible through a chronologically and precisely defined Anthropocene. Only then will it become possible to place climate change forecasts in the planetary context of earlier spatial and temporal scales.

A third possibility is the evidence of a significant change in the dynamics of the Earth system. While the first two criteria represent a practical tool of stratigraphy for classifying layers, the Earth System Sciences (ESS) show a geohistorical change in the interaction of the planet's spheres (see Anthropocene 2 in the glossary). Only these refer specifically to the long-term interaction of humans with their environment (Schellnhuber 1999). Some argue that only a definition of the Anthropocene according to this third criterion would be relevant for cultural and social science questions, such as the dissolution of the distinction between natural history and cultural history, or ontological questions about humans as a "geofactor"(for example, see Bauer and Ellis 2018:210).

In regard to the reception of cultural science, whether affirmative or critical, it is often mistakenly assumed that geoscientists unanimously endorsed the introduction of the Anthropocene. But this is clearly not the case as there is a minority in the AWG that primarily advocates for a geologically, archaeologically, and historically expanded concept of the Anthropocene (*diachronous Anthropocene*, *diachronic Anthropocene*). This is the sum *of all* empirically verifiable human influences that the *Homo sapiens* has ever had on the geosphere (diachronic), and only regionally (see Anthropocene 3 in the glossary).

It should be noted that the criteria for formalization described above represent an ideal construct that stratigraphy does not always follow precisely in geological practice. This also explains why it is taking so long for the AWG to decide on the Anthropocene's formalization. There are fundamental concerns from geology itself. To date, the Anthropocene has not been formalized, and there is still debate as to whether it is on par with the Pleistocene (2,5. Mio.—11.700 yrs. BP) and the Holocene or rather whether it forms a subgroup of the Holocene (Walker et al. 2012;

Leinfelder 2018a, b). Unlike the representatives of the *Anthropocene Working Group*, the majority of whom are *not* geologists, the concept is viewed rather cautiously by geologists and stratigraphers and is rarely promoted at geoscientific conferences (Finney 2018: 217).

1.4.6 Formal Epoch and an Alternative—A Geological 'Event'

The formal introduction was rejected by the Subcommission on Quaternary Stratigraphy (SQS) in March 2024 (Witze 2024; Zong 2024). This happened despite the fact that objections raised by experts (Finney 2014) were refuted in detail after several years of studies (Waters et al. 2023a, b; Zalasiewicz et al. 2024a, b, c). The official reasons given for the rejection are as follows: Firstly, the first anthropogenic impacts on the geosphere set in long before the middle of the twentieth century. Secondly, a new unit in the official geological time scale GTS would restrict the Holocene due to its lower limit. This would mean that the Anthropocene covers a time span of less than a single human lifetime (so far), whereas the usual units of the GTS span thousands or even millions of years. Thirdly, human impacts on global systems are trans-temporal and also spatially and temporally variable. Their onset cannot be adequately represented by an isochronous horizon, which reflects a single point in time (IUGS 2024). Opponents of the formalisation of the Anthropocene as an epoch also emphasise that the decisive anthropogenic layer (stratotype sequence) in Crawford Lake in Ontario, Canada, proposed as a gold nail in 2023, is only 15 cm thick (Walker et al. 2024: 2).

Fundamentally, many geologists are simply not convinced that the incision is so huge as to justify such a high-ranking unit of an epoch in Earth's history . Powerful critics from geology argue that the use of the syllable "-cene" has the effect of automatically seeing the Anthropocene as an epoch or series, i.e. not even considering lower time units. After all, one could consider labelling it as an era with the ending "ian" instead of "-cene" (Gibbard and Walker 2014: 32; Walker et al. 2015; Head and Gibbard 2015: 21). Critics of the formalization argue that the Anthropocene can be seen as a cultural rather than a geological concept (Klein 2015; Gibbard et al. 2021; Witze 2024: 250; Autin and Holbrook 2012).

Christian Schwägerl, who was instrumental in popularising the concept in Germany and knew Crutzen well, reports that what Crutzen himself found most important about his proposal was that it could trigger intensive debates about the relationship between humans and the Earth. "He felt that the risk of formally recognising the Anthropocene was that the debate would paralyse and humanity would become accustomed to its massive influence on the biosphere and the climate" (Schwägerl and Urban 2024). The victorious faction of the rejectionists also emphasises strongly that all these reasons for rejection do not mean that human societies would not massively and rapidly change the planet (Ellis 2024: 1–3; Walker et al. 2024: 2). This is also the conclusion of the official statement of the International

Society of Geological Sciences, stating the concept should only be used informally, but

> … the Anthropocene as a concept will continue to be widely used not only by Earth and environmental scientists, but also by social scientists, politicians and economists, as well as by the public at large. As such, it will remain an invaluable descriptor in human-environment interactions (IUGS 2024).

Looking back from the future, the Anthropocene could therefore prove to be less an era than a transitional period, a transitional phase or rather a geological event, some argue, just as the 'great extinction event' in the Permian was according to current knowledge (Gilbert, in Haraway et al. 2016: 540–541). From a purely stratigraphic perspective, the Holocene as the warm period in which we live today does not constitute an epoch, as earlier warm periods are not separated out as separate epochs (Mönnig 2016, as an overview of the Holocene Roberts 2014).

In a nutshell, the geological criticism essentially states that the term (a) does not correspond to the usual naming practices of chronostratigraphy; it is (b) inconsistent with other epochs of the Cenozoic; (c) its etymology is flawed in several respects. The informal character should be stylistically emphasised, for example by using inverted commas or by writing the term in lower case (Ruddiman et al. 2015: 39, Rull 2018a: 4, 2018b). Representatives of the German Stratigraphic Commission (DSK) expressed a similar view in 2016:

> On the one hand, it is a matter of formally establishing an Anthropocene like all other stratigraphic units with a GSSP and Golden Spike. The DSK considers this to make little sense, even if the unit could become much more popular if it were included in textbooks. As an in-formal term, it is already on everyone's lips, so that the Working Group [of the 'Anthropocene' of the Subcommission on Quaternary Stratigraphy of the ISC] is presumably also driven by this public impact with its project. However, we must be careful not to mix scientific concepts with political worldviews. So perhaps it is better to place the Anthropocene in the field of geoethics rather than on stratigraphic tables' (Mönnig 2016).

The Anthropocene may prove to be less of an epoch in the geological time frame than a comparatively short event. A more recent third view among geologists, as an alternative to formalisation or non-formalisation, therefore proposes that the Anthropocene should not be regarded as a high-ranking formal series/epoch (i.e. a chronostratigraphic unit and corresponding geochronological unit), but rather as a geological event. This would correspond to the major transformative events of Earth's history, such as the 'Great Oxygenation' 2.4–2.1 billion years ago. Such 'events' are not included as stratigraphic units in the geological time table: the Anthropocene would thus 'only' be an informal non-stratigraphic term (Gibbard et al. 2021; Finney and Gibbard 2023; Walker et al. 2024). From a human perspective, the Anthropocene appears to be a catastrophe, but in geological terms it has so far only lasted a moment. Only if it actually lasts a very long time would it resemble earlier drastic geological transformations, such as that at the end of the Permian 252 million years ago, when 96 per cent of all marine species became extinct, or at the end of the Cretaceous 66 million years ago. The argument against this proposal is that geological events cover very long periods of time of very different durations (Waters et al. 2022, 2023a, b).

It can be said that the Anthropocene's own position shows an anthropocentric view of the Earth's history, one that is more precisely centred on the present and therefore presentist ... a perhaps all too human view. The assumption, which initially appears to be purely geological, that humans will continue to be geologically powerful, would also have to be underpinned by social theory, because the current growth-orientated way of life cannot represent a stable permanent state (Ostheimer 2016: 38, cf. Sieferle 2000, first 1997: 160). The anticipatory review with the geoanthropologist Amy at the beginning of this book showed the speculative elements in the concept of the Anthropocene. Taking a future geologists' perspective, after all, we cannot know exactly what will be left of human activity in the distant future and in what form (Santana 2019: 1077–1088).

What will remain permanently of asphalt surfaces, plastic bottles, teabags, T-shirts, Covid 19 masks, ballpoint pens, aluminum cans and broiler bones? Which parts of our currently produced mixture of sediment grains and other natural deposits (e.g. shells, wood) held together by molten plastic, called "plastiglomerates" will persist for so long to be Anthropocene markers? (Fig. 1.5). If we do the thought experiment and assume, unlike in the story with or future geoanthropologist, that humanity disappears from Earth's history, '... how permanent would the traces we have left behind in the *Earth system* be?' (Sakkas 2021: 161, emphasis CA). The notorious plastic bags in the rubbish swirls of the oceans, the sealing of surfaces and the huge open-cast mining holes would soon have disappeared or at least can no longer be attributed to a clear origin. Even the plutonium from the fallout with a

Fig. 1.5 Plastiglomerate from Kamilo Beach, Hawai'i, displayed at "One Planet" Exibition, Museon, The Hague, The Netherlands. (Photo: Aaikevanoord; https://commons.wikimedia.org/wiki/File:Plastiglomerate_Museon.jpg)

20,000-year half-life would have almost completely decayed in 200,000 years. To put it bluntly, all of this—seen on a global and geological scale—is 'ephemeral trivia' (Müller 2021: 3) compared to the Alpidic folding, for example.

However, all of this can only be definitively assessed in the distant future, for example, by the cultural geologist in the Quintary mentioned at the beginning. A premature introduction could therefore run the risk of presenting a hasty interpretation of the current global change as a trend in long-term geological change. The Anthropocene thesis would fall victim to an ill-considered exceptionalism or exaggerated presentism. On the other hand, not introducing it would miss the geologically and historically verifiable realization of the force that the *unprecedented* and possibly irreversible anthropogenic change represents. Apart from all these rather methodological-technical objections, the ultimately decisive reason for the rejection is probably that geologists traditionally feel responsible for the deep past, less for the recent past, hardly for the present and almost not at all for the future.

The speculative element in the argument in favour of the Anthropocene as a geological unit of time (Table 1.2) cannot be denied. Apart from all the more methodological-technical objections, the ultimately decisive reason for the rejection in 2024 is probably that geologists traditionally feel responsible for the deep past, less for the recent past, hardly for the present and almost not at all for the future. Nevertheless, my (current!) conclusion on geological classification is that the arguments in favour of formalisation together are stronger than the counter-arguments.

The evidence of several synchronous time markers around the year 1952 is particularly important. The arguments repeatedly voiced in geological circles that such a formalisation would run counter to previous "well-established" practices do not really seem to hold water. In a careful analysis, the historian of science Eugenio Luciano shows that the previous procedures of stratigraphers were anything but consistent and still are today. This becomes clear when analysing the International Stratigraphic Guide (Salvador 1994; Murphy and Salvador 2000) and its application (Luciano 2022a: 32–36). Conceptual diffuseness and inconsistency are evident in chronostratigraphic literature and practice both in the characterisation, definition

Table 1.2 Chronostratigraphy of the Quarternary with proposed Anthropocene added (BP = years before 2000 CE)

System/ period	Series/epoch	Stage/age	Onset; golden spike GSSP	Primary geological marker
Quarternary	*Anthropocene*	Crawfordian	1952	Plutonium 239, plus multiple indicators; "Great Acceleration"
	Holocene	Meghalayan	4250 BP	Paleoclimatic event 4.2 ka
		Northgrippian	8236 BP	Paleoclimatic event 8.2 ka
		Greenlandian	11.700 BP	Paleoclimatic warming
	Pleistocene	(to be named)	129.000 BP	(termination discussed)
		Chibanian	0.774 Ma	Paleomagnetic change
		Calabrian	1.8 Ma	Paleomagnetic change
		Gelasian	2,58 Ma	Paleomagnetic change

Author's own work

and naming of stratigraphic units—three aspects that are often confused (e.g. in Bonneuil 2015). Thus, some system units are named according to temporal position, some according to lithological character, others according to geo-graphic locations. Some even bear the names of early human cultures in Wales (Ordovician, Silurian). The preservation of historically established names and a line of tolerance and flexibility deliberately pursued by the guide with regard to criteria of definition and characterisation clearly dominate over terminological consistency (Luciano 2022a: 34).

Even the use of the term "well-established" in relation to the introduction is unclear and should be evaluated more strategically. If one takes the example of the word "Anthropocene" itself, "well established" can firstly mean that the term is popular. Wikipedia has (as of August 2024) keywords for the Anthropocene in 52 languages. As explained in the following chapter, it is widely used in the media and, for example, in the visual arts. Secondly, it is well-established in that it is widely used linguistically and institutionally. There are many translations of the term, such as *anthropocène* (French), *anthropoceno* (Spanish), *Anthropoceen* (Dutch), Mannöld (Icelandic) and Антрапацэн (Russian). It has also been translated into non-European languages, e.g. Chinese (人类世), Japanese (人新世) and as an *Antroposen* into Bahasa Indonesia. Thirdly, "well-established" can mean that the word is used in concrete research in the natural sciences, especially in the earth system sciences and, for example, the water sciences (Luciano 2022a: 35–36). This is all the more true for the humanities, social sciences and cultural studies, which are particularly emphasised in this book.

The effects of human activity can lead to us being trapped in the landscape dynamics that we ourselves have (co-)created. In history, this can be seen in the responses of rocks, soils, and whole landscapes: different temporalities collide—geological deep time and human time. Typically, social systems respond by opening themselves to deep time or by trying to slow it down. Irvine demonstrates this using the example of how peatlands are dealt with in England ("fixing flux," Irvine 2020: 66).

In the Anthropocene, the sciences as part of modern society have partially opened themselves up to deep time. In this respect, it is not only a professional and technical question whether the institutions of geology will commit to formalizing the Anthropocene. No, it is also a cultural question within geology (Zalasiewicz et al. 2021a, b: 6; Will 2021: 143–201). Scientifically, the topic developed in an exchange between a rapidly emerging interdisciplinary community of Earth scientists on the one hand, and an established and disciplinary community of geologists on the other. This also includes matters of organizational sociology and the sociology of knowledge within the geosciences, because the AWG is an honorary institution of geology where not only non-geologists play an important role, but where minority positions can also make themselves heard. Apart from the existential anthropological significance of the Anthropocene as an issue, this content-related and knowledge-related nexus as such can represent one of the interesting objects of the globally interested cultural and social sciences. This brings me to the cultural resonance of the Anthropocene.

Chapter 2
Cultural Resonance—Loss of Orientation, Fears, and Hope

Abstract This chapter presents the fascinating discovery of the Anthropocene in the year 2000 and follows the diverse paths of its rapid popularization. I must explain that the Anthropocene is not simply old wine in new bottles, and that the Anthropocene is much more than human-made climate change: it concerns large parts of the geosphere. The chapter discusses a variety of proposals for the beginning of the Anthropocene in human history, which date back to many thousands of years from today. These historical and archaeological debates about periodization contain a hitherto untapped social and cultural-theoretical potential for research on cultural change and the question of the main factors of long-term cultural revolution.

> *The prophecy of the "end of the world" must therefore be announced performatively so that it does not materialize.*
>
> Deborah Danowski & Eduardo Viveiros de Castro (2019: 108)

> *Never before have more people believed that the ecological crisis is global, and never before has it been less clear what this actually means.*
>
> Frank Uekötter (2020: 635).

The word "Anthropocene" is known in a wide range of disciplines and has also found its way into non-scientific fields. Several of the Sustainable Development Goals (*SDGs*) are built around core issues of the Anthropocene, such as global emissions, the degradation of ecosystems, and dependence on fossil fuels. The UNESCO dedicated an entire issue of its in-house magazine *Courier* to the topic (UNESCO 2018a, 2018b). In the mass media, documentaries such as "Anthropocene: the Human Epoch" (Baichwal et al. 2018) are being shown, and even feature films are drawing public interest to the topic. The attention given to the Anthropocene has led to many new visualizations of the state of the world (Mirzoeff 2014) and has also stimulated the visual arts (Horn and Bergthaller 2022: 117–138). There are podcasts dedicated to the topic, for example, John Greene's "The Anthropocene reviewed" (Green 2021). Furthermore, the Anthropocene has led to new and experimental text forms (for example, Tsing 2015; Tsing et al. 2017; Krogh 2020). In short, the Anthropocene is increasingly becoming a charismatic mega-concept (Davis and Turpin 2015: 16).

C. Antweiler, *Anthropology in the Anthropocene*, Anthropocene – Humanities and Social Sciences, https://doi.org/10.1007/978-3-031-74591-1_2

2.1 Multiple Births—Epochal Breaks and Popular Culture

In the 1990s, we had no vocabulary to discuss the emerging trends in humanity's relationship to the environment.
Lourdes Arizpe-Schlosser 2019: 288

Not a day goes by in the 2010s without some humanities scholars becoming quite exercised about the term Anthropocene.
Timothy Morton (2014: 1)

The term 'Anthropocene' is known in a wide range of disciplines and has also found its way into non-scientific fields. Several of the Global Sustainable Goals (SDGs) are built around core Anthropocene issues, such as global emissions, the degradation of ecosystems and dependence on fossil fuels. UNESCO has dedicated an entire issue of its in-house magazine Courier to the topic (UNESCO 2018a, b). The mass media are showing documentaries such as 'Anthropocene: the Human Epoch' (Baichwal et al. 2018), a three-part series from Terra X and even feature films are focussing public interest on the topic.

The attention paid to the Anthropocene has led to many new visualisations of the state of the world (Mirzoeff 2014) and is also booming in the visual arts (Horn and Bergthaller 2022: 117–138). There are dedicated podcasts on the topic, e.g. John Greene's 'The Anthropocene reviewed' (Green 2021). The Anthropocene has also led to new and experimental text forms (for example Tsing 2015; Tsing et al. 2017; Krogh 2020). In short, the Anthropocene is increasingly becoming a charismatic mega-concept (Davis and Turpin 2015: 16).

2.2 A Multiple Birth—Epochal Breaks and Popular Culture

The world's largest linked research dataset *Dimensions* counts a whopping 186,291 publications, 633 datasets, 1519 funding grants, 84 patents and 1581 policy documents using the term "Anthropocene" under the input "anthropocene" on 13 July 2024 since 2000.The word "Anthropocene" has experienced a meteoric rise (see Table 2.1). A search engine for the chronology of mentions in scientific publications shows that the English Word "Anthrpocene" and English adjective "anthropocenic" as well as the German word "Anthropozän" have increased significantly since around 2010 (Ngram Viewer, accessed 13.7.2024, cf. Krämer 2016). We can find a very broad field of users beyond science, which the geographer Noel Castree jokingly refers to as the *Anthropo(s)cene* (Castree 2014a, b, c).

However, the flourishing of a word and a discourse career, which is quite evident (Matejovski 2016: 9–10), does not automatically mean that there is also a proliferation of the concept and its significance. In many cases, the word is used more for marketing purposes (for example, see Hüther et al. 2020). The word often simply serves as a replacement for "environment" or is used as a mere show term, even in the scientific context. For example, a major article on the subject of a necessary

Table 2.1 Proliferation of the term Anthropocene: institutionalization in the sciences and reception in public and popular culture

Occurrence	Fields, disciplines, institutionalization	Examples, media, authors and newer sources
Nineteenth century	Similar formulation in terms of terminology	Stoppani 1889: "nuova forca tellurica"
1980s	The word "Anthropocene"	Stoermer (verbally and informally)
2000	The term "Anthropocene"	Crutzen and Stoermer 2000
2000	History, environmental history, *global history*	Radkau 2000, McNeill 2000 (2000), Chakrabarty 2009, Hughes 2009, McNeill and Mauldin 2015, Headrick 2021
2004	Earth system sciences	Steffen et al. 2004: Meta-analysis
2007	Multidisciplinary approach	Steffen et al. 2007
2008	International institutions	*Stockholm Resilience Center* 2011
2008	Geography	Ehlers 2008
2009	Geology and Paleontology climate change research	*Anthropocene Working Group* (*AWG*) 2009, Rockström et al. 2009
2010	Humanities and cultural studies, social sciences	Among others, Urry 2010
2010	Social media	*Ted talks*
From 2010	Interdisciplinary Anthropocene journals	*The Anthropocene review, Anthropocene, Anthropocenes—Human, inhuman, Posthuman, environmental humanities, Earth's future, Elementa, science in the Anthropocene, Anthropocene science*
2011	Interdisciplinary science	*Philosophical Transactions of the Royal Society* Special Edition (Williams and Zalasiewicz 2011), Conference "The Anthropocene. A New Epoch in Geological Time" 2011
2011	General public	*The Economist* Cover Story 2011, *National Geographic Magazine* (Kolbert 2011)
2012–14	New hybrid disciplines	*Environmental humanities, integrative humanities, Sustainability science*
2014	Inclusion in general dictionaries	*Oxford English Dictionary*
2014	Jurisprudence	Kersten 2014, Pottage 2019, Vidas et al. 2020
2015	Catholic Church	*Laudato si'*, Pope Francis (& Schellnhuber)
2015–2017	Art studies	Davis and Turpin 2015
2016	General acceptance as an informal term	World Geological Congress, Cape Town
2017	Religious studies	Deane-Drummond et al. 2017, Steiner 2018
2018	Feature-length films	Baichwal et al. 2018: "Anthropocene"
2018	Special encyclopedias	DellaSala and Goldstein 2018 (2280 pages), Krogh 2020, Alexandre et al. 2020, Howe and Pandian 2020, Fourault 2020
2021	1st Atlas of the Anthropocene	Gemenne et al. 2021

(continued)

Table 2.1 (continued)

Occurrence	Fields, disciplines, institutionalization	Examples, media, authors and newer sources
2021	1st Museum Curator for the Anthropocene (Nicole Heller)	*Anthropocene center, Carnegie Museum of Natural History*, Pittsburgh
2022	1st chair for the history of the Anthropocene (Debjani Bhattacharyya)	Swiss Federal Institute of technology, Zurich
2024	Rejection of formalization as an epoch	*International Subcommission on Quarternary Stratigrapphy (SQS)*

global environmental policy (*Earth governance*) in *Science* (Biermann et al. 2012) carries the word Anthropocene in the title, but it does not appear even once in the text (see, however, Biermann 2014). A book on environmental economics with "Anthropocene" in the title only mentions the word on three of its pages (Brown and Timmerman 2015: 194, 295, 375).

2.2.1 The Meteoric Rise of a Term

How should the Anthropocene be classified—epistemologically and in terms of social sciencesociologically? Does the Anthropocene offer a new perspective on the world, or a "distinguishing lens" that allows different disciplines to re-examine their concepts of nature and humanity (Bauer and Ellis 2018: 210)? Is it truly about new insights, or rather about seeing and naming something more clearly, or even about recognizing something already known (Horn and Bergthaller 2022:125, Mentz 2019)? Is the discussion about the Anthropocene about previously unknown facts, about a politically important "attention anchor" for humanity, or is it about the obfuscation of accountability for catastrophic environments? r are we ultimately dealing with a new interpretive framework for how modern society describes itself? To answer such questions, we must deal with the proliferation of the idea and the word "Anthropocene," because this concerns both the development of the geosphere and the recent history of human societies.

2.2.2 Earth System Science—The Discovery of Rapid Earth Change

Earth System Science, considered to be an interdisciplinary metascience, is the product of the distant view of the Earth from space, which allowed us to see the planet as a whole. From the perspective of Earth System Sciences, the Earth—more precisely, its shell, the geosphere—forms a complex system dominated by the oceans. Individual sciences, such as biogeography or landscape ecology (which leaves out the oceans), will notoriously underestimate the Anthropocene problem because the Earth system is more than a collection of ecosystems (Hamilton 2016a: 94). The idea of an integrated Earth system emerged in the 1960s and 1970s in connection with the concept of *Gaia* as conceived by James Lovelock (Lovelock 1979, 2020; see Sect. 3.2). It offers a holistic perspective on the planet as an integrated system. More specifically, this means that the tomato we eat, for example, cannot be understood without reference to soils, rocks, ice, water, and air—and all this over millions of years (Thomas et al. 2020: x).

Earth System Science is a kind of long-term and large-scale extension of ecology. The term *Earth System Science* emerged in the late 1980s when NASA began to focus on the depletion of the ozone layer and more generally on climate change (National Research Council 1986: Ehlers and Krafft 2010; Lenton 2016: 1, 5). The Earth system refers to *one* system, but it is a very large system and an extraordinarily complex one. Complex systems have two basic characteristics, namely (a) emergent properties at the system level that are not evident in their components, and (b) the so-called attractors, that is, certain well-defined states. In particular, recent biodiversity research has provided insights into large-scale *and* long-term networks. From the perspective of the Earth sciences in view of what seems to be an overly comprehensive claim, it should be noted: "There cannot go any smaller" (Mulch and Zizka 2021: 18–21).

Even before the postulate of the Anthropocene, the *Ea*rth sciences were already able to prove that the spherical surfaces, which together form the geosphere, interact strongly with each other: the spheres of the air (atmosphere, stratosphere), rocks (geosphere, in the strict sense), soils (pedosphere), water (hydrosphere), ice (cryosphere), and life (biosphere of about 20 km thickness). They are mutually constitutive parts of the *one* Earth system. Recent geological research, however, has made it clear that parts of the Earth system are also largely decoupled from each other, such as the *decoupling* of parts of the lithosphere as the upper part of the Earth's mantle from the asthenosphere.

Similar to organic evolution, the entire Earth system also shows different forms of change on different time frames: individual irreversible change events ("revolution"), long-term trend-like or directional change ("evolution"), and oscillating change, that is, changes that modify the direction several times within certain frames, such as during ice ages.. The majority of human evolution took place in the Pleistocene between 2.5 million years ago and around 11,700 years ago. The Pleistocene climate was a period of great climatic instability with many ice ages and interglacial periods as well as dry and wet phases. According to the "variability-selection hypothesis", heterogeneous environments and drastic fluctuations in the environment—and not a specific type of habitat—have decisively determined the framework of the evolution of human cognition, especially human adaptive flexibility (Potts 2013; Meneganzin et al. 2020: 2–3). Currently, the geosphere is on a rapid path out of the comparatively stable Pleistocene–Holocene transition with a global temperature rise of eight to 11°C at that time (11,700 BC) and the likewise relatively stable Holocene or interglacial period. Since the middle of the twentieth century, the geosphere has been moving out of the glacial–interglacial boundary cycle of the late Quaternary without having yet developed its own attractor, as paleobiological studies have shown (Zalasiewicz et al. 2018a, b: 221–222).

As studies in the history of science have demonstrated, the understanding of rapid change in the geosphere itself does not stem from the Anthropocene discussion itself. As early as the mid-1960s, that is, at least 35 years before Crutzen's intervention, it became clear from Earth System Science, which emerged in the context of space travel, that the geosphere has the potential to move extremely quickly from one state to another, as seen in geological time scales. Four central insights were important for this, which were established in the geosciences between 1966 and 1973 (Brooke 2014: 23, 25–36, see Turner 2011): (a) the confirmation of plate tectonics, (b) the significance of extraterrestrial impacts for the Earth's history, (c) the relevance of (also) sudden or disruptive change in the evolution of life (punctuated equilibria), and, above all, (d) the insight into the existence of the Earth as an integrated system, which first manifested itself in by James Lovelock's Gaia hypothesis.

The "old" or rather static geology was greatly energized by the discovery of plate tectonics, a paradigm shift that I myself experienced during my geology and paleontology studies from 1975 onwards. Nevertheless, the scientific focus of none of the four new perspectives was on humanity. This capacity of the climate and other sub-spheres of the geosphere for abrupt state-shifting predates the existence of humanity by billions of years and is therefore independent of our existence. None of the four processes requires the presence of humans (Clark and Szerszynski 2021: 21). The problem in the Anthropocene is that humanity as a species now also has the ability to (most likely) trigger such abrupt system changes in the geosphere. These insights culminated in 1999, a year before Crutzen's proclamation of the Anthropocene, in Hans Joachim Schellnhuber's interim conclusion of the International Geosphere-Biosphere Program (1987–2015), in which he stated that we were experiencing a "second Copernican revolution", "namely… that we may have irreversibly crossed critical limits on a planetary scale" (Schellnhuber 1999: C23).

2.2.3 The Year 2000: Crutzen's Intervention and McNeill's Historical Diagnosis

When did the discussion start about the fact that we are now living in a geological epoch in which humanity is a significant factor in planetary change? The first broadly effective mention of the idea was a more or less spontaneous statement by atmospheric chemist Paul Crutzen at a conference in Cuernavaca near Mexico City on February 22–25, 2000 (Crutzen 2000; Crutzen and Stoermer 2000: 17). There, a small group of two dozen experts discussed global environmental changes in the Holocene, the current relatively climate-stable section of the Earth's history, which began 11,700 years ago. Conference chair Will Steffen and his colleagues kept mentioning the term Holocene, which increasingly annoyed one of them:

> Finally, he burst out: 'No! We are no longer in the Holocene. We are in the …,' he thought for a moment, '… Anthropocene!' The room fell silent. Crutzen had obviously struck a nerve. The term came up again and again during the rest of the conference (Zalasiewicz 2017: 52).

In this way, or something similar, is how the beginning of the debate is always portrayed. This founding story is usually simply accepted as a given fact, but one can ask to what extent this founding story itself constitutes a narrative or—in my opinion, quite an exaggeration—should even be seen as the genesis of the myth (Hoiß 2017: 25, Unmüßig 2021). Witnesses at the conference recalled that the ideas culminated in Crutzen's proposal had already been discussed since the mid-1980s (Arizpe-Schlosser 2019: 268). Crutzen, himself neither a geologist nor a historian, suggested that the Anthropocene began at the end of the eighteenth century. However, he emphasised that a precise dating would be rather coincidental. These chronological questions about the beginning of the Anthropocene were soon to lead to heated discussions.

A movement working towards an interdisciplinary Earth System Science had already existed since the International Geophysical Year IGY 1957–1958. In addition to geophysics, the disciplines of meteorology, oceanography, and glaciology were also represented, while biology, for example, was not yet included. Biology and other disciplines were only added around 30 years later, within the framework of the *International Geosphere-Biosphere Programme* (*IGBP*) of the *International Council of Scientific Unions* (*ICSU*, now the *World Science Organization*), from 1986 onwards. This has been funded through the research program *Human Dimensions of Global Environmental Change* of the *International Science Council* (*ISSC*) since 1988.

In this respect, it is important to know that Crutzen's outburst in Cuernavaca was a response to a presentation of the results of the project *PAGES* (*Past Global Changes*). This was in the context of the IGPB, and the intervention "happened" at the annual meeting of the scientific committee of the IGBP. Despite the seemingly geological nature of the name "Anthropocene," the context of Crutzen's interjection was thus an Earth science context and hardly that of evolutionary biology, geology, or even stratigraphy.

In any case, Crutzen's interjection was an intervention in the literal sense. He had reacted to the contradiction of constantly talking about an epoch in which humans had factually caused such extensive changes to the geosphere without making any mention of them in the epochal designation. While Earth system scientists have increasingly embraced the concept, their older idea of a steady change from the Holocene to the Anthropocene was transformed into an image of a clear and rapid change in the state of the Earth (Zalasiewicz et al. 2021a, b: 8).

In terms of climate geology, the Holocene is a warm interglacial or a warm period of the current Quaternary ice age, which began 2.6 million years ago. In evolutionary terms, mankind developed during this long period of temperate climate, which was unprecedented in the history of the earth. With 'Anthropocene', Crutzen found a proxy that summarised several fundamental transformations of the geosphere as an overall system in a short formula. In view of the fact that conferences on global change were still being held around 2000 where human activities were not mentioned at all, the hope was that with 'Anthropocene' there was now a term that could bring together the natural and social sciences on the basis of a human issue. In 2001, a major conference was held in Amsterdam: the 'Challenges of a Changing Earth'. With the term "Anthropocene," Crutzen had found a *proxy* that brought together, in a short formula, the several fundamental transformations of the geosphere as a complete system. Lourdes Arizpe Schlosser was involved in such meetings as a cultural anthropologist and remembers:

> Somehow, I felt that the veil that had separated social scientists from biophysical scientists had suddenly become more transparent (Arizpe-Schlosser 2019: 269).

As Crutzen quickly realized after the conference, Eugene Stoermer, a limnologist and researcher of marine diatoms, had already been using the term Anthropocene informally since the early 1980s, using it in lectures at the University of Michigan and at the Iowa Lakeside Lab. However, as one of Stoermer's former students recalls, he used the term informally and intuitively for human influences on the geosphere: „I recall Gene Stoermer using the term Anthropocene, but in such a way that it did not even seem new, or even novel. It was another way that he expressed what he had observed for decades." (Luciano 2022a: 30).

The biologists Francis Putz and Andrew Samways had proposed the term "Homogenocene" (also "Homogocene") in 1998 and 1999, i.e. very shortly before Crutzen's intervention, in order to summarise the effects of the massive global exchange of animal species since the discoveries ("Columbian exchange", Crosby 2003, 2015: 96ff, 116ff.) and the resulting standardisation of species composition (Putz 1998; Samways 1999: 65). In 1999, one year before Crutzen's intervention, the biologist Andrew Samways proposed the term "Homogenocene" (also "Homogocene") to describe the effects of the massive worldwide exchange of animal species and the resulting homogenization of the living world. For example, biodiversity is reduced by the approximately 10,000 species spatially displaced through their transport in the ballast tanks of ships (*neobiota*). In some parts of the world, this process began as early as the beginning of the seventeenth century (Samways 1999: 65, Mann 2011: 17, 2024, Seebens et al. 2018). Due to the global connection of all habitats, from a geological perspective we are living in a new, low-barrier Pangaea (McKinney 2005). This was the primordial continent before it split into individual plates. Although we do not have a physical supercontinent today, we do have a biotically largely interconnected world of life.

Andrew Refkin, an American science journalist, had even argued in a book about global warming in 1992 that the world had entered a geological interval that could be referred to as an "Anthropocene." Refkin was a well-read and visionary science journalist, but Crutzen had become famous as the director of the Max Planck Institute for Chemistry in Mainz (1980–2000), primarily for his thesis that a large nuclear war could trigger a nuclear winter and wipe out all animal and plant life.

As the "savior of the ozone layer" and the most cited scientist in the world at the time, it was he who successfully popularized the term "Anthropocene" with his reputation as a Nobel laureate. In the "prehistory" of postmodernity, an alternative to the epochal term "Atomic Age" was thus found (Schmieder 2014: 45–47).

Shortly after the Cuernavaca conference, Crutzen contacted Stoermer, and together they published the first short paper in the *IGPB Newsletter* (Crutzen and Stoermer 2000)—just in time for the turn of the millennium. However, the term's popularity suddenly increased with a one-sided, – but in my opinion anything but one-sided—article by Crutzen in *Nature*, which had the powerful title "Geology of Mankind" (Crutzen 2002). There he stated the goal as: "… to assign the term 'Anthropocene' to the present, in many ways human-dominated, geological epoch, supplementing the Holocene" (Crutzen 2002: 19). The meeting in Cuernavaca was not a conference of geologists, but of Earth system scientists. Earth scientists had used the computing capacity of mainframe computers available since around 1990 to represent the Earth system, especially the atmosphere, in quantitative models and to simulate long-term processes. It increasingly became apparent that the norms of the Holocene were being abandoned—a "regime change" within the global climate system.

Crutzen was not a geologist, and geologists first had to realize that Crutzen's intervention had generated so much literature in the Earth sciences and in biology that the term was probably not just a passing scientific trend after all. In addition, Crutzen and other Earth scientists had cited environmental parameters as evidence, such as new chemical components in the atmosphere or the extinction of species. Crutzen subsequently used the concept in many scientific papers, which had a high impact, as the book by Benner et al. and especially the bibliometric analysis by Brauch show (Benner and Crutzen 2021; Brauch 2022).

A global perspective on human environmental effects was in the air at the turn of the millennium. It had long been known that humans had been causing local environmental disasters and, in some cases, regional environmental change for a long time. Now, however, the focus was on global environmental change and globally orientated environmental history was gathering pace. Some representatives took a very long-term perspective and argued that the history of human cultures and civilisations was a history of environmental disasters. The system-wide and structural effects of human 'macroparasitic' activity would go back at least 5000 years (Chew 2000). Also in 2000, the major study 'Nature and Power' was published in Germany by the German environmental historian Joachim Radkau (Radkau 2000). Radkau focussed on the political contexts of environmental change and showed that the awareness of and concern about rapid environmental change did not just begin with the environmental movement. Even during early industrialisation there was a strong concern about drastic environmental effects. Social awareness of the environment was growing and practical movements, such as the practical network of the hygiene movement, existed to promote healthy public life. These environmental historical works saw a very long longue durée and argued with similar historical data as sparser proponents of an 'early Anthropocene'.

However, these important environmental history studies did not yet argue with geological concepts, deep time and globally synchronous time markers. John R. McNeill, a pioneer of ecologically extended global history, came closer to this step also in 2000, but independently of Crutzen. Based on historical findings from a variety of sources, he identified a dramatic increase in a number of phenomena in the global environment around the middle of the twentieth century. McNeill speculated that a historian at the end of the twenty-first century looking back on the twentieth century would probably place less emphasis on the two world wars, fascism, communism, and the women's movement than the fundamentally changed relationship of humans to the environment (McNeill 2000: 17). McNeill then went on to work with Earth scientists, such as Steffen and Crutzen. He is permanently engaged with stratigraphic questions and became *a* member of the AWG as a *historian*, where he stands for the historically correct localization of geoscientific knowledge (Will 2021: 222). Today, 20 years later, environmental historians, who were first made aware of the global dimension of their topic by McNeill, are surprised at how much faster McNeill's assumption is likely to come true, according to current knowledge (Marks 2020: xi).

In the same year, 2000, Kofi Annan launched a multi-year investigation for the United Nations, which delved into the global transformation of the environment; the investigation was named the "Millennium Ecosystem Assessment" (MA). The aforementioned science journalist Revkin subsequently played an important role in spreading the concept of the Anthropocene in the English-speaking world. Revkin is a talented communicator between science and the public. He runs a very successful blog called "Dot Earth" and was chosen by *Time Magazine* as one of the world's top 25 bloggers. In 2003 and 2013, Revkin received the "National Academies Communication Award" from the National Academy of Sciences, the National Academy of Engineering, and the Institute of Medicine. These awards attest to his abilities as a multiplier or, to put it critically, his ability to transform scientific findings into narratives that appeal to the public and can be used in mass media (Hoiß 2017: 25).

2.2.4 A Geological Category Without Geologists?

As early as 2001, climatologists and chemists led by Will Steffen and Hans Joachim Schellnhuber began collecting data for a geoscientifically underpinned synthesis of quantitative studies on trends in long-term global change from 1750 onwards.. The aim was a compilation and a new interpretation of data that had been collected for other purposes. This meta-study compared the change trends of 12 Earth system variables, such as atmospheric gases, biodiversity, ocean acidification, and nitrogen levels, with 12 human parameters, such as population increase, economic growth, resource consumption, dam construction, fertilizer production, waste generation, and artificial radionuclides. The results showed in two graphic charts an extremely rapid increase in almost all the trends from the middle of the twentieth century onwards (Steffen et al. 2004). The curves all rose slowly, only to rise abruptly around 1950 (J-curves or "hockey stick curves" after Michael Mann).

This sudden and simultaneously drastic increase in many parameters was interpreted as a "second Anthropocene"—a second phase following the onset of colonial resource exploitation, including the industrial revolution. After the CO2 concentration in the atmosphere had risen linearly until the middle of the twentieth century, it increased dynamically, rapidly, largely exponentially and in some cases irreversibly from around 1950 onwards ('Keeling curve'). Based on the accumulating geoscientific data, this global trend reversal was increasingly recognized as a qualitative leap and was then dubbed the "Great Acceleration" at the Dahlem Conference in 2005 (Steffen et al. 2007, updated in Fig. 2.1). The underlying data were largely confirmed in an updated study by scientists from the IGPB in collaboration with environmental historians in 2011 (Steffen et al. 2011a, b: 4, 6–7, Steffen 2019). On *Google Scholar*, the number of hits for the search term "Anthropocene" doubled from 12,900 in October 2014 to 28,400 in September 2016, and at 241,000 hits as of this writing in May 2021 (Braje 2015: 374, Arizpe-Schlosser 2019: 277).

The key insight from these data was that this is not simply a multiplication of individual environmental trends. As early as 1995, a group of environmental historians used historical data to diagnose a rapid increase in global electricity demand and the diverse associated environmental consequences for the years between 1949 and 1966 as the '1950s syndrome', although at that time this related to industrialised countries (Pfister 1995). The new geoscientific perspective now showed even more clearly how closely the developments in Earth's history and in human history, which once operated on completely different time scales, are intertwined (Haardt 2022: 127). The multiple "human footprints" call for a reorientation of environmental and also historical thinking in terms of content. It now also concerns more than the sheer magnitude of human-made effects, such as the fact that the number of dams has risen from almost zero in the late nineteenth century to around 25,000 (Duflo and Pande 2007; Merchant 2020a, b: 4). It is about a categorical change, namely that of the geosphere *system*. This systemic leap calls for a fundamentally new way of thinking in the field of international politics, and about the place of political economy in its relationship to the Earth system (Dryzek and Pickering 2019: 4–5, Arias-Maldonado et al. 2019; Simangan 2020a, b).

Nevertheless, unlike climate scientists and Earth system scientists, geologists as field scientists are only interested in the archive of rocks. They want to determine geospheric change based on fossils, the structures of layers and facies characteristics, and the chemical traces in rocks (Thomas et al. 2020: xi, 46–48). Fossils and also "trace fossils" play a role here, where nothing of the organism itself is preserved, except for an imprint. Robert Macfarlane summarises this in a beautiful formulation: "Trace fossils are space solidified by vanished structures, in which absence becomes a sign" (Macfarlane 2016: 98). This also applies to technofossils, such as cavity fillings as future traces of former tunnels. Through the totality of such petrified *proxies*, the history of the Earth is documented. The focus of stratigraphic interest is less on the historical event itself, such as a volcanic eruption or an

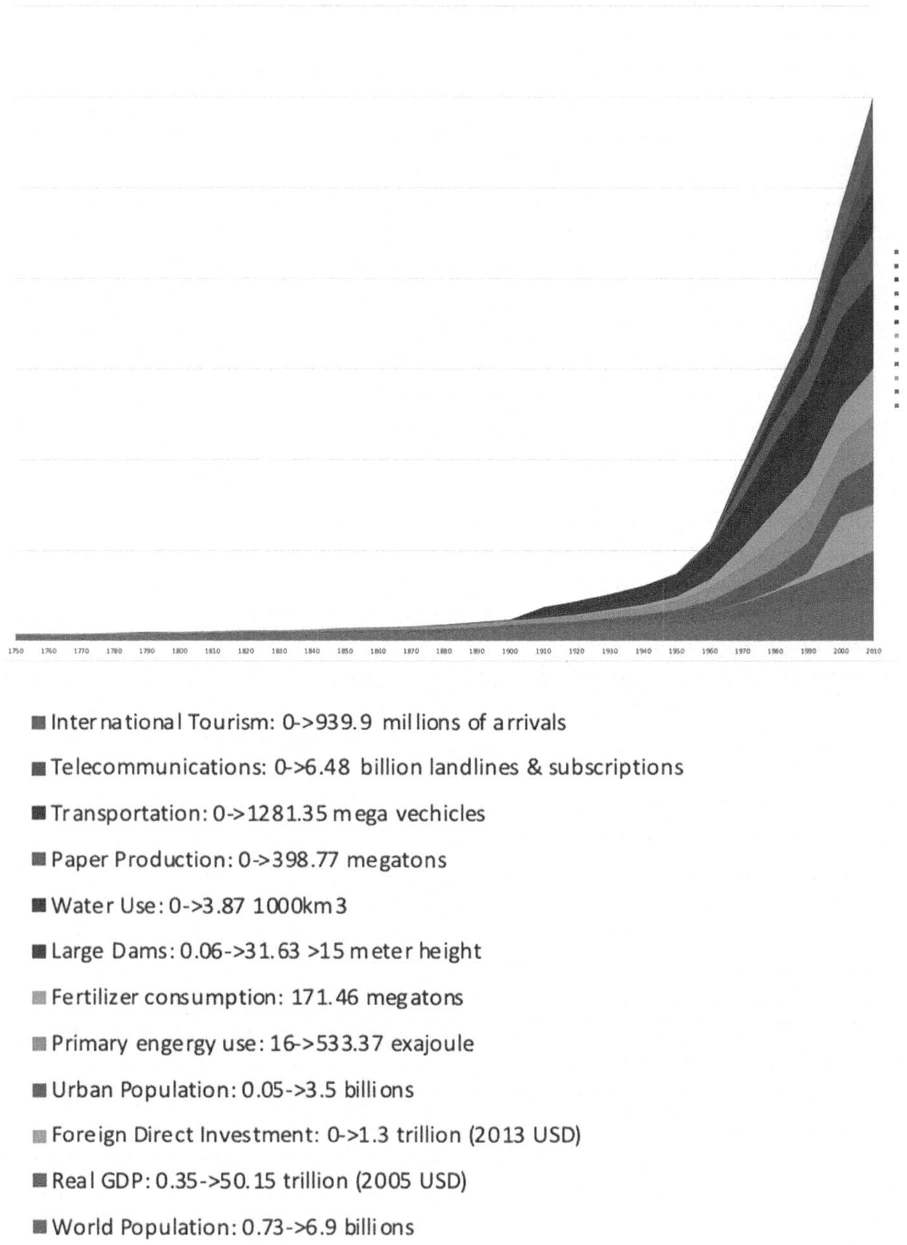

Fig. 2.1 Socioeconomic trends 1750–2010: the "Great Acceleration" from 1950 onwards https://commons.wikimedia.org/wiki/File:Anthropocene-GreatAccelerationSocioEconomicTrends-1750-2010.png; Brian MacKinnon CC BY-SA 4.0

earthquake, than on the evidence base for a time boundary in the rock, which serves as the basis for the chronostratigraphic time scale. The "rock" can also be an ice core that provides us with a snapshot of the global atmosphere of the past through traces of carbon dioxide or methane (Yan et al. 2019).

2.2.5 Planetary Boundaries vs. Boundaries Due to Resource Scarcity

The slow journey of the Anthropocene from an informal term to a formal time category only gained momentum around 2009. A major factor for this was the concept of "planetary boundaries" by resilience researcher Johan Rockström and climatologist Will Steffen. Planetary boundaries are parameters that must be adhered to for the continued secure existence of humanity, a "safe operating space for humanity" of maximum loads and material flows The physicist and climate researcher Schellnhuber had already put forward the idea in *Nature* in 1999, but now it has been empirically implemented (Schellnhuber 1999: C23, Rockström et al. 2009, updated Steffen et al. 2015a, b, 2018). Thirty leading environmental scientists developed a comprehensive summary of the interrelationships that reveal the factual limitations of the Earth. The planetary boundaries are evaluated in terms of nine dimensions (for example, land use), which together are indispensable for the stability and resilience of the geosphere and the biosphere, and thus ensure the survival of humans. In 2023, Rockström's team from the Stockholm Resilience Center was able to show that six out of nine thresholds have already been exceeded today, so that a return to a global environmental regime before 1950 or even before the eighteenth century no longer seems possible (Richardson et al. 2023).

Behind the term is the precautionary principle ("better safe than sorry"): Our uncertainty about the possible consequences of human action must not be a justification for a further *business-as-usual*. Planetary boundaries are important because they dramatically reveal environmental damage and at the same time show the possibilities of how we could prevent further disasters by staying within the safe zone. This has also led to the popularization of the concept (Rökström and Klum 2015). On the other hand, the concept of planetary boundaries is problematic because it makes the boundaries appear immutable. Moreover, the authors explicitly state that they are non-negotiable, for factual reasons. The concept could be constructively developed further to consider how the boundaries could be shifted through technological as well as social and political innovations (Dryzek and Pickering 2019: 8). However, the concept of the Anthropocene raises the question of whether we can return to stable Holocene conditions at all, more so than the stability-oriented idea of planetary boundaries.

Limited resources and "limits of growth" are rather non-technical terms for predicted upper limits, and they have been discussed in relation to the planet since the work of the *Club of Rome* ("limits of growth," Meadows et al. 1972, 2012).

Contrary to the assumptions of the *Club of Rome*, we now know through insights into the dynamics of the Anthropocene that the pressing problem is not the limited resources, but the effects of expansive resource use on a limited Earth: the limits of the Earth are therefore central. At present, efforts are being made in many places to shorten the fossil fuel era. However, this is by no means due to a shortage of energy raw materials or too high costs of extraction and use. Horizontal drilling methods and hydraulic fracking have increased reserves: peak oil is being postponed. No, we have to end the fossil era because of the environmental effects (Smil 2021: 272 f.).The resource collapse predicted by Meadows et al. has not yet occurred, it can be postponed—and this is precisely the political problem, because it means that the catastrophic situation is not seen or is obscured. From the perspective of the Anthropocene, more than 90% of oil, gas, and coal reserves should therefore no longer be fully utilized: "There are more available resources than we would like" (Schneidewind 2019: 128).

At first glance, this concept of planetary boundaries is reminiscent of the concept of ecological carrying capacity and the talk of limited resources or limits of growth. Ecological carrying capacity is a term introduced in ecology for the maximum number of organisms that a given ecosystem can sustainably feed. This is already the case for some animals that depend on how the landscape is altered by other animals, such as beavers (*Castor fiber*). However, this assumes that the system is not fundamentally altered by humans. In the case of humans, the problem is that population size alone is not decisive as per capita consumption can and will increase. At its core, planetary boundaries are a completely different concept that is not fixated on limited resources, but focused on the interconnection between human societies and the Earth system:

> The 'new' planetary boundaries are the problems caused by human action: the effects on the oceans, the atmosphere, species and soil. Humanity is still overstretching the planet, not primarily by reaching the limits of *resource availability*, but rather by approaching or even transgressing the *limits of anthropogenic disturbance absorption and ecological resilience* – what the planet can absorb (Pálsson et al. 2013: 6, emphasis in original).

One problem with talking about planetary boundaries and tipping points is that it could be interpreted—out of ignorance or intentionally—to mean that there is no cause for concern below the respective limits, which could therefore exhaust the thresholds (Haardt 2022: 137). Other critics, such as Kate Raworth, have argued that the concept of planetary boundaries does not take into account concrete human livelihoods and, above all, inequality, and thus represents biophysical essentialism. Accordingly, the biophysical upper limits must be supplemented by social lower limits that cannot be undercut. Instead of Rockström's "safe space for human development", it is about a "safe and just space" that no country in the world has yet offered for its entire population ("Doughnut Model", Raworth 2018).

From the point of view of evolutionary ecology, one could object to the pessimistic orientation of the concept of planetary boundaries beyond the fact that there was never a "safe operating space" in the Holocene. Flood disasters, extreme droughts, epidemics and social uprisings as well as colonial exploitation have existed for a long time. Instead of conjuring up the global crisis in an alarmist way, the focus must now be on doing something politically for real progress in human-nature relations through worldwide cooperation (Ellis 2023: 5, 8). Even though the idea of planetary boundaries arose from the debate on the Anthropocene, a concept of the Anthropocene that is actually geologically conceived raises the question of whether we can return to stable Holocene conditions at all even more strongly than the ultimately stability-oriented idea of planetary boundaries.

On Google Scholar, the number of hits for the search term "anthropocene" doubled from 12,900 in October 2014 to 28,400 in September 2016 and currently stands at 241,000 hits in May 2021 (Braje 2015: 374, Arizpe Schlosser 2019: 277). The term Anthropocene has its origins in an earth-scientific thesis that only slowly became an object of geological investigation from 2000 onwards (Will 2021: 81, 93–104).

2.2.6 Delayed Appearance of Geologists—The Stratigraphic Turn

In view of the origin of the term from Earth system science, geologists were not formally involved in the discussion until almost 10 years after Crutzen's initiative. The first concrete step towards this was a conference of the Geological Society of the London Stratigraphic Commission in Burlington House in May 2006. While this was only a national body, its work resulted in an authoritative discussion paper (Zalasiewicz et al. 2008a, b). which in hindsight initiated the "straitigraphic turn" of the Anthropocene (Zalasiewicz et al. 2018a; Luciano 2022a: 30). This said that the Anthropocene could have beena serious candidate for unity in the official geological time scale, the Geological Time Scale. This was the trigger for the aforementioned Anthropocene Working Group (AWG), which was established in 2009. The process of dealing with the Anthropocene was timed significantly differently than the usual sequence envisages (Will 2021: 107–126). Usually, the units of the geological "stone clock" of chronostratigraphy are obtained after a long-term and comprehensive study of rock layers representing certain geological eras. In accordance with its origins in Earth system science, the concept of the Anthropocene, on the other hand, was developed primarily with reference to atmospheric change and thus initially without a clear reference to rock layers.

This explains why it was only due to the increasing perception of the fundamental limits of the earth in political circles and in the wake of the scientific success of the term Anthropocene that the Anthropocene Working Group (AWG) was founded in 2009 as part of the Sub-Commission on Quaternary Stratigraphy within the International Commission on Stratigraphy (ICS). As the official geological "world authority" of stratigraphy, the ICS is responsible for the classification of geological ages. Accordingly, the self-imposed task of the AWG as a regularly convening committee of experts was to examine dating proposals for the Anthropocene, to conduct its own stratigraphic investigations, and to develop a plausible periodization for the more recent sections of the Quaternary "System." The group, which currently comprises 37 scientists, including Revkin, is thus preparing the groundwork for the formal recognition of the Anthropocene by the highest authority, the *International Union of Geological Sciences* (*IUGS*), as an epoch on the geological timescale. The Commission of Stratigraphers and the International Commission of Geologists will then have to vote on this.

This is also sociologically interesting. The main responsibility for advancing the argument for the Anthropocene as a unit of time in the Earth's history has now shifted from the historically young Earth System Sciences to the representatives of the older disciplines, mainstream geology and paleontology (Clark and Szerszynski 2021: 17–19, Gordon 2021: 55–71). This is noteworthy as their stratigraphic commissions work according to the strictest rules and demand quasi "forensic evidence." Two aspects are of central importance to geologists—firstly, whether the Anthropocene can actually be established as a *synchronous* unit, and secondly, at what hierarchical level the Anthropocene should be set: is it an epoch, an age, or even a period? If the Anthropocene were to be defined as a period, the most far-reaching proposal, the Holocene (as the most recent series in the Quaternary stage) would have ended in the middle of the twentieth century, as some have suggested. The Holocene would then have been an extremely short interlude in the Earth's history, replaced by an Anthropocene, a series that would probably be significantly shorter than the Holocene given the rate of current growth.

In the layers that *temporally* represent the Anthropocene, naturally deposited *and* anthropogenic sediments of the same age lie side by side, for example, on the ocean floor or in lakes. The natural components are therefore also part of the Anthropocene, especially if they contain small particles that were produced by humans elsewhere. The natural components of the layers deposited in the Anthropocene are often thinner than the anthropogenic components. Nevertheless, they can be the result of continuous deposition, such as at the bottom of freshwater lakes. The completely preserved and thus the most detailed deposits of the Anthropocene, just as in geology in general, are often found in remote regions (Waters et al. 2018; Thomas et al. 2020: 59, Renn 2023).

While chronostratigraphic units must be strictly synchronous, lithostratigraphic and biostratigraphic units can also be diachronic, just like time units in archaeology and history. The Stone Age, for example, begins and ends at significantly different times in different regions of the world, as does the boundary between prehistory and history, which—at least according to the classic criterion of written evidence—lies in some areas of Oceania in the twentieth century. These different approaches are often forgotten or lead to misunderstandings. Since its inception, geologists were not the only ones involved in the AWG, but also, for the first time in such a commission, Earth system scientists. The participation of archaeologists, historians, and, because of the formal implications, a lawyer, was also new. For a working group in the International Commission on Stratigraphy, this was something new and thus provides an insight into the importance of humans and the need for a multidisciplinary approach through institutionalization. No unit of time in geology is remotely as political as the Anthropocene. The question of formalizing the Anthropocene did not only provide a fertile basis for multidisciplinary professional discussions among the "time lords," as one science author aptly called them. There were also heated debates and even personal clashes (Gordon 2021: 55–71).

The term "anthropogenic" ("human-made") does not necessarily correspond to a chronographic unit called Anthropocene in geology, that is, a unit of *time*. Not *every* Anthropocene layer must exclusively contain traces from human activity, and not every anthropogenic soil is also part of the Anthropocene. For example, if we look under the asphalt surface in London, we might find a Roman roof tile in the base layer that is older than two thousand years, shards of medieval pottery in the middle, and above it a mixture of concrete residue and plastic waste from the post-war period. In this case, all these layers would be anthropogenic, and they could be seen as part of the "archaeosphere" as the totality of human-altered land (see Sect. 3.3). From a geological perspective, more precisely from a chronostratigraphic point of view, only its uppermost layer would represent the Anthropocene (Thomas et al. 2020: 57, see Oschmann 2018; Williams et al. 2020). It is important to keep emphasizing that the goal of the AGW is chronostratigraphic in nature:

> We emphasize here that the task of the Anthropocene Working Group (AWG) is not to provide another prism through which to reinterpret human history and environmental impact, but rather to identify a practical strata and time marker as point of reference in the formal classification of geological time (Zalasiewicz et al. 2019a, b: 1).

The proposal to designate such a short phase as a separate *geological* period is a very far-reaching idea, because it postulates that this phase is comparable to other geological epochs such as the Pliocene or Pleistocene, or the Holocene—or even periods such as the Paleogene or Neogene. Because of this brevity, some geologists argue against its formalization or for formalizing the Anthropocene not as an epoch, but only as a unit of lesser rank, such as an age or stage of the Holocene. According to one argument, a few decades would only leave a few centimeters of sediment in the geological record. And that would not justify a top-ranking geological unit. After all, periods such as the Middle Ages or the "Little Ice Age" are not part of the formal geological calendar either (Walker et al. 2015). There is always the danger of taking on a presentist perspective, one that is too present-oriented:

> Even the proclamation of the Holocene was actually an illusion of perspective; a forgivable one, because the things that are right under our noses, with their apparent size, concern us more than the seemingly smaller ones in the background. A thumb held up is enough to cover the sun. This says a lot about our perception and nothing about the sun. (…) But the Anthropocene completely lacks the most important thing, the dignity of duration (Müller 2021: 5,6).

What is sometimes forgotten by non-geologists is that the suffix "-cene" in geology is traditionally reserved for units with the status of an "epoch," such as the Eocene or the Oligocene, and not for subordinate units such as a "stage" or a "subdivision." Normally, such periods in geology or stratigraphy are identified *retrospectively* based on index fossils (Gebhardt 2016: 39). Hence, while in the case of the Anthropocene we are accompanying the process of the emergence of a geological age itself within the span of a human lifetime, such proclamations of such an epoch will not usually take place until a very long time *ex* post, namely many millions of years post factum! These are normally retrospective rather than a future-oriented determination of facts. and this alone makes the Anthropocene a reflexive concept (Kersten 2014: 381). In this sense, the Anthropocene thesis is a necessarily hypothetical determination of future developments, *science fiction* in the literal sense. Thus, the characteristic inclusions in ice cores that are used today as Anthropocene markers represent guide fossils in environmental archives, which will soon disappear with the advancement of global warming (Bubenzer et al. 2021: 28).

Compared to conventional geological time periods, the dictum of the Anthropocene is a hypothesis, or more precisely: a *claim about the past situated in the future*—a retrodiction (as a special form of abduction) about the present moment. According to the thesis, a geologist in the Quintary would be able to recognize the onset of the epoch as an anthropogenically marked layer in sediments. She would see, for example, that many technofossils appear stratigraphically with high industrialization, while the biofossils of many animal species suddenly cease. She might also be able to determine stratigraphically—after the extinction of the index fossil *Homo sapiens*—the end of the "human era." This particularity of a prospective rather than a retroactive determination means that the approach used for the Anthropocene cannot remain a factual description. The Anthropocene is thus a concept that is inherently reflexive (Kersten 2014: 381).

In August 2016, the name Anthropocene was first officially adopted by the International Geological Society (IGS), although in the end it was still disputed as to when this epoch began and which marker should be taken as a criterion for its beginning; the proposal for its beginning was the year 1950, with the accumulation of radioactive elements in the soil. The current status on the formal introduction of the Anthropocene into the geological chronostratigraphy of a new epoch is as follows: Instead of replacing the Holocene—that is, the current epoch—with the Anthropocene, or at least renaming its last phase as the Anthropocene, the commission only introduced a more precise subdivision of the Holocene series in July 2018. After the Greenlandian and the Northgrippian, the last 4250 years, beginning with a dryer phase, have now been separated as a stage under the name *Meghalayan* (Leinfelder 2018a, b; ICS 2020).

Geologists tend to think in long time frames, while stratigraphers are, for good reasons, somewhat bureaucratic in the naming of layer stacks. The geological time scale is the backbone not only of stratigraphy and paleontology, but of geology as a whole. It should be stable, allow for worldwide comparisons, and provide an internationally understandable basis that is binding between generations. Any revision of this temporal framework is slow to be implemented because a broad empirical basis is required, and this implies many technical questions, which I will address in Sect. 6.6.

One example is the fact that artificial minerals are not formally recognized as minerals in the classification of the *International Mineralogical Association*. This raises the question of whether they can still be recognized as stratigraphic markers. In addition, the overwhelming support of over 60% of geologists is required at *every* level of the decision-making process (Will 2021: 71, Fig. 6). Such a decision can take a very long time for stratigraphers as "guardians of geological time" (Haardt 2022: 118). Some boundaries between time periods have not been formalized to this day, such as the base of the Cretaceous as a system and the beginning of the Cretaceous period. As for the Holocene, 52 years passed between the proposal of the term (Lyell 1833) and its formalization (1885) and a whopping 102 years to the current formalization (Bajohr 2019: 63, Jobin 2021: 786).

A few geologists are generally against formalization or are in favor of a postponement. On the one hand, they argue that the layers of the Anthropocene are simply too thin to provide a long-term, stratigraphically useful trace. Others say that the peak of human influence on the geosphere may not have been reached yet, and that the Anthropocene should only be formalized when a crystal-clear picture has emerged in the rocks. The criticisms from geoscientists are more fundamental, arguing that a formalization should be forbidden because the phenomenon is more of a political statement or a phenomenon of popular culture (Autin and Holbrook 2012; Finney and Edwards 2016).

Given the reluctance of the geologists, it makes sense that the word "Anthropocene" is not used on the website of the *International Commission on Stratigraphy*. Critics can see this as a sign "… that this is a political term with attention-grabbing implications" (Herrmann 2016: 50, see Herrmann 2014: 44–45, 49, 2015: 7). However, its formalization has probably only been postponed in the short term (Bajohr 2019: 63). This hesitation has not harmed the academic career of the concept in any way. This will remain the case even after the IUIG's decision to reject formalisation in 2024. There is now a voluminous body of literature, dedicated book series, professorships, and since the 2010s, several journals specialized in the Anthropocene (see media guide in the appendix).

Although the term in its current sense was coined in 2000, it has actually only been used extensively in the sciences since about 2010 and has spread very quickly, as Krämer shows in an analysis of 770 essays from the *Web of* Science (Krämer 2016). However, this first strongly affected the natural sciences: as late as 2016, Christoph Görg noted that apart from the historical sciences, there were only a few contributions from the social and cultural sciences to the debate on the Anthropocene (Görg 2016: 11). This was slightly exaggerated at the time and has fundamentally changed since then. A broad field of intellectual debate has developed around the concept of the Anthropocene. Almost every discipline is at least marginally concerned with the topic.

The term has spread very quickly in all directions. However, this *Anthropo-Scene* (Lorimer 2017: 117, 131, Zalasiewicz et al. 2021a, b: 8–11) consists mainly of political decision-makers, artists, writers, and cultural studies academics, and hardly any from the general public. The spread of the idea of the Anthropocene was by no means driven solely by epistemological interests. Some important factors include the emergence of system models after the Second World War and during the Cold War, which intensified from 1980 onwards, the formation of an interdisciplinary science dedicated to climate change (often referred to briefly as *Climate Science*), concerns about the global spread of radionuclides as fallout from above-ground nuclear weapons tests (about 1500 to date), and the popularization of images of the planet from space in the 1970s (Hamilton and Grinevald 2015; Mathews 2020: 68).

2.2.7 Non-academic Inflation and Discourse Networks

Since around 2015, the non-academic life of the Anthropocene has not only expanded, but has also multiplied and branched out in a gorgonian manner. The unequal yet almost global dissemination in the mass media has played an important role in this. In view of anthropogenic climate change, there is a growing awareness of the problems caused by human-induced changes to the planet, and so it is hardly surprising that the word is on everyone's lips. Almost no one remains indifferent in the face of the topic of the Anthropocene because climate change has made people aware of the global problems caused by humans because it affects the surface of the *entire* planet, because it concerns *all* of humanity, and because it is a *long-term* threat. The topic also has an emotional appeal because the theses on the Anthropocene are rather tenacious: often strong stuff. The Anthropocene affects the surface of the entire planet; it is about the whole of humanity and a long-term threat. The term summarises a mega-problem and therefore many people feel an overwhelming threat, a characteristic of apocalyptic attitudes.

> The original thesis has long since evolved into an interdisciplinary network of discourse that is difficult to grasp, in which the Anthropocene is a bridging concept between different sciences, a cross-sectional task for science and society, and a term of reflection for the relationship between humans and nature (Dürbeck 2018b: 11, see Dürbeck 2015).

He concept comes from the natural sciences and was thus introduced into the humanities and cultural studies from the outside. However, the appropriation of the term Anthropocene is by no means limited to the humanities and cultural studies. The topic of the reciprocal relationship between humans and nature and the buzzword Anthropocene achieved a broad resonance in the creative fields about 10 years after the term was first introduced (see Dürbeck 2018a: 12, Dürbeck and Hüpkes 2022). The thematization of the Anthropocene received a strong boost in 2011 and 2012, when an essay by Elizabeth Kolbert appeared in the *National Geographic Magazine*, the most widely read popular science magazine in the world; the essay focused on species extinction under the title "Into the Anthropocene" (Kolbert 2011, see Kolbert 2014, 2021). With this perspective, UN Secretary-General Ban Ki-Moon opened the Rio Environmental Summit in 2012 with a video bearing the same title and showing images of the globe along with 24 graphs on the effects of rapid global economic expansion, the "great acceleration" from the middle of the twentieth century, in the form of the famous "hockey stick curves" (Rojas 2013), which had far-reaching consequences for public perception. Numerous keynote speeches have since been held under the same expansive title, and as late as 2018, a special issue of the UNESCO Courier bore the title as well (UNESCO 2018a). In the same year, the British publication *The Economist*, in its "Leaders Section" on May 11, 2011, published an essay with the optimistic title "Welcome to the Anthropocene" (The Economist 2011). The leading science magazine *Nature* published an editorial in 2011 stating that the Anthropocene could …

> (…) encourage a *mindset* that will be important not only to fully understand the transformation now occurring but to *take action and control it* (Anonymous 2011: 254, emphasis CA).

Overall, it can be said that from 2010 onwards, more and more people have become aware that the Anthropocene calls for responsibility and action, an aspect that Crutzen had emphasized from the beginning. The Anthropocene is obviously a very stimulating and effective catchword, as evidenced by its widespread use (Hoiß 2017). Even though the word still is unknown to most laypeople, not only scientists but also large parts of the population worldwide are aware that we are living in a new era (Hann 2016a: 3). This has been evident since around 2010 through presentations in TED Talks and the attention given in mass media, for example, in podcasts such as *Generation Anthropocene* from Stanford University (Osborne et al. 2013), and resource pools such as *The Age of Humans. Living in the Anthropocene*, created for interested laypeople by the Smithsonian Institution (Waters 2016).

There are now not only popular books and journalistic online magazines, such as *Anthropocene: Innovation in the Human Age*, but also a dazzling range of popular formats, such as theater productions, novels, and comics on the subject. In the arts and creative industries, the term is used primarily because it promises the chance to initiate change (Trischler and Will 2020: 237, Will 2021: 277–288). The topic has even spawned cookbooks that use the Anthropocene in an attempt to reconcile sustainability as a goal in consumer society with the reality of a world population that continues to grow (Rockström 2019). This often creative reception is already the subject of its own field of literary and cultural studies research (for example, see Turpin 2013).

Blake & Gilman write that the Anthropocene has managed to leave the world of the ivory tower like few other academic concepts, such as existentialism and postmodernism (Blake and Gilman 2024: 253). However, its rapidly increasing popularity (Trischler 2013a, b; Brauch 2022; Zottola and De Majo 2022) should not obscure the fact that the concept is predominantly widespread, in the broader sense, in the two fields of science and art. The concept has barely made it into the general public, as quantitative analyses by media scientist Leslie Sklair have shown (Sklair 2019, 2021a, b, see Ellis et al. 2016b: 193, Lorimer 2017; Toepfer 2018).

The overly broad use of the word "Anthropocene" is definitely a problem, which I will return to in Chap. 4. Many authors use the word to rhetorically inflate any content by attaching the word to sentences or titles à la "… in the Anthropocene," when they actually only mean "nowadays" or "in the globalized world." All of this should lead to more precise use of the term, but not to a blanket dismissal, for example, when today's Anthropocene literature is presented as a variant of spontaneous philosophies by scientists, as "… treatises that scientists write after work … " (Hoffmann 2020: 81.

2.2.8 Intensive Popularization in German-Speaking Countries

Although the topic of the Anthropocene is still little or not at all known to most laypeople in the German-speaking world, it is quite present in the media (for example, Lesch and Kamphausen 2017; Hüther et al. 2020). A decisive factor in its early popularization in German-speaking countries was a book by science journalist Christian Schwägerl entitled "Menschenzeit" (Schwägerl 2012, first 2010). Schwägerl had already been tackling environmental issues for 20 years as an editor at *Der Spiegel* and for other leading media. He consistently advocated for the topic and also published articles together with scientists, for example, with paleobiologists (Williams et al. 2015). Contrary to purely apocalyptic prospects, he emphasized the opportunities for conscious human action against the Anthropocene dynamics. He focused on political design tasks and spoke of the positive possibilities of a "biofuturism," in which science and technology are put at the service of the biosphere (Schwägerl 2016). The book was presented to the public in 2010 by Achim Steiner, the head of the UN Environment Programme, who praised it as an "intellectually challenging book" and the "work of a father explaining to his children what is happening to our Earth." Schwägerl's book was then also translated into English (Schwägerl 2014, 2024), and he succeeded in winning over Paul Crutzen, the great godfather of the new geological age, to support the popularization of the topic in the German-speaking world (Crutzen and Schwägerl 2011; Schwägerl 2016; Crutzen 2019a, b).

When Crutzen made the term known in geoscientific circles in the year2000, he was working in the Department of Atmospheric Chemistry at the Max Planck Institute for Chemistry in Mainz. He was known worldwide as an atmospheric chemist and as a CFC specialist for his work on the then-highly topical problem of the hole in the ozone layer (Crutzen and Brauch 2014). Even before he popularized the term Anthropocene, Crutzen had not only made groundbreaking contributions to climate chemistry, for which he was awarded the Nobel Prize, but he had also written texts on human-made climate change in the 1980s that were accessible to laypeople. Crutzen wrote these popular science texts at the height of the Cold War, often together with colleagues or journalists, and some of his books were published in German translation (Graedel and Crutzen 1996; Crutzen et al. 2011; Crutzen 2019a, b).

A third reason for the strong popularity of the Anthropocene term in the German media is the scientific and also popular science activities of the geobiologist Reinhard Leinfelder, the former director of the Berlin Museum of Natural History. Leinfelder recognized the importance of educating the general public about the topic, and he saw the didactic potential of the Anthropocene as a catchy narrative (see Hoiß 2019). He initiated the popularization of the topic in various formats, in essays (Leinfelder 2011, 2012, 2015a, b, 2019; Leinfelder and Schwaderer 2020), comics and graphic novels (Hamann et al. 2013; Hamann et al. 2014; Leinfelder et al. 2015), and even a cookbook (Leinfelder et al. 2016). In his own "Anthropocene" blog, Leinfelder uses texts, graphics, and even self-drawn sketches to convey the topic in an understandable form for young people (Leinfelder 2018b, 2019). A special tool that Leinfelder, often together with Schwägerl, repeatedly uses in his publications is the use of metaphors, such as the image of the Earth as a "spaceship without an emergency exit." In addition to his own graphics and hand-drawn sketches, he also uses his own neologisms, such as "human Earth" and "ourworld" (catchy, as the word in German, "Unswelt," is very similar to the word for environment, "Umwelt"), which are catchy and thought-provoking. While the British palaeobiologist Jan Zalasiewicz speaks of humans as a "predator," Leinfelder presents the image of humans as "world gardeners" and demands: "We should treat the Earth as if it were a foundation" (Leinfelder 2017b, 2020).

These popularization activities also extend into politics. This is particularly evident in the activities of the permanent German Advisory Council on Global Change (WBGU). This influential *think tank*, which advises politicians, is pursuing the project of a comprehensive restructuring of technology, economy, and society, a "great transformation" (WBGU 2011a: 5). The WBGU is also speaking of the "third great transformation." According to this, the first great transformation was marked by the beginning of agriculture and urban development (Neolithic Revolution) around 11,000 years ago, while the second great transformation began with the use of fossil fuels at the start of the nineteenth century (Polanyi 1944). The innovation that was consequential for the geosphere in the early twentieth century was the Haber-Bosch process, the large-scale industrial extraction of nitrogen from the air, which was patented in 1908. However, as a process for the synthesis of ammonia, it consumes so much energy that it can only be implemented with the use of fossil fuels, an example of a mechanism that we need to move away from through the third transformation. The WBGU propagates this through flagship reports and short summaries, all of which are accessible in full text as pdf files. The dissemination of ideas is also promoted, for example, through a non-fiction comics, which aims to make the main report accessible to laypeople (Hamann et al. 2013).

Uwe Schneidewind, a member of the WBGU, the *Club of Rome*, and President of the Wuppertal Institute for Climate, Environment and Energy, and who plays an important role in public discussion, reiterates that the Anthropocene marks the necessary leap forward in civilization that requires a global social contract as a new "business foundation," in short, "the upcoming third 'great transformation'of human history in the twenty-first century" (Schneidewind 2019: 134, similar to Schellnhuber 1999). The idea of the Anthropocene is referred to in publications, such as the extensive reports of the WBGU, to justify the necessity and urgency of this fundamental restructuring of the economy and society. However, it is noticeable that scientific contributions on the Anthropocene are only used very selectively in the main reports of the WBGU (WBGU 2011a: 66, 2016, 2020). I will address more fundamental criticisms of the projected great transformation, which see this as a depoliticizing project, in Sects. 4.4 and 4.5.

One of the paradoxes in the discourse surrounding the Anthropocene is that a term, whose establishment is disputed by natural scientists, has initially been accepted as a fact by the cultural sciences, especially in Germany (Görg 2016: 9).

The focus of the early cultural exploratory movements was initially not on discourse, as was the case in postmodernism in the 1980s and 1990s, but on material dependencies: "Matter does matter." The hesitation of natural scientists did not prevent feuilletons and museums from surveying the new age even before it was recognized in the geosciences. The humanities and cultural sciences then quickly turned their attention to the discourse on the Anthropocene (Klingan et al. 2015; Springer and Turpin 2015).

The "Haus der Weltkulturen" (HKW) in Berlin launched a major Anthropocene project in 2013. Under the patronage of Crutzen and in collaboration with the Max Planck Institute for the History of Science, an interdisciplinary and also multidisciplinary "Anthropocene Project—Basic Cultural Research using the Means of Art and Science" was held at the haus der Weltkulturen (House of World Cultures) in Berlin in 2013–2014 (Möllers 2013; Möllers et al. 2015; Haus der Kulturen der Welt, House of World Cultures 2013, 2014). The Berlin Anthropocene project was continued in the "Technosphere" project held in 2015–2019 and was accompanied by the "Anthropocene Lectures" in 2017–2018. Many of these activities culminated in a large exhibition at the Deutsches Museum in Munich in 2014–2016, which was programmatically headlined with the notorious title from *The Economist*, "Welcome to the Anthropocene. Our responsibility for the future of the Earth" at the Deutsches Museum in Munich (Möllers and Schwägerl 2015). In Austria, the first European coin for the Anthropocene was issued in 2018, a 25-euro coin. HKW also financed the worldwide search for candidates for the Golden Spike, organised together with the Anthropocene Working Group (AWG), until 2023. The central idea was to take this as an opportunity to examine the scientific, historical and cultural 'construction' of the Anthropocene and to communicate this primarily in an audience-orientated, artistic and experimental format (Zalasiewicz et al. 2023: 318, Rosol et al. 2023).

The research activities in the German-speaking world have also led to exhibitions. Many of the early activities culminated in a major exhibition at the Deutsches Museum in Munich in 2014–2016, which was programmatically entitled 'Welcome to the Anthropocene. Our Responsibility for the Future of the Earth at the Deutsches Museum in Munich (Möllers and Schwägerl 2015). This was the world's first major exhibition on the topic and it focussed on the role of technology and infrastructure (for a critique, see Jørgensen and Jørgensen 2016). At the Centre for Art and Media (ZKM) in Karlsruhe, Bruno Latour was involved in a long-term art project that brought together science and the visual arts and culminated in an exhibition in 2020 (https://zkm.de/de/bruno-latour-ueber-critical-zones). In 2021, the 20th Major State Exhibition in Baden-Württemberg was dedicated to the topic under the title "Anthropocene—Age? Turning point? The future?". In the Deutsches Museum in Munich, revcently an exhibition of my photos organised together with Fabienne Will ran in 2024 under the title 'We are the Anthropocene. The horror and beauty of global change' (Antweiler and Will 2024).

Recently, there have been several major cultural or social science research projects in Germany that have focused on the Anthropocene as a topic or at least strongly touched on it. The activities and publications of the *Rachel Carson Center*, which deals with the global environment from a social and cultural science perspective and has worked on the Anthropocene early on (for example, Trischler 2013a, b), were pioneers. A DFG project at the University of Vechta dealt with "Narratives of the Anthropocene in Science and Literature. Structures, Themes, Poetics" from a literary and media science perspective in 2017–2020 (Dürbeck 2015, 2018a, b, Dürbeck and Nesselhauf 2019; Dürbeck and Hüpkes 2022; Mauelshagen 2019, 2020; Probst 2020).

In 2019–2021, a Hamburg DFG research group explored the "Futures of Sustainability" from a predominantly sociological perspective and addressed the question of how we can make sense of the Anthropocene from the perspective of social theory (Adloff and Neckel 2020; Adloff et al. 2020; Adloff and Busse 2021). Colleagues at the University of Cologne, my *alma mater*, are exploring the connections between flora, fauna, and humans in the transboundary landscape in southern Africa in a current project entitled "Rewilding the Anthropocene."A research focus at the Max Planck Institute for the History of Science has been dealing with "Knowledge in and of the Anthropocene" since 2017 (MPI 2017). A more recent initiative of the MPI intends to introduce the new research field of "geoanthropology" as an Earth-human science of the Anthropocene. In 2023, the "Max Planck Institute for Geoanthropology" was founded in Jena. Geoanthropology is the study of the totality of processes, mechanisms and natural-historical and cultural-historical paths that lead human societies into the Anthropocene (Renn 2021: 5, cf. Renn 2020: 375–376). This partly resembles a required 'geosociology' (Bammé 2016; Schroer 2019, 2022). German social and historical sciences concerned with the Anthropocene are becoming increasingly international. For example, the CALAS Maria Sibylla Merian Centre at Bielefeld University is promoting research into the Anthropocene in Latin America, for which a six-volume handbook is being published in collaboration with colleagues there (Kaltmeier et al. 2024a).

2.3 Anthropocene—Really a New Perspective?

> *We are all struggling for an appropriate language to help us analyze the changing human and ecological condition.*
> *Simo Nicholson and Sikina Jinnah* (2016: 4)

There is often the misunderstanding that "Anthropocene" is a simple descriptive term for what used to be called "environmental damage." This is by no means the case. Unlike previous environmental terms, "environment" in the Anthropocene view of earth system sciences is no longer the local living environment, but the entire ecosphere as the basis of all life (Warde et al. 2018). The question of what the Anthropocene should actually be understood as is as controversial as it is unclear. Initially, three strands can be distinguished: the geological concept, the Earth science concept, and the cultural science idea, as well as the above-discussed appropriation of the term in the media, popular culture, art, and the public sphere.

However, in the concrete practice of the multipolar debate on the Anthropocene, these strands are closely and discursively related to each other (Trischler 2016a, b, Trischler and Will 2019: 236).

What exactly is meant is often diffuse or remains open: Is Anthropocene a geological-stratigraphic phase, as it is technically seen, or does the Anthropocene represent a more extensive geoscientific finding? Or is the Anthropocene more of a geological speculation that looks back from an *assumed* future development: in other words, *science fiction*, in the strict sense of the term. In addition to descriptive components, the concept often also involves normative settings. The following two statements, one from geobiology and one from literary studies, do not provide a definition in the technical sense, but they get to the heart of the concept behind the word "Anthropocene" and at the same time reveal important nuances:

> The Anthropocene as a concept is first and foremost a scientific hypothesis, which states that the changes initiated by humans are already reflected in a geologically visible form and are of a sufficiently long-term nature to locate them on the timescale of the Earth's history (Schwägerl and Leinfelder 2014: 234).

> The term 'Anthropocene' denotes a new geological age in which humanity has the dominant geophysical influence on the Earth system and from which human responsibility for the future of the planet is derived. The concept also contains a call to redefine the position of humans in relation to nature and the cosmos and to deal responsibly with limited natural resources (Dürbeck 2018a: 11).

The Anthropocene as a postulated fact is about humanity substantially altering the near-surface spheres of the planet, primarily through the effects of technology. The sum total of human activity is bringing about a profound global change that permeates a large part of the geosphere. This change has already progressed so far that there is not a single ecosystem on the planet without an anthropogenic component. In short, humanity has become a profound geo-factor.

2.3.1 Anthropocene—A Contribution to the Organisation of the Facts and the Idea

The guiding thesis is repeated: the accelerated or abrupt change seen in geological time periods leaves traces in emerging layers of sediments and this justifies the introduction of a separate geo-chronological time period ("-cene", "new" or "recent age"). The proposal that has been hotly debated since then is to define the most recent part of the Holocene ("completely new age"), popularly the "post-ice age," as its own era in the stratigraphic system of the International Commission on Stratigraphy (ICS) due to the accumulated impact of human activity on the Earth system. It should be noted here that the Holocene epoch began just under 12,000 years ago and is thus, at least in geological terms, a very short phase of the Earth's history.

To what extent do the concepts relate to the geological facts? Does the Anthropocene describe the human way of life in modern times, or is it an open key concept that bundles problems and questions, as is the case with the word resilience (Bröckling 2017; Chandler 2018, 2020)? Is the idea of the Anthropocene a genuinely new concept or merely a semantic shift (Bruns 2019: 56–58)? Does the Anthropocene provide an insightful metaphor for uncertainty and anxiety, or also for creative problem-solving? Or are we looking at a conceptually disguised authoritarian ideology (as Baskin 2015, Haraway, in Haraway et al. 2016: 536, Bruns 2019)? Are we perhaps just dealing with a powerful but rather empty buzzword or an overloaded catch-all term?

A rough distinction can be made between the extremely different interpretations of the concept in the natural sciences and the humanities (Zalasiewicz et al. 2021a, b: 9, Fig. 3). In the natural sciences, the Anthropocene is understood as a descriptive or analytical concept at the Earth system level, while humanities scholars and cultural scientists rather tend to understand it as a normative ("What should we know?") or narrative concept ("How did we get here?"), or as a narrative-normative combination that names those responsible ("Why did we get here?"), or a call to action ("What do we need to do now?"). Tables 2.2. and 2.3 show the actual diversity of conceptions of a word that generates such strong resonances that it is even causing a "conceptual earthquake" (Rowan 2014: 447, Egner and Zeil 2019: 15, 28–30).

Table 2.2 Polyvalence of the Anthropocene (1): Understanding as a new Earth system phase and geochronological or historical epoch (descriptive, analytical, Anthropocene 1, Anthropocene 2, and Anthropocene 3)

Anthropocene is to be understood as ...	Author/s
Earth system: Geophysical evidence of a new status (*"new state", "new earth", "transformed earth"*), (earth system sciences)	Crutzen 2000, Chakrabarty 2009, Leinfelder 2012, Ziegler 2019: 278, Syvitsky et al. 2020, Renn 2020, Zalasiewicz et al. 2021a, b: 6–8
Epoch: Geochronologically and archaeologically an emergent epoch (geology)	*Anthropocene Working Group* (AWG), Dryzek and Pickering 2019: 2, Zalasiewicz et al. 2021a, b: 1
Epoch: Geological Holocene is over (Geology, Paleobiology)	Zalasiewicz et al. 2021a, b: 1
Epoch: Biogeographical, dissolution of the biotic uniqueness of places (place)	Lomolino 2020: 19
Epoch: Geosociocultural phase	Bajohr 2019: 64
Event instead of a formal epoch	Gibbard et al. 2021, Finney and Gibbard 2023; in contrast, Waters et al. 2022, 2023a, b
New "universal" history (*geohistory*)	Chakrabarty 2009: 221
Cumulative impact of civilization on the global environment	Syvitski 2012, Syvitsky et al. 2020
Historiographic period, historical materialization of the human–environment relationship	Bauer and Bhan 2018: 17
Problem, social and/or ecological	Wapner 2012, Lövbrand et al. 2015, Mathews 2020: 67, 77

Table 2.3 Polyvalence of the Anthropocene (2): Understanding as cultural idea or concept (as a narrative and/or normative goal)

Anthropozän is to be seen as …	Authors
Story, narrative, narration, *master narrative* (often apocalyptic, e.g., on environmental destruction, many variants)	Masco 2010, Moore 2015a: 27, Dürbeck 2015, 2018a, b, Hoiß 2017: 17–25
Metanarrative	Dürbeck 2018a, b, Haardt 2022: 158
Naturalistic master narrative of humanity, distanced "from nowhere" (geocratic grand narrative)	Bonneuil 2015: 20, 29, Bonneuil and Fressoz 2016: 45–96
Appeal, 'wake-up call', formulation of a problem that is intended to generate responsibility, a 'rallying cry' that is intended to motivate	Crutzen 2002, Purdy 2019, Horn and Reitz 2020
New universal historiography	Chakrabarty 2009: 221
Collective term for global crisis symptoms, bundle term	Horn and Reitz 2020, Goeke 2022: 88
Integrative concept für global change	Zalasiewicz et al. 2015: 198–199 (AWG)
Central thought system (in the earth sciences)	Horn and Reitz 2020
Denkrahmen, Meta-Konzept, „new glasses"	Ostheimer 2016: 34; Bornemann et al. 2019, Ellis 2018
An event instead of a thing, new human situation (*condition*) and also political event	Bonneuil and Fressoz 2016: xiii, 24
Dilemma (*collective predicament*), existential crisi	Magni 2019, Thomas et al. 2020: 3, 196
Cultural concept, term (*designation*), cultural idea, idea as a social fact, emerging idea	Kersten 2013: 40, Lövbrand et al. 2015: 212, Moore 2015a: 27, Purdy 2015: 16
Construct	Kruse et al. 2011, Neckel 2020: 157
Zentrales Denksystem (der Erdwissenschaften)	Lövbrand et al. 2015
Hypothesis	Egner and Zeil 2019: 16
Geohistory, *geostory* (vs. *geohistory*), narrativity	Latour 2017a, b, Arizpe Schlosser 2019: 267
Time concept for the present, also implicitly for the past and future	Moore 2015a: 27, 41, Rabinow 2008
Catchphrase, *buzzword*, empty phrase (for inevitable change by humans)	Moore 2015a: 27, 41, Malhi 2017: 93, Uekötter 2020: 643, 2021, Herrmann 2014: 44–45, 2021, pers. Mitt.
Boundary object sensu Star and Griesemer 1989	Lorimer 2017: 117ff.
Empty signifier ("A." means everything or nothing)	Moore 2015: 600; Bubenzer et al. 2021: 29
Fulguration figure and apotropaic word	Herrmann 2015: 7
Social category	Moore 2015a: 27
Mode of societal self-description (of society)	Werber 2014: passim; Goeke 2022
World formula	Lippuner et al. 2015: 5–7
Reflection term	Dürbeck 2015: 113, Lippuner et al. 2015: 14

(continued)

Table 2.3 (continued)

Process that reflects on itself	Renn and Scherer 2017a, b, Scherer 2015, 2017
Claim to truth as an eco-authority	Howe 2014
Political statement	Gibbard and Walker 2014, Herrmann 2014: 44–49, Finney and Edwards 2016: 8.9
Anti-politics machine	Krauß 2015a: 59–60, 74
Absolute Metaphor of a discourse industry (on industrial policy criticism)	Metejovski 2016: 9, 14
Counter-discourse (against globalization)	Dibley 2012
Planetary imagination	Moore 2015a: 27, 2015b; Adloff et al. 2020
Geopolitical imagination Geopolitische imagination	Lövbrand et al. 2020, Dalby 2020
Science fiction (unusual time and space)	Swanson et al. 2015: 149
Hyperobject	Morton 2013, Tsing et al. 2019: 190
Paradigm shift, new way of thinking	Baskin 2015: 9, Lewis and Maslin 2018 Bhan: 108, 114, Krämer 2016, kritisch, Bauer and Bhan 2018: 4
Evolving paradigm, paradigm to be improved	Butzer and Endfield 2012, Bauer and Bhan 2018: 12
Discursive arena, discursive space, conceptual terrain	Gibson and Venkatesvar 2015: 6, 9, Chua and Fair 2019, Moore 2015a: 41
Scientific turn (geologic turn)	Bonneuil 2015: 19–31
Problem framing, problem space Problem formulation, intended to generate responsibility	Moore 2015a: 27, 41
Key term (e.g., for diagnosing the present)	Werber 2014
Crisis diagnosis, crisis rhetoric	Rockström et al. 2009, Moore 2015a: 35, Matejovski 2016, Egner and Groß 2019
Threshold concept or boundary concept	Bonneuil and Fressoz 2016: xiii, Clark 2015: 21, 176–178, Görg 2016: 10 Cera 2022: 129
Negotiation zone (*trading zone*) for central questions about the present and future of humanity on earth and the related evidence production and assurance	Will 2021
Bridge concept, platform for alliances of sciences, cross-cutting task	Dürbeck 2015: 107, Brondizio et al. 2016, Arizpe Schlosser 2019: 276
Depoliticizing narrative, *mainstream story*	Lövbrand et al. 2015: 212
Ideology, ideological project (e.g., enforcement of authoritarian policies or neoliberalism)	Baskin 2015, Visconti 2014: 382, Wilson 2016: 84–86, 95, Malm 2016, 2021, Mathews 2020: 70; Finke 2022
Label, distraction, semantic shift (e.g., distraction from inequality)	Malm and Hornborg 2014, Demos 2017, Bruns 2019
Domesticating concept against experiences and expectations that are perceived as scary	Taddei et al. 2022: 7

(continued)

Table 2.3 (continued)

Pop-cultural phenomenon in cultural battles for the environment	Autin and Holbrook 2012
Scholastic worldview	Neckel 2020: 162–167
Theodicy (for the "good Anthropocene")	Hamilton 2015
Political motivator for action, and generator for the assumption of broad responsibility and advocacy for environmental justice	Gibson and Venkatesvar 2015: 6, Görg 2015, 2016: 10, Bauer and Bhan 2018: 6?, Clark and Szerszynski 2021: 3
Potent label or weapon (for creating awareness, for implementing sustainability)	Gibson and Venkatesvar 2015: 7, Moore 2015a: 27
Productive idea (especially against dualisms, such as nature/culture and subject/object)	Gibson and Venkatesvar 2015: 7
Eye-opener for an attitude of openness and curiosity towards others	Rose 2013
Anchor of hope (for an endangered planet)	Gibson and Venkatesvar 2015: 18–19, Cunsolo and Landmann 2017, Folke et al. 2021: 851–853
Historical-philosophical term to characterize the current epoch, caesura; plus a call to action	Schüttpelz, personal communication 2021
Palimpsest (A. As a foil to be described again and again)	Scherer 2022

In view of the sheer endless variety of ways of understanding the Anthropocene, proposals for categorisation have been made. Bonneuil (2015: 18–29) also distinguishes between four variants, which he refers to as 'narratives' (cf. in more detail in 3.4): (a) the naturalist narrative (human species as a geofactor), (b) the post-natural narrative (development of a hybrid techno-nature), (c) the ecocatastrophic narrative (possible collapse of civilisation and the living world) and (d) the eco-marxist narrative (Anthropocene as the result of a history of inequality). Buhr also makes a fourfold distinction between the phenomenal situation, the concept, the message of urgency and the question of the 'ontological sutuation' of man (Buhr 2022: 312–313).

Dibley makes a distinction of seven conceptual understandings within a variety of conceptual understandings seven main forms: (1) epoch name and humans as geofactor (2) marking a temporal change, 3) a risky condition that limits human freedom, 4) dissolution of the separation of society and nature, (5) the new close relationship between the earth and humans as 'earthlings', (6) a new concept of the earth as a system, and (7) the inevitable change of ideas of the future beyond progress (Dibley 2012).

These and other typologies (e.g. Toivanen et al. 2017; Lorimer 2017) of aspects of or perspectives on the Anthropocene bring conceptual order, but they overlook the fundamental difference between geological and other understandings. Toivanen and colleagues distinguish between four 'approaches', as they call them, (a) geological, (b) biological, (c) social and (d) cultural Anthropocene (Toivanen et al. 2017: 187–192). However, they also combine geology and Earth system science.

In order to do justice to the different interests, data and also methods of different approaches, but in particular between geology and Earth system science, and at the same time not to get out of hand, I propose the following four basic understandings of the term 'Anthropocene':

1. A first understanding (Anthropocene 1) is guided by Earth science and refers to the geohistorical caesura in the status of the Earth system (*new state*) from around the middle of the twentieth century, *mainly* due to the effects of human activity as an important geofactor.
2. A second understanding (Anthropocene 2) is geologically influenced—the stratigraphic Anthropocene. This refers to the sum of events during the very short and currently ongoing geochronological "epoch." As a geological "series," this includes *all* deposits during this time interval (synchronous), *regardless* of whether they are of anthropogenic, partly natural, or entirely natural origin. This is the dominant view in the AWG.
3. A third, temporally extended, and spatially more limited understanding (Anthropocene 3) is primarily archaeological and historical and can be described as the diachronic Anthropocene. It refers to the sum *of all* empirically verifiable influences of *Homo sapiens* on the geosphere across all time periods, even if they only manifest themselves regionally. This is the view of the minority and resembles the concept of the archaeosphere.
4. A fourth understanding (Anthropocene 4) comes from the humanities, cultural studies, and social sciences. This is a synthesis term for the consequences of the break caused by human-induced environmental change and also includes corresponding reflections, especially on the responsibility and status of humans relating to or within nature. This understanding is used extremely inconsistently for the meta-level up to the analytical levels of Anthropocene 1, 2, and 3 (cf. Zalasiewicz et al. 2021a, b).

2.3.2 Precursors of the Concept—Not Old Wine in New Bottles

The diagnosis of a human-made and permanent as well as possibly irreversible change in the Earth as *one system* fundamentally differs from the precursors of the Anthropocene concept (Hamilton 2015; Hamilton and Grinevald 2015; Will 2021: 12–13). However, this does not mean that there were no precursors for certain aspects of the concept. One spiritual father of the Anthropocene is the biologist Alfred Russel Wallace (1823–1913), a contemporary of Darwin. As early as 1898, Wallace pointed out pointed out the "plundering of the Earth" in popular scientific works and branded it as an equally reckless and irresponsible treatment of nature, not so much in the struggle for existence, but in the "struggle for prosperity." In another work, he substantiated this through descriptions of the clearing of tropical forests for coffee plantations in Southeast Asia (Wallace 1910, 2019, see Kutschera 2013).

However, the influence of humans on the geosphere has been discussed for a long time, at least since the seventeenth century. Long before de Buffon and Hutton, Émile Rousseau, René Descartes, and Francis Bacon wrote about the strong human influence in changing the Earth, even if they did not see its global extent (Grinevald 2012; Hamilton and Grinevald 2015; Dürbeck 2018b: 4). Within the genesis of ideas about climate, a historically young idea, there was a concept of climate from the late nineteenth century that included cultural factors and also the first considerations of global human influences (Horn 2016). The question is in what respect the questions about the human–nature relationship are really new. Bernd Herrmannn, as a sharp critic of the Anthropocene concept, says:

> As if the entire Western reasoning about humans and their relationship to nature had not existed before the introduction of the Anthropocene concept, this term *seems* to have suddenly made us aware of humankind's creative power (Herrmann 2016: 46, emphasis CA).

An awareness the fact that humans are permanently changing their environment on a large scale and that natural resources are becoming scarce is something that developed independently of Europe in China and the Asian part of Russia as early as the eighteenth century. This also applies to the perception of the loss of species caused by humans, for example, through increasing deforestation of forests in China (Marks 2015, 2020: 118). These non-European precursors of an awareness of the Anthropocene only became known through new studies in global history, which aim to explore world history in a less Eurocentric way. Numerous studies in this research field highlight the role of Asia in the emergence of the modern world, which I will take up again in Chap. 7. However, the "Anthropocene" has precursors not only conceptually, but also terminologically (Steffen et al. 2011a, b; Davis 2011; Davies 2016: 43, Bonneuil and Fressoz 2016: 4, Guyot-Téphanie 2020).

The neologism "Anthropozoic," which is linguistically similar to the Anthropocene, was first used by Thomas Jenkyn from Wales (1854), by the used by the Irishman Samuel Haughton (1865). In the face of the industrial revolution, the Italian geologist and clergyman Antonio Stoppani (1824–1892) in particular argued in his work *Corso di geologia* (*era anthropozoica*, Stoppani 1873) pronouncedly in favour of man being a new 'telluric power' equal to the great forces of nature. It is still sometimes used today, for example, by the zoologist Hubert Markl (1986, 1998: 23) and by the paleoclimatologist Broecker, who also explicitly speaks of the "Anthropozoic era," which expands the meaning to an epoch (Broecker 1987). The early scholars were both geologists and theologians or clergymen. In the 1870s, there were terms largely synonymous with the Anthropozoic era that emerged as a name for the age of humans or for the post-glacial period, which was referred to by Charles Lyell 1833 as "Recent" before the term Holocene was established. An example is the term "Psychozoic" given to the time of the dawning of human consciousness by Julian Huxley, the founder of the synthetic theory of evolution. Huxley also suggested elevating humans to the rank of their own kingdom of life, one he called "Psychozoa," because of the unique features of their central nervous system (Huxley 1966, see Oeser 1987: 52, Reszitnyk 2020).

Precursors similar in content to the Anthropocene emerged in the 1920s. In 1922, the Russian geologist Alexei Petrovich Pavlov coined the term "Anthropogen" ("-gen" is a formal ending for a geological period) for the period beginning with the appearance of humans, which more or less corresponds to today's Quaternary (Rull 2017: 1067, 2018). Central are the works of the Russian Vladimir I. Vernadskij (1863–1945), the founder of biogeochemistry, and his term "noosphere." In his philosophical writings, he developed a three-part concept of the geosphere, biosphere, and noosphere. He understood the "noosphere" as a sphere of human thought, a worldwide linkage of culture and shared ideas, which can be explained with the use of newer approaches to cumulative cultural evolution. In contrast to the others, he understood it as an Earth sphere that emerges from the actions of rational humans and gradually permeates the biosphere (Vernadskij 1997). With regard to the biosphere, Vernadskij drew on the works of the founder of this term, the geologist Eduard Sueß (1831–1914), and otherwise on those of the Bergson student Edouard Le Roy and the Jesuit priest and paleontologist Pierre Teilhard de Chardin In Soviet geology, the word "anthropogenic" was sporadically used as a synonym for the Quaternary after the Second World War (Shanster 1973; Gerasimov 1979). However, this term was used without reference to anthropogenic effects (Thomas et al. 2020: 44).

The tenor of some of these concepts was quite positive, in stark contrast to the current almost consistently apocalyptic narratives on the Anthropocene. The latter overlook the fact that at the core of the concept of apocalypse as a revelation of systemic grievances, hope for a better world is always dialectically included (Horn 2014; Fladvad 2021: 158). Instead of scientific predictions, some apocalypses offer meaningful 'historical overviews in future tense' (Ostheimer 2016: 46) as anticipatory reviews of today's environmental situation. This has now become a popular book genre. On climate change, the historian of science Naomi Oreskes and Eric Conway have dared to take a cautionary look back from the future with 'The Collapse of Western Civilisation' (Oreskes and Conway 2014).

In the first half of the twentieth century, in the wake of increased industrialization, an awareness of the dilemmas created by cumulative human activity emerged. The sheer scale of the effects of human action was now being noticed. An early precursor is the American diplomat, writer, and conservationist George Perkins Marsh (1801–1882), the "father of American conservation." His work, "Man and Nature, or, Physical Geography as Modified by Human Action" (Marsh 2021, orig. 1864), established environmental geography as a subordinate branch of geography. This was an early voice and is still an influential one today (for example, see Olwig, in Haraway et al. 2016: 550).

The impact of humans on ecosystems was recognized early on by geologist Robert Sherlock and plant ecologist Tansley. Tansley criticized the concept of succession in plant communities because human influence was not seen there. In 1923, he introduced the term "anthropogenic" into the scientific literature; according to Meriam-Webster, the term has existed since around 1883 (Tansley 1923: 48). In a later work, he made a clear prediction of a coming Anthropocene:

> Limited at first to the regions where civilization originally developed, this destructive activity has spread during recent centuries, and at an increasing rate, all over the face of the globe except where human life has not yet succeeded in supporting itself. It seems likely that in less than another century none but the most inhospitable regions – some of the more extreme deserts, the high mountains and the arctic tundra –will have escaped. Even these may eventually come, partially if not completely, under the human yoke (Tansley 1939: 128).

Nonetheless, for a long time, voices that considered the effects of human activity to be minor dominated. Edward Wilber Berry, for example, thought that the supposed assertions of the noosphere proponents were a remnant of medieval holocentric thinking (Thomas et al. 2020: 45). In the post-war period, which was shaped by modernist and technological ideas, the environment was then again narrowed down to an object of the natural sciences (Pálsson et al. 2013: 4). But even then, there were clear voices on the global extent of man-made change. In an early environmental bestseller entitled "Our Plundered Planet" , the biologist Fairfield Osborn Jr., President of the New York Zoological Society, wrote as early as 1948: "It's man's earth now. One wonders what obligations may accompany this infinite possession" (Osborne 1948: 66).

It was only in the immediate post-war period that global effects became obvious. Awareness of human-induced global change then increased significantly during the Cold War between the 1950s and 1980s. Nuclear bomb tests and the knowledge gained from space missions about global change were decisive factors (Horn and Bergthaller 2022: 102). Of central importance were the environmental movement that emerged in the 1960s and the resulting studies by the *Club of Rome*, which focused on the future development of the Earth and the limits of growth (Meadows 1972). These developments culminated in the International Geosphere-Biosphere Program (*IGBP*) in the late twentieth century.

In the geosciences, there is a great deal of research on this topic in geography, whose central subject with regard to the lithosphere—in contrast to geology—is precisely the Earth's surface. "Man's impact on the Earth" has been an important research topic in physical geography and geomorphology, but also in human geography since at least the 1960s (Simmons 1996; Anderson et al. 2013; Goudie and Viles 2016; Goudie 2019). Since the 1980s, there has also been an increase in the amount of works on the large-scale and even the global effects of human activity, mostly under the aforementioned catchphrase "global change" (Glaser et al. 2014; Flitner 2014; Messner 2016).

Experiences with global crises and the increasing awareness of the role of humans in the emergence of such crises led to crisis diagnoses: crisis of the environment, crisis of capitalism, crisis of development, crisis of rationality, and even: crisis of humankind. However, the idea that runs through all of these diagnoses of the times thought in terms of trends, in terms of increasing developments in the twentieth century, as can also be seen in the concept of global change. These human-made imprints and transformations of the land surface cover became evident, namely visually documented and publicly perceived, through satellite images and satellite image atlases (for example, Eisl et al. 2011).

From around 1995, an *Earth System Science* developed that was comprehensively oriented towards the geosciences, and in which geography, geology, and geophysics came together. Here, the concept of global change was used heavily. Social and cultural science research on the interrelationship between society and the environment was established in and between many disciplines from the 1960s onwards. Environmental economics developed in economics and environmental sociology in sociology (for an overview, see Kropp and Sonneberger 2021). The humanities began to devote themselves more intensively to environmental topics from the 1970s onwards. Beginning in the 1980s, this was reflected in fields such as environmental history and historical ecology (Isenberg 2014; Haumann 2019; Kupper 2021), environmental philosophy, and especially environmental ethics (Ott 2020; Widdau 2021) and the educational sciences (Wulf 2020a, b; Wobser 2024). Literary studies and media studies have dealt with the topic particularly intensively, where *ecocriticism* has become established (see Chap. 3).

Archaeologists and historically oriented ecologists and geographers have long investigated the role of social conditions in the formation of landscapes (Crumey 1994). In the 1980s, political ecology emerged, which can be understood as an environmentally related complement to political economy. Here it was worked out that not only individuals and collectives, but also institutions and discourses should be considered as influencing factors for environments. Through constructivist-influenced studies, political ecology significantly contributed to the recognition of soil erosion and general environmental degradation not as purely natural processes, but as being caused by political factors (Blaikie 1985; Fairhead and Leach 1996). A complement to this are cultural geographic studies that show how natural disasters are also largely social disasters in their origin. A variation of the discussions on global environmental processes is Jason Moore's concept of *World Ecology* (Moore 2020).

At the same time, many scientific fields developed that were focused on sustainability. The supposed contradictions between nature and culture had already been resolved in the late 1900s in the concept of 'socio-ecological systems'. Aand from the 2000s onwards, research on the resilience of ecosystems and post-disruption collectives (resilience, Folke 2006, 2016; 2021) also increased later in cultural anthropology and archaeology (Bollig 2014; Bradtmüller et al. 2017). It was only in the wake of the Anthropocene discussion around 2012 that many of these fields culminated in the institutionalization of the *Environmental Humanities*, for which still no fitting German term has been established (Wilke and Johnstone 2017; Bergthaller and Mortensen 2018). Even though there is extensive literature on the Anthropocene (I have about 2700 articles on my computer), the topic has hardly made it into the canon of established topics in the social sciences. This applies even to the sub-areas of the social sciences that are thematically close to the Anthropocene, like sociology, where the concept was offered a 'cautious welcome' in environmental sociology (Lidskog and Waterton 2016). The number of mentions in more recent comprehensive manuals can be taken as an indication In the *Routledge Handbook of Environmental Anthropology*, the Anthropocene appears briefly on 14 pages of text (Kopnina and Shore-Ouimet 2017), on two pages of the two-volume *Handbook of*

Cultural Sociology (Moebius et al. 2019), and on six text pages of the *Handbook of Environmental Sociology* (Caniglia et al. 2021).

A number of authors described a phenomenon *avant la lettre*, which comes close to today's understanding of the Anthropocene. The biologist and philosopher Erhard Oeser wrote a book in 1987 under the title "Psychozoikum," in which he adopted Julian Huxley's term. One chapter is entitled: "The Age of Man as a Catastrophe of Natural History" (Oeser 1987: 51). Oeser speaks of monoculture agriculture and mass livestock farming as serving only one purpose: the "feeding of the most voracious predator that has ever inhabited the Earth." Oeser emphasizes that the break, the actual catastrophe of the whole of natural history, can especially be seen in the change in the speed of all processes determined by humans (similar to Rosa 2021). While Oeser mainly names the biological break, another author in 1988, at least 12 years before Crutzen, already captured this break in specific *geological* terms:

> We are acting on a geological and biological *order of magnitude*. (…) the *anthropogenic shock* that is overwhelming the Earth is of an order of magnitude beyond anything previously known in human historical and cultural development. As we have indicated, only those geological and biological changes of the past that have taken hundreds of millions of years for their accomplishment can be referred to as having *any comparable order of magnitude* (Berry 1998: 206, my emphasis)

My interim conclusion is that the term bundles existing ideas but also brings something really new. On the one hand, "Anthropocene" is a term that has the ability to congenially bundle pre-existing ideas and environmental thinking styles (Goeke 2022: 97–100). On the other hand, it also contains something substantially new conceptually: geological and deep time thinking and an Earth system perspective.

2.4 Turning Points and Breaks—When Did Humanity Become Geological?

> *Intentionally or not, we are all exhausting the Earth.*
> Robert Marks (2020: 235)

> *We are conducting an experiment with the entire Earth system, which we still understand very inadequately.*
> Jürgen Renn (2021: 3)

The current geological consensus is that the Anthropocene began in the middle of the twentieth century, around 1950 to be precise. According to the dominant view in the *Anthropocene Working Group*, the beginning of the Anthropocene epoch should be placed in the recent past, namely with the dropping of the first atomic bomb in 1945 or even better with the first hydrogen bomb "Ivy Mike" in 1952 (Zalasiewicz and Wing 2024c; Zalasiewicz et al. 2024a, b, c). The radioactive fallout lead to a worldwide horizon of artificial radionuclides. However, there is a wealth of alternative proposals, especially from geology, geography, archaeology, and environmental history, or from studies on social and cultural memory (*memory studies*,

Craps et al. 2018). A rough distinction can be made between proposals for a beginning in the last 250 years or so and others that go back to the early history of agriculture ("young Anthropocene" vs. "old Anthropocene," Kawa 2016; Hoelle and Kawa 2021: 656–657).

Where exactly the chronological boundary, i.e. the beginning of this period, is to be set is, however, disputed, and this question has been a major topic of scientific publications on the Anthropocene. The fundamental chronological decision is therefore whether to let the Anthropocene begin with the appearance of humans and especially with the first permanent settlements or only from the time when the cumulative effects of human trade are detectable worldwide—*anthropisation* vs. systemic change (Wallenhorst 2023: 432). This can lead critics of the term, for example, those from the field of environmental history, to say: "The Anthropocene is currently based on perceived boundaries, not on objective ones" (Robin 2013; Herrmann 2014: 49, 2016: 49). In a letter to the editor of Nature, Noel Castree goes so far as to say that an objective determination of the beginning of the Anthropocene is not possible in principle. Specific value judgements about when quantitative change turns into qualitative transformation are necessarily 'scientified' (Castree 2017: 289). Nevertheless, a research topic can be the periodizations themselves, as political motives also play a role here in addition to scientific facts. If a later point in time is brought into play as the beginning of the Anthropocene, a resulting sense of urgency to act can be generated (Görg 2016: 10). More important, however, is the fact that the question of the beginning not only raises the scientific question of origins, but also that of authorship. This results in controversies about who bears the responsibility for the crisis. With this in mind, I will systematically go through the facts and questions here.

Additionally, I will go into these proposals in detail for special reasons. First, the arguments for different times of the beginning of the Anthropocene raise many theoretically important as well as empirically open questions in anthropology. Second, the decision of when to speak of the Anthropocene in historical terms is very consequential for the political assessment of the current ecological situation of the world and, consequently, for the assessment of our possibilities for action. Determining the historical beginning can lead us to realistic imaginations or visions of society in the future (Schroer 2017, 2020; Adloff 2020). What forms of sustainable collectives are emerging today in response to the understanding of Anthropocene effects, and what forms of Anthropocene society are realistically conceivable? Third, the respective approach to the temporal beginning of the Anthropocene has considerable consequences for the question of what the humanities can contribute—a central theme in this book.

More specifically, the question of which branches of anthropology in the broad sense, such as American four-field anthropology, are relevant to the topic of the Anthropocene. The assumptions about the beginning have implications for cultural anthropology, that is, for cultural and social anthropology. The possible contribution of cultural anthropology as traditionally present-oriented and micro-oriented science varies depending on the assumed beginning of the Anthropocene. It makes a big difference whether the Anthropocene began 11,000 years ago or in 1945, which would correspond to the beginning of cultural anthropology as a globally widespread science. This also raises the question of which subject areas and methods of anthropology can contribute what and how much to the study of the Anthropocene.

2.4.1 Diverging Academic Interests in Periodization and Notorious Misunderstandings

> *The current lack of agreement on a tart date and which marker to use should not detract from the Anthropocene as a concept.*
> Anonymus (Nature) (2024: 466)

With the strong focus on the question of the beginning of this Anthropocene phase, different academic interests collide, and misunderstandings arise. Firstly, it is often forgotten that in the original conception by Crutzen and Stoermer, *it was not* about periodization nor even the exact dating of the beginning of this Anthropocene phase, but about drawing attention to the magnitude of this change. When it comes to the question of dating, different academic fields collide. (a) While geologists want to align the periodization with the effects visible in the deposits, some archaeologists advocate setting the beginning according to the causes of global anthropogenic change (Smith and Zeder 2013: 11). (b) While geologists place global influences at the center, archaeologists are often more interested in the local and regional impacts of humans. Even more fundamental is the fact that (c) geologists strive for setting synchronous period boundaries, while archaeological and historical periods are diachronic: they begin and end in different parts of the world at different times (for an overview, see Pare 2008). I will address this fundamental problem in Chap. 7.

The crucial point of the debate, in my opinion, is not the beginning per se, but rather the fundamental questions about humanity raised by the Anthropocene phenomenon as such, and the questions that the geological and historical beginning of this phase raise for the understanding of natural history-cum-cultural history. Firstly, in a particularly concrete way, the disputed beginning raises the question of whether the idea of the Anthropocene inappropriately generalizes concepts and chronologies that are based on European experiences, data, land use patterns, and temperate climates to the whole planet, as postcolonial historian Dipesh Chakrabarty has argued (Chakrabarty 2012). This is especially true for the research conducted under the banner of a planetary-oriented and primarily scientific Earth System Science (or *Earth-Systems Science*, Wainwright 2009). It should be noted that Western Europe only makes up 7% of the Earth's land surface. Seen in this light, agriculture in England or Ireland, for example, would be a small and scarcely representative part of the agrarian world, as if it were a province (Morrison 2015: 79, 2018).

Soil science, which became an academic discipline in the middle of the nineteenth century, also sees soils as an "archive" of human history that bears witness to the history of the landscape and the people. Soil scientists identify soils that have been permanently shaped by economic activity. These include so-called "plaggen soils," which are classically found in the Netherlands, Denmark, and Northern

Germany. For generations, farmers added grass "plaggen" cut elsewhere to the naturally sandy and nutrient-poor soil, which was enriched with the urine and manure of domestic animals until a layer of arable land was formed. "Wölbäcker" from the Middle Ages, with their characteristic wavy shape due to the plowing technique used at that time, can still be found in many places in Europe today that are protected as wastelands through reforestation (Bartz and Sperk 2015: 10). Plaggen-like anthrosols can be found in rice-growing areas of Southeast Asia.

Agriculture began as a regionally dominant economic system outside Europe thousands of years earlier, about 10,000 years ago. Agriculture, which has a strong and lasting impact on the environment, has existed in some areas since around 8000 BC, such as wet rice cultivation in East Asia. As a result, entire large landscapes in these parts of the world have been "domesticated" to an extent that is equivalent to modern monocultures. In more recent times, technosols have been added, soils, which consist mainly of materials such as concrete, glass and bricks, rubble, household waste, and industrial waste. Consequently, some climate models may significantly *under*estimate the changes in land cover in earlier periods. This applies, for example, to a simulation model that was originally intended to simulate vegetation in land use in Europe over the last 3000 years and its extension to the entire land area of the Earth and the period from 8500 years BP to 1850 (Kaplan et al. 2009; Kaplan et al. 2011, according to Morrison 2015: 80). This made a "tropical correction" necessary, which shows the problem of unquestioningly assuming European conditions as a baseline.

From a general methodological perspective, I believe this shows that postcolonialist-motivated criticism can indeed also be constructive for supposedly purely scientific studies. It can, for example, lead us to critically rethink variables, such as the "potential natural vegetation" used in agricultural geography. Certainly, dating is not the central question, especially from the perspective of cultural studies and the humanities. But the demand for decentralization already shows that the question of when the Anthropocene began is about much more than just chronological or stratigraphic questions.

2.4.2 *"Great Acceleration" from the Middle of the Twentieth Century*

So how new is the "new" age, "when did humans become geological" (Paál 2016: 2)? The growing interest in the Anthropocene has led to a wide range of proposals. According to the currently dominant view, the beginning of the Anthropocene epoch should be placed in the middle of the twentieth century. Around the middle of the century, there was a rapid growth in many of the measurable parameters of humanity. A precise dating proposal is the year 1945, more precisely July 16, 1945, with the test of the first bomb, which was based on the fusion of plutonium. This was determined through the consensus between Earth scientists and environmental historians (Steffen et al. 2007; 2011a, b). The "Trinity Test on July, 16th, 1945 (Fig. 2.2)

Fig. 2.2 "Trinity Test": first detonation of a nuclear weapon, July 16th, 1945 https://en.wikipedia.org/wiki/Anthropocene#/media/File:Trinity_Test_Fireball_16ms.jpg

and the drops on Hiroshima and Nagasaki shortly afterwards August 6, 1945, and August 9, 1945; initiated the atomic age. The two bombs on Japan resulted in over 100,000 deaths. The radioactive fallout marked the beginning of the widespread distribution of artificial radionuclides. However, this spread did not yet lead to a marker that could actually be detected worldwide, which is why the stratigraphic community ultimately did not support this proposal. An alternative that is currently being discussed is the year 1950, when the first globally persistent industrial chemicals were detected in ice cores. The explosion of the first hydrogen bomb over the Eniwetok Atoll on November 1, 1952, which produced a much higher fallout, also occurred during this period. In the wake of the Cold War, over 2000 nuclear weapons tests were carried out, 500 of which were above ground; this has led to another proposal favored today—the year 1964, the peak of radioactive and permanently detectable fallout (Thomas et al. 2020: 64).

During this time, the aforementioned industrial phase of the "great acceleration" began with a huge increase in population, increased water consumption, enormous dam construction, and a sharp rise in the production of plastic, aluminum, and concrete. From 1945 to 1950, to name just a few of the parameters, there was a drastic increase in soil sealing with concrete and asphalt, the spread of permanent organic chemicals, the spread of microplastics in the seas, and ocean acidification (Aragonite, Steffen et al. 2015a, b; Zalasiewicz et al. 2008a, b, 2015, 2017a, c, 2019a, b; Lewis and Maslin 2018: 222–223). Furthermore, from that time onwards, there was a rapid global increase in the production and use of synthetic fertilizers, the production of which requires fossil fuels. The use of synthetic fertilizers also made it possible to feed billions more people. In short, nitrogen-based artificial fertilizer provided nourishment by means of fossil energy (Christian 2018: 297). The focus of this phase, however, is the drastic decrease in the diversity of ecosystems, species, and genes. The diversity of life, biodiversity in the broadest sense, which is at the center here, exists on several causally interlinked levels. It is related to cultural diversity and is therefore ultimately a specific human-made construct that emerged in the 1980s and was promoted by the recently deceased evolutionary biologist Edward O. Wilson (1929–2021) (Wilson 2016). The basis for the diversity of ecosystems is created by the variation in the genetic material (genetic variation, genomic diversity) and their products (species or, more precisely, biodiversity) in interaction with the living and the non-living (abiotic) environments.

The marking around the middle of the century is based on the diversity of these data on the one hand, and an Earth system view of the planet on the other. However, this periodization also coincides with methodologically independent and much earlier works from climate history (Sieferle 1982, 2020, see Wanner 2016 and Pfister and Wanner 2021). In addition, after the Second World War, there was a sudden increase in the world's population and massive changes to ecosystems that had a *permanent* impact on the surface of the planet (Carey 2016; Waters et al. 2016). The remains of roast chickens (*Gallus gallus domesticus* as an anthropogenic species) that have been eaten worldwide will be a geological indicator for future geologists, alongside the ubiquitous plastic micro-particles (Bennett et al. 2018).

From a historical perspective, it should be emphasised that the Industrial Revolution, like the Neolithic Revolution, had a slow start in a small region and it took time for global climate effects to become apparent (Brooke 2014: 479). It is important to note once again that the determination of the great acceleration as the beginning of the Anthropocene is not a periodization made from a historical perspective, but rather from the perspective of geology based on extremely long periods of time. The basic statement by no means excludes the realization emphasized by critics (Görg et al. 2020: 49–51), that the acceleration is different in intensity and has different manifestations in different regions of the world.

The central argument for choosing the geological period at the end of the Second World War is the *global* accumulation and almost *synchronous* increase in *many* of these human-made effects. From a historical perspective, however, it should be emphasized that the Industrial Revolution, like the Neolithic Revolution, had a slow start in a small region, and that it took time for global effects to become apparent (Brooke 2014: 479). It makes a big difference whether you want the Anthropocene to begin when (1) the human capacity for strong environmental change emerged, when (2) the first effects on environments began or (3) when the cumulative effects began worldwide (see Table 2.4).

2.4.3 Industrialization and Colonization

There are about ten more proposals for the temporal boundary between the Holocene and the Anthropocene, of which Table 2.4 shows a selection (for a chronological presentation of the debate: Will 2021: 169). The first and classic proposal comes

Table 2.4 Periodization proposals for the beginning of the Anthropocene

Origin phase (alternative stratigraphic data in brackets)	Event and the possibly most important stratigraphic markers	Source
At any time in the future	Still unclear	Wallenhorst 2023: 392
1960s, e.g., 1964	Effects of atomic bomb tests, Bomb Spike, economic and social *overheating*	Lewis and Maslin 2015a, Eriksen 2016
1952	First thermonuclear hydrogen bomb "H-Bomb" *Ivy Mike*, Eniwetok Atoll, Pacific	Anthropocene Working Group 2023: 4–5
1940s and 1950s (1945, 1952)	Spread of artificial isotopes in sediments, caused by atomic detonations, *Trinity Test*, Plutonium 239	Zalesiewicz et al. 2010. McNeill and Engelke 2013: 208, 2016. AWG 2023
1750–1800 (1760, 1784)	Increase in methane and carbon dioxide due to the industrial revolution, steam engine	Crutzen 2000, 2002, Tickell 2011
1712	First steam-powered pump	Lovelock 2020
1610	Fly ash from coal combustion, *Orbis Spike*	Davis and Todd 2017
1492	Transatlantic exchange between the „Old World“ and the „New World“, *Columbian exchange*	Lewis and Maslin 2015a, b, Koch et al. 2019
1000 BP	Increased soil erosion worldwide	Bubenzer et al. 2021: 28
3000 to 2000 BP	Organic waste and anthropogenic soils as evidence of human ecosystem engineering, methane budget	Certini and Scalenghe 2011, Fuller et al. 2011, Mitchell 2013
3000 to 500 BP	Anthropogenic soils (diachronic, no clear stratigraphic marker)	Lewis and Maslin 2015b
5000–4000 BP	Methane increase due to increased wet rice cultivation and livestock farming, traces in ice cores	Fuller et al. 2011 Ruddiman et al. 2020
8000–5000 BP (7000 BP)	Methane and carbon dioxide increase due to deforestation and wet rice cultivation, "early Anthropocene"	Ruddiman 2003, 2013, Ruddiman et al. 2020
11,000–9000 BP	Human settlement, deforestation, human constructions, and animal domestication	Smith and Zeder 2013
13,800 BP (50,000–10,000)	Mass extinction of the Quaternary megafauna due to hunting and anthropogenic vegetation change (large mammal fossils) plus natural climate change	Braje 2015, Braje et al. 2017, Foley et al. 2014

from Crutzen and Stoermer, who placed the beginning at time of the Industrial Revolution, more precisely in the year 1800 (Crutzen 2000). During this period, the levels of anthropogenic greenhouse gases in ice cores rose dramatically. The decisive factors are the invention of the coal-powered steam engine, the sudden increase in the exploitation of mineral resources, the use of coal, and generally, the industrialization in parts of Europe (Crutzen 2002; Zalasiewicz 2008a). The resulting disruption of biotic, geological, and climate-chemical cycles is usually explained by the massive use of technology based on a conscious decision by the bourgeoisie. In contrast, some proponents of the term Technocene argue that it is an unintended consequence of profit-seeking and greed (Cunha 2015a). From the perspective of this periodization, the crisis of the Anthropocene is above all a crisis of the climate, as the "geological mineral layer" is mainly found in the atmosphere.

A slightly earlier alternative is the late eighteenth century, at the start of the industrial use of fossil fuels, especially coal. The year 1760, with the first detectable traces of fly ash from coal combustion, is currently being discussed as a stratigraphic marker. This was the period when human impact equaled that of other geological factors (Tickell 2011). The focus here is industrial capitalism. This coincides with periodization proposals, such as those made in historical sociology and by economic historians long before the concept of the Anthropocene emerged (SONA 2021). Ernest Gellner placed the beginning of state nationalism in this phase (Gellner 1983), and Edward Wrigley characterized England's transformation at that time as one from an "organic economy" to a growth economy (Wrigley 2016: 7–18).

In a 1944 book on market societies, the economic historian Karl Polanyi described, from a holistic perspective, the conflictual transition from an agrarian society to an industrial society as a fundamental "great transformation" (Polanyi 1944). Large civilizations with their hitherto embedded economy collapsed with the enforcement of abstract capitalist principles in the nineteenth century. Both proposals for the beginning of the Anthropocene, the early nineteenth and the late eighteenth centuries, correspond to the worldwide change in economic growth. Worldwide economic growth was low until the beginning of the nineteenth century: it was about 0.05% per year or 6% per century. This growth reflected the steady increase in population, so it was not per capita economic growth.

Sustained and broad growth in many measurable dimensions, such as water consumption, fertilizer use, urbanization, transport infrastructure, and dam construction, is historically a relatively new phenomenon (Schmelzer and Vetter 2019: 46–47). Such comprehensive growth only began in the nineteenth century and then only accelerated dramatically from the middle of the twentieth century, during the aforementioned "great transformation." If one takes this periodization with the beginning of the Anthropocene in the industrial age as the "first Anthropocene," the break in the middle of the twentieth century with the large increase in emissions can be referred to as the "second stage of the Anthropocene" (Steffen et al. 2007: 616, Merchant 2020a, b: 14).

The "long sixteenth century" or the early seventeenth century has been proposed as a significantly earlier alternative for the onset of the Anthropocene (Lewis and Maslin 2015a, b). After the discovery of America in 1492, new production relationships and a truly transcontinental exchange between the 'Old World' and the 'New World' emerged for the first time. This resulted in enormous changes in ecosystems almost worldwide.. As part of the "Columbian Exchange" (Crosby 2003, 2015: 96ff, 116ff.) in the fifteenth and sixteenth centuries, shipping and trade led to a strong exchange of plants and animals—such as potatoes, tomatoes, cattle, and sugarcane—between the New World on the one hand, and Europe, Africa, and Asia on the other. This also led to a biotic homogenisation of some ecosystems, which left traces in rock layers.This macro-scale and therefore evolutionary trajectory-changing transformation can be demonstrated both archaeologically and through a CO_2 minimum in ice cores, the so-called *Orbis Spike*, around 1610 (Boivin et al. 2016; Lewis and Maslin 2015b; Koch et al. 2019). These phenomena are more sharply formulated by some proponents of the Capitalocene concept, who associate the beginning of the Anthropocene with early capitalism. During the time of Europe's global expansion in the long sixteenth century, Europeans established a violence-based hegemony of pre-industrial capitalism (Moore 2016; Yusoff 2018). However, recent studies have shown that the environmental effects of the *Columbian Exchange* started at different times in different areas of the tropics and varied greatly in quality (Hamilton et al. 2021).

In the wake of colonial expansion, there was a noticeable decrease in CO_2 concentration in the two Americas due to unused fields and farmland as well as reforestation due to the decrease in indigenous populations, from 50–60 million to about six million as a consequence of diseases and genocide (Lewis and Maslin 2015a, b, 2018). Archaeologists studying species extinction and species displacement in the modern period support this timeframe, but sometimes refer to the entire period between 1400 and the early 1800s (for example, Lightfoot et al. 2013: 112). This contradicts a clear geologically synchronous approach, but is of interest from a cultural studies perspective because this also represents the period of the encounter with non-European cultures and the resulting change in concepts of nature (Arizpe Schlosser 2019: 277).

Authors critical of colonialism arrive at a similar periodization (Davis and Todd 2017). The *Columbian Exchange* phase saw the beginnings of colonial exploitation or the unequal exchange between powerful countries and the regions of global population centers in Asia, Africa, and Latin America, the "majority world." Davis & Todd argue that the ecocidal logic of the Anthropocene is a result of specific decisions that can trace their cause to colonization, namely the violent displacement of people and the powerful distortion of landscapes ("forcing a landscape," Davis and Todd 2017: 763–769, Baldwin and Erickson 2021). Irvine deepens this by linking specific spatial and temporal attitudes towards land and country with a colonial past. Authors who are part of the *Black feminism* movement see in the dominant Anthropocene terms not only an obscuring of coloniality, but also a racial exclusion of Africans from the concept of humanity by assigning Black people to nature. Thus, the geographer Kathryn Yusoff set the beginning of the Anthropocene to the middle of the fifteenth century, specifically the year 1452, with the arrival of the first slaves at the Portuguese sugarcane plantations in Madeira (Yusoff 2018: 33–35, 106, Luke 2018).

This is an example of how periodizations are also strongly linked to basic political convictions and weightings with respect to the assumed causation and thus today's responsibility (Chua and Fair 2019: 3, 13, 19). However one phrases it, whether as an "exchange" or "encounter" or as an "act of violence": something new is happening with the approach of Lewis and Maslin, because here geological knowledge becomes a testimony to the global effects of violence in human history. Thus, the approach of Lewis and Maslin bridges the gap between those contributions from the natural and social sciences that emphasize the role of colonialism in the genesis of the Anthropocene (Folkers 2020: 594).

2.4.4 Early or "Deep" Anthropocene and the Importance of Historical Data

The ability of humans to drastically change the environment dates back to early times. In pre-modern times, there have been local to regional environmental changes due to the use of fire since at least 800,000 years ago. A reduction in megafauna came with the displacement of grassland and habitat fragmentation (with the influence of natural climate change) in the late Pleistocene since probably 400,000 years ago. Strong greenhouse gas emissions due to land use are documented since around 8000 BC and large-scale deforestation in parts of China and Mediterranean Europe since ancient times and strongly since the early Middle Ages (Smil 2019: 205–207).

One of the three proposals originally made by the *Anthropocene Working Group* goes much further back, namely to the time between 3700 and 2000 years BP. The tangible traces here are landscape changes, habitat change of animals, and increased carbon dioxide levels due to land-use change (Zalasiewicz et al. 2015: 198). Some proposals emphasize the heavy deforestation and the emergence of rice cultivation in East Asia around 5000–9500 BCE. The beginnings of rice farming are unfortunately much less researched than those of wheat cultivation or sheep farming in the classic area of the Neolithic Revolution in the "Fertile Crescent." The actual beginnings may therefore be even earlier (Kaube 217: 168). Especially the technologically advanced agricultural economies under the "biological ancient regime," such as England and China, which had already largely deforested their land area by 1800 (Li 2009 for China, Balter 2013; Archaeoglobe Project 2019; Marks 2020: 232).

Intensive agriculture led to sudden increases in carbon dioxide from 7000 B.P. and methane from 5000 B.P. Early wet rice cultivation in eastern and southern Asia was probably responsible for about 80% of the early methane content in the atmosphere (Fuller et al. 2011; Stephens et al. 2019). All this possibly postponed a coming cooling, as the palaeoclimatologist William Ruddiman tried to show (Ruddiman 2003, 2010). Ruddiman's thesis that the early elevated levels of CO_2 and CH_4 can only be explained by the spread of agriculture is controversial, because there are also other factors that can abruptly disrupt the coherence of the climate. In addition, there is the general insight into the "punctualistic" behavior of evolving systems (Brooke 2014: 286–287, 440–442). However, it was also substantiated for the period of European expansion: The CO2 level decreased noticeably between 1550 and 1650 due to reduced land use in Latin America as a result of the thousands of deaths of many indigenous people due to measles (Nevle et al. 2011). The hypothesis is also being tested in models and is probably currently the best empirically substantiated of the proposals for the early Anthropocene (Thomas 2020: 52, 74). The thesis has also been tested in models and is currently the best empirically validated from among the proposals for the early Anthropocene (Thomas 2020: 52, 74). From a geological and archaeological point of view, one problem is that human intervention started very early (as early as the Paleolithic), but for a very long time, they were only associated with local or at most regional traces. Only from the Neolithic period onwards does this change with regional and possibly even global effects (Ruddiman 2013).

One problem is that we have little reliable knowledge about the population sizes and densities of early human communities. According to palaeoanthropological findings and models, the groups in the early history of humans probably consisted of between 150 and 200 people. According to archaeological data and models, hunter-gatherer groups in the Palaeolithic period in Europe usually consisted of a few hundred to a few thousand people (800 to 3300), towards the end of the Palaeolithic, i.e. between 14,000 and 11,600 years BC, on average around 6600 at most (Schmidt and Zimmermann 2019; Schmidt et al. 2021). Until the beginning of the Holocene, humans lived in a world of low population density and small communities.

An earlier proposed threshold of human-induced land change is thus the emergence of agriculture (Fig. 2.3) at around 8000 years BP (Ruddiman 2003, criticism by Wanner 2016: 164), or the formation of cities (Smith and Zeder 2013; Lewis and Maslin 2015a, b, 2018; Bauer and Ellis 2018: 224, West 2017: 198).

Fig. 2.3 Intensive rice agriculture in Central Java, Indonesia. Foto: Giri Wijayanto https://commons.wikimedia.org/wiki/File:Penggembala_Bebek.jpg licensed under the Creative Commons Attribution-Share Alike 4.0 International license

Even environmental historians who reject the term Anthropocene see the onset of global human influence on the geosphere as coinciding with the beginning of agriculture (Herrmann 2014: 49, see Sieferle 2020, and from an anthropological and archaeological perspective, see Graeber and Wengrow 2021). These processes were associated with a strong deforestation, especially in parts of Eurasia, and this change is detectable through fossil pollen and phytoliths. However, some human activities, for which there is archeological evidence, have only left limited stratigraphically usable traces, such as the early development of anthropogenic soils from around 3000 BCE (Certini and Scalenghe 2011). From an archaeological perspective, the Holocene largely coincides with the decisive phase of human history, the significant landscape change in the wake of the domestication of plants and animals, the so-called Neolithic Revolution. Morrison puts it succinctly:

> We must come to terms with and better understand the *anthropogenesis of the entire Holocene* before we can evaluate the novelty or significance of present-day human impacts (Morrison 2015: 80, emphasis mine).

Today we know that humans, even without agriculture, were biotic and geological agents for long periods, especially through the widespread use of fire. Fire was the first tool used by humans to change landscapes extensively and permanently, or to create an artificial econiche (Glikson 2013; Glikson and Groves 2016; Soentgen 2021a). James Scott therefore suggests that the Anthropocene should begin with the use of fire as a weak or "thin Anthropocene" (Scott 2018: 19). The controlled use of fire dates back to at least 800,000 years BP (Goren-Inbar et al. 2004), possibly up to 1.7 million years ago, long before *Homo sapiens*. The effects on local landscapes go back to almost 100,000 years (Thomson et al. 2021). However, their effects remained small on a global scale due to the low population numbers until about 10,000 years BP, when the systematic use of fire in agriculture began. The world population was then around 2–4 million people, less than a thousandth of today's population. Then, the Neolithic Revolution began with the emergence of permanently inhabited settlements around 12,000 BP. States, which only emerged in Mesopotamia from around 6000 BP, represent the second central trigger of the early Anthropocene for Scott. Permanent settlements, cultivated plants, and cattle farming are the central motors of landscape change. It was only with "state landscaping" by early states that the thin Anthropocene became a "thick" or "strong Anthropocene," a development that then experienced another leap with industrialization and globalization (*thickening Anthropocene*, Scott 2018: 18–19).

Newer positions, especially from geography, archaeology, the climate sciences, and global history as well as *Big History* go even further back historically and speak of the deep Anthropocene (Stephens and Ellis 2020). The *Archaeoglobe* Project, for which 250 prehistoric archaeologists compiled data, came to this conclusion:

> An empirical global assessment of land use from 10,000 years before the present (yr B.P.) to 1850 CE reveals a planet largely transformed by hunter-gatherers, farmers, and pastoralists by 3000 years ago, considerably earlier than the dates in the land-use reconstructions commonly used by Earth scientists (Archaeoglobe 2019: 1, see 3, Fig. 3).

Numerous findings speak in favor of an early Anthropocene. Recent syntheses of the available data show that beyond all temporal differences, a large part of the geosphere was transformed by hunters, gatherers, farmers, and herdsmen from around 3000 years ago. Nevertheless, hunter-gatherer societies brought about major changes in the flora and fauna: depending on the region, 20 to 70% of the large fauna became extinct (Wagreich and Draganits 2018; Wagreich et al. 2019; Stephens et al. 2019). Asian rainforests and even the Amazon rainforest, the icon of "untouched nature," are largely a product of reforestation following human intervention (Balée 2013; WinklerPrins 2014). Amazonian black Earth soils (*tierra preta*) are probably largely the result of human intervention, that is, anthrosols (Glaser and Woods 2004). For social scientists and historical ecologists, this may not be surprising. However, as geographer Nathan Sayre aptly pointed out, long-term human influences pose a real challenge from the perspective of *mainstream* ecology. The combined findings of archaeology and paleoecology look like this (see Archaeoglobe Project 2019):

> Even 12,000 y ago, nearly three quarters of Earth's land was inhabited and therefore shaped by human societies, including more than 95 % of temperate and 90 % of tropical woodlands. Lands now characterized as 'natural,' 'intact' and 'wild' generally exhibit long histories of use, as do protected areas and Indigenous lands, and current global patterns of vertebrate species richness and key biodiversity areas are more strongly associated with past patterns of land use than with present ones in regional landscapes now characterized as natural (Ellis et al. 2021: 1).

These new realities manifesting as the Anthropocene raise new questions, particularly for environmentalists: the question arises as to what kind of environment we want to protect if there has long been no wilderness, no *pristine nature* (Reichholf 2011; Sayre 2012: 61, 63, see Cronon 1995, 1996).

All of this shows how important historical facts are for the topic. The introduction of the Anthropocene makes perfect sense *if* it can be proven that humans have significantly contributed to ending the Ice Age (Paál 2016: 2). With the assumption of the very early onset of the Anthropocene, only a small phase of the Holocene would remain. If the Anthropocene were to take over the entire (already short) Holocene, the Anthropocene would directly follow the Pleistocene. In effect, the Holocene would be renamed as the Anthropocene. On the other hand, the introduction of the Anthropocene as a stratigraphic unit would seem unnecessary. In my opinion, however, it still makes a difference in the weighting of content whether we name this period of the last 12,000 years or so as the "Holocene" or the "Anthropocene." Some researchers push the beginning even further back, into the Pleistocene, and even suggest the likely human-caused extinction of the mammoths around 13,800 years before as a turning point (Doughty et al. 2010; Lewis and Maslin 2018: 25–41). Other authors even go back as far as 50,000 years ago, a period of a strong species extinction of large mammals ("overkill" of the megafauna, Martin and Wright 1967; Erlandsson and Braje 2014; Braje 2015) that can be proven in the fossil record. However, this may be primarily due to natural climate and vegetation change. The theses of a very early Anthropocene or a very deep history are diametrically opposed to suggestions that the abrupt global geological change is very recent. Without relying heavily on theses on the Anthropocene, the anthropologist Thomas Hylland Eriksen places an epochal break even closer to our present: the end of the Cold War (Eriksen 2016). I will come back to this in Sect. 5.3.

2.4.5 Mediating Proposals

Unlike industrialization and global radiation, most of these upheavals were regionally limited and also occurred with some delay rather than globally *and* simultaneously. A number of mediating proposals from the historical sciences locate the roots of global environmental change at some point between early complex societies in Eurasia and Africa and the early modern period, for example in the sixteenth century (White 1967; Chew 2000; Mosley 2024). However, these studies do not refer to supposed traces in the rock, but work with historical sources or prehistoric data (Haardt 2022: 122–123). These theses of a very early Anthropocene or a very deep history are diametrically opposed to proposals that the abrupt global geo-change was very recent. Without relying heavily on theses on the Anthropocene, the anthropologist Thomas Hylland Eriksen set an epochal break even closer to our present: the end of the Cold War (Eriksen 2016). I will return to this in Sect. 5.3.

Contrary to many assumptions among anthropologists (as, for example, the tenor in Bauer and Ellis 2018), a geological-stratigraphic break in time does not mean that there were no consequences of human activity for the Earth system before the Anthropocene. A stratigraphic boundary is not a black-and-white matter, as geologists and paleobiologists have emphasized (Zalasiewicz et al. 2018a, b: 220–221). *Anthropogenic* changes are therefore by no means synonymous with *Anthropocene* change. From a geological, that is, deep-time perspective of time and duration, *all* of the proposed starting points would fall into one (Dartnell 2018, see Bjornerud 2018: 129–131). I alluded to this with the experiences of the cultural paleontologist in the Quintary in the preface. We should distinguish between the question of the formal introduction of a geostratigraphic time unit called the "Anthropocene" on the one hand, and debates on "Anthropocene" effects on the geosphere on the other. Unfortunately, unlike in English, the German script does not allow for the possibility to distinguish between a capitalized *Anthropocene* (*the* Anthropocene) and one written in lowercase (*an* anthropocene) (see, for example, Zalasiewicz et al. 2017b: 219, Braje 2018: 215–216). The drastic Anthropocene geo-change is also scientifically important and politically relevant, even if it is not a matter of formalizing the time unit.

Contrary to the tenor of the study by Bauer & Bhan, which suggested that the Anthropocene, regardless of whether it is understood from a natural science or cultural studies perspective, is about a chronologically determined period ("chronological designation," Bauer and Bhan 2018: 9–10), it should be noted that the original motive of Crutzen and Stoermer—despite the geological neologism Anthropocene—*was not* the search for the geohistorical start of the period. Contrary to popular perception, the Anthropocene was *not* originally introduced as a stratigraphic concept. The focus on the beginning of the Anthropocene, which still dominates scientific discussions today, easily makes us forget this fact. The central message of Crutzen & Stoermer was to use a new epochal concept to draw attention to the abrupt human-induced environmental change, our responsibility, and the importance of sustainable management for the resulting problems (Braje 2018: 215–216, Kaplan 2018: 217, Zalasiewicz et al. 2018a, b: 221).

With the participation of Crutzen, German colleagues, especially from prehistory and early history, have proposed a mediating periodization: the Palaeoanthropocene (Foley et al. 2014: 83, Gronenborn 2024). This would be the period of limited human intervention in the landscape ecosystem in subspaces of the geosphere. The Palaeoanthropocene would therefore begin with the first barely noticeable environmental interventions that came with the emergence of hominids, show active interventions already in the Paleolithic yield the first massive effects around 1750 or 1780, and end around 1850 with the massive transformations brought about by the Industrial Revolution, thus initiating the Anthropocene in the truest sense. This proposal aims to do justice to several insights: humans have been intensively shaping the landscape ecosystem in many areas of the Earth for a long time, they reach a new quality with agriculture but only with industrialization and mechanization did this turn into a *global* and in *geological* time perspective *synchronous* geospheric change.

The concept of the Palaeoanthropocene opens up important insights: In a logarithmic scale (Foley et al. 2014: 84), three roughly equal areas of Earth history emerge: 1. The phase without humans (4.56 billion years to 1–three million years BP, 2. the Palaeoanthropocene (from the appearance of humans, or the Neolithic at the latest, until 1850), and 3. the Anthropocene (beginning with the globally synchronous effects around 1850 or later. Only in this logarithmic representation does it become clear that *all* the time proposals discussed for the Anthropocene—that is, as a globally synchronous period of time—from a geological and linear (not logarithmic) time perspective, coincide at one point! The Palaeoanthropocene is therefore explicitly not to be understood as a synchronous geological epoch of Earth-wide change, but as the accumulation of peaks of human influence on a local or regional scale, which accumulated and then only in the last 200 years or so, with the Anthropocene, transitioned into a real geological period, which one could then, in my opinion, actually call the "Neoanthropocene."

Overall, the concept of the Palaeoanthropocene also shows the respective relevance of the data from climatology, biogeochemistry, archaeology, paleoanthropology, and paleoclimatology for different periods within a broadly interdisciplinary research area (Foley et al. 2014: 84, similar Ellis et al. 2013). Strictly speaking, for the Anthropocene, climate history (Cronin 2009; Brooke 2014; Mauelshagen 2012, 2013, 2019, 2023; Summerhayes 2020) and other sciences, which are the focus of this book, would need to be added. However, the term Palaeoanthropocene should not lead to confusion with the idea of the "early Anthropocene." The thesis of the early Anthropocene postulates the early onset of *globally synchronous* change, sometime during the Neolithic (Ruddiman 2016a, b), while the concept of the Palaeoanthropocene sets the beginning of the Anthropocene with industrialization, which is why I think the term "Proto-Anthropocene" would be more appropriate.

I would like to emphasize that none of the concepts of an "early Anthropocene"are compatible with the requirements for a formal geochronological unit in the geological time scale (GTS). Instead, they frame the entire Holocene or part of it, and possibly also parts of the Pleistocene, to recognize the long history of humans in reshaping the global environment (Zalasiewicz et al. 2021a, b: 13). With their *diachronically* understood origin findings for the Anthropocene, none of them can be used as chronostratigraphic markers:

> Seemingly *counterintuitively*, despite human modification of the planet being most clearly expressed in artificial deposits associated with the archaeosphere, no candidate GSSP is currently being investigated in such deposits, despite their richness in anthropogenic evidence (…), because of their typically *punctuated*, *patchy*, and *locally disturbed* accumulation (Zalasiewicz et al. 2021a, b: 4, emphasis mine).

Conversely, it is clear that an Anthropocene set chronostratigraphically in the middle of the twentieth century *per definitionem* excludes the preceding thousands of years of human influence on the Earth, a point repeatedly emphasized by the minority in the AWG. However, contrary to common misunderstanding, this does *not* mean that the geohistorically new Anthropocene Earth dynamics (*new state*) would be analytically decoupled from historical precursors and continuities. The diachronic onset of human influences can be integrated and analyzed within a timeframe that allows for precisely defined and consistent comparisons. The new geospheric dynamics show continuities and are based on path dependencies that go back to the nineteenth century and, in some cases, much further. In the geological-chronostratigraphic approach, a sudden change ("great acceleration") within a system that is continuously *changing* over a long period of time is recorded as a *synchronous* time marker by means of clear markers or geological *signals* (Williams et al. 2016).

Based on his Marxist (!) criticism of the Capitalocene concept (see 4.7), John Bellamy Foster proposes the following chronological boundaries. The Anthopocene begins as a geological epoch in the middle of the twentieth century with its first age, which he calls the 'Capitalinian Age'. This replaces the last age of the Holocene, the Meghalayan (Foster 2022: 466–472). This is a chronological solution that equally takes into account (a) the geological deep time perspective, (b) the great acceleration with monopoly capitalism, plastics, petrochemistry and radionuclides from the middle of the twentieth century and also (c) the much earlier world-systemic origin of historical capitalism in the sixteenth century.

2.4.6 Arguments for and Against a Chronostratigraphic Determination

The chronostratigraphic procedure of locating an acceleration point within a long-term change is equivalent to geostratigraphic practice, such as the groundwork for the Cambrian and Silurian systems in the geological timescale. Thus, the base of the Meghalayan in the Holocene series at 4250 years BP emphasizes a point within a strong as well as a social transition that took place over 250 years (Zalasiewicz et al. 2021a, b: 10). The debate about the historical onset of global human effects implies the central arguments for and against a globally synchronous determination. The arguments were broadly discussed in the journal *Progress in Human Geography*. Ruddiman concisely named the arguments against it (2018: 451): 1. Humans extensively altered the geosphere long before 1900. 2. A strict chronostratigraphic determination of a beginning would exclude all changes that were "merely" regional or continental or whose beginning was staggered, such as wet rice cultivation in Asia or deforestation in the prairies of North America. 3. The classic approach pursued by the *Anthropocene Working Group* is not actually followed by the majority of field-oriented researchers concerned with more recent geological phases. All of this speaks in favor of a rather informal and flexible use of the Anthropocene as a period term.

The proponents of the formal definition in the AWG argue, on the other hand, that a formal version is perfectly compatible with a diachronic onset of anthropogenic influence. Their argument is (ad 1) that the effect size increased slowly and thus clearly differs from the rapid change in the twentieth century, (ad 2) that the stratigraphic nomenclature certainly allows for pre-industrial markers in the Holocene, and (ad 3) that classic chronostratigraphy, with its clear time units, is meaningful and indispensable (Zalasiewicz et al. 2019a, b: 320, 329–330, Zalasiewicz et al. 2021a, b: 4–6). Despite all the diversity of the proposals, at the moment, the *Anthropocene Working Group* is only intensively considering two time phases for the onset of the *geologically* understood Anthropocene: first, the proposed dates in the twentieth century (1945, 1950, 1964), and second, those in the late eighteenth century (Lewis and Maslin 2015a, b).

The group led by the archaeologist Edgeworth, on the other hand, supports Ruddiman's argument. Chronostratigraphy is the right tool for geological periods, but not for prehistoric and historical ones. Edgeworth et al. argue that the splitting of Holocene stratigraphic sequences and continuous transformation processes into discrete time units would directly obscure the peculiarities of human impact. The typical chronostratigraphic sequences would accordingly miss the human impact (Edgeworth et al. 2015, 2019: 343, 341). The two positions ultimately result from the differences in disciplinary approach, different period concepts, synchronous vs. diachronic (time-transgressive), and also the disparate time scales.

If the Anthropocene is seen as a series of disruptive events and not as a geological period, one can assume a coevolution of geological and human processes (*Earth system trends* and *socio-economic trends*) within the framework of a "history of the whole system" (Bonneuil and Fressoz 2016:10–11, similar to Jasanoff 2010). Important surges in the increase in human geospheric effects can then be recorded, such as the thresholds around 14,000 years BP, 3000 BP, and then around 1510, 1650, 1750, 1800, 1870, 1945, 1950, and 1964.

The only major synthesis in book form to date that represents an empirical approach in this sense was published by environmental historian John Brooke. In his work, he combined data from Earth System Science, climatology, climate history, and other environmental history as well as medical history. Brooke arrived at the following periodization of the Anthropocene into seven climate phases, from the Bronze Age to the present day (Brooke 2014: 408, 553–558): 1. Beginning of anthropogenic climate effects in the Bronze Age and the sudden increase in methane in the eighteenth century with the spread of agriculture in China; 2. The 1840s up to the 1850s, the "first industrial revolution" with early emissions dominated by carbon dioxide; 3. From 1880 to 1910, the "second industrial revolution" with a dramatic increase in greenhouse gases; 4. From 1910 to 1950, with multi-causal global warming; 5. From 1950 to 1970, with a slight cooling; 6. From 1970 to about 1995, with warming despite the cooling effects of major volcanic eruptions and economic depression; 7. From 2000 onwards, with a slight reduction in the increase of global temperatures.

Research on all these thresholds is important and is largely still lacking. Nevertheless, I think that the core message of the geologically understood Anthropocene concept, that is, the "conceptual earthquake" (Egner and Zeil 2019: 28–30), only becomes clear when we see all these thresholds as phases of the "Anthropocene." From a geological perspective, for example, from the point of view of Amy, the geologist in the Quintary, all these dates coincide in time. On a geological and planetary scale, regional distinction and temporal differences disappear. What will remain is a series of *structurally similar* deposits that date back to *a few thousand* years and occur at various places around the globe. It will be obvious that they were formed by an interconnected global process of exploitation of the Earth (for example, Lawrence et al. 2016: 1360). It is only from this distant deep-time perspective, which goes far beyond the usual historical time scales, that it becomes apparent how suddenly and unpredictably humans have changed the geosphere.

Chapter 3
End-Time Stories—Mostly Dramatic Framings

Abstract This chapter is dedicated to the strong metaphors and mostly alarming narratives about the Anthropocene. The ideas concerning the Anthropocene are related to conceptions of the Earth, globe, and planet, which were strongly influenced by space travel. At the center are images of the sick planet on the one hand, and immeasurably large numbers, alarming curves, and drastic diagrams on the other. There is a wide panorama of powerful narratives that simultaneously, and often implicitly, carry political ideals and moral concepts. The chapter shows how the Anthropocene is linked with certain images of humanity in the discourse, and that these metaphors represent a social fact that influences societal change.

> *It's not a matter of fact as much as a way of organizing facts to highlight a certain importance that they carry.*
> Jedediah Purdy (2015: 2)

> *Human beings participate in history both as actors and as narrators. (…). In vernacular use, history means both the facts of the matter and a narrative of those facts, both „ what happened" and "that which is said to have happened."*
> Philippe Trouillot (1995: 2)

In the discussion about the Anthropocene, scientific findings are combined with a universal narrative and a mainly negative linguistic picture, particularly through dystopian metaphors and—in my opinion, understandably—alarmist and melodramatic narratives. End stories dominate the discourse (Sakkas 2021: 154): We hear about the "climate emergency," the threat to our "natural heritage," the world "borrowed from our children," and a "stolen future." Many representations of the planet's future are in the form of narratives about the "end of nature." The same applies to statements about Western or modern civilization.

In view of the Anthropocene, some authors argue that the biggest problem is admitting that modern civilization is *already dead*. This "civilization" also includes all the achievements and hopes of modernity, such as freedom, self-determination, mobility, knowledge, and the vision of equal access for all to all of these. Recognising the Anthropocene would mean accepting a rupture, that this old world is dead because

C. Antweiler, *Anthropology in the Anthropocene*, Anthropocene – Humanities and Social Sciences, https://doi.org/10.1007/978-3-031-74591-1_3

we are already living in a new one. Now it is about dealing with it, about "learning to die" (Scranton 2015: 23). Critics criticise that such extreme ways of talking about the inevitability of "dying" could make the problem seem no longer seriously addressable. The central motive of creating responsibility could dissolve in a depoliticising way (Purdy 2015: 4–5). For Scranton, however, "death" refers to the foreseeable decline of civilisation, By "death", however, Scranton means the foreseeable decline of civilisation. When ecosystems that form the basis of our lives collapse, the current techno-economic system, which is fixated on production and consumption, will fail.. And here the argument points to the urgency of rational, rapid and fundamental change away from the current imperial relationship to nature (Read and Alexander 2019: 11, 101–106). Recognizing the Anthropocene means accepting a break, that this old world is dead because we are already living in a new one. Now it is a matter of dealing with it, of "learning to die" (Scranton 2015: 23). Such extreme positions, however, make the problem appear too insurmountable and dissolve the central motive of generating responsibility in a depoliticizing way (Purdy 2015: 4–5).

Much less common are Anthropocene narratives that frame the Anthropocene in a value-based and possibly also encouraging narrative or utopia (as Leinfelder 2017a does). Since the pessimistic narratives dominate so strongly, we should not forget that we need good reasons to empower cultures and societies worldwide to work against the Anthropocene. Happiness research has shown how important most people deem the factors for enriching life beyond income and material growth. This could be part of a plausible narrative for the energy transition, as the technology and environmental sociologist and transformation researcher Ortwin Renn has suggested: "We need narratives that show how life can become better with the necessary changes"(Renn 2021: 28).

In addition to the narrative style, texts on the Anthropocene are teeming with metaphors. We read about "planetary guardrails," a "threatened Earth," and an "overheated planet." I see such metaphors and narratives not only as an anthropologically interesting topic, but as a powerful part of social reality and especially of scientific evidence-building and societal relationships with nature. Linguistic images, narratives, and fantasies are causally important influencing factors right up to the creation of material realities:

> While an attention to narrative may appear to abandon a commitment to material processes, I argue that such stories help constitute ecologies and socialities by bringing them into new kinds of socioecological and ecobiopolitical (earthly) relation (Moore 2015a: 39).

3.1 Alarm and Dystopia—Environmental Narratives with Mobilization Potential

We live in a time of catastrophic peaks and curve inversions.
Deborah Danowski & Eduardo Viveiros de Castro (2019: 20)

The Anthropocene hypothesis must be understood as a contribution to the self-description of society.
Niels Werber (2014: 245)

The Anthropocene has generated a particular form of experience that, for many people, evokes a feeling of helplessness, especially in the face of climate change. At the end of the twentieth century, uncertainty about the future of the world generated a specific feeling of threat to our freedom (Stoner and Melathopoulos 2015). This feeling of constriction is not only an individual one, but it is also reflected in the image that societies have of themselves. Here, the Anthropocene constitutes just one of several occasions for catastrophic images, albeit a particularly monstrous one (Ostheimer 2016: 45–50, Hinrichsen et al. 2020, Guiliani 2021: 83, 140–193).

3.1.1 Self-Descriptions of Society

Narratives and imaginations about nature and society, as well as the environment of society, in the Anthropocene must first of all be seen as self-descriptive categories of the world or society in the face of rapid change. From a social science perspective, however, the Anthropocene is not an Anthropocene automatism. Despite all the natural restrictions on human activity, which the Anthropocene also makes clear, there is no Anthropocene determinism. Imaginations and categorizations play a decisive role in determining which claims to autonomy are granted, which potentials of creativity are unleashed, and which standards of criticism are considered legitimate (Adloff et al. 2020¸, see also Oppermann 2023).

Ideas and images about the relationship of society to nature (and culture) and to humanity have a considerable influence on social developments, political decisions, and, above all, on visions, future concepts, and thus on long-term *policies*. A new content analysis of specialist articles on the Anthropocene revealed three imaginations of the Anthropocene: The world situation is typically seen as one of compromised habitability (*endangered*), dependent on coexistence (*entangled*), and based on resource exploitation (*extractivist*, Lövbrand et al. 2020: 3–5). The narratives analyzed in this chapter are not only interesting as such, but are also eminently practical. Such imaginaries are geopolitically relevant because they undermine conventional binary categories such as inside vs. outside, north vs. south, and “us” vs. “them” in the face of an interconnected world.

Both the personal and the public perception of catastrophes or disasters primarily concern sudden changes. Industrial disasters in particular are imagined as short-term events (Kirsch 2018: 243). Disasters are prototypically associated with, for example, the explosion and the poisonous cloud emission in Bhopal, the nuclear meltdown in Chernobyl, the *Exxon Valdez* shipwreck, and the oil spill in the sea off Alaska. However, given the spread of fear-based narratives, it should be noted that many “end-of-the-world metaphysics” go back to general worldviews and are therefore only indirectly related to the physical event of the planetary catastrophe (Danowski and de Castro 2019: 10).

While the Anthropocene cannot be experienced by the individual or presents itself as a slow-motion process, from a geological perspective, it represents a geological catastrophe. Against this background, the catastrophe narratives in which the

Anthropocene is often told must be evaluated differently. They may initially appear exaggerated and sensationalistic, for example, when talking about the "death of the planet"or the "death of species." From a deep-time perspective, however, the Anthropocene actually represents a catastrophe of the geosphere, which is evident in the *geological* formulation "sixth extinction of species." Dystopian allegories and narratives (DeLoughrey 2019) thus constitute the central starting point for raising public awareness about an extensive crisis, for making politically urgent trend-setting decisions, and for education.

Images and narratives not only determine scientific thinking, but also have practical and political implications. This is evident throughout history, where we can find guiding ideas about the environment of individual regions that are explicitly or implicitly linked to a call for necessary action. Purdy (2015) illustrated this using the example of North America, and Dewey summarized four broad approaches that continue to shape the life, laws, and policies in North America today:

> (…) *providential* (colonial period–1800s, Errand into the Wilderness/Manifest Destiny), *Romantic* (1800s–1900s, transcendentalism/scenic preservation, Henry David Thoreau and John Muir), *utilitarian* (1900s, progressive technocratic resource management, Gifford Pinchot and Theodore Roosevelt), and *ecological* (postwar environmentalism, Rachel Carson, 1960s–1970s federal environmental laws) (Dewey 2016: 1329, emphasis mine).

These different constructions of nature did not only show the dominant perception of the environment at the time, but also postulated certain human needs and legitimized political measures based on them. However, nature was almost always seen as a sphere separate from humans. This separation is seen in a markedly different way under the Anthropocene concept, but the importance given to narratives, imaginations, and metaphors is even greater than in the environmental discussion (see Radkau 2011; Dalby 2017; Uekötter 2020).

3.1.2 "Humanity in Ruins"—Apocalypses and Exaggeration

In the 1970s, environmental awareness and increasing globalization fostered the feeling that one, as an individual, was responsible for the planet. This also changed the meaning of the word environment. The environment was no longer just the concrete surroundings of organisms, but a political and societal field. Now, the environment was not only a matter of political responsibility, but also of individual morality. This influenced consumer behavior, and the industry responded, especially in advertising and marketing, for example, through recyclable packaging (Thomas et al. 2020: 179).

On the other hand, the stories told in Anthropocene articles are big, and they are getting even bigger, "… and perhaps getting bigger too quickly so that they want to virtually be theories of everything," as feminist philosopher of science Donna Jean Haraway has noted (Haraway, in Haraway et al. 2016: 561). Moreover, we are currently seeing a trend of "Anthropocene fiction" (see Horn and Bergthaller 2022:

118), and there are also non-fiction books, for example, about environmental problems or climate change, that fly under the banner of the "Anthropocene" without addressing the concept. The use of *catch-all* words and the underlying ideas is insightful as such, because the discussion about the Anthropocene and the emergence of a supposedly omnicompetent word, and thus a necessarily vague concept, represents a new social fact. This fact can be examined from a humanities and cultural studies perspective.

Texts on the Anthropocene are teeming with word images, often containing very powerful metaphors, which are known to be an important means of thought and can serve as symbolic markers that draw attention (cf. Fig. 3.1). In terms of the factual clarification of issues, metaphors become problematic when they are substantively absurd, when they promote anthropomorphic ideas, or when they become mere empty phrases. We can ask ourselves, for example, which concepts from the Earth sciences, such as "great acceleration," "threshold," or "tipping points," tell us about the actual causes (Pálsson et al. 2013: 7). But texts that examine or criticize the Anthropocene as a concept also work with environmental narratives and metaphors. Pertinent examples can be found in texts by Haraway and her student Tsing (see Hoppe 2021).

In a much-quoted phrase, Tsing speaks of us living in the ruins of capitalism, defining ruins as "spaces abandoned for the production of assets" (Tsing 2015: 6). This phrase is then used much more generally in other publications (Tsing 2015: 40–41, 205–213, Gan et al. 2018; Tsing et al. 2020: 193). Haraway and Tsing argue

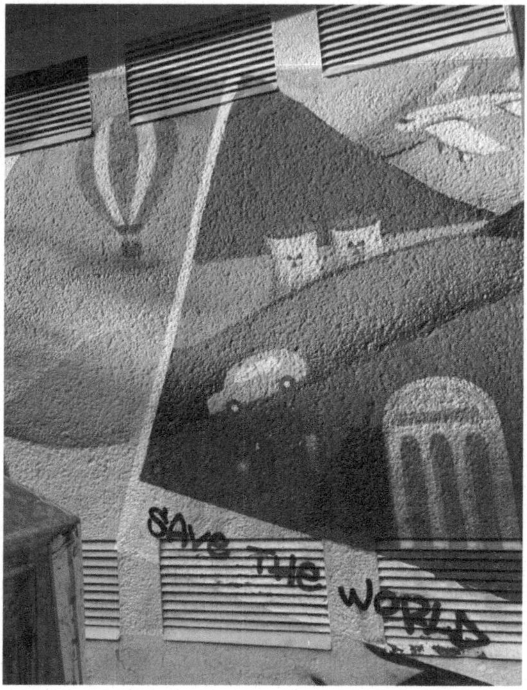

Fig. 3.1 "Save the World": Utopia and Dystopia; graffito in Cologne, Germany. (Photo: author)

in favor of overcoming the separation between humans and other living beings, and consequently also speak of the fact that not only humans, but also other living beings are "on the run." From this perspective, deterritorialization and environmental flight are fundamental phenomena of both the living and the social (Folkers 2017: 378). In a geological present characterized by capitalist "ruins," refuges for refugees of all kinds become a necessity for survival. What threatens in the Anthropocene is thus a

> (…) wiping out of most of the refugia from which diverse species assemblages (with or without people) can be reconstituted after major events (like desertification, or clear cutting, or, or …) Right now, the Earth is full of refugees, human and not, without refuge (Haraway 2016: 100).

The currently dominant imaginations of the future are polarised between 'the world is already lost' and 'we can do it'. Widespread catastrophism stands against techno-utopianism. However, there are several variants within alarmist narratives and dystopian imaginaries (Fladvad 2021: 154–158). On the one hand, there is the classic variant of catastrophism, which is based on the natural sciences but does not challenge the status quo or the capitalist system. One example is the narrative of impending climate collapse and the corresponding 'ecologies of fear'. Critics see this approach as depoliticising and an expression of naive techno-utopianism. Recently, however, there has been another form of endtime ideas that has emerged in recent years in the context of activist groups, in particular Extinction Rebellion (XR,; cf. Read and Alexander 2019). This catastrophism is also based on scientific findings supplemented by indignation about the lack of effect of environmental policy measures. What is new, however, is the clear naming of the political causes of the anthropocene crisis and the aim of opening up a space for genuinely alternative politics and economics. This new 'collapsology' works with forms of protest, for example through public mourning rituals, so-called 'Die Ins', which aim to draw attention to the proximity of humans and animals. It is about'catastrophe visualisation' (von Redecker 2020: 92).

3.2 Globe, Planet, and *Gaia*—World-Views Full of Resonance

> *In just a century or two, we have been thrust unawares into the role of pilots who are supposed to steer the planet without really knowing which instruments to keep an eye on, which buttons to press, and where to land.*
>
> Christian (2018: 292)

The choice of images and metaphors is consequential, but even more descriptive words convey certain ideas and also valuations. This is shown by studies on the image politics of climate change (Schneider 2018). In the case of the Anthropocene, this applies to very important words, starting with the names of the Earth. The word "Earth" is initially descriptive, while the word "world" is more general and ambiguous, because the word can also refer to the living world of organisms or the cultural living world of humans. With the world population likely to rise to ten billion, German promoters of the Anthropocene idea have predicted the emergence of a "human Earth," an Earth where human needs and human-made infrastructure dominate the Earth system (Schwägerl and Leinfelder 2014: 234). Critics have complained about the overestimation that comes with this term, "human Earth," which is often used by Anthropocenists; I will be discussing this further in Sects. 4.4 and 4.5. At the same time, the word Earth can be too strongly associated with the perception of oneness (Gibson and Venkateswar 2015: 1).

3.2.1 *Globe and Planetary Thinking*—**Mondialisation** *and* **Planétarisation**

A term linked to the word and the concept of "Earth" is mondialization (French: *mondialisation*). The term is not simply the French translation of the word globalization, but it especially conveys the critical aspects of globalization. Mondialization is intended to highlight the diversity of political entities and regimes that have sought regional dominance throughout history, even if they never achieved worldwide hegemony (Augé 2017: 123, see also Beau and Larrère 2022). An alternative to using "Earth" and "world" is to speak of the "globe." The word "globe" focuses on politics, history, human expansion and dominance as well as economic globalization. A critical point against the concept of the globe is that—similar to the concept of humanity—it ignores socio-ecological inequalities, such as access to drinking water. The literally global perspective tends to make spatially undifferentiated statements (Gebhard 2016), and critics instead recommend "… adopting a general inequality perspective in and on the Anthropocene" (Bruns 2019: 59, 2020).

The term "planet"ostensibly refers less to the human-centred construction of the world, to political aspects and to human dominance (which the globe shows us), but rather to the special position of the planet Earth as an earth system in the solar system that makes life possible (Chakrabarty 2019; Blake and Gilman 2024: 71–75). Planets dominate an orbital region around a star. Nature, size, but also the limitations of the Earth come into view here. In images of the "blue planet," the loneliness of the "home planet"is also highlighted. It therefore makes a difference in perception when we speak of either "global change"or "planetary change." While many of the forces of globalization do not actually affect the entire globe, that is, they are not really global, the term planet points to truly worldwide forces and assemblages. Such an outlook opens up theoretical perspectives on planetary thinking (Szerszynski 2019; Blake and Gilman 2024: 71–101). But it can also reveal empirical views on relationships between humans and other species (*inter-species relations, multi-species entanglements*) and post-human perspectives.

On the other hand, "planet," just like "world," constitutes a global icon, with which worldwide change is presented to the public and which is used as a symbol for research programs and projects. It shapes the contentual problem framing in institutions and the partially ritualized global conferences, which serves as a fruitful field for anthropological studies (Little 1995; Rojas 2013, and Krauß 2015a: 68, see esp. also Jasanoff 2004a, b). The term "planetarization" (French: *planétarisation*), which is mainly used in French contributions, is used as an alternative to the concept of globalization, whereby it refers to the limits of political and technological globalization. Together with the term mondialization, planetarization can be analytically valuable in its second—ecological—meaning. Planetarization refers to the Earth as a physical body, which, as the common property of humanity and all living beings, is threatened and thus raises questions about a global community. This global community consists of "earthlings," all of whom must be taken into account in their individual, collectively cultural, and generic (*génerique*) dimensions (Augé 2017: 26, 84, 88, see Pálsson 2020).

The everyday word "global" ignores important differences (Chakrabarty 2021: 71–80, Blake and Gilman 2024: 71–101). The global in globalization is primarily man-made, intentional and historically recent. The global of global change, on the other hand, affects and is affected by humans, but it includes non-human things and living beings. This global as an interdependent whole was already there before humans and will probably still be there long after humanity. Therefore, the following suggestion regarding the choice of words in relation to climate change makes sense:

> The human-centric *global* of globalization can remain the *globe* or *global*; the Earth-centric *global* of global climate change is better understood as the *planet* or the *Planetary* (Blake and Gilman 2024: 73, emphasis orig.)

3.2.2 The Power of Images from Space—"Blue Planet" and "Spaceship Earth"

In the age of visual media, ideas and images of the Earth are closely linked to visual experiences that are present worldwide. When we speak of the "Blue Planet," this refers rather soberly to the significance of the oceans, which cover 70 percent of the Earth's surface, or more generally to humankind's dependence on water. Much more fundamental, however, is the visual reference to the appearance of the planet from space. The image "Earthrise," taken by Bill Anders from the lunar orbit during the Apollo 8 mission on December 24, 1968, which shows the Earth rising as seen from the Moon, and the photo "Blue Marble," taken by geologist Harrison Schmitt from Apollo 17 on December 7, 1972, this time with a view of the entire planet (Fig. 3.2), were the most printed photos of the twentieth century (Poole 2008; Lazier 2011; LeCain 2017: 311–322). These photos replaced the iconic images of World War II and the Vietnam War, which had been prevalent until then. As seen from space, the blue marble against the black velvet background impressively demonstrated both the wholeness of the planet and the loneliness of the Earth in space.

Fig. 3.2 "Blue Marble" photo taken from Apollo 17 https://commons.wikimedia.org/wiki/File:The_Blue_Marble_(remastered).jpg?uselang=de. NASA copyright policy states: "NASA material is not protected by copyright unless noted"

From a symbolic and art studies perspective, these images of the circle inside the square do not only evoke a sense of harmony, but they also bring a whole string of associations into play, such as globalization and the supposed absence of humans. Thus, it is no exaggeration to say that these images are veritable icons of the twentieth century (Bredekamp 2011; Glaubrecht 2019: 18). As detailed qualitative analyses show, the crucial aspect is probably the external and distant view of the world and thus of humanity, because it shows many things and at the same time also points to a number of things that are not visible (Hoggenmüller 2016). These images from space showed from a new—distanced—perspective where we come from and what keeps humanity alive. At the same time, however, *Earthrise* also shows the significance of established perspectives. The photo is the result of an observation perspective we are used to, in this case an observation-dependent construction from the moving *Apollo* spacecraft (Goeke 2022: 100). On the moon, there are no earth rises comparable to our sunrise, because it is in bound rotation with the earth.

These photos are omnipresent to this day, and together with the old metaphor of "Mother Earth" that they revived (for example, Caputi 2020), they were enormously influential in shaping the guiding principle of one world, instead of the concept of three worlds, as geographer Dennis Cosgrove presented in comprehensive studies (Cosgrove 1994, 2001). However, this conceals quite different guiding ideas (*One World* vs. *Whole Earth*, Nitzke and Pethes 2017). The visual revelation of the unity of the planet from a distant view from space also marked the birth of Earth System Science, which views the Earth system as a self-regulating system (*Earth System Science*, Lenton 2016: 1). Geoscientists have since created a wealth of visualizations of global environmental change (De Vries and Goudsblom 2012, for image examples, see Buttimer 2015).

However, this reorientation of perception began even before the first of such recordings. The story of the image of the "Blue Planet"began in the context of the Californian counterculture, where romantic and technophile ideas from the cybernetics of the 1960s came together and continued to the concepts of system and self-management in network capitalism, which remain globally effective today (Diederichsen and Franke 2013). In 1966, the futurist thinker Steward Brand distributed buttons in San Francisco that read: "Why haven't we seen a photograph of the whole Earth yet?" Brand repeatedly urged NASA to release the images to the public, which eventually happened. Brand was a counterculture networker at the time and shortly afterwards published a famous mail-order catalog called "Whole Earth. Access to Tools," whose cover was adorned with the image of the Earth (Brand 1968, see Turner 2006; Buchmann and Bunz 2013). With the subtitle "Access to Tools," Brand wanted to provide access to seminal tools, both practical and conceptual. There were, for example, building instructions for dome houses inspired by the design of Buckminster Fuller, but also texts on a holistic worldview. It became one of the cult books of the 1960s and 1970s worldwide.

These space images of the planet, first described in 1987 as a dramatic "Overview Effect" (White 2021), continued to be enormously influential for the awareness of a threatened environment or even an endangered world. This has also been reflected in the worldwide interest in satellite images of global change, such as in the Instagram account "Daily Overview" (Grant 2016). This perspective reflects common interests rather than a struggle for power between individual nations—quite different from what was actually the case during the Cold War. This was how a photo, which emerged in the context of the military-industrial complex, became an icon of the ecology movement and counterculture (Diederichsen and Franke 2013).

In the Anthropocene discourse, the whole planet is also occasionally seen as an "Earth Island" (Moore 2015a: 37). This could be traced back to the perception of the Earth from space. With "Earth Island," we are dealing with a double metaphor, because "island" itself is a word to which many ideas are attached, often romanticizing ones. Islands are often seen metonymically as representative of the planet or as a pars pro toto "laboratory" (Antweiler 1991; Moore 2015a, b, 2019: 16, Chandler and Pugh 2020). Talk of the "blue planet" or even "our blue planet"

(Haber 1965) emphasises the important role of water, but is even more anthropocentric. Talk of the "home planet" (Sagan 1994; Kelley 1988; Nitzke and Pethes 2017; Christian 2018: 335) points even more strongly to the fact that humanity's home is only here so far, that it can only be here (so far), thus forming a community of fate.

> The resulting image conveys the fragility of the terrestrial ecosystem. The perception of the Earth as a vulnerable home represents a clear break with the previous perceptions (Krause and Reitz 2020).

What is hardly mentioned by the commentaries on the space images is the fact that human influence on the geosphere is hardly visible in these early images. Even the Great Wall of China—contrary to popular belief—is hardly recognizable, if at all, in images showing the whole planet. The enormous human impact is all the more evident in composite night shots of the planet, which—depending on your perception—show the beautiful lights… or the light pollution of the planet. In such photos, had they been taken a few hundred years ago, we would have seen only complete darkness. These glowing chains of "night lights" originate mainly in cities or in areas where rainforests have been cleared by fire. They reflect the rapid pace of urbanization and illustrate what some call the "Technocene" or "Urbanocene" (West 2017: 220).

In detail as well as in its entire extent, human influence on the geosphere only becomes clear in images taken from space stations, artificial satellites, airplanes, high locations, and recently, from drones (for example, Dech et al. 2008; Grant 2016; Steinmetz 2020; Grant and Dougherty 2020). There we often see human-made landscapes often filling the picture and extending up to the horizon. It is there where they are typically manifested—at least for me, in their beauty—as the notoriously standardizing elements of human-driven change (Burtynsky et al. 2018; Baichwal et al. 2018; Steinmetz 2020; Lang and Mauch 2024; Antweiler and Will 2024). Visuality implies attention. People also want their activities to be visible from afar. Thus, the city of Dubai, which was planned on the drawing board, was deliberately designed so that it would be visible on *Google Earth* at normal resolution (Graham 2018: vii–xiv).

In the same spirit as the blue planet, there is also the metaphor of the Earth as a spaceship, popularized by the designer and architect Richard Buckminster Fuller in 1963 and by the economist Kenneth Boulding in 1966 ("Spaceship Earth," Boulding 2006a, 2006b, Fuller 1963, vgl. Höhler 2015). In deep historical terms, humans as "Homo migrans" have repeatedly moved on or moved away whenever the environment changed or was overused. Given the densely populated and largely used surface of the Earth, we must abandon the image of the "unlimited expanse" and internalize the image of the "closed spherical surface." If the Earth now appears as a "closed world of the future"(Boulding 2006a, b: 15), the conquering cowboy must now be replaced by an astronaut who is steering and maintaining a small and fragile spaceship in a hostile environment (Schlaudt 2019: 79).

> I've often heard people say: 'I wonder what it would feel like to be on board a spaceship', and the answer is very simple. What does it feel like? That's all we have ever experienced. We are all astronauts on a little spaceship called Earth (Fuller 1963: 43).

When Buckminster Fuller declares every human being an astronaut here, it becomes clear that the metaphorical spaceship has already been filled in different ways by Buckminster Fuller and Boulding (Höhler and Luks 2006a, 2015). Crutzen himself and popularizers like Schwägerl and Leinfelder have used the spaceship image with the phrase "spaceship without an emergency exit" (Crutzen et al. 2011). This particularly refers to the need for a *life support system* in a hostile environment, but it can also be read as a statement that there are always technical solutions. The spaceship image implies that humans can steer the planet, which is a problematic assumption based on all current knowledge about the geosphere and biosphere (Wilson 2016: 187). Critics of Fuller's 'Operating Manula for Spaceship Earth' see in the metaphor of the spaceship and the astronaut perspective by no means an overcoming of anthropocentrism, but rather a cybernetic-technical power fantasy, a planning gigantism (Jansen 2023: 118ff., 132–133).

Such metaphors are stimulating, but they also have their limits because they often contain several components, allow several connotations, and encourage associations that were not originally intended. Andreas Folkers sarcastically diagnosed the impending expertocracy by stating that the "command bridge" of Spaceship Earth is occupied by climate experts and Earth system scientists (Folkers 2020: 598). Köhler used the example of the spaceship metaphor to illustrate how contradictory and implicitly political such an image can be:

> In the face of increasing ecological risks and dwindling resources, the metaphor 'Spaceship Earth' is used to draw attention to the vulnerability of Planet Earth and at the same time to its finite nature. This is for the most part normatively linked to the urgent appeal that *only through immediate action* can the survival of humanity be ensured. In the use of this metaphor, collective as well as tendentially authoritarian *demands* for a societal reorientation in the face of the ecological crisis and *at the same time the belief* in technological progress and the controllability of nature overlap in a contradictory way (Köhler 2015: 245, emphasis mine).

In the Anthropocene, the Earth is no longer a robust entity created for eternity, but rather appears to be under threat and worthy of preservation and protection. In this view, it is no longer about a struggle for power between individual peoples—as was factually the case during the Cold War. The new perspective is also reflected in individual perceptions, such as in a changed concept of identity. The metaphorical use of the word "terrestrial" by science and technology sociologist Bruno Latour is also oriented towards the concept of planetary thinking. Latour sees "the terrestrial" as a political actor, as a terrain of life that would need to be reclaimed in order to overcome the previous lack of "…grounding, reality and consistent materiality" (Latour 2018: 50). This use of the word may be disconcerting for geoscientists since they use the term "terrestrial" for solid parts of the geosphere, as opposed to marine zones, the seas. Latour says that humanity as a whole does not exist and therefore advocates for a "decomposition" of the global (Latour 2014). With the term "the terrestrial," Latour wants to dissolve the polarity between human geography and physical geography and, above all, the geopolitics that are partly based on it.

The concept of *Gaia*, which has (often forgotten) predecessors (Crutzen 2019a; Hüpkes 2020a), is currently being discussed in a manner that originated from James Lovelock, a chemist, and Lynn Margulis, a microbiologist (Lovelock 1979; Margulis 2017; Lovelock 2018). The basic assertion is that the development of the atmosphere was not only shaped by cosmological conditions and geological factors, but also significantly by the development of bacteria, plants, and animals. The thesis essentially states that the Earth's atmosphere could only develop in *coevolution* with the living world and tectonics, and *vice versa*. Feedback between the atmosphere and biosphere has made parts of the geosphere habitable for higher life forms (habitability, Chakrabarty 2019, 2020b: 37–41, 2021: 83–85). Unlike the term "Earth," the planet is seen here practically as a large autonomous organism, as a superorganism that organizes itself (on Margulis cf. Clarke 2020 and Kerner 2024: 83–129). The idea of homeostasis, a dynamic equilibrium found in normally healthy organisms, plays a central role here. Just as homeostasis can be disturbed in organisms and is manifested as fever, the Earth can also become "sick."

In Lovelock & Margulis, Gaia appears as a system of self-organizing units with many feedback loops and constantly emerging (emergent) properties. In contrast to the Anthropocene concept, *Gaia* emphasizes a new way of thinking about space rather than a different experience of historicity, a "new time of time" (Danowski and de Castro 2019: 101). Popularizations of the *Gaia* idea, but also anthropological essays, suggest seeing *Gaia* as a person, as many formulations in Danowski and de Castro (2019), 101–120). At a conference on the "Thousand Names of Gaia," it was concluded that the term *Gaia* emphasizes the autopoietic self-creation of the Earth. This is similar to the concept of societies as being quasi-large organisms that organize themselves and have environments. *Gaia* is considered by many to be less Eurocentric than the idea of the Anthropocene, and *Gaia* is deemed to be more "anthropological" compared to the term Anthropocene (Danowski et al. 2014, Bubandt, in Haraway et al. 2016: 547). We will still need to clarify what "anthropological" can specifically mean here.

Twenty years before Crutzen, philosophers called for a geologization of morality (for example, Deleuze and Guattari 1992). When the Earth is understood as *Gaia*, this often results in moral or political demands being attached to it. This can be seen, for example, in Latour's metaphorical use of the word *Gaia*, but even more so when extended Gaia concepts are seen as a counter-concept to the Anthropocene, thus becoming even more metaphorical and normative. Isabelle Stengers speaks of the "interference" *of Gaia* into the human world, or an "intrusion" into the creation of meaning, or the eventful "intrusion" of a transcendence into our history. She says that *Gaia* constitutes an "operation" in the sense of the global anthropogenic effect that turns "one" into a "we." Stengers derives from this the program of a different society and a slowed-down science (Stengers 2008, 2009, *ralentissement des sciences* 2013: 135).

3.3 Earth Spheres and Critical Zone—The Human Skin of the Earth

We have made enough concrete to cover the whole surface of the Earth in a layer 2 mm thick.
Mark Maslin (2021: 48)

The discourse on the "Earth system" (*system Earth*) has significantly been shaped by these new visual experiences through images from space. The institutionalization of the Earth System Sciences took place, as just mentioned, in the middle of the 1980s in the wake of NASA's interest in anthropogenic ozone reduction, among other things (Steffen et al. 2020: 50). On the one hand, the Earth system forms an Earth system scientific model of the geosphere as an integrated system. On the other hand, the term can be seen as core metaphor for the interconnectedness of all our bodies, not only with the biosphere, but all of our embedding in the geosphere. For the social sciences, this (*pace* Hamilton 2014, 2017) need not mean an end. Rather, it can serve as an eye-opener to change their concepts and extrapolations of possibilities and imaginations of the desirable. In particular, the form of change in an evolving large system (*world of becoming*) can serve as a productive confusion and challenge for the social sciences (Ingold and Pálsson 2013).

3.3.1 Earth System, Instability and Depth of Intervention

The younger Earth sciences have shown that it is part of the *normality* of such systems that after long periods of equilibrium, sudden reversals can occur—on a geological time scale—and after such a period of imbalance, a new stability at a different level can occur. The dominant theories in the social and political sciences, on the other hand, usually assume a stable situation, or one that only changes evenly or in the short term (Connolly 2011: 150, after Pálsson et al. 2013: 4). This also applies to most concepts of resilience, that is, the ability of a system to absorb disturbances and withstand them while maintaining its basic structure. Here lies a problem for all concepts that are primarily oriented toward continuity and are based on scientific as well as societal experiences in the Holocene ("holocene ideas," Dryzek and Pickering 2019: 65–67, 107–108), a challenge that I will discuss in the context of current criticism on concepts of sustainability in Sect. 6.5.

The quasi-medical diagnosis of the current state of the planet reads: The health of the Earth is threatened, *Gaia* has a fever, she is dying of thirst or is reeling, especially if the doctor, humanity, does nothing about it. With the description of the state come signposts and claims (critically on this Werber 2014: 245). To use medical terminology: With the diagnosis comes the medicine along with a detailed package insert. This can be observed in very different voices, such as in the reports of the German think tank WBGU on the necessary "great transformation," but also in the programmatic works of Latour. In Latour's case, this extends to the demand for a constitution for networked beings, ultimately a "parliament" for the entire assemblages of the superorganism. An alternative to this is to argue for a "public sphere" that includes things and animals, but which only arises through an awareness of a *common* problem and an interest in avoiding suffering, and is therefore limited to humans (Bennett 2020). I will address this issue in Sect. 4.5 on depoliticization.

It should not be forgotten that there are also dynamic metaphors, such as that of "heating up" (Eriksen 2016) or that of "the long thaw" (Archer 2009). Recent postulates of "planetary thinking" state: "Thinking planetary means taking the Earth seriously as a planet, from the Earth's core to interstellar space, from the nanosecond to deep time, from the elementary particle to the Earth's mass" (Hanusch et al. 2021). What is easily forgotten when talking about the Earth as a whole planet, however, is the fact that anthropogenic change "only" affects the outer skin of the Earth, namely the geosphere and the biosphere, which is around 20 kilometres thick.

The biosphere is connected to the other envelopes of the geosphere, i.e. the atmosphere, hydrosphere and lithosphere, but today's civilisation only has a really strong influence on some of these linked components, especially those in the thin biosphere (Smil 2021: 208). We have no influence at all on the fundamental properties that make life on Earth possible. The position of the Earth in a relatively quiet part of our galaxy, the enormous solar energy generated by nuclear fusion, the shape, tilt and rotation of the Earth and its eccentric orbit around the Sun (seasons and ice ages) are all completely beyond human control (Smil 2021: 208). With regard to the concept of *Gaia*, therefore, it should not be forgotten that *Gaia*, in the sense of Lovelock and Margullis, *does not* mean all the spheres of the planet, but only the outer spheres, especially the biosphere and the atmosphere, and the spheres of culture and technology that are additionally mixed with these in the Anthropocene.

3.3.2 Critical zone and Verticality

In the face of the Anthropocene, climate and Earth scientist Tim Lenton distinguishes between a surface Earth system, which makes life processes possible, and the much more extensive rest of the Earth (*inner Earth*). It is this thin layer, this extremely thin shell with its specific, astonishing, and often still-unknown systemic properties, that is the focus of today's Earth System Science. The Earth system extends from the upper limit of the atmosphere down to the ground and the upper lithosphere, with the lower boundary being more blurred or set differently depending on the time scale (Lenton 2016: 17, 29).

This fits with the even more reduced geoscientific concept of the thin and near-surface "critical zone" (Arènes et al. 2018; Arènes 2022: 17–24, 306–323, Latour 2018: 133, Latour and Weibel 2020; Critical Zone n.d.; Etelain 2023). This thin and vulnerable zone forms an area of dynamic interaction between the atmosphere, hydrosphere, biosphere, and the uppermost part of the lithosphere. It extends from the treetops down to the unweathered rock or the base of the aquifers. This boundary layer is characterized by life, is highly complex, and can reach more than 30 meters deep (von Blanckenburg 2021). In contrast to deep geology and the zone above, human activity has especially severe destructive effects in this layer. The concept of the critical zone is similar in content to the other concepts of sphere, especially the "anthroposphere" and Tansley's landscape concept of "ecosystem" (Richter Jr. and Billings 2015; Richter Jr. 2020). For an understanding of the Anthropocene, it is important that the term "critical zone" emphasizes the ecological and political relevance of soils beyond their purely agricultural significance (Montanarella and Panagos 2015).

From the perspective of material culture, however, the critical zone would need to be extended both upwards and downwards. Most images of the anthropogenic layer are still too strongly influenced by "flat" everyday geographical thinking (Graham 2018). The legacy of horizontal rather than vertical thinking is also evident in geographical visualizations of human activity, such as the maps of the *Urban Theory Lab* at Harvard University. However, humans penetrate up to almost four kilometers deep in the form of bunkers, mines, traffic tunnels, and drillings and up to one kilometer high through high-rise buildings. This verticality is increased by drones, helicopters, airplanes, long-range weapons, satellites, and space debris: the atmosphere is not only shaped by human-produced exhaust gases.

We need a more vertically informed human science, such as that which exists empirically in GPS-oriented geography to some extent and as developed programmatically by the philosopher Peter Sloterdijk in his "Spheres" project. We need sciences that examine the entire space between deep-sea trenches and the middle atmosphere, all fields in which the *anthropization* (Magni 2019) takes place. For cultural anthropology, this would be a more vertical and more three-dimensional documentation of material culture, which I will discuss in Sect. 5.7.

What terms do we have that are both spatially oriented and emphasize that human activity "only" affects the Earth's surface and not the entire planet? The obvious terms are those that correspond to the spheres distinguished in the geosciences. An early term for a culturally shaped Earth shell is noosphere, which refers to the sphere of communicative exchange, especially of knowledge, which is equivalent to the spheres of the living world and the Earth. This term, coined by the Jesuit philosopher Teilhard de Chardin and the Russian geologist Vernadskij, particularly emphasizes the enormous terrestrial creative power of human societies. Ehlers proposes the term anthroposphere (Ehlers 2015: 27, 29–31, Cornell et al. 2012; Steffen et al. 2020). The anthroposphere would thus be the entirety of the thoroughly and permanently changed parts of the geosphere, which form the "… nodal point and integrating link between nature and society, between humans and their environment" (Ehlers 2015: 29, trans. CA).

3.3.3 *Technospere and Technocene*

An alternative to the anthroposphere as a concept is the proposal of a "technosphere" by Friedrich Rapp, a philosopher, and Peter Haff, a geophysicist who is a member of the Anthropocene Working Group (AWG) (Rapp 1994; Haff 2013, 2018; Klingan and Rosol 2019). As a thin but dense shell of our existence effects, the technosphere constitutes a full equivalent of the other spheres, such as the hydrosphere, and primarily represents a physical system that includes not only technology per se, but the whole range, from large-scale extraction systems to domesticated animals to the laptop on my desk. The technosphere includes some completely new phenomena, such as the transport of solids over long distances. In human history so far, it has included only the small-scale movements of solids via rivers, soil flows, avalanches, shifting dunes, wind and ocean currents, and marine turbidity currents on the continental slope as well as movements resulting from animal migration. While these movements mostly followed natural energy gradients, mainly gravity and wind directions, they are now joined by intentional mass movements to specific destinations, which take place along planned routes and by means of specific human infrastructures (paths, roads, railways, pipelines) (Wuscher et al. 2020).

The technosphere encompasses almost all human communities, and it is diverse and massive. A special feature of the technosphere compared to the other spheres of the geosphere is that it is based on non-renewable energy. However, a second feature is even more important for the Anthropocene. The technosphere also contains many unintended and barely controllable components such as huge amounts of waste and globally distributed microplastics as well as chemical traces in waters and rocks and displaced animals that become invasive elsewhere (Seebens et al. 2018). A large portion of the waste products from the technosphere, such as reactor waste, cannot be recycled by the geosphere, biosphere, or atmosphere. Some construction waste, for example, can somehow be recycled but remains in the technosphere (Hecht 2018: 218–222).

In contrast to earlier individual proto-technospheric, technically overlaid landscapes, the technosphere according to Haff is a globally connected system. This technosphere is not a separately created phenomenon, but an integral part of the emergent Earth system. Within it, it forms a sphere that operates outside the control of individuals and, to some extent, also human collectives, whereby complex innovations such as airplanes and mass storage build heavily on earlier technologies and coevolve with the social system. In this sense, humanity is only partially the creator or director, but rather a part of the system on whose existence we depend (Haff 2013, 2018). This autonomy of the technosphere then results in its own demands on human social systems. Proponents of the concept, such as the group around Zalasiewicz, see the concept as a productive non-anthropocentric (!) prism, through which the Anthropocene can be viewed in a special way and which would allow for a strictly quantitative assessment of technology (Zalasiewicz et al. 2017a, b, Thomas et al. 2020: 105).

This view, guided by the concept of the technosphere, leads to remarkable findings If you add up human-produced and transported materials and distribute them evenly over the earth's surface, there is an average of around 50 kilograms per square metre, a fact that should be repeated here because it initially seems unbelievable (Zalasiewicz et al. 2014a, 2017b, c: 15–23). The most common new type of rock in the technosphere is concrete. The concrete produced alone, evenly distributed, would cover the entire surface of the earth with a two-millimetre-thick layer (Waters et al. 2016; Waters and Zalasiewicz 2018). This becomes clear when you realise, for example, that more concrete has been built in China every 3 years recently than in the USA over the entire last century (Smil 2019: 189; Fig. 3.3).

In my opinion, this demonstrates the ingenuity of an initially confusing geological perspective. Many objects are materially preserved and will thus become future fossils: technofossils that will surpass biodiversity in number and range within a short geological period of time. While the diversity of living species, namely the biodiversity, is decreasing, the technofossil diversity is increasing sharply, as shown, for example, by the number of book *titles* worldwide at approximately 100 million. The number of potentially fossilizable techno-"species," which could serve as index fossils for the cultural paleontologist in the Quintary mentioned in the preface, has currently been determined to be 130 million.

However, the thesis that humans are thus part of an uncontrollable and materially reduced sphere of technology has also provoked criticism, for example, from

Fig. 3.3 High-rise buildings in Pudong, Shanghai, China. (Photo: author)

historians of technology, who find the term Technocene more appropriate because it is less strictly limited to materiality (Trischler and Will 2020: 238–239). Accordingly, concepts of technology have been developed that include animals and plants and see technology as an integral part of ecosystems (Jørgensen 2014). This can be connected with the concepts of niche construction and environmental heritage, which will be discussed later in this work.

The harshest criticism, however, comes, as might be expected, from the ranks of cultural studies. Whereas an overestimation of the Anthropos has been noted elsewhere (Mathews 2020), scholars from this area are complaining that the focus on humanity is wavering, indeed they even see a degradation of the human being (for example, Szerszynski 2017). I would criticize the fact that human technology is not actually a sphere, but rather a network, a network with dense nodes, loose connections, and also several holes with little connectivity. Moreover, the term technosphere also obscures the coevolution of technical, cultural, and even biotic systems and anthropogenic niche construction.

3.3.4 Archäosphere und Humanosphere

One concept of spheres that is particularly more suitable for cultural anthropology, in my opinion, is that of the archaeosphere (Edgeworth 2014, 2018b: 21–24). The archaeosphere is the totality of human-altered land and soil. It distinguishes the parts that have been *significantly* shaped by human activity from the layers underneath that have remained undisturbed by humans. According to recent studies, it proves to be a living and active complex of deposits with numerous interactions, influences, and impacts on other parts of ecological systems. The concept is reminiscent of the avant-la-lettre work of the Austrian geologist Eduard Suess, who studied the soil of Vienna as a "rubble cover" in the nineteenth century and already included the lifestyle of 'bourgeois society' in the title (Suess 1862). The archaeosphere is partly created by humans and is constantly being reshaped and expanded, but it is so significant and durable in the Anthropocene that it should be seen as an independent unit (Edgeworth 2018b: 19).

As such a unit, the archaeosphere can, in my opinion, be seen as a meaningful addition to the concepts of the "critical zone" and "verticality" (Graham 2018), the ethnographic significance of which I will discuss in Sect. 5.7. As a unit, the archaeosphere is more horizontal and global, while the critical zone is more of a network of the living, and verticality is the sum of interconnected clusters in the height and depth of the Earth. Edgeworth goes so far—in Latour's perspective—as to see the archaeosphere as an accumulation of a multitude of small effects that add up to a global force that will have a distributed material agency in the future (Edgeworth 2018b: 19, similar to Van Oyen 2018). This goes too far, in my opinion, because of the exaggerated concept of agency, which states that things can act.

Perhaps another term may help here, one that limits the role of the humans but includes the living world. Ishikawa introduced the term "humanosphere," which emerged from a Japanese discussion of Asian and African examples (Ishikawa, in Haraway et al. 2016: 542–543). This term is supposed to emphasize the non-uniformity of the globe and take human effects into account, but attribute less weight to humans. While the geosphere has existed for about 4.5 million years and the biosphere for around four million years, the humanosphere would have existed for about 200,000 years. Inspired by the situation in the tropics (Lopez et al. 2013), he emphasized that this sequence and the articulation of the spheres are important, as human societies are dependent on the existence of the earlier spheres. With this, Ishikawa positions the humanosphere much earlier in time than the earliest proposals for the Anthropocene. With respect to the scope, Ishikawa's ideas come close to an argument by Leinfelder, who said that the Anthropocene is nothing more than the human sphere of influence, a spherical layer corresponding to the biosphere and the geosphere (Leinfelder 2017b).

The term humanosphere has been criticized for its focus on humans. However, Ishikawa has also linked this version of a spherical perception with the Japanese concept of *shinra bansho*, "all things of the universe" or "the entire creation between heaven and Earth," of which humans form only a small part. This is more reminiscent of concepts of biomass, networks, entanglement, rhizomes, or the Malay expression "body hair of the Earth (*bulu gumi,* meratus), as Tsing aptly noted (Tsing, in Haraway et al. 2016: 543). Haraway emphatically concludes from the encounter of Asian and Western terms that this is useful for a critique of what she sees as the inevitable but "arrogant" Anthropocene term and an argument for the perspective of the more-than-human:

:

> See, I think people like us have an obligation to propose these words for naming our urgent conjuncture, and not to be dazzled and tame in the face of the proposal of these other terms that are maybe useful in ways. I think folks like us, who are really over-educated have an obligation not to let elites get away with another simplification, that I think is part of the problem with the Anthropocene in the first place (Haraway, in Haraway et al. 2016: 545, see 548).

However, because of their material basis, namely their geoscientific, biological, and material-cultural foundation, these three concepts and that of the technosphere are, in my view, much more productive for theory and empiricism than, for example, the currently more prominent term of the "more-than-human." This enormously popular approach places humans on a level with other living beings and can be found, for example, in Abram, Haraway, Tsing, and Puig de la Bellacasa and in some new materialisms, as well as the many new neologisms, such as "para-sites," which I analyze in Sects. 5.3 and 5.5. What is surprising, in view of the Anthropocene concept's strong relation to geology, is the fact that geological phenomena such as sedimentation, erosion, stratification, faulting, or the thermal transformation of rocks (metamorphosis) are rather rarely used as a metaphors for analysis in scientific works on the Anthropocene (with exceptions, for example, Horn and Bergthaller 2022 and Thomas and Zalasiewicz 2020), quite unlike literary texts (see, for example, Schnyder 2020).

3.4 A New Reordering of the World—Powerful Narratives and Moral Visualisation

> *(...) the Anthropocene is good to think.*
> Lourdes Arizpe Schlosser (2019: 283)

> *Its effects, to the anthropos, are visible only through the data, diagrams, and visualizations of science.*
> Hannes Bajohr (2019: 71)

Texts on the Anthropocene often appear as narrations, so that they can be studied as narratives. Contrary to its use as a buzzword, I chose to use the word narrative to refer to the meaningful stories that often have a political or socially mobilizing effect and show a clear pattern: protagonists, chain of events, plot, causes, effects, a moral as well as a specific spatial and temporal structure. These are examined here as collectively effective narratives. Narratives can be found in history and historical scholarship and are especially strongly in public stories. The idea that the Anthropocene shares with other historiographies is that it is a present-oriented narrative. What makes Anthropocene narratives special, however, is that they are not only related to present interests, problems, and concerns, but are also future-oriented (Trischler and Will 2020: 241):

> Anthropocene narratives (...) connect a *long history of the present* with partly implicit, partly explicit, but in any case highly normative assumptions, also and especially with regard to the question of which role is ascribed to which technologies for solving *problems of a planetary dimension* identified in the present *and projected into the future* (Trischler and Will 2020: 241, emphasis CA).

For one thing, I consider the examination of Anthropocene narratives as such, as a subject matter, to be interesting from a cultural studies perspective. The narratives are especially important in this sense because a large part of the criticisms of the idea of the Anthropocene is related to the dominant scientific and technologically shaped narratives about the Anthropocene in that they are depoliticizing (see Chap. 4). The French historians of ideas Bonneuil and Fressoz got to the heart of the critical perspective by stating that their goal is a future-oriented deconstruction of the Anthropocene narrative:

> It means deconstructing *the* official account in its *managerial* and *non-conflictual* variants and forging new narratives for the Anthropocene and thus new imagineries. Rethinking the past to open up the future (Bonneuil and Fressoz 2016: xiii).

This implicitly contains a third reason as to why narratives about the Anthropocene are worth studying, namely the widespread misconception that it is *a single* narrative. Bonneuil & Fressoz emphasize this by viewing the Anthropocene as an "event" which triggers quite various narrations. Upon closer inspection, there are multiple narrative forms and a wealth of specific topoi, images, and metaphors relating to the Anthropocene. Bonneuil outlines and criticizes four narratives of the Anthropocene. These are (1) the dominant *naturalistic* narration of the Earth System Sciences, a narrative of humans as a powerful species that, however, ignores societal dynamics.

The second (2) is the *postnaturalistic* narrative, which dissolves the boundary between nature and culture, or proclaims the end of nature. The (3) *eco-catastrophic* crisis narration sees an impending collapse and calls for a turnaround away from growth. The (4) *eco-Marxist* narrative emphasizes the intimate link between the (metabolic) environmental crisis and capitalism. The (5) *eco-feminist* narrative is similar to the eco-Marxist one in its focus on inequality and power, but emphasizes the gender dimension (Bonneuil and Fressoz 2016: 23–29, Grusin 2017; Ebron and Tsing 2017: 683, Glabau 2017; Gramlich 2020) and also allows for a methodological approach through intersectional environmentalism (Angerer and Gramlich 2020; Walton 2020).

In Table 3.1, I have compiled the most important characteristics of the narratives. The literary scholar Gabriele Dürbeck distinguishes five central narratives, and she distills a superior metanarrative (Dürbeck 2018a: 13–17, 2018b: 7–15). Of practical and political significance is the fact that the above-mentioned metaphors and visual images can have such a strong impact on thought and action. The narratives contain a different, often strong potential for mobilization to action (Dürbeck 2021) that Dürbeck was able to articulate well, which is why I quote her extensively here:

> The *catastrophe narrative*, with the image of a 'world without us', places the victims at the center and derives from this the urgency of a radical rethinking and changed action. The *judgement narrative*, in its different manifestations, names the perpetrators or culprits (Europe since the industrial revolution, capitalism, the industrialized countries since 1950), refers to the victims in the Global South who are hit much harder by climate change, and emphasizes the 'common but differentiated responsibility' or fair burden-sharing. The *narrative of the great transformation* is based on the discourse of ecological modernization and endorses technology and enlightenment as solutions to the problem by promoting the reduction of the causes of environmental destruction and the measures for reasonable adaptation to changing environmental conditions, and by calling for a radical 'cultural change' for the transformation into sustainable societies. The *(bio-)technological narrative* also strives for efficient solutions and improved nature conservation, with the plot emphasizing the importance of power mechanisms and technological elites and representing a neo-Promethean approach to nature as an object. Finally, the *interdependence narrative* presents a self-awareness plot and an insight into the mutual dependence of humans and nature, thus relating reflexively to the other Anthropocene narratives. It is based on a systemic concept of nature, with humans as part of a network in exchange with other species, where, from a critical perspective, humanity is not conceived as an abstract unit, but differentiated in its local possibilities for action and creative scope (Dürbeck 2018a: 17, emphasis mine).

In addition to emphasizing a threat, an apocalyptic narrative can also result in concrete instructions for action… and often does. One example is the *Gifford Lectures* by Bruno Latour (Latour 2017b). The potency of narratives and the metrics used is exemplified by the work of the expert panel WBGU for a sustainable society, where the Anthropocene is repeatedly invoked (WBGU 2016: 33, 66, 98, 288, 2016: 283, 2020: 22, 41, 245). Here, a combination of the Anthropocene narrative with the specific concept of planetary boundaries is used to promote a *social* goal, the socio-ecological restructuring of society, the "great transformation". It is precisely this intertwining of normative goals with hard science that has met with massive criticism, which I will examine in detail later (Kersten 2014; Krauß 2015a: 69–70, Altvater 2013, 2017, 2018, on the intertwining in climate change research, see

Table 3.1 Narratives of the Anthropocene

	Plot	Prota-gonist/s	Time persp.	Metaphors, Topoi, Alternative Concepts	Mobilization Potential
Catastrophe narrative	Apocalypse (pessimistic)	Transgressors, victims, humanity	Neolithic, 1800	Sick planet, awakening, Technocene	Appeal to responsibility, radical rethinking
Judgement narrative	Colonial subjugation, capitalist globalization, *Whodunnit* (pessimistic)	Perpetrators, culprits, industrial nations, capitalism	1610;1950	Indictment, liability, Capitalocene	Liability, diff. Responsibility, fair burden-sharing, pressure
Narrative of the great transformation	Modern escape is partly possible through consumption change and *geo-engineering* (optimistic), possibly *happy end*	Enlightened humans, technology, Progress, radical cultural change	1950	Human era, garden earth, earth gardener, us–world, spaceship earth	Ethical mission, *stewardship*, adaptation and mitigation, enlightenment
(Bio)techno-logical narrative	Markets and technology will sort it out (optimistic)	Humanity as Prometheus, elites	1945–1950	Technocene, great Anthropocene	Green revolution 2.0
Interdependence narrative	Nature/culture with unintended network effects	Humanity or nomadic human-non-human arrangements	Neo-lithic	Posthuman “we”, nature/ cultures, deep time	Self-awareness, *geo-stories*
Naturalistic narrative	Humanity changes the world and adapts	Human as a quasi-natural force	Open	Enlightenment	Reason, rationality, science
Post-narrative	We solve problems through self-technologization	AI, algorithms	Ca. 2000	„Good Anthropocene“„Novacene“	Technophiles and optimistic people
Eco-catastrophic narrative	Apocalypse (pessimistic)	Transgressors, victims, humanity	1800	Sick planet	Responsibility, radical rethinking
Eco-Marxist narrative	Capitalists exploit the earth	Classes, elites	ca. 1850	Fossil capital,"Capitalocene“	Class interests
Eco-feminist narrative	Emancipation	Patriarchy, victims of male violence	Trans-historical	Partly (white) men (?)	Servitude

Bauriedl 2015; Krauß 2015b; Von Storch and Krauß 2013). Bruns works out how the originally descriptive-analytical Anthropocene idea, as well as the idea of planetary boundaries and tipping points (Lenton et al. 2019), as a scientific narrative of justification, legitimizes the transformation as a largely *prescriptive* social project by means of factual constraints and without explicitly referring to the political, because the argument is:

> Anthropogenically induced global environmental change, and in particular the already-exceeded or approaching tipping points, *require* the navigation of human–environment relations within planetary boundaries. This navigation (which thus addresses questions of collective decision-making (governance) pursues one *goal* (sustainability) (Bruns 2019: 56, emphasis mine).

A comparative look also reveals, in addition to the many differences, similarities between the narratives. While Bonneuil emphasizes the differences and the critical potential, Lövbrandt and co-authors and Dürbeck elaborate on the continuous lines and the respective potential of each narrative to mobilize people to think or act differently. Lövbrandt et al. diagnose a combination of three perspectives as a common narrative: the "post-natural," the "post-social," and the "post-political" (Lövbrand et al. 2015: 212–215). The post-natural perspective unites nature and society instead of opposing them. The post-social perspective speaks of humans or anthropos instead of diversity and local dynamics. The post-political perspective mixes apocalyptic rhetoric with a defense of the status quo. Dürbeck sees similar similarities:

> Despite all their differences, the five narratives have a common structure, namely the (partly critical) reference to the endangerment of the world by humanity as a plot, a deep-time perspective on the past and the future, a planetary frame of reference, a dissolution of the categorical boundaries between nature and culture in the horizon of the Earth system concept, and finally, the thematization of ethical responsibility for the reduction of further environmental destruction and the survival of human civilization (Dürbeck 2018a:17, see 2017b: 15–17).

Görg rightly notes that many of these narratives are not new. For many years, theses as well as basic assumptions and diagnoses of the times have been discussed in debates about the ecological crisis and the development of nature relations (for example, Becker et al. 2006). Also in view of multiple paradoxes, such as apocalyptic scenarios with simultaneous inactivity, on a philosophical level, this is partly a renewed discussion about the "Dialectic of Enlightenment" (Görg 2016: 12–13, 2019). I consider it all the more important to precisely examine the narratives and the specific metaphors, topoi, and visual images in order to clarify the particularity of the Anthropocene as an idea.

One particularity of Anthropocene narratives is that they are often associated with the mention of large numbers, in which they resemble earlier demographic narratives about the world's population (Engle Merry 2011). Measurements, statistics, drastic examples and progression curves, and constantly new figures play a big role in the publications. This applies not only to descriptive literature, but also very much to policy-oriented texts on the subject. In economic texts, figures generally have a high persuasive value (McNeill 2017). In texts on the Anthropocene, geological facts and figures are additionally linked with dystopian exaggerations.

Record values prove the influence of humans, and this is often accompanied by a catastrophic choice of words, such as "Everyone can see how the polar ice caps are melting and how ecosystems are going to the dogs."

If one interprets the Anthropocene as a self-descriptive formula for society, the figures serve to substantiate *certain* self-descriptions with evidence as compared to others. This allows the contingency of the respective description formula to be forgotten or deliberately ignored (Werber 2014: 244). But this is also where the more fundamental problem of abstraction promoted by money economy comes into play, namely the temptation to treat the most diverse realities as equivalent by means of numbers and curves (Horkheimer and Adorno 2006). The numbers in dominant narratives about the Anthropocene are not just any, they are the numbers of geologists. They are particularly important because the Anthropocene is juxtaposed with other epochs of the Earth's history. These other periods last tens of thousands of years, such as the Holocene, or many millions of years, like the Pleistocene.

These numbers are impressive as "stunning figures," so much so that even a constructivist-oriented scientist like Bruno Latour, who otherwise problematizes "facts," argues strongly with the use of figures, such as in the *Gifford Lectures* of 2013 (Latour 2017b, see Werber 2014: 241). There, Latour compares human-generated energy with the forces of plate tectonics. Latour also repeatedly refers to the sheer size of the Earth and the massiveness of the problems and speaks eloquently of humanity's "telluric" power, which is equivalent to the forces of nature (for example, Latour 2018: 25). This brings us to the justified—as well as the less justified to absurd—criticisms of the concept, idea, and rhetoric of the Anthropocene, which are systematically analyzed in Chap. 4. Dürbeck summarizes concisely:

> The Anthropocene thus appears as a narrative that (a) understands humanity as a geophysical force, (b) casts a planetary perspective on the global environmental crisis, (c) has a deep time dimension, (d) assumes a close interrelationship, that is, non-separability of nature and culture, and (e) derives from this an ethical responsibility of humans for the Earth system (Dürbeck 2018b: 4).

3.5 Images of Nature and Humanity—Between the Misanthropocene and the "Mature Anthropocene"

> *The late modern public sphere is in a deplorably bad state in the face of the complexity of the Anthropocene.*
>
> Bernd Stegemann (2021: 12).

Popular texts on the "human era," but often also scientific texts on the Anthropocene, frequently work with an alarmist vocabulary and convey a negative outlook for the future—dystopias wherever you look. Even in the climate change debate, a "clear break" in human history has long been part of the discourse, and there is often talk of "end times" (Orr et al. 2015: 162). We read even more dystopian metaphors about the Anthropocene, from a "mega-crisis" up to a "downfall," "apocalypse," and the "end of the world" (Danowski and de Castro 2019).

3.5.1 Alarmism and Strict Deadlines

However, the awareness that the planet is under existential ecological threat goes back to a phase before the recognition of anthropogenic climate change. It can be historically traced back to nuclear science in the USA and the perception of the effects of atomic bombs (Masco 2010). Atomic bombs revived the always-latently present catastrophism or eschatological attitudes towards human history. They made it *empirically* clear for the first time that humans have the potential to plunge the geosphere into a catastrophe. This awareness of a possible worldwide, human-induced catastrophe was then promoted almost simultaneously by the realization of a possible global "sixth extinction event," which also geologically challenges any assumption of uniformity.

In relation to the Anthropocene, body metaphors and linguistic images from the field of disease are particularly common, as already seen in the debate on climate change and resource consumption ("footprint," "fingerprint"). The Earth "is thirsty," overheated, "has a fever," is out of balance, or is staggering towards a "collapse." The planet appears as overheated as the pace of its change (*overheating*, Eriksen 2016). While *planetary health* mostly concerns the health problems caused by the Anthropocene (Hoffmann 2020), some sustainability and development researchers say that human health must be promoted along with planetary health (Haines and Frumkin 2021). Humans are seen as the main cause of this disease of the Earth, appearing as a "virus" or even as a "monster." We humans, or humanity as such, have become the problem (Scherer 2017, 2022).

The clinical picture is also reflected in the "syndrome concept" of the German Advisory Council on Global Change (WBGU). This is used to characterize certain patterns of environmental degradation (Glaser et al. 2012: 200–205). Here, the concept of syndrome, borrowed from pathology, is used as a term for typical constellations of several symptoms. The development of the world is threatening to become an accident, and the worst can only be prevented by "planetary guardrail", i.e. political goals to limit change in order to enable long-term survival (Leinfelder 2012). In sum, a particular form of negative anthropology prevails, according to which the Anthropocene appears as a "Misanthropocene" (de Castro and Eduardo 2019: 296, Horn and Bergthaller 2022: 42, 124, Bajohr 2019).

3.5.2 Earth Management and the "Good Anthropocene"

Positive utopias or nostalgic metaphors are much less common, such as when humans in the Anthropocene are made to appear as "gardeners" or as "shapers" or as nurturing human beings, as stewards or advocates, for example, for the living world or nature (see, for example, Knauß 2018). Positive or optimistic metaphors and words claiming power are becoming more frequent, such as the talk on the role of humans in a *trusteeship* (Pielke and Roger 2014). They are found mainly in those writings from the Earth sciences that highly estimate humanity's ability to find answers to the Anthropocene (McPhearson et al. 2021). They derive hope for a targeted "rescue" of the Earth from the Anthropocene catastrophe, and speak of the "good Anthropocene" in the sense of an *ecomodernism* (Asafu-Adjaye et al. 2015: 1, 4). Critics of technophile and optimistic ecomodernism argue that these rather technophile and optimistic voices, claiming that they are technicist. Then again, there should also be positive voices that go beyond technology (*techné*), such as poetics that emphasizes the principle of creative vitality. This can likewise build on insights into the conditions of common land in economics as well as on knowledge of symbiotic forms in biology (*poiesis*, *enlivenment*, Weber 2019: 10–14).

A deep historical or geologically deep-time perspective or a spatially broad view of *Big History* could help us to see the positive aspects of the human success story while remaining realistic. The "bad Anthropocene" of enormous inequality and massive environmental pollution is contrasted—at least from a human perspective—by a "good Anthropocene." For the first time in history, billions of ordinary people achieved a better life. Thus, the use of synthetic fertilizer not only allowed the aforementioned reduction of hunger, but also the emergence of a broad global middle class, a completely new phenomenon in the history of mankind (Christian 2018: 317). Astrobiologist David Greenspoon's idea of the "mature Anthropocene" is a mixture of realistic insight into negative aspects and positive aspects. He sees opportunities for humanity, which has already been on the brink of extinction at least once in the course of evolution. A guiding principle of the mature Anthropocene would describe characteristics of a future world that preserve the best of the "good Anthropocene" and at the same time identify measures that halt or reduce trends towards the "bad Anthropocene" (Greenspoon 2016: 222–244, Christian 2018: 333).

Some extreme proponents of a good Anthropocene consider anti-capitalist orientations to be the greatest obstacle to a development towards the good Anthropocene, most notably the co-founders of the Californian think tank *Breakthrough Institute* Michael Shellenberger and Ted Nordhaus (Shellenberger and Nordhaus 2011). Stewart Brand, who fueled the perception of the planet as a networked world in the 1980s with the *Whole Earth Catalog*, also argues in the direction of a good Anthropocene. As an entrepreneur, he now optimistically sees a technological future. Optimistic or pessimistic views of the future certainly also always depend on the personality of the authors. Critics, on the other hand, see these representatives as technophile advocates of a booming late industrial capitalism or a "green capitalism." They see them as prophets of a "cornucopia eschatology," of which there are also left-wing, "accelerationist" variants (Danowski and de Castro 2019: 63, 65, 143–146). But, despite all the criticism, what needs to be considered is whether a moderate "environmental humanism" would also lead politically further than a fearful alarmism (Shellenberger 2020: 274–279, 285).

Ideas of the good Anthropocene are often associated with ideas of geoengineering (*geo-engineering*, *geoclimatic engineering* (Wiertz 2015a, b; Preston 2019; Meiske 2021; Will 2021: 184–189). Geoengineering refers to 'earth management' measures that use technologies or—more rarely—new institutions to deliberately change processes in the earth system in such a way that the environmental crisis can be counteracted. This would be possible on different scales, that of the planet, at the level of national energy strategies, to regional green technologies and down to the level of households (for critiques see Baskin 2019; Jansen 2023: 156–244 and extensively Schlemm 2023). They are often perceived as optimistic, but they also stem from the realization that individual behavioral changes can hardly do anything against Anthropocene trends. The perceived way out here is technological rather than political.

Despite their name, however, these measures usually do not refer to technical responses to the Anthropocene *per se*, but to the climate (Haardt 2022: 148–155). These include the artificial generation of rain by releasing particles into the atmosphere, the management of solar radiation using reflectors, the underground storage of carbon dioxide, or the large-scale use of bacteria to clean the oceans. Most proposals represent *geoclimatic engineering* rather than *geo-engineering*: the geospheric, that is, Anthropocene, dimension is left out (Thomas et al. 2020: 181 but there are exceptions (for example, Kolbert 2021: 157–217). Firstly, there is hardly any mention of the many dimensions of the Earth system involved. Secondly, hardly any reference is made to geological deep-time, for example, in decarbonization technologies. Thirdly, although unplanned repercussions are discussed, hardly any attention is paid to the potential political implications. It is precisely these technological approaches that are therefore primarily under fire as "technological fixes," which are branded as "technicist" (see Chap. 45).

Ideas for earth management are often perceived as optimistic, but they also stem from the realisation that individual changes in behaviour alone can hardly do anything to counteract anthropocene trends. The central argument is urgency and the perceived solution here is technological rather than political. It should not be forgotten, however, that there are also other forms of earth management, such as the idea of large-scale renaturalisation of landscapes as a countermeasure to species extinction or even the call by evolutionary biologist Edward O. Wilson to declare half of the earth untouchable nature reserves (Wilson 2016). There is also the idea of 'pleistocene renaturalisation'. This would be a promotion of those present-day species in wildlife parks that are similar to Pleistocene species in their ecological function (elephants replace mammoths). What connects these ideas with the technological ones in the narrower sense is the idea of large-scale intervention for a 'good Anthropocene'.

Such a positive and simultaneously technological utopia is also James Lovelock's newer vision of a post-human evolution of humanity. Lovelock sees a new age, the "Novacene," a post-Anthropocene future of technological innovations, especially by means of artificial intelligence, which we as humanity would enter after the Anthropocene (Lovelock 2018). This approach strongly deviates from Lovelock's earlier ecological work on the *Gaia* system. Instead, it adopts from the earlier considerations on self-organization an optimism about governance, which is now based more on technology than on organisms.

Given the planetary theme and the fact that the debate was initiated by an atmospheric chemist, Paul Crutzen, it is not surprising that many metaphors come from the Earth sciences, especially from geology and paleontology. The formulation of human agency as a “geological force” is fundamental (Crutzen and Stoermer 2000). At first sight this seems to be a metaphor for the unprecedented force of human activity, but it is also a statement of the fact of the quantitatively strong and lasting geological impact of human activity. More precisely, however, these are rather geographical and geomorphological effects, because they only affect the Earth’s surface (geosphere) and not deeper geology of the Earth. Contrary to some assumptions, this does not mean that the other geological forces are becoming less important (Hooke 2000; Davies 2016).

Deep time (Sects. 1.4 and 6.6) is also a metaphorical description of the enormously long geological time perspective compared to human and historical standards. Metaphorically, the geologically underpinned talk of stratigraphies, metamorphisms, and faults can be useful. These three geological terms can be used to effectively summarize: firstly, different layers of the debate, such as factual versus normative statements; secondly, changes in central concepts, like those relating to humans and nature; and thirdly, faults, such as the divergence of the micro- and macro-scales of human action. In my opinion, the most fruitful use of these geological terms can be found in the works of the literary scholars Horn and Bergthaller and the anthropologist Irvine (Horn and Bergthaller 2022: 25–58, 59–138, 139–212, Irvine 2020: 106–128).

The concept of humankind plays a particularly important role in explaining human action in the Anthropocene. For example, Donna Haraway used Hannah Arendt’s thoughts on Eichmann’s thoughtlessness during the deportations he supervised (Arendt 1963: 285) to explain why people systematically destroy their environment (see Hoppe 2021). As with Eichmann, the focus is not on stupidity, but on the lack of a sense of connection with the material world and the living world:

> Here was someone (Eichmann) who could not be a wayfarer, could not entangle, could not track the lines of living and dying, could not cultivate response-ability, could not make present to itself what it is doing, could not live in consequences or with consequence, could not compost. Function mattered, duty mattered, but the world did not matter for Eichmann. The world does not matter in ordinary thoughtlessness. The hollowed-out spaces are all filled with assessing information, determining friends and enemies, and doing busy jobs, negativity, the hollowing out of such positivity, is missed, an astonishing abandonment of thinking. This quality was not an emotional lack, a lack of compassion, although surely that was true of Eichmann, but a deeper surrender to what I would call *immateriality, inconsequentiality, or, in Arendt’s and also my idiom, thoughtlessness* (Haraway 2016: 36, emphasis and addition mine).

Haraway thus implies, unlike Arendt herself, that everyone has the ability or tendency to significantly damage environments (see the criticism in Park 2020: 97 f.). Many authors ask what the insights into the reality of the Anthropocene mean for our concept of humanity, that is, to what extent a concept of the human condition must do justice to the fundamentally new situation. Reference is often made to an earlier text by Hannah Arendt, namely her thoughts on a changed image of the *human condition* (Arendt 1998, first published in 1958, see Szerszynski 2003).

Arendt’s central argument was that people’s connection to the world had been unsettled by modernization, and that the prevailing situation then was characterized by alienation from the common human as well as from nature. These thoughts can be related to the Anthropocene. What does the Anthropocene mean for concepts of "Anthropocene societies" in the future? How can the novelty of the "human condition" in the sense of Arendt be grasped in the Anthropocene, and what does it consist of?

> Overall, the 'naive' belief in human 'mastery' of the forces of nature must be tempered by responsibility and humility, by respect for other agents, cultural differences, and other disciplines. This is the crux of Anthropocene society, the new human condition (Pálsson et al. 2013: 11).

Just as Stone Age societies had specific human–environment relationships, we have in fact been living in *Anthropocene societies* for some time now (Dalby 2007). An argument of the recent eco-feminist debate is that this fundamentally new situation (*new human condition*) requires a profound change in politics, science, and thinking in general (Schneiderman 2016, 2017; Guin and Ursula 2020, as a science fiction classic Guin and Ursula 1947). In concrete terms, this would mean completely new systems of care, for example, a family-like identification with humanity and concern for the biosphere and geosphere as a whole, or solidarity with all forms of life, "… a new mode of humanity and a new mode of belonging" (Pálsson et al. 2013: 11, similar to Haraway 2016).

Chapter 4
Critique—Strengths and Weaknesses of Anthropocene Thinking

Abstract This chapter asks about the strengths and weaknesses of the idea of the Anthropocene. While the geological-stratigraphic criticisms were discussed in the first chapter, this chapter deals with criticisms relating to the social, ethical and political relevance of the hypothesis. A clear benefit is that the natural sciences are now actually working together with the humanities and cultural studies on a very concrete topic. A problem with the term Anthropocene is its increasing diffusiveness. Important criticisms suggest that the idea of the Anthropocene overestimates the power of humans and is anthropocentric. More fundamental criticisms see the idea as Eurocentric and criticize its blindness to the diversity of the perpetrators and also the inequality of the victims. Critics object to its depoliticizing tendency or simply consider the Anthropocene to be a distraction from the capitalist causes of the world crisis. Accordingly, a multitude of alternative terms are being discussed. If we look at the scientific landscape, we have now arrived in the "Neologismocene." My own criticism is that many contributions wrongly reduce the Anthropocene to anthropogenic climate change or misuse the word for the rhetorical enhancement of known concepts or other topics.

The term means: today, natural relations are always also 'relations of domination'.
Michael Müller (2021b: 107)

As we have entered the Anthropocene,
strong continuity looks problematic.
John Dryzek and Jonathan Pickering (2019: 27)

What is the benefit of such a loaded term as "Anthropocene" compared to terms, such as "human-shaped Earth system" or "anthroposphere"? These and other proposed alternatives, such as *humanized Earth system* (Rull 2016), would de-emphasize the controversial chronostratigraphic periodization and more honestly mark our uncertainty about the future development of the geosphere. The benefits of using the term Anthropocene is seen across academia, from the Earth sciences to *literary criticism,* either as an *Earth system* term with fundamental significance for the future of humanity in the geosphere (comparable to Anthropocene 1), as a

C. Antweiler, *Anthropology in the Anthropocene*, Anthropocene – Humanities and Social Sciences, https://doi.org/10.1007/978-3-031-74591-1_4

geological period designation (Anthropocene 2), or as an indication of the historically long-lasting transformation of the geosphere (Anthropocene 3), and as a *concept* that stimulates cultural and natural theory as well as history (Anthropocene 4), or as a term with an educational impact, a potential to be critical of society (for example, Müller 2021a), or as a guide for improved *governance* or even for a "democratic Anthropocene" (Purdy 2015). This can be represented as a multi-level approach, with each of the levels having a proximity to different disciplines: the Earth system level, the geological-stratigraphic level, and the "consequential meta-level" and new form of social criticism (Anthropocene 4).

A practical potential is seen, for example, in an educational effect, a socio-critical potential (e.g. Müller 2021a) or as a signpost for improved governance or even for a 'democratic Anthropocene' (Purdy 2015). This can be represented as a multi-level approach, with the levels each having a proximity to different disciplines: the Earth system level, the geological-stratigraphic level and the "consequential meta-level" (Leinfelder 2019: 26–35, Zalasiewicz et al. 2020a, b). Like these positive views, criticism is also strongly linked to fundamental values and normative ideas. In view of the inflation of the use of the word 'Anthropocene', the following sharp criticism is at least worth considering before we ask about the benefits:

> The popularity of the Anthropocene concept does not imply its acceptance in a scientific-systematic sense. (…) Unlike comparable designations for geological epochs, the term Anthropocene, according to the will of its main propagandists, primarily contains deontic properties (Herrmann 2016: 49).

> Instead, the use of the term rather has apotropaic traits, as if the naming of a phenomenon could limit, mitigate, or avert its detrimental properties. It is therefore no coincidence that it is often associated with a term that also has high PR properties: 'sustainability' (Herrmann 2016: 49).

Criticisms from the field of geology were dealt with in the first chapter. This chapter will focus on criticism of the Anthropocene concept from the humanities and social sciences. The philosophical critique is essentially an amalgam of historical-philosophical catastrophism and anthropological negativism. Central philosophical findings are that we reify our self and world references, unrealise original world references, succumb to an instru-mentally narrowed reason and, as a new Prometheus, increase ourselves into a totalitarian, technology-fixated worldlessness (Sakkas 2021: 153). In the German-speaking world, these include early critical theory (Hannah Arendt, Günther Anders), the post-Maximist critique of unlimited growth and capitalism and the environmental movement since around 1970.

Objections, however, go far beyond such critiques: they see it as basically modernistic and thus aim to politicise and fundamentalise the concept. In addition, mass media catastrophism and 'quite normal' insecurities and fears in the face of climate change should also be mentioned. Markus Wissen and Ulrich Brand summarise the basic tenor of many criticisms of the grand narrative of the Anthropocene:

> The powerful Anthropocene narrative conceptualises the ecological crisis as a man-made one. as a man-made crisis. In doing so, it neglects the power-based social mediation of human impact on nature' (Wissen and Brand 2022: 263).es die herrschaftsförmige gesellschaftliche Vermittlung des menschlichen Einwirkens auf Natur (Wissen and Brand 2022: 263).

4.1 Benefits of the Term—Great and "Real" Multidisciplinarity?

The Anthropocene not only challenges the dispositions of our survival, rather, it also challenges our understanding of the role of science.
Werner Krauß (2015b: 4)

We are not always talking about the same thing, or talking about it in the same way.
Julia Thomas et al. (2020: 13)

The term Anthropocene bridges geological and historical time (Renn 2021: 3). So it is not surprising that some, albeit rather few, representatives of the social sciences and the humanities also explicitly use the geological term (Conversi 2020: 3–4). These include representatives of international political science, international law and geopolitics, but also, for example, Julia Adeney Thomas, Dipesh Chakrabarty, and Jürgen Renn as historians, Bruno Latour as a sociologist, and myself. A core insight from the findings on the systemic nature of the Anthropocene is that the future depends heavily on present decisions.

> The Anthropocene not only challenges our notions of 'nature' and 'human,' but also the foundations of political decisions and democratic procedures, which go neither with the geological time scales nor with the urgency of decelerating resource consumption – they seem unsuitable for both time scales (Görg 2016: 13).

4.1.1 Path Dependency and Science

Our power to change our environment seems to be increasing much faster than our understanding of the effects of these changes or our capacity to change our economies.
David Christian et al. (2014: 285)

Path dependencies, especially a "pathological path dependency," which does not recognize the feedback from the geosphere, are dangerous. The main political problem is that ecologically decoupled path dependencies result from information filters. The consequence is the continued existence of "Holocene" institutions, which are oriented towards continuity and are thus unable to react to the sudden change in parts of the Anthropocene Earth system. This applies in particular to international institutions (Dryzek and Pickering 2019: 22–24, 27; Chandler 2018; Chandler et al. 2021). The key insight is that the Earth system and human systems are interconnected, but there is no unified approach to understanding them, nor is there a single comprehensive planetary history. We will have to live with constant dynamics and will consequently never find a permanent political solution, such as Naomi Klein's call to abolish the global capitalist economy in the face of climate change, which "changes everything" (Klein 2015). A view informed by the Anthropocene rather than the Holocene instead relies on continuous global learning from mistakes:

> The Anthropocene is not just something bad to be lamented and avoided as far as possible. Nor is it something to be embraced, mastered, and celebrated. Rather it is *inescapable* and must be negotiated. (…) The Anthropocene is something that humanity must *continually* learn and relearn to live with. (…) The Anthropocene is now for better or worse humanity's chronic condition, a constant presence (Dryzek and Pickering 2019: 10, similar Eriksen 2018).

Just as there are many ways to look back at the causes of the Anthropocene, there are also more than one way to look into the future (Thomas et al. 2020: xi; Thomas 2022). The Anthropocene affects many topics and problem areas at different scales and thus requires multidisciplinary knowledge. Above all, an explicitly multidisciplinary perspective, as distinguished above from a merely sub-disciplinary one, considers the diversity of the involved disciplines as a strength. For responses to the Anthropocene, the diversity of the issues involved and also that of the scales is to be seen as a strength. It requires global knowledge as well as regionally oriented and local knowledge. One open question is whether knowledge that is explicitly non-hierarchically organized and networked but not coordinated is sufficient, as Tsing (2020) and Thomas et al. (2020) believe, or whether we need more consilient knowledge, in the sense of the evolutionary biologist Edward O. Wilson, as Ellis (2020) and I believe. Among the comparatively few anthropologists who have long engaged in intensive exchange with natural scientists, some see the main benefit of the term in its potential for interdisciplinary collaboration beyond pure programmatic thinking (Arizpe-Schlosser 2019: 276).

The historians of ideas Bonneuil and Fressoz have pointed out a fundamental problem regarding the role of knowledge and science in the Anthropocene. When the Anthropocene is presented as a scientifically authorized catastrophe narrative, humanity is virtually elevated to the status of subjects of Earth's history. This can lead to an overestimation of humanity, as most cultural studies critics complain about (see Sect. 4.3). But this is only one danger. Even more, this "geocratic grand narrative," as the authors call it (Bonneuil and Fressoz 2016: 45–96), could have the effect of relativizing human responsibility or culpability according to the motto "without knowledge, there is no culprit." We only know about the extent of the problem through ecology and geosystems research. According to the critics, earlier societies are portrayed as indifferent to environmental destruction. Accordingly, Earth system scientists could now stage themselves as central actors in Earth history and as saviors.

Latour provided an example of such logic when he applied a cybernetic twist to Haraway's idea of "response-ability" (Folkers and Marquardt 2017, 2018; Folkers 2020: 597). In one of her many creative word games, Haraway had said that "response-ability" only occurs when conditions are met that enable one to truly respond to problems rather than just react (Haraway 2016: 16, 36). Latour argued that as humanity, we would only become fully responsible to the extent that we have a technically enabled feedback about the environmental effects of human action (Latour 2017a, b: 29, 2018: 92–95). As a consequence, Latour, together with the geosystems scientist Tim Lenton, postulated a kind of Earth system Gaia 2.0, which "in-forms" and regulates itself by means of a complex knowledge infrastructure *and* is "conscious" of this (Lenton and Latour 2016).

4.1.2 Necessary Multidisciplinarity vs. Interdisciplinarity

Our best interdisciplinary moments are when we are most practical, as it were, being led by a shared curiosity about the world.
Nils Bubandt (in Haraway et al. 2016: 552)

The Anthropocene represents a recent culmination of the effects of human activity, some of which have a long history, such as the controlled use of fire, agriculture, urban culture, and plantation economy, and others that are extremely recent from a geological perspective, such as globalization, migration, and world tourism. Since the Anthropocene as a phenomenon is not only multifaceted, but also involves a wide variety of scales, that is, it is "multiscalar," it is relevant here that the disciplines also use different spatial and temporal scales as a basis. The diversity of empirical evidence mentioned earlier as such forms a strong argument for the reality of the sudden change in the geosphere. To be precise, the cultural markers—plastic parts, technofossils, and the broiler chicken—that we all know from everyday life are "secondary markers" for geologists. Their ubiquity has also contributed to opening up the humanities and cultural sciences as well as the public to an originally technical and stratigraphic field of investigation (Will 2021: 43, 170–179).

To research the Anthropocene, it is therefore not only various humanities and social sciences that need to work together, or various natural sciences among themselves. What is required now is the cooperation of many disciplines on both sides of the divide, for example, when it comes to socio-ecological dynamics and resilience in urban areas (Brondizio et al. 2016; Soentgen 2021b). Thus, the phenomenon of ecological novelty in complex systems, a fundamental characteristic of such systems, makes such broad cooperation indispensable (Kueffer 2013: 27; Toivanen et al. 2017; Pries 2021). Understanding the Anthropocene requires not only an interdisciplinary, but also a *multidisciplinary* perspective (Thomas et al. 2020: ix–x). the terms interdisciplinarity and multidisciplinarity are often used synonymously, but there are differences in the cooperation and in the objectives.

Interdisciplinary research tries to synthesize research and harmonize approaches. It is about coordinated research that creates knowledge that is as coherent as possible. The program of the evolutionary biologist Edward O. Wilson that strives for a scientific unit of *consilience*, which is generally rejected by social scientists, could be classified as such (Wilson 1998). This program fits in with biocentrically arguing approaches, but it was largely rejected by historians as well as cultural and social scientists because it ultimately leads to giving precedence to one perspective, such as, in Wilson's case, the evolutionary approach (Thomas 2014).

Multidisciplinary research, on the other hand, aims to create networks of knowledge. It emphasizes more strongly that representatives from very different disciplines come together to work on a question, each with *their own lens*. In multidisciplinary approaches, it is assumed that the involved disciplines use specific questions, theories, and methods. In addition, there are political and ethical aspects that will never lead to a correct answer. Thus, tensions are to be expected, and unlike in the *consilience* approach, there is no expectation that the collaboration will lead to a unified picture. The ultimate goal here is to dissolve the boundaries between sciences, which is why this approach is sometimes also called "transdisciplinary," although one should actually speak of meta-disciplinary work or a "pandisciplinary" approach (Davis 2011: 130–134).

Alexander von Humboldt (1769–1859), who was historically active before the great division of the sciences, can be seen as a forerunner here. He had a cross-disciplinary view and at the same time anticipated both a global and a local perspective, as is appropriate for the Anthropocene. From this position, Humboldt collected data on various parameters and phenomena such as temperature, air pressure, and altitude, on rocks, plants, animals, and humans. Across these levels, he was interested in unity, diversity, and change. As the historian of ideas Julia Thomas put it: he wanted both data and stories (Thomas et al. 2020: 1). Humboldt saw cultural history as linked with the natural-historical change in the landscape and explicitly traced cultural history far back in history. In doing so, he focused on humans as a transformer and also reflected on their role as a researcher of nature. Guided by his concept of the "cosmos," Humboldt not only collected a wide variety of factual data, but also did so at different scale levels, from the micro to the macro level (von Humboldt 2014). Thus, Humboldt's approach represents an early *big data* approach, as is inevitable for understanding some complex interrelationships (Grana-Behrens 2021: 3–5, 22).

Many people see the special benefit of the idea of the Anthropocene bringing together different disciplines. The topic of the Anthropocene downright demands the collaboration of the sciences; it calls for genuine multidisciplinarity instead of a friendly neighborhood approach. There is often talk here of the opportunity to form an "alliance of the sciences" (Gilbert, in Haraway et al. 2016: 552, similar Skinner 2021). This is not just about facts and data. Scientists from very different disciplines are called upon to "critically scrutinize their established narratives, histories, and stories and review their theoretical, methodological, and conceptual foundations" (Trischler 2016a: 273–274). The special potential lies in the fact that this is not just about a temporally and spatially large phenomenon, but also addresses globally relevant problems in regard to the future organization of humanity.

> The multipolar debate (…) is being conducted, on the one hand, in a disciplinary container (…) On the other hand, it is forcefully breaking down disciplinary boundaries and developing in interdisciplinary and transdisciplinary formats (Trischler and Will 2020: 237).

4.1.3 Encounter of Disparate Scientific Cultures and Misunderstandings

> *There is more at stake in the Anthropocene than a simple addition of natural sciences and those concerned with anthropos.*
>
> Werner Krauß (: 74)

A special feature of the Anthropocene discussion is that the call for the integration of disciplines necessary for the topic comes loudest from the natural sciences and partly came up in a clear manner even before the concept was launched by Crutzen and Stoermer (Vitousek et al. 1997). The invitation coming from the natural sciences offers the chance to bring the "two cultures" together. I see here above all the opportunity to establish a *great interdisciplinarity*, that is, a form of cooperation in which the natural sciences work together with humanities disciplines. A more recent example of a fruitful bridge from the natural sciences is the 'anthropocene science' approach. The aim is to supplement the earth system sciences with more future-orientated scenarios. This involves systematically asking how the space of potential societal visions of the future is reduced to the much smaller space of realised visions. How is the potential normally reduced to the actual potential (*worlding potential*) via (a) inability to perceive, (b) refusal to look and (c) cultural values that channel the thinking space? Building on this, how can the potential be expanded again through creative work on realistic scenarios (Ellis and Haff 2009; Keys et al. 2023: 3, 13)? These questions concern both long periods of past history and the distant future. The Anthropocene deals with a significantly longer time frame than what is usually covered by the natural sciences in the climate debate (Pálsson et al. 2013: 1). The *Anthropocene Working Group* of the International Geological Union sees potential in the precise determination of the onset:

> (…) may open possibilities of *historical fields other than Earth History (geology)* to more easily engage with the emerging interdisciplinary science base of the Anthropocene (Zalasiewicz et al. 2015: 201, emphasis mine).

A crucial point of collaboration is the mutual understanding between different scientific cultures. This is primarily about goals, methods, thinking styles, evidence, and forms of evidence. Will systematically outlined the contrast between geology and history by means of the Anthropocene debate (Will 2021: 143–201, 230–276). In addition, there are differences in the discussion behavior. For example, natural scientists see themselves more as contributors of complementary elements, not as representatives of divergent perspectives, unlike many humanities scholars and cultural scientists (Horn, pers. Comm. 2021). Furthermore, there are not only differing terminologies, but also different forms of scientific writing. Humanities scholars often quote positions from the literature verbatim because they are concerned about the specific meaning of words, and these often refer to the specific use by other authors and thus establish connections. This seems strange to many natural scientists (Thomas et al. 2020: xii). These differences also run through the relevant literature on the Anthropocene. Longer direct quotes would hardly survive even the first review in *Science* or *Nature*, for example, but also in the subject-relevant journal *Anthropocene* or *Earth's Future*, while they are commonplace in *environmental humanities* and cultural studies works on the Anthropocene.

Misunderstandings arise especially when the same words from everyday language have been adopted differently in the sciences. One example is the notoriously difficult word "culture," another is "Earth." When natural scientists talk about the "Earth" (capitalized), they usually mean the planet in our solar system.

For humanities scholars and historians, the earth (lowercase) rather represents the landscape and living environment of people and collectives. When geologists and paleontologists speak of "revolution," they mean changes that are infinitely slow from a human perspective and therefore not perceptible, which differs from the historical understanding of the word (Schulz 2020). Evolutionary catastrophes or "crises of evolution," such as the five major mass extinction events ("Big Five"), were processes that took millions of years (Stanley 1987; MacLeod 2013; Hannah 2021). This also applies to the "sudden" increases in certain organism groups during so-called evolutionary radiations. The oldest radical change "event" (sic!) about 550 million years ago, between the Ediacaran period and the Cambrian, with the beginning of multicellular life lasted about 30 million years.

4.1.4 Multi-scalar Institutions and Collaboration

For the implementation of interdisciplinary knowledge in order to obtain solutions, we need not only cross-regional organizations, but also institutions at various levels (*multiscalar institutions*, Thomas et al. 2020: 195). The economist Elinor Ostrom sees a structural similarity here in terms of adaptability between evolving systems and human organizations: "Institutional diversity may be as important as biological diversity for our long-term survival" (Ostrom 1999: 278). This is particularly true when large solutions cannot simply be scaled down to small spaces or living environments, or vice versa. Responses to the Anthropocene require an orientation towards common goals and agreements that are voluntarily reached and can therefore build on mutual trust. Social cohesion is of central importance for the adaptability of complex social systems, but a form of social cohesion that allows for diversity and uses it as a strength (Isendahl 2010; Antweiler et al. 2019).

In addition to many programmatic calls, the Anthropocene has in recent years also led to numerous networks and concrete projects involving broad collaboration between ethnologists and other scientists. This has not only brought together scientists from a wide range of disciplines, but also artists, representatives of civil society, and environmental activists. Examples of such broad collaboration include the aforementioned *Feral Atlas* project and other projects led by Anna Loewenhaupt Tsing and the *Plantionocene Project* at the University of Wisconsin (Tsing et al. 2020, 2024; Moore et al. 2021). The Rivers of the *Anthropocene* project at Indiana University in collaboration with Purdue University is also multidisciplinary (Kelly 2019). A particularly productive project is the primarily historical-ecological and archaeological project *Integrated History and Future of People on Earth*, with the beautiful acronym *IHOPE*. In terms of content, this is a historical project that spells out the fusion of geological and historical temporality in an exemplary way and includes historians, among others, but was not initiated by the historical community (Constanza et al. 2007; Robbins 2007: 1704; Robin et al. 2013; Crumley 2019). In this—still rare—direction, there is also the current collaboration between a climatologist with a flair for history and a historian well-versed in climatology (Pfister and Wanner 2021).

In Europe, in addition to the mentioned German activities and networks, there is the group *Aarhus University Research on the Anthropocene (AURA),* the *Vienna Anthropocene Network* from the University of Vienna, and the *Humans and Other Living Beings EASA Network* (*HOLB*) of the European Society of Social Anthropology (https://www.easaonline.org /networks/holb). Under the title *Overheating: The Three Crises of Globalisation*, a research group led by Eriksen dealt with the experience and handling of anthropogenic crises by groups of people (Eriksen and Stensrud 2018; Stensrud and Eriksen 2019). At the University of Cologne, a project on water dynamics in Anthropocene-shaped deltas is underway (*Volatile Waters and the Hydrosocial Anthropocene in Major River Deltas, DELTA*, Krause 2018). In Asia, the *Research Institute for Humanity and Nature (RIHN)* in Tokyo, the South Korean *Center for Anthropocene Studies KAIST*, and the *Taiwan Anthropocene Network* founded in 2021 also deserve mention. There are also a number of other projects (Mathews 2020: 73).

The idea of the Anthropocene is a provocation in several respects. It invites us to go beyond the fields integrated into a broad ecology, such as the life sciences, social sciences, cultural studies, and economics, by cooperating with the geosciences and the technical sciences (Table 7, see Trischler and Will 2020: 240–241). The provocation applies to both the geosciences and the social sciences, but the elements of the Anthropocene theses that are perceived as provocative differ from discipline to discipline (Will 2021: 79–84, 203–209). However, the inviting call made by the natural sciences to historians and the humanities and cultural sciences is also viewed critically by them. Some sociologists, for example, emphasize that there is a risk of a reductive “natural scientificization” of global problems, which in fact have historical-social causes (Neckel 2020: 162).

It is an eminently political question as to what form the cooperation between the disciplines should take (Brondizio et al. 2016). With regard to the social sciences and cultural studies, and especially the relatively small discipline of cultural anthropology as compared to the Earth sciences, some tough questions arise: Should cultural anthropology (a) act as a social science addition that provides data on local impacts or (b) play a conducive role in popularizing the issue or the societal challenge among political decision-makers or mediators of a sense of responsibility to the general public? Or, something that is clearly different—substantively more demanding and at the same time political—should it (c) guide a critical, actually reflexive, and thus always tense dialogue between the disciplines (Krauß : 60, 61)? These largely open questions are discussed in the chapter on cultural anthropology (Chap. 5).

It is fitting that some see the emergence of the discussion on the Anthropocene as a revelation of the failure of conventional natural sciences. In any case, it is a sign of the failure of the idea that forms of human life and the Earth’s biological and geochemical cycles can ever be studied in isolation. The Anthropocene makes it possible to enrich the often extremely detailed research (*molecularization*) in the life sciences and Earth sciences (Rose 2013, b). Perhaps, as humanities scholars and cultural scientists in particular hope, the global hierarchy of the sciences can be broken down.

4.1.5 Fruitful Irritations—Anthropo-Scientific Openings

The idea of the Anthropocene is heuristically inspiring and promotes ecological thinking in completely new areas, a "general ecology" (Lippuner et al. 2015: 9). From a cultural science perspective, the objectively necessary collaboration with natural scientists can have the concrete advantage that a discussion about concrete research and its results would take place, which is often more fruitful than endless epistemic debates (see Bubandt in Haraway et al. 2016: 548, 552). The debate about the Anthropocene forces us to take new perspectives on supposedly familiar objects. For example, waste appears as something other than just dead remains or matter in the wrong place. "Wild animals" that cannot survive permanently without human intervention, such as populations of elephants or tigers, appear as what they actually are: pets (cf. "petificaton", Cera 2020: 34–36, 2023). Since a large part of today's primates live permanently in anthropogenically modified environments, or more specifically, together with humans, primatologists have established their own ethnoprimatology. Primatology historically has cultural-anthropological roots, and the research unit of these ape researchers are human-primate communities. When working "among fellow primates," the methods used are relatively close to cultural anthropology (Riley 2020: 67, 76–98; Hartigan 2021; Sommer 2021).

From the perspective of science research, especially the sociology of science, the question arises as to which basic forms of organized science are developing in the wake of an awareness of the Anthropocene. After the earlier *Little Science* and the large-scale, industrially oriented, and often state-orchestrated *Big Science* (from 1940s to 1970s) as well as the technically dominated *Technoscience* (from 1980s to 1990s), to what extent is the formation of a truly inter- and transdisciplinary *Anthroscience* (Filipi 2011: 8) possible, one that revolves around the problem of sustainable life in changing environments?

Such a science, I hope, would then be "anthroposcient," knowing the human being, without becoming anthropocentric (cf. Table 4.1). This also applies to the historical sciences, which are increasingly expansive in terms of space and time. The planetary perspective of the idea of the Anthropocene is suitable for going well beyond the "flight altitude of eagles" (Osterhammel 2017) that global history assumes. The planetary view actually demands and promotes what can be denoted "planetary thinking." A planetary perspective involves a lot of challenges for the social sciences, but—not to be forgotten—also for the geosciences (Hamilton 2013; Clark and Szerszynski 2021). The question is whether their eagle eyes will remain as sharp with a planetary and therefore necessarily distanced view. In terms of

Table 4.1 Diversity of research directions with potential for the anthropological exploration of the Anthropocene

Direction	Representatives (examples)	Leading theories, basic sciences	Methods	Anthropocene-relevant aspects (selection)
Interdisciplinary Disaster Research (*disaster studies, collapse st.*)	Middleton (2017) and Oliver-Smith and Hoffmann (2020)	Systems science, human geography	Observation, participant observation, document analysis, simulation	Perception, sociocultural *coping* with abrupt change
Human Ecology	Herrmann et al. (2021)	Natural history, life requirements	Synthesis	Basic needs, Anthropocene effects
Environmental History, International/ Global Environmental History	Mauch and Pfister (2009) and Isendahl and Stump (2019)	Ecology, human ecology	Parallelization of documents with scientific data	Environmental crises in historical chronology
Historical Ecology	Balée (2006)	Globalization theory, colonial theories	Source analysis (texts, images)	Political-economic shaping of the landscape
Climate history	Mauelshagen (2023) and Pfister and Wanner (2021)	Historical theory	Source analysis (texts, images)	Historical documentation of trends
Global History	Conrad (2016) and Osterhammel (2020)	World system theory, Postcolonialism	Document analysis, diachronic comparison	History of Frontiers
Anthropological Archaeology	Butzer and Endfield (2012), Barton et al. (2012), and Scarborough (2018)	Historical sciences, *material culture*, practice theories	Ground findings reconstruction, Retrodiction	Early local cultural Landscape shaping, civilizational collapses
Global Archaeology	Smith (2020) and Stephens et al. (2019)	Globalization theory, domestication	Comparative regional analysis, meta-studies	Early global and regional cultural Landscape shaping
Historical Anthropology	Mauelshagen (2019)	Philosophy, history, biology	Source analysis, Retrodiction	Environmental crises in relation to being human
Philosophy	Federau (2017), Raffnsøe (2016), Welsch (2019), Rolston III (2020a), Belardinelli (2022), and Cera (2023)	Logic, ethics	Analysis, deduction, inference	Environmental ethics, anthropocentrism, biocentrism

(continued)

Table 4.1 (continued)

Direction	Representatives (examples)	Leading theories, basic sciences	Methods	Anthropocene-relevant aspects (selection)
Educational Sciences	Lampert and Niebert (2019) and Wulf (2020b)	Theories of understanding, cognition	Learning experiment, survey observation	Learning of concepts for stability and instability
Environmental macrohistory	Diamond (2005), Headrick (2021)	Ecosystem research	Natural experiments	Transgenerational channeling through geography
Landscape Geography	Kramer (2018) and Bätzing (2020)	Debates in the German-speaking landscape	Cultural and natural geographic procedures	Political references to landscape, rurality
Human-Animal Studies (*HAS*)	Wirth et al. (2016), Krebber (2019), and DeMello (2021)	Actor-network theory	Interpretation, self-observation,action-Analysis, relationship analysis	Human-animal relationship, animal use, pets, animals as Research perspective
Science and Technology Studies (*STS*)	Latour (2017a, b)	Actor-network theory	Observation	Realities of the research situation
Cultural Science and Material Studies	Malafouris (2013, 2019)	Practice theories	Objects, archaeological finds	Reference object—Cognition—Action
Complexity Science	Berkes (2003), Lansing and Cox (2019)	Systems science, cybernetics, physics	Observation, simulation	Unintended consequences, non-linear dynamics
Sustainability Transformation Research	Bollig (2014), Spangenberg (2011), and Renn (2021)	Systems science, communication science	Simulation of the overall handling of non-linearity	Resilience, societal awareness
Behavioral Ecology	Casimir (2021), Turner (2021), and Schill et al. (2021)	Evolutionary biology, behavioral research (ethology), Biocultural *constraints*	Observation, modeling, experiment	Basic needs, niche formation, Universals
Truly Evolutionary Approach	Richerson and Boyd (2018) and Laland (2017)	Cultural selection, niche construction, coevolution	Diachronic comparison, simulation, *reverse engineering*	Long-term niche construction

science policy, the Anthropocene makes it abundantly clear that in times of post-factual uncertainty, a strictly knowledge-based understanding of “perceived truths” is needed.

4.2 Strengths of the Anthropocene Idea Compared to Related Concepts

> *Precisely because overheating processes are now planetary and epidemic, the conditions for a genuinely (sic!) conversation look better than at any time in the past.*
> Thomas Hylland Eriksen (2016: 156)

Even before the Anthropocene idea emerged in the geosciences, there was already an interest in long-term environmental change the historical sciences. The term “environmental history” deals with long-term changes in the environment, such as the historically documented climate change, and, for example, “little ice ages” (Dukes 2011; Herrmann 2016; Behringer 2022). From an Anthropocene perspective, however, it is not just the environment that is changing, but also humans themselves. The Anthropocene thesis is based on very specific, fundamental relationships and, above all, on global and long-term interactions. The Anthropocene involves potentially irreversible changes, the onset of which can be marked historically (Haumann 2019; Renn 2021: 4–6).

A prominent example of a form of environmental history that focuses on anthropogenic change is the work of environmental historian William Cronon. Using the example of the American Midwest and particularly the city of Chicago, he combines economic history with environmental history and shows how socio-ecological relationships are manifested historically in the context of market institutions in environments that have changed over the long term. It becomes apparent that neither the city itself nor its surroundings can be understood as human or natural entities. Overall, human environmental history can be depicted as a tendency towards the infinite movement of frontiers (Richards 2003; Osterhammel 2020: 465–564), but earlier notions of the advancement of a cultural *last frontier* against nature appear too narrow (Cronon 1991). These boundaries are primarily *frontiers* of resource extraction, but also of a more extensive commodification of nature that is promoted by state law against local interests, as, for example, in Southeast Asia (Tsing 1993, 2005a, b; Geiger 2009; Li 2014; Kelly and Peluso 2015). More recent process-oriented research emphasizes that *frontiers* sometimes also dissolve again (Acciaioli and Sabharwal 2017).

The Anthropocene idea represents the human–nature relationship differently than in classical ecology (human–environment). Humans are not an external force, and culture is a normal part of the biosphere. Accordingly, natural disasters are no longer, as in the past, disasters that affect societies from the outside. No, today we ourselves are part of the cause of many disasters (Bruns 2020). Humanity itself, as a geological actor, is part of the Anthropocene and thus leaves an imprint in nature, but so do its dependence on the environment and the unintentional and uncontrollable effects of its actions. Crises in the Anthropocene are a multiple natural catastrophe and social catastrophe at the same time. In this respect, the concept involves both a centering on humans (geofactor) and decentralization (Clark 2015: 25). This means that nature can no longer be economically ignored or externalized. Conceptually, the Anthropocene as an epoch is something different from the "technological age" or the "history of civilization" of humankind. I have already mentioned that the Anthropocene approach, in contrast to earlier cultural ecology and human ecology, emphasizes that *every* ecosystem has humans or the effects of human activity as a component. Unlike the theories of late modern globalization, an explicitly *geo*biological and *geo*historical perspective is adopted. With the Anthropocene, two different kinds of present come together, so to speak. The now of human history is closely interwoven with the long now of geological and biological time. Human geoactive action connects the histories of the earth, of life and of modern civilisation (Chakrabarty 2009, thesis 3).

In a human society, all human activity in one area of life affects other areas of human life. This interaction is what makes up a significant part of human culture. This is a commonplace of the humanities, cultural studies, and social sciences. Cultural anthropology in particular has emphasized this holistic nature, at the latest since Tylor and Boas and their holistic concept of culture. The insights of the discussion on postmodernism also tells us all this. The Anthropocene thesis extends this holism in the sense that it identifies connections not only in the sphere of culture (like cultural anthropology) or within nature (like ecology), but also in both, especially the connections between these two clusters of natural and cultural phenomena. To link this to change and time scales, Chakrabarty's thesis that the problem of mismatching "temporalities" is the fruitful approach to understanding the Anthropocene is highly relevant. We need to analyse deep history and "recorded history" together, using historical sources and data as well as geological traces.

The development of humankind follows certain trends and directions, but not a clear historical path. We cannot speak of a clear sequence, such as one that goes from societies living close to nature to traditional ones, up to modern societies. This also has implications for all chronologies, as the situation of later periods did not fully and seamlessly replace that of the preceding periods. Modernity as a period cannot be clearly distinguished from the pre-modern era. Modern societies contain traditional elements, and even today there are—albeit few—communities that live in a non-capitalist way and are not fully integrated into the global system, for example, in the Amazon lowlands, Central Africa, and in the mountain areas of Myanmar in Southeast Asia.

Something similar is stated by Darwin's theory of evolution, which argues for a historical conception of evolution: natural history. The accurate picture of evolution is not that of an arrow, a staircase, or a ladder (*ladder stereotype*, Greenwood 1984), but that of the multi-branched tree or the complex branching bush. Thus, modern evolutionary theory does not oppose Darwinian theories of higher development deemed as natural and conceived as a necessary evolution (social evolution)—quite unlike the dominant public perception of Darwinism. Contrary to the popular conception of evolution, modern evolutionary theory primarily deals with the *mechanisms* of evolution (microevolution) and less with long-term development (macroevolution). But when it does, in contrast to social evolutionism since the nineteenth century, it tries to *explain* (average) increases in complexity in the development of the living world rather than assuming them to be necessary. This requires the inclusion of insights from complexity sciences (for example, Lansing and Cox 2019; Wilson 2019). This applies particularly to the increase in complexity during the emergence of primates, the lineage of hominids, and especially of *Homo sapiens*.

Anthropocene research differs in that it sees the trends in the development of the natural living world and that of the cultures of Homo sapiens as fundamentally intertwined. Anthropocene research names clearly identifiable trends for the phase of strong human influence. One of the best-known graphs in Anthropocene publications is the aforementioned compilation of clear trends in the form of sharply and often abruptly rising curves, such as that of carbon dioxide in the atmosphere, the human population, the sealed surface of the Earth—the famous hockey stick curves.

If we asked whether the Anthropocene is not simply another, more academic word for the human-made contribution to rapid climate change, we would have to say unequivocally: No! The Anthropocene is about much more than climate, namely also about soils, rivers, seas, and organisms, including humans, and the diagnosis is much more complex than that of anthropogenic climate change itself (Thomas 2019). This also applies, albeit to a limited extent, to what is probably the clearest precursor of the A-thesis, the *Gaia* concept mentioned above. The Anthropocene deals with the influence of humans as special beings and particularly powerful organisms. Moreover, it is no longer "just" about the atmosphere, but about a comprehensive anthropogenic change to the entire surface of the Earth's, the geosphere, including the global energy balance.

From the perspective of cultural studies and sociology, the Anthropocene is both a fact-based real concept as well as a consequential fundamental metaphor for the permanent perception of crisis. The Anthropocene is a means of controlling attention; the Anthropocene represents a formula for observing the world and a self-description mode of society. The Anthropocene can be seen as a semantic that captures the basic dynamics of a society, similar to diagnoses such as "late modern society," "knowledge society," or "risk society" for individual societies. Understood as a "world observation formula", the Anthropocene often contains a world observation imperative and thus—implicitly or explicitly—often also a call for the fundamental transformation of the economy and society; in this case it is a transformation imperative (Goeke 2022: 101–103).

What are the strengths of the Anthropocene concept compared to other approaches to environmental change that are currently being discussed? What distinguishes the Anthropocene thesis in particular from concepts that address similar issues? All these approaches have provided important insights and fruitful perspectives for understanding the Anthropocene, but the Anthropocene concept goes beyond that or is less extreme. Unlike systems theory approaches to autopoiesis, it is assumed that there is also a—multiply populated—outside to the system. In contrast to the theory of world society (Stichweh 2018) technological factors are emphasized more.

In contrast to world system theory (for example, Wallerstein 1988, 2004; Zinkina et al. 2019), a stronger bio- and geo-reference is made. In contrast to actor network theory (most recently, Latour 2018), a more conventional, but also less diffuse actor concept is used. Actors are seen as consciously acting individuals or collective actors, while material entities, such as objects or infrastructures, are seen as "only" causally significant. Assemblage theories (for example, Ong and Collier 2005; Ogden et al. 2013; DeLanda 2016; Hastrup 2016) about a materially multidimensional world are close to Anthropocene approaches and are used in them, but the Anthropocene is about a much more macro-scale material view of the human world.

The above-mentioned term "global change" (Glaser 2014; Flitner 2014) came about by thinking in terms of global environmental change, which emerged in the mid-1950s. In 1955, a renowned Wenner-Gren Foundation symposium on the "Role of Humanity as a Changer of the Face of the World" took place at Princeton University, which resulted in an extensive publication of 1236 pages (Thomas 1956; Steffen et al. 2011b: 844). This conference provided the groundwork for the establishment of an interdisciplinary study of the global environment. Until then, people had mostly spoken of individual communities embedded in their respective environments under the title "man and his habitat." By the end of the 1980s, two distinctly different understandings of global change had emerged. On the one hand, as an *anthropocentric* term, it referred to global change caused by human influences over a period of decades. On the other hand, there was a more *geocentric* understanding that focused on the interactions of the various sub-spheres of the geosphere over much longer periods of time (Arizpe-Schlosser 2019: 268–269).

A terminological variant that emphasizes diverse change is the current term "anthropogenic global change" (Moore : 34). Although global change emphasizes the worldwide extent of change, the planet is not only changing on a global scale, but also regionally, locally, and variably over time. An example of this is the "invasion" of animals that stems from anthropogenic change (neobiota). Quality of life in particular is also changing on a local and regional scale due to global interventions (Kueffer 2013: 21). Criticism of the exclusively global perspective has also led to the elaboration of regional and local boundaries within the Earth System Sciences (Dearing et al. 2014).

The very term itself represents a "disruption," a disruption in the cacophony of the better-known crisis diagnoses, such as the alarmist talk of "climate collapse" or the "heat death of the Earth" (Horn and Bergthaller 2020). We are reaching a threshold that is difficult to grasp through purely cultural representations. In contrast to the talk of "global change," the word Anthropocene more clearly marks the exit of the Holocene, through which we may have already passed. The Anthropocene gives this threshold a name; it also creates the notion of a boundary concept at the interface of science and politics, which also challenges previously adequate concepts (*threshold concept*, Clark 2015; Bonneuil and Fressoz 2016: xiii; Cera 2022: 129; see boundary concept, Görg 2016: 10, see boundary object, Star and Griesemer 1989).

In my opinion, all of this taken together in fact makes the Anthropocene a new concept rather than just another innovation in the constant change in political buzzwords or scientific fashions and academic terminological acrobatics. However, this does not rule out the fact that terminological acrobatics are unfolding on the word itself, which I will problematize in Sect. 4.7.

4.3 General Criticism—Diffuseness and Atlantocentrism

> *In the social sciences, and thus also in human geography, we do not actually need the term Anthropocene.*
>
> Hans Gebhardt (2016: 39)

> *(the word Anthropocene) has been tossed into debate more frequently than it has been explained or defined.*
>
> Davies (2016: 2)

The term "anthropogenic" shares with the Anthropocene term the problem of being too abstract as a category. A fundamental objection is that the word "Anthropocene" as the "age of humankind" already etymologically ignores non-human living beings and thus falsely presents the phenomenon as an exclusive result of human activity. Therefore, it can also lead to hasty conclusions (Krämer 2016: 24–44). If there is no longer an ecosystem unaffected by humans, everything and anything can be too quickly considered anthropogenic. One example is dioxin (TCDD), which, for a long time, was regarded as a compound of elements that did not occur in nature and was therefore exclusively anthropogenic. However, this was only true until naturally produced dioxin was found (Sayre 2012: 67).

As a science journalist, Gabór Paál has rightly criticized the fact that the non-scientific popularization of the topic in newspapers, museums, and art is leading to a softening of the concept. "The thesis that we are living in a new geological age is scientifically appealing, but threatens to fray into platitudes" (Paál 2016: 4, cf. Zottola and De Majo 2022). Paál sees the danger that the idea of the Anthropocene becoming a popular scientific topos, similar to Mandelbrodt's famous butterfly effect on climate, which meteorologists tend to smile about. The term threatens to flatten out and become synonymous with "the technological age," "the history of civilization," or even "the great problems of our time."

Due to the range of understandings shown so far, it is rightly criticized that the term is becoming increasingly vague. It shares this with the concept of sustainability. This is mainly related to how the concept is introduced to the general public. There is a risk that a vague guiding metaphor and a dramatizing framing will hinder a rational discussion. Above all, dramatizing metaphors would, according to the criticism, all too easily presuppose negative evaluations and create a "dominant discourse" in public debate, for example, against industry. A hypermoralization makes rational argumentation difficult and tends to excessively reduce the complexity of the global environment and also that of ecological argumentation (Matejovski 2016: 3, 6). The polymath James Lovelock also fears a conceptual flattening and calls for precise definition, because otherwise:

> (…) this clear and useful term is in danger of losing resolution in the background of vague academic niceties and amorphous thought about ecological sin (Lovelock 2014, cited after Visconti 2014: 383).

In addition to the temporal definition, a thematic delimitation is also needed. Paál goes further and advocates the strict separation of the scientific concept from ethical considerations of responsibility. In order to commit to sustainability issues and to strengthen ecological responsibility, the Anthropocene concept is no more necessary than the *Gaia* idea. These questions would later be the topics of a separate geoethics (Paál 2016: 5). No direct responsibility or policy follows from the findings. An exaggerated or excessive popularization weakens the science. In contrast, some popularizers, above all Reinhold Leinfelder, see the educational potential and deliberately link scientific findings with moral appeals (Leinfelder 2015a; Leinfelder & Schwaderer 2020).

Despite the criticism of its vagueness, Paál concedes, and I agree, that the Anthropocene concept can sensitize the public to the fact that human-made global change is about much more than just "climate change." This raises the question of whether we, from a scientific perspective, even need the term. This is the line of criticism voiced by the geographer Hans Gebhardt. He argued that the term Anthropocene often becomes a buzzword in a "… strategic 'discourse game' that focuses on disciplinary interests" (Gebhardt 2016: 35, similar Herrmann 2016; Bubenzer et al. 2021: 28). The criticism of some sociologists is similarly harsh. Taken together, I believe this means that the Anthropocene raises fundamental questions about meaningful societal communication on science, in the sense of a more acute form of a *Public Understanding of Science* (Matejovski 2016: 3).

The term is as controversial as it is vague. As for the sharpness of the debates, however, nothing is left to be desired. In a discussion of a recent essay in the journal *Current Anthropology* (Bauer and Ellis 2018), there are numerous statements to the effect that the crucial point has been overlooked, that the authors' interests have not been understood, that their own arguments have been "misunderstood" or even "avoided" by their discussion partners, that their esteemed colleagues have "sardonically averted" something (Bauer and Ellis 2018: 223) or have "decided to ignore xyz and continue with their massive misrepresentation" (Finney 2018: 219). The arguments there partly hinge on disciplines, and some see the misguided idea of a "… battlefield with the Anthropocene as a singular trophy to be fought over and won or lost" (Zalasiewicz et al. 2018a: 222). The many mutual accusations not only show misunderstanding, but reveal fears that geoscientists are now taking over the work of historians and humanities scholars.

Fundamental pro and contra voices on the term Anthropocene can be found across disciplines, as arguments from around 2015 show (Table 4.2). Such a table shows well how controversial the topic is; however, it cannot show the weighting of the arguments. In such an interdisciplinary field of research, I think it makes sense to organize the criticisms according to the arguments. Differences of opinion exist not only between disciplines in general, but also between the co-authors of an essay (such as Bauer and Ellis 2018: 225), and they exist within disciplines. On this basis,

Table 4.2 Cursory pros and cons of the Anthropocene concept in the early debate phase

Discipline	Pro-Anthropocene	Contra-Anthropocene
Geology	Consistent, adequate (human as a geological factor)	Quaternary geology is sufficient as it describes anthropogenic sediments; Anthropocene (A.) is too short for a geological epoch
Geography	Integrates human, economic, and physical geography	A. Is synonymous with geography
Archaeology	Basic consideration of human artifacts	Early human influence before the eighteenth century not considered
Biology	Realistic and consistent	Humans and animals always interfere with the earth
Ecology	Good metaphor for biodiversity threat	Biotic factors have long been considered
History	A. Integrates historical processes in nature and culture	Historical data are more suitable than conversion into geological data
Cultural studies	Culture is considered important	A. Is western and lacks gender conformity
Sociology	A. Integrates societal processes; emphasizes social construction	A. Lacks discursiveness, controversy, and disruptive changes bring more
Cultural Anthropology	– not listed –	– not listed –
Political science	A. Is political	A. Is too apolitical
Philosophy	A. Is not anthropocentric, holistic, intrinsic value of nature	A. Is anthropocentric
Religious Studies	Presumption of geology	Human claim to creation, lack of humility
Environmental NGOs	A. Is integrative, strengthens responsibility, creates protection concepts	A. Dilutes the concept of sustainability

the debate in cultural anthropology, which does not appear at all in this table or is probably subsumed under cultural studies, can then be tackled. Gebhardt also says, in my opinion not conclusively, that the term is simply unnecessary because the Anthropocene as a phenomenon coincides with the subject of modern geography, especially political geography:

> In the social sciences, and thus also in human geography, we do not actually need the term Anthropocene. In a phase of 'consuming the planet to excess' (Urry 2010), we should rather concern ourselves with a political geography that critically reflects on spatially relevant conflicts in the society–environment system of an economy fueled by neoliberal thinking and action (Gebhardt 2016: 39).

Most sharp criticisms focus on the political consequences of thinking in terms and categories of the Anthropocene. In particular, the uncritical adoption of the periodization carries the risk of naturalization and teleologization (Bauer and Ellis 2018: 224). The change to the Anthropocene appears to be the inevitable result of human evolution, and this would counteract an inclusive environmental policy that seeks to mitigate the damage. Critics often have the impression that the geoscientific approach obscures the decisive causes of the global environmental crisis, especially through the word "Anthropocene" itself. However, this objection overlooks the fact that geoscientists, especially geologists—unlike historians and ethnologists—do not usually claim to clarify the causes. They primarily ask about the cumulative effects or are focused on the geological time division, as, for example, in the AWG (Will 2021: 209, 215, see Kaika 2018).

A central objection to the 'Anthropocene' is directed against the homogenising gesture of talking about humans as geofactors. According to a central objection, the conception of the Anthropocene would, through its emphasis on the human species, obscure or "silence" the social and political inequality between the perpetrators and the victims (*silencing*, Trouillot 1995, similarly Bonneil and Fressoz 2016). This would be reinforced by words such as "humanity," "geophysical force," and phrases stating that humans "dominate" nature. The authors therefore advocate an explicitly critical analysis and commentary of the historical process with an emphasis on entanglements between social conditions, materials and non-human life as well as a focus on the possibilities of socially influencing environmentally relevant actions. The grand narrative of the Anthropocene, according to the core of the critique, tends to make existing transformative practices and possibilities for action invisible, but in return, all-encompassing solutions by 'humans' are almost implied (Hoppe 2022: 2). Furthermore, possibilities for action beyond Western innovations are not even considered. The authors therefore advocate an explicitly critical analysis or commentary on the historical process with an emphasis on entanglements between social conditions, materials, non-human life, and a focus on the possibilities of socially influencing environmentally relevant action.

The term is very ambiguous and has therefore been described not only as "controversial," but also as "problematic," and even as "unaccountably polysemous" (Swanson et al. 2015) or as "polluted" (Bubandt, in Haraway et al. 2016: 548). Nevertheless, even some of the harsh critics explicitly see the need to use the term for strategic or etymological reasons (such as Haraway and Tsing, in Haraway et al. 2016: 539, 541). This is particularly true in view of the contradiction that humanity is expected to solve the problems caused by humans themselves (Tsing, in Haraway et al. 2016: 541). Haraway also sees at least the usefulness of the word, propagating two critical directions for the humanities and the social sciences. On the one hand, concepts of the Anthropocene should be used for productive uncertainty by renaming them, for example, into *Plantationocene* or *Chthulocene*. Haraway explicitly advocates the multiplication of Anthropocene-shaped terms. Having said that, a different substantive content for the term Anthropocene should make the phenomenon the subject of discussion. From a science-critical and feminist perspective, she defines Anthropocene as "... the necessity of tragic domination of the secular project of phallic man" (Haraway, in Haraway et al. 2016: 28).

Some argue that the concept opens up a "space of possibility" for inter- and transdisciplinary geoscientific research (Bruns 2019: 56). Others go further and say that it is so multifaceted and undeveloped that it offers playful potential for thinking

(Swanson et al. 2015). According to Haraway, anthropologists and other humanities scholars and social scientists should counteract the contributions of the natural sciences with "speculative fabulation." But does this constitute a real deconstruction? In 2016, Hann was able to determine something that is still true today, namely that ...

> ... anthropologists are currently joining in the debates launched by natural scientists with a curious mixture of environmentalist commitment and playful imaginative reflection at multiple levels, ranging from conceptual critique to global political economy (Hann 2016a: 18).

A fundamental criticism of the idea of the Anthropocene is that the proponents of the idea repeatedly claim to perforate the boundaries between nature and society or culture, but actually cement them (Malm and Hornborg 2014; Malm 2021). The authors attribute this to the dominance of the natural sciences in the discussion. It is argued that the dichotomy would be reproduced by climate scientists in particular, even though natural scientists, such as Richard Lewontin, had already pointed out the permeability of the boundary between organism and environment (Bauer and Bhan 2018: 2). Bauer & Bhan counter the dichotomy by arguing that cultural anthropology has proven many times that a nature-society divide is anything but universal. However, Bauer & Bhan do not distinguish between analytical external views and internal views in different cultures. Such arguments, which see great potential in the idea and discourse of the Anthropocene, are countered by criticisms asserting that works on the Anthropocene often represent conceptual noise that is not capable of theorization:

> Sociologically, talk of the 'Anthropocene' has so far largely remained an empty formula, but its considerable impact demands to be examined with the whole sociological toolkit for an analysis of the social construction of reality. After all, the hybridization of nature and culture is also a process that has been driven by the effects and side effects, by the intended and unintended consequences of social action, so that for this reason alone, sociology has not become obsolete in the Anthropocene (Neckel 2020: 166).

4.4 Ahistorical periodisation—The 'Great Divide'

> *We are all visitors on the Earth.*
> Carolyn Merchant (2020a: xiii)

In Sect. 4.1, geological criticisms of the Anthropocene were presented, which mainly criticize the low number of truly synchronous markers. This chapter deals with criticisms that predominantly come from the humanities, cultural studies, and social sciences (Table 4.3). To be able to classify these, it is important to note that the representatives of these sciences predominantly use a broad concept of the Anthropocene, which differs significantly from the Anthropocene as an Earth system and as a geological epoch (see Anthropocene 4 with Anthropocene 1, 2, and 3 in the glossary). They use "Anthropocene" as a synthesis term for the consequences of the break caused by human-induced environmental change and the corresponding reflections, especially on the responsibility and status of humans.

Table 4.3 Systematisation of criticisms of the dominant Anthropocene concepts

Criticised points of dominant Anthropocene research	Authors (example)	Critical apraisal of critiques
1 Orientation towards the human species and humanity		
View of the human being as special: Thus anthropocentric, universalizing, totalizing (exceptionalism, speciesism)	Haraway, Tsing	(relative) human peculiarity is well substantiated in terms of both science and philosophy (cf. primatology, philosophical anthropology)
Humanity as a "geofactor" overestimates the power of humans to act (omnipotence, hubris, human mastery)	Nixon, Haraway, Cera	Anthropocene, however, emphasizes powerlessness (irreversible change, peak points partly reached)
Humanity reference ignores proportions of causes, e.g. colonialism, capitalism, wealthy vs. poor (de-politicisation)	Todd Whyte, Moore	Partially true
Humanity-based approach effectively levels out different levels of responsibility today (de-politicisation)	Swyngedouw, Moore	Partially true
Reference to humanity ignores different current and future vulnerabilities	Many Anthropo-logists	Partially true, but decreasing
Reference to humans ignores non-human beings (regarding causes and consequences) (anthropocentrism)	Tsing	Not necessarily, cf. natural science-ecology, human ecology
2 Dominant periodization: Onset "great acceleration" mid-twentieth century		
Onset thesis in the 1950s ignores historical causes: colonialism and capitalism	Maslin	No, but with less weight, because only regionally;
Geological-stratigraphic demand for synchronous epochal boundaries ignores diachronic historical beginnings	Bauer, Ellis	Only excluded in geochronology; from a geol. Point of view, all onset suggestions are almost simultaneous
Dramatization, dystopias, end-time orientation prevent positive future thinking	Haraway	Empirically open; dramatization may also spur on, possibly give hope
Geological time framing hides human history	NN	No; Geochronology already connects cultural and geological history
Directed concept of time fades out other concepts of time or concepts of history	Benjamin, Cultural an-thropology	Concept directional time is in fact a near-universal (beyond other local time conceptualisations)
The duration of the A. (70–12,000 years) is much too short and too present-related for a geological epoch; too actualist (presentism)	Majority of geologists	As a geological alternative, the Anthropocene offers itself as a geological "event" instead of an epoch

(continued)

Table 4.3 (continued)

Criticised points of dominant Anthropocene research	Authors (example)	Critical apraisal of critiques
3 Natural Science Basis and Dominance of Western Science		
Western scientific basis separates nature from culture (dualism)	Haraway	No, dominant sciences are monistic: Humans are (a specific) part of nature
Western scientific basis ignores non-Western concepts	Mostly Anthropology	"Western" science is a polygenic global system of knowledge
Western scientific basis shows factors and effects of Anthropocene in a narrowing way (reductionism)	Hulme	Does not apply to e.g. evolutionary biology, ecology, complexity science and ethnology
Narrowing of the solution approaches to technologically modernized mastery of nature	Wissen & Brand	Yes, largely applicable to US discussion
Theories and scientific data are concentrated on Europe and America (eurocentrism, Atlantocentrism)	Bergthaller, Horn, Antweiler	Yes, as before Atlanto-centric debate, theory and data
Earth-based orientation ignores regional differences in causes and consequences	Maslin, an-thropology	Partially true, but regional variation already mentioned by Crutzen (2002)
Earth system science ignores cultural and social sciences	Pálsson	Yes, but social sciences are being heard to some extent
Earth system science as a leading science ignores local and indigenous knowledge	Cultural Anthropology	Local or Indigenous knowledge offers only local solutions
Geology as a basic science works with colonialist and racist concepts	Yussoff	Partially true regarding colonialists. But criticism is too narrowly linguistic

The word is used extremely inconsistently for the meta-level and even for the analytical levels. Within an integrative overall concept ("The integrative Anthropocene concept sensu lato"), this term usage can be distinguished from the two analytical levels of the Earth system and the epoch as a *consequential meta-level* or as *the responsible anthropocene* (Zalasiewicz et al. 2021a, b: 9, Fig. 3).

What are the most important, more explicit criticisms of the postulates of the existence of an Anthropocene? Here there are criticisms of details on the one hand, and fundamental objections on the other. Against the epochal claims by geoscientists, which are motivated by the Anthropocene, some historians and many humanities scholars emphasize that the periodization of human history is their business (Hamilton et al. 2015). It can also be argued against the claims of the Earth sciences that universal and global historical periodizations, which sometimes concern long periods of time, such as in Osterhammel (Osterhammel 2020), are not even used by geoscientists in their debates about the beginning of the Anthropocene.

In view of the currently favored thesis among geoscientists that the Anthropocene began in the middle of the twentieth century, one would expect debates on the beginning of globalization à la "Globalization since 50, 500, or 5000 Years" to be considered (for example, Frank and Gills 1996). In fact, the dominant plausibilization of the "age of humans" (sic!) seems to do largely without knowledge of human history, social order, and culture. With the exception of Erve Ellis and William McNeill, for example, the proponents of the Anthropocene are rather indifferent to sociological concepts of formation (Schmieder 2014: 47). With respect to this geologically based plausibilization of the Anthropocene, Werber sarcastically concludes:

> With secularization, the restructuring of historical semantics, the differentiation of functional systems, structural changes in the public sphere, the French Revolution, or the discovery of universal values, with cosmopolitanism or world traffic, the evidence of this epoch, which is apparently so important for humankind, has nothing to do with any of these (…) This view is based on a geological theory of epochs and its evidence, thus escaping the incessant social and cultural science dispute about the historicity of society (Werber 2014: 243, 245)

Here, the contrast between geoscientific thinking in large spaces and long periods of time and a historical view becomes very clear. At the same time, it becomes apparent that some cultural scientists do not realize that geology is a very special natural science, namely a historically oriented one. Instead, they place geology with the other natural sciences. What Werber overlooks, for example, is that the protagonists of the Anthropocene, whom he refers to as "Cartesian geologists," in fact hardly come from geology itself. They notably come from *other* geosciences, such as climatology, Earth System Science, and soil science.The criticism that viewing the environment from the perspective of deep geological time would make colonialism appear as an insignificant detail on a planetary scale (Ferdinand 2022) is also unfounded. A deep time perspective does not call into question a historically developing sense of the Anthropocene.

It is also often forgotten that Crutzen himself sought an early exchange with William McNeill, who also drew attention to the rapid influence of humans in the year 2000, but as a historian. In a rather comprehensive book, McNeill contributed data of a completely different nature to that of Crutzen, and the two then also published together, including one of the central texts on the Anthropocene (McNeill and Engelke 2013). What is rightly criticized is that only a few works from the geosciences explicitly include cultural history, but even this only applies to the mainstream. An exception is the detailed work of Erle Ellis (esp. Ellis 2015) and the IHOPE group, which explicitly tries to brings together geology, Earth system science and findings of historical sciences (Constanza et al. 2007; Robin 2013: 20). This is also reflected in the latest introductions and systematic overviews (Ellis 2018; Thomas et al. 2020; Will 2021: 218–225), and now also in recent English-language publications on world history (Fernández-Armesto 2019, especially Christian 2019).

The central issue here is one's attitude towards the use of knowledge from outside the discipline, as permitted by the AWG (Will 2021: 127–138). From geology itself comes the objection that the hypothetical and prospective rather than the retrospective approach is methodologically atypical from a stratigraphic perspective. It is feared that the periodization of Earth history, the unique selling point, the *arcanum* of stratigraphy, would now be taken over by the humanities and cultural sciences in an almost imperialistic manner (Gibbard and Walker 2014).

Finney & Edwards criticize this as geologists and complain that the Anthropocene is similar to the broad concept of the Renaissance in that it has several meanings and leads to different periodizations, depending on the interest (Finney and Edwards 2016: 8–9). As historians, Warde, Robin & Sörlin rightly counter that this critical argument itself does what Finney & Edwards accuse the AWG of doing, namely mobilizing non-specialist knowledge to reinforce their thesis (Warde et al. 2017; Robin 2013: 20).

In addition, some geoscientists fear a politicization of the Earth sciences (Finney and Edwards 2016), while others point out that important ideas and key issues in geology have always been negotiated in a political and social manner (Warde et al. 2018). Since the establishment of geology at the beginning of the nineteenth century, and increasingly since Romanticism, literature has been fascinated by the temporal scaling of geological processes, and this has been intensified by the Anthropocene (for example, Völker 2021).

The social sciences also object to the hypothetical approach to periodization. The Anthropocene as a new geological age is not old enough to have already left a clear stratigraphic mark (*geologic record*) (Pottage 2019). The stratigraphic decision can actually only be made in the distant future. That is why the imaginary figure of a "future geologist" is always used (Yusoff 2016: 4). In the case of this book, I imagined Amy with my optimistic assumption that the Anthropocene could end, but that there would still be people because humanity would have managed to take a turn for the better. However, if one speculates pessimistically, the only way to look back is through a post-humanoid geologist or through AI based on technofossils. In my opinion, these criticisms overlook the fact that such markers *already* exist from a *geological* perspective, especially plastic particles and radionuclides.

From the perspective of cultural anthropology and environmental history, periodization as such has also been criticized. The criticism was directed at the unconditional will to find a time marker for the onset of the Anthropocene that is valid for the entire planet. A real diachronically gradual change of environments induced by humans would be arbitrary and separated by a sharp step, the *Anthropocene Divide* (Bauer and Ellis 2018: 209). However, the effects of the activity of mere "ecological humans" have never been ascribed to "fully geophysical" humans. We would need a much "grayer answer" to the question of when humans became geophysical agents, because there has certainly been a quantitative leap in the geo-effects produced by humans, but no qualitative change (Bauer and Bhan 2018: 15).

The idea that the planet was free of human influence in the late Holocene, before industrialization, is a myth. The boundary between the Holocene and the Anthropocene, no matter where it was drawn, would introduce a binary separation between a "before" and an "after," and it would also divide the disciplines. In fact, as Bauer & Ellis put it, the increase in human effects would be distributed diachronously instead of synchronously (or isochronously). Such an assumption would hinder an understanding of the human contribution to the transformation of the Earth's functional systems and the complex links (*entanglements*) that exist within them (Bauer and Ellis 2018: 223). In short: humans would have already changed ecosystems on a broader scale long before the present (Bauer 2016: 409; Morrison 2015).

This coincides with the objections of some geologists: On different continents, cultures have intervened in the natural environment at different times. In America or Australia, demonstrable traces have appeared later than in Southern Europe in the Near East. Furthermore, human intervention has slowly increased since the end of the last Ice Age, so that a sharp boundary does not exist. This refers to the epoch of the Holocene, which does not require a demarcation with the Anthropocene. However, this criticism by Bauer & Ellis is unwarranted: they are throwing the baby out with the bathwater. First of all, they are overlooking the simple fact that demarcations in no way deny continuous transitions:

> A popular example of a logical demarcation problem that discredits the character of evolutionary changes is the assertion that if humans were to be traced back to ape-like ancestors in terms of phylogeny, there would also have been a case in which a 'last ape must have given birth to the first human'. – It is immediately apparent that periodizations within continuously varying feature complexes are based on agreements that are (or must be) arbitrarily set (Herrmann 2016: 47).

The emphasis on early and also larger human influences on the environment is correct, but Bauer & Ellis do not distinguish between anthropogenic change in individual landscapes or regions and *systemic* change in the geosphere (Thomas et al. 2020: 119). They lack insight into what an Earth system is, into the fact that individual changes, even strong ones, are something different from a *systemic transformation* of the geosphere. All species change the environment and thus the living world of other organisms. But only a few change the geosphere systemically *and* globally *and* permanently. The current level of carbon dioxide is higher than it has ever been in the last three million years.

Bauer & Bhans'criticism aligns with the criticism already mentioned in the discussion about the onset, that the idea of and especially the research on the Anthropocene are Eurocentric. From a postcolonial perspective, the idea of the Anthropocene appears Eurocentric in the sense that existing models from Europe—based on experiences in temperate zones limited to Western Europe and the history of land use there—are being "expanded" as unmarked categories. As a result, other regions and variants appear as exceptions. In philosophy, this is called "nostrocentrism." One author summarises this view succinctly: 'It's not the anthropocene, it's the white supremacy scene' (Mirzoeff 2018, cf. White (M)Anthropocene, Di Chiro 2017). The European perspective and the focus on the new carry the risk of not seeing that the Anthropocene actually began much earlier, was more complex in its causes, and had more diverse trajectories (Morrison 2015: 78–79).

Some critics see this danger not only with natural scientists, who would naturalize human social relationships in models, but also and especially with cultural scientists. They find the idea of the Anthropocene exciting and attractive because of the strong emphasis on human agency and run the risk of "converting" to it with flying colors (Morrison 2015: 76). The obvious "novelty" of human agency in the Anthropocene is just that: that it is obvious. With the resulting call for the "decentering" of research on the Anthropocene, the chronological criticism is intertwined with the thesis that the concept of the Anthropocene extends colonialist discourses into a postcolonial world. I will take this up again in Chap. 7 with the question of an Asian Anthropocene.

However, it should be noted that there are indeed works by representatives of the Anthropocene that emphasize the gradual onset and also the strong, varying, regional effects. In particular, these are publications on which natural scientists and historians have worked together and which Bauer & Ellis also mention. When Steffen et al. wrote: "the current epoch in which humans (...) have become a global geophysical force" and that their goal is to investigate the "evolution of humans and our societies from hunter-gatherers to a global geophysical force" (Steffen et al. 2007: 614), I do not see this, *pace* Ellis & Bauer, as binary before-and-after thinking because evolution allows for gradual change, indeed it is at its core a notion of gradual change. There are studies in Anthropocene research that show diverse anthropogenic imprinting in the Holocene (for example, Gibbard and Walker 2014).

In addition, humanities scholars and historians also carry out periodizations, and they do so in relation to the Anthropocene. It has been discussed whether it is possible to account for a historical transition from "biological" or "ecological action" to "geological action," which only began with the use of fossil fuels. Chakrabarty then introduced the idea of "disjunctive" forms of human action into the discussion (Chakrabarty 2012). Bauer & Ellis do admit the alarming rise of human impact with the industrial revolution and its intensification since the 1950s, but they emphasize, with arguments from archaeology, that humans have always changed their environment, for example, through the use of fire and the displacement of animals. Ellis himself is involved in significant publications on periodization (Braje 2018: 216), but still concludes together with Bauer:

> Nevertheless, an anthropocene periodization that begins at these points fundamentally obfuscates qualitative similarities and historical linkages with the dynamics of human-environmental relationships in previous periods (Bauer and Ellis 2018: 210).

Overall, a mutual adoption of perspectives between disciplines is required. From an external cultural studies perspective, it is surprising that a large part of the contributions from the natural sciences have so far focused on discussing the beginning of Anthropocene change (Dürbeck 2018a: 12, 2018b). Since the Anthropocene was "gifted" by geoscientists (Latour 2017a, b), it is based on their epistemic premises (Bruns 2019: 55). Periodization and dating are, of course, always a core interest in geology and paleontology as *historical* natural sciences. The environmental historian and philosopher Carolyn Merchant nicely sums up the benefit of the Anthropocene as an epoch concept:

> I argue that the concept of the Anthropocene goes beyond earlier concepts and periodizations such as preindustrial, colonial, industrial, modern, and postmodern by presenting a clear and forceful characterization of the future crisis humankind faces (Merchant 2020a: xi).

4.5 Ideology—Depoliticization, Anthropocentrism, Gender Blindness

There is no common world, and yet it has to be composed, nonetheless.
Bruno Latour (2017a, b: 12)

From an Earth science perspective, one key reason why we find ourselves in the Anthropocene is the enormous increase in the consumption of raw materials and its effects on the environment since the middle of the twentieth century. Many contributions to the Anthropocene therefore deal with the economy, and especially with growth and the *runaway processes* typical of unchecked capitalist systems, as exemplified by the uncontrolled trade in fictitious commodities, the basis of the major financial crises. Global growth currently relies on the use of cheap, non-renewable energy. This is ultimately a destructive process that causes lasting damage to the environment and is reflected in the increasing scarcity of raw materials. The current generation is living at the expense of future generations, which is why sustainability and generational responsibility are central themes in all development debates. A large part of the fundamental criticisms of the idea of the Anthropocene argues from a perspective that is critical of society and capitalism.

4.5.1 Background—Criticisms of Growthism *and Plutocracy*

At the center of the fundamental criticisms of the Anthropocene concept is a critique of the neoliberal thinking of its main representatives in the geosciences. This often unacknowledged neoliberal attitude is accompanied by depoliticization. To characterize a mode of production that is based on growth and increase—or one diagnosed with a basic neoliberal attitude that many propagators of the idea of the Anthropocene have –critics sometimes sarcastically speak of the "Growthocene" (Malm 2016) or of us being in the "toxic grip" of the "growth algorithm" (Ebron and Tsing 2017: 683). Depoliticization treats the Anthropocene as a problem to be solved by technology and management, thus avoiding political debates about fundamental values, social justice, and the conditions for long-term well-being (Eriksen 2016: 131; Svampa 2020). Critical social scientists and environmental activists see the danger that past human interventions in the natural environment will be used as a justification to establish new environmental control mechanisms under the guise of "repair" in a targeted and more ambitious manner. The proponents of the Anthropocene would cleverly invoke the end of nature and thus create a fatalistic attitude in order to then elevate humankind to the role of "thermostat" of the planet (Hamilton 2013; Crist 2013, 2020; see Corlett, b; Büscher and Fletcher 2020).

Due to the continuous relevance of this background to the criticisms, I will first briefly explain the fundamental criticism of growth thinking, which mainly comes from the cultural and social sciences. Earlier civilizations also grew in an environmentally damaging way, but economic growth per capita and expansion as guiding principles are historically inseparable from the capitalist economy. The accumulation of capital began with risk capital ventures in the early capitalism of discovery and expansion voyages, and especially with the trading companies of colonial times, which developed into joint-stock companies (Schmelzer and Vetter 2019: 100–110; Scheidler 2021; Marks 2020: 69–101). An orientation towards growth and a market economy characterizes an economic policy that has been particularly strongly since the 1980s with the implementation of *Reaganomics* and *Thatcherism*. Neoliberalism is a variant of market-oriented economic principles based on the free market, free trade, and private property, for which the state provides only the appropriate framework. This has also largely applied to Japan since the 1960s, and again increasingly so since the 1980s. The central goal of "growthism" was to increase the gross domestic product (GDP).

There were very few voices against the global growth orientation after the Second World War, foremost among them was Mahatma Gandhi (1869–1949). The growth orientation characterizes American and globally widespread consumerism, but it is also a consistent feature of centrally planned economies, such as in Soviet productionism with the net material product (NMP) as a standard and in Vietnam since the *Doi Moi* reforms of 1986. Growth also characterizes the developmentalism of most post-colonial economies in the Global South. It is no coincidence that the increase in CO_2 associated with the growth of the former developing countries coincided with the emergence of the term *Global South* (Fuhr 2021). The world economy became 14 times larger in the twentieth century than it was in 1900 (Marks 2020: 226–227).

This fundamental criticism of approaches to the Anthropocene can be attributed to an older fundamental criticism of universalist economic models. In his aforementioned seminal work on the "great transformation", Karl Polanyi argued that every economic system is substantially embedded in life-worlds. As a non-Marxist socialist, Polanyi argued against so-called formalist approaches to economic theory, which later led to the famous debate on formalism vs. substantivism in economic anthropology. Thus, he was considered the most important critic of neoliberalism *avant la lettre*. (Eriksen 2016: 20; Hann 2016a). Every form of economic activity, so the argument goes, is built on a "human economy," on social conditions, for example, the relational forms of reciprocity and the norm of solidarity (Hann and Hart 2011). Social substance is as fundamental to every economy as Tönnies' community(-ness) is to every society. Accordingly, every economic system should also be normatively oriented towards concrete sociality rather than individual interests. Polanyi did not reject market principles, but argued against their extension into genuinely social spheres of life. This means a rejection of a comprehensive commodification of labor and other narrowed forms of mainstream economics (Polanyi 1944).

It is precisely this social embeddedness of its own economic principles that modern growth-oriented capitalism has lost sight of since the nineteenth century; in capitalism, the social dependency of its own system has been forgotten. Although Polanyi prematurely predicted the decline of such capitalism, the "success" of neoliberal economies since the 1980s has only been possible on the basis of state support and—central to our topic—the persistently excessive exploitation of raw materials in view of the Earth's limits (Schneidewind and Mitarbeiter 2019: 2018).

Another, albeit unintended, effect of growth-oriented economic and development policy is the increasing inequality between countries, or between regions and between socioeconomic classes in individual countries, especially since the turn of the millennium. Income inequality and even more so wealth differences have been growing in recent years. Even communist countries like China and former communist ones such as Russia have largely abandoned the planned economy and command economy and have adopted market principles. The dominant concepts of economic policy pursue an approach of unlimited growth that runs counter to insights into the Anthropocene (Thomas et al. 2020: 139–144).

Insights into the reality of the Anthropocene now undermine the leading assumptions of currently dominant economic policy on a much broader empirical basis than Polanyi's, that is, spatially broader and temporally deeper. They do so in a much more fundamental way than the realization that current climate change is largely anthropogenic in nature. However, mainstream economics still largely ignores the insights of the Anthropocene, insights into realities of our geosphere that can only be controlled to a limited extent. This is particularly important because economics is arguably the field of science that guides politics like no other. The guiding assumptions of orthodox economic models determine to a large extent the perception of threats, risks, uncertainty, and vulnerability. They influence ideas about "good life" and the future of societies. And they shape forms of control, regulation, and governance. In short, economics is the "mother tongue" of politics, the language of public life, and the worldview that significantly shapes society, but does not avoid the profound "degradation" of the geosphere and also the profound "depravation of man" (Raworth 2018: 5).

From the perspective of the Anthropocene, the error of orthodox economic approaches lies in the following points. First, it is assumed that "nature" is an externality that does not need to be considered in economic calculations. This includes inputs into the economy, such as raw materials, drinking water, and air, but also outputs such as waste and environmental gases. Second, there is an assumption about time, that development will proceed evenly into the future. Accordingly, it is assumed that current growth will solve future problems because costs will decrease in the future (*discounting*). This makes it advisable to postpone current measures against the effects of the Anthropocene. This is one of the clashes of scales so typical of the Anthropocene (Eriksen 2016: 29, 131–156).

Orthodox economists admit that raw materials could become scarce, but they optimistically believe in replacement through market forces and innovation. It is not recognized that the Earth system is not only entering a crisis in the Anthropocene, but is experiencing a real break in development (*rupture*, Hamilton 2016b). Third, it is assumed that all key economic variables can be quantified and, in particular, monetized. The cardinal example is the gross national product, which only takes into account goods and services that are marketable. Thomas et al. summarized the criticism well:

> In sum, mainstream economics treats 'the economy' as separable from the environment, does not acknowledge disruptions in the overall trajectory of growth, and believes in metrics, thereby limiting the grasp of nonquantified, nonmonetary values (Thomas et al. 2020: 144).

Neoclassical approaches do not include natural resources in their market calculations, while alternative approaches to economics insist on seeing the environment as an integral part of the economy. Even within the dominant economics paradigm, the limits of human growth on a finite planet are recognized. Crucial to this were Aurelio Peccei (1908–1984) and the system dynamicist Dennis Meadows at MIT, and the work of the *Club of Rome* from 1968, which quantitatively modeled future developments (Meadows et al. 1972, 2012).

4.5.2 Pathological Path Dependency Through Holocene "Success"

The portrayal of neoliberalism by critics, however, overlooks the fact that the neoliberal position, unlike classical liberal approaches, does not start from the *homo oeconomicus*, but from *imperfect* markets and *limited* human insight and therefore places emphasis on regulations. In view of the criticism of growth, which I share, it should be noted that this focus on growth, besides widespread inequality worldwide, has also shown phenomenal successes. These are sometimes not recognized by critics, and for a long time I did not want to acknowledge them either. However, as recent studies based on very different historical data and new empirical evidence consistently show, rapid progress has been made in terms of concrete living conditions in important areas such as nutrition, security against war, violence, murders, accidents as well as health, life expectancy, infant mortality, and education (Rosling 2018; Pinker 2018; West 2017: 185; Behringer 2022). However, the epochal transitions in population growth, agriculture, nutrition and energy use were bought at the price of enormous environmental effects, which are hardly highlighted by Behringer, Rosling, Pinker, for example (Smil 2019: 439–448, 2021: 205–243, 253–260; Tauger 2021: 50, 92).

From a long-term perspective of social evolution, it can be said that humanity's political institutions have coped well with some problems. In the economic sphere, a number of communities that depend on certain resources have successfully developed rules to prevent overuse of common resources (*commons*). Through collective effort, international violence has been largely contained in the political field, and in the development field, some infectious diseases have almost been eradicated through *public health*. This is not a given as many economic and political systems have collapsed—sometimes after a long period of existence. Polanyi had prematurely predicted the decline of a disembedded capitalism. This positive outcome also applies to the reduction of economic and gender inequality, at least until the increased implementation of strengthened neoliberal principles from around 2010.

However, this success only applies to certain areas and is based on political institutions that emerged in response to Holocene experiences and not Anthropocene conditions. Moreover, states were successful in those areas that concerned their core business, such as the welfare state of the late nineteenth century in response to pressure from workers' interests (Dryzek and Pickering 2019: 20–23). In the longer term, capitalist economies and nation states have had their greatest success in generating sustained growth… and that is now the problem. The "success" of neoliberal economies since the 1980s has only been possible on the basis of state support and—central to our topic—the continuously excessive exploitation of raw materials. The downside of this success is the exorbitant consumption of energy and raw materials, especially of resources that—in human time scales—are not renewable and, what is more dramatic from an Anthropocene perspective, lead to irreversible damage to the geosphere.

4.5.3 Industrial Interests and Scientific Skepticism

Leading economists did not believe the gloomy predictions of their colleagues. This was reinforced by criticisms of the scientific basis, which stated that there was no professional consensus on climate change issues. These attitudes were instrumentally used for propaganda by networks between industry, *think tanks*, and individual scientists, as sociological studies have qualitatively and quantitatively demonstrated (Oreskes and Conway 2010). The sociologist Farrell demonstrated, through a computerized text analysis of 40,000 documents between 1993 and 2013, how leading economic elites systematically orchestrated an "alternative" discourse through lobbying, *think tanks*, advisors, and politicians (Farrell 2015, see also Mann 2021).

Another factor that may have played a role in public doubts about facts and thus the postponement of concrete measures was the fundamental criticism of science by social constructivism, especially in the form of the *Science and Technology Studies* (*STS*). The basic message was that scientific statements are social constructs rather than representations of reality. Science is based on culture-specific assumptions, social contexts, linguistic conventions, and institutional, often powerful structures. Therefore, one should always be critical of knowledge gained in this way. However, the "… flip side of this critical concept of knowledge is a relativism that can be used to question scientific knowledge in general" (Thiemeyer 2020: 71). Some concluded from this that scientific statements have no other value than (other) statements of faith. And accordingly, scientists in general, especially natural scientists, were disparaged in the *Science Wars*.

Twenty years after his ethnographic-sociological field studies on "science in action" at the *Salk Institute for Biological Studies* in San Diego, Bruno Latour, the *STS*'s key figure, admitted self-critically in an interview that he was not against science at the time; it just felt good to disparage the natural scientists a bit. He attributed this to his excessive youthful enthusiasm (DeVries 2017). Thomas et al. argue that this exaggerated criticism of science from the cultural sciences played a role in the non-perception of Anthropocene change and the resulting inactivity:

> These attacks on science were damaging. Unintentionally, the interests of business and politics, social constructivists, and neoclassical economists converged to delay actions recommended by the Club of Rome and others alarmed by the evidence of environmental degradation (Thomas et al. 2020: 147).

Whether the critical concept of knowledge and the emancipative idea of science criticism have actually played such a strong role in the denial of anthropogenic climate change, as Thomas et al. have suggested, and the extent to which they produce anti-enlightenment attitudes and other "unappetizing kinships" such as conspiracy theories (Latour 2004) would, in my opinion, still need to be empirically clarified. Thiemeyer characterizes the role of radical science criticism in the climate change debate in a more considered way:

> The populists and deniers of climate change are an *extreme manifestation* of a skepticism and ignorance of science, the arguments of whom are derived *not least* from an epistemology that has questioned the claim to universal validity of scientific knowledge *with good reason* (Thiemeyer 2020: 78, emphasis mine).

Nevertheless, some recent contributions from cultural anthropology point to the dangers of an exaggerated social constructivism in the face of the Anthropocene. A leitmotif of many anthropological contributions is the demand to historicize and contextualize scientific findings or concepts (Pálsson et al. 2013). However, this should not go so far as to consider threshold values or tipping points merely as a "social construction," as even anthropologists, who are not at all suspicious of thoughtlessly following scientific models, object (Danowski and Viveiros de Castro 2019: 17).

4.5.4 Alternative Economic Forms

Nonetheless, there is indeed a wealth of approaches to alternative forms of economic policy, which should at least be mentioned here (according to Thomas et al. 2020: 149–161). In simplified terms, two basic economic approaches can be distinguished here: environmental economics and ecological economics. The aim of both approaches is to introduce nature into economic theories and models, and both use the term "ecosystem services," but they differ fundamentally in terms of the conclusions they draw from the reality of the Anthropocene, as they do in their proposed solutions.

Environmental economics treats nature as a subsystem of the economic system and its aim is to rely on market forces for the protection of the environment. Environmental economics is thus a branch of neoclassical economics. The value of the environment must be determined economically in order to make it economically visible and thus enable growth while maintaining sustainable development. Market forces would ensure that the economy "dematerializes" through innovation.

Thematically, CO_2 emissions are a major focus for environmental economists. Environmental economics thus aims to integrate nature into the market without significantly changing the fundamental values and political goals. A crucial assumption in relation to the Anthropocene is that the services of nature, just like the results of human activity, are seen as "products" that are in principle interchangeable.

Ecological economics, on the other hand, treats the economy as a subsystem of nature and relies on social objectives and political decisions for environmental preservation, such as the conscious limitation of growth. Representatives of ecological economics insist that the "planetary boundaries" represent actual limits, which is evident in the fact that many of these limits have already been reached. Programs such as "green growth" or "sustainable growth" are therefore unrealistic because the planet is limited. While the goal of environmental economics is to subsume nature to the market, the aim of ecological economics is to subordinate the market first to nature, and second to politics.

Thematically, ecological economics, such as that of Robert Costanza, is more broadly oriented, which makes the approach more relevant to the Anthropocene than environmental economics. In addition to CO_2 emissions, issues such as land use, water scarcity, and the collapse of entire habitats are also addressed (Brown and Timmerman 2015; Figueroa 2017). This stance argues that improved technology and more conscious consumer decisions by individuals are important but not sufficient, and that it is therefore a matter of changing values and new forms of political *governance*. This is in line with a consensus among many critics of the Anthropocene idea, but also of leading representatives (such as Steffen et al. b). The approach is situated outside the economic mainstream and is strongly associated with civil society programs and attempts at post-growth or *degrowth* (Schmelzer and Vetter 2019; Kallis et al. 2020; Bruns 2020).

However, there are certainly connections and overlaps. Costanza and colleagues originally introduced the concept of ecosystem services in the 1990s as an eye-opening metaphor to criticize market fundamentalism. The term was later absorbed in environmental economics, but used for something completely different (Thomas et al. 2020: 150). While an ecosystem service for ecological economics is the contribution of an environmental compartment to a larger ecosystem, for environmental economists it is the quantifiable benefit that humans derive from an ecosystem. From this it was deduced that it is the market that can protect the environment from excessive use, not environmental policy. For example, the market would ensure that the consumption of fossil fuels would decrease and that wind power and solar power would be introduced.

From the perspective of ecological economics, the following insight is more important than ever in the Anthropocene: The market cannot value goods and services that cannot be substituted and for which there is therefore no exchange value. Without daily drinking water, we as humans can no more live than humanity can exist in the long term without the climate-regulating effect of phytoplankton or rainforests (Thomas et al. 2020: 154). In addition, companies can hardly afford truly comprehensive innovations due to the financial risk involved. There is also a fundamental temporal *mismatch* between the metabolisms of the market and nature.

The metabolism of the market and the metabolism of life are intertwined, but the metabolism of the market is faster: too fast (Hamilton 2015: 35; see Sieferle et al. 2011). Ecological economists emphasize (a) that societies cannot be theorized separately from nature, (b) that natural systems have growth limits, (c) that we should strive for well-being instead of growth, (d) that social equality increases average well-being, and (e) that societies that pursue equitable sharing cause less anthropogenic damage (partly according to Thomas et al. 2020: 150).

Even though the Anthropocene is objectively something much more comprehensive than climate change, the latter is the paradigmatic example and therefore much of the research on the Anthropocene is focused on it. Accordingly, criticisms are first mostly related to anthropogenic climate change, and second, they are critical of a capitalist orientation of the economy that permeates all levels of life, and third, critics mostly follow ideals of human equality and justice. In relation to political measures aimed at curbing the Anthropocene, it must be noted that protection against further anthropocenic change does not easily align with goals of global equality. This is evident in the climate: the lower standard of living in poorer countries and regions, caused by inequality and injustices, is currently keeping global CO_2 emissions within limits:

> Let's imagine the counterfactual reality of a more economically just world, with the same number of people and based on the exploitation of cheap fossil fuels. Such a world would undoubtedly be more egalitarian and fairer, at least in terms of income and wealth distribution – but the climate crisis would be even worse! Ironically, we owe it to the poor, that is, to uneven and unfair development, that we do not emit even more greenhouse gases into the biosphere than we already do (Chakrabarty 2014: 11, trans. Horn 2017: 17).

Fulfilling the legitimate demands for a higher standard of living would thus exacerbate Anthropocene trends. This is already evident in Asian countries such as China, Thailand, and Indonesia, where the consuming middle class is growing rapidly. The demand for global social and ecological justice is objectively as well as morally justified. Contrary to the pathos with which justice is often put forward as a goal, as in the United Nations, it is not an easy moral remedy against the Anthropocene.

4.5.5 Exceptionalism and Feasibility Ideology

The term "Anthropocene" reflects the species category of humans. A central criticism of the rhetoric of the Anthropocene concerns the extreme emphasis on human agency. Critics argue that the dominant grand narrative of humans as the greatest destroyer, but also savior, must be set against counter-narratives under the motto "undoing the Anthropocene" (Barca 2020: 18–58). Even geology has criticized the fact that the very population of industrial societies designates itself as the cause of a geological epoch, although the permanence of human influences on the Earth is by no means assured. The concept of the Anthropocene as a (geological) "age of humankind" initially entails the danger of human overestimation. It could be seen in an ascending series after the "age of fish," the "age of reptiles," and that of mammals, which would correspond to the outdated view of the "Great Chain of Being" in biology (Gilbert, in Haraway et al. 2016: 540, see Gould's critique 1987). The argument for a separate epoch is oriented towards geological measurements and particularly focuses on the aforementioned "stunning figures" from the natural sciences, such as the impressive "hockey stick curves." From this perspective, the Anthropocene thesis overestimates the importance of humans by focussing on people and cultures and current conditions. In terms of scope, that is, spatially and geosystemically, the actual effects of humans are significantly less than in such an anthropocentric view.

Critics argue that humankind now comes across as a Titan or Prometheus, or a world created by the *Homo faber*, controllable (*mastery*) and also self-accountable. The role model is Goethe, who speaks of the "human-made Earth," of which humans can say, "it is my Earth" (Werber 2014: 243). From the critics' point of view, historians, for example, say that the environments on our planet are now so strongly anthropogenically shaped that human beings can no longer be compared with other genera and species. Critics consider this to be speciesism or exceptionalism (Uhrqvist and Lövbrand 2014; Mathews 2020: 69). Prototypical of this view are the conference reports from the Earth System Sciences of the early 2000s. There, humanity appears as a (sic!) subsystem of the Earth system, as a generalized natural force of global change. This was graphically reflected in the Bretherton diagram (National Research Council 1988: 29–30; Pálsson 2020: 55), which I will discuss in Sect. 7.6. The exaggerated view of human uniqueness also appears as a dystopian narrative when humans are seen in a victim role, which is why Kersten calls for a "de-escalation" in order to be able to see positives in human complexity once again, instead of the deplorable human species (Kersten 2013: 51).

Another criticism is that the Anthropocene as an idea normatively emphasizes human appropriation of nature instead of its integration into a comprehensive living environment. The approach promotes a delusion of feasibility (*human mastery*, Bubandt, in Haraway et al. 2016: 545; Mathews 2020). If humanity is seen as the cause of global change, then humankind and the human-made science are now also seen as the cure. Instead of humanization, this approach pushes for hominization. The approach is caught up in modernist thinking, which is evident in technicist and perfectionist solution proposals.

Critics mainly interpret ideas of geoengineering and the "good Anthropocene" as part of an arrogant approach of the *technological fix*, as a vision of a euphoric creative power. The self-proclaimed "earthmasters" see a civilization that is compatible with nature, but which is in fact as technoid as it is politically unrealistic (Launder and Thompson 2010; Brand 2010; Hamilton 2013; Wiertz 2015b; Dickel 2021; Neckel 2021: 161). However, this criticism tends to overlook the fact that technological responses to the Anthropocene do not only come from capitalist or industry-friendly circles, but often also from civil society (examples in Thomas et al. 2020: 181–186). Moreover, not all proponents of technological responses see them as a "solution" (*fix*) that would otherwise allow us to continue with *business-as-usual*.

In connection with this, anthropologists in particular criticize the fact that nature is conceptually understood as externalizable in the dominant understanding of the

Anthropocene (Moore 2015a). Nature would thus be thought of as accessible for technological interventions and would therefore be *made* accessible. Instead, it is also argued, nature should be seen as "unfathomable" (Bauer and Bhan 2018: 6). Of course, such a conception of nature will not exactly make cooperation with natural scientists easy, a point which I will discuss in more detail in Sect. 5.5 using a critique of the concept of the *patchy Anthropocene*. Haraway struck a similar chord when she said that the concept of ecosystem services is similar in that a utility-oriented model presents the Earth as an *accounting system*, as a tool for capitalizing the planet. This goes hand in hand with a distanced and globalist view of the Earth, and this also shows the implicit geopolitical positioning of the concept in the era after the Cold War and after the space travel hype (Haraway, in Haraway et al. 2016: 538). However, Haraway did not say which of the above-presented conceptions of ecosystem services she was highlighting: Was she referring to services for ecosystems or only for humans?

Some critics diagnose the *supposedly* value-neutral, purely descriptive concept as an expression of this specific model of civilization, namely a "feasibility ideology," which ultimately culminates in post- or transhumanism (Loh 2018). An anthropologically analyzed example of the overly political use of the term Anthropocene criticized here is provided by forms of politics that Chloe Ahmann has characterized as "subjunctive politics" based on debates about waste disposal, gas emissions, and the energy industry in Baltimore, USA. Typical of such Anthropocene-framed politics is the use of conditional language registers that reduce an initially broad field of potential future developments to a few alternatives that appear "actionable" (Ahmann 2019: 11).

The counter-proposal is a "new human ecology," the task of which, according to theologian Jürgen Manemann, who is mainly concerned with the debate within Germany, would be to fundamentally renew societies and work towards creative structures in order to develop basic capabilities that allow for a humane life despite all catastrophes (Manemann 2014). Critics complain that "Anthropocenists" see humans as a Titan, Prometheus, and as a "world builder." The idea of human beings or humanity as a quasi-global subject runs the risk of thinking in purely utilitarian terms and of drawing certain political conclusions almost automatically. Instead of the hominization of the world demanded in the face of the climate catastrophe, the focus must be on the profound humanization of humankind. Spaces must be created for others and other things, because otherwise the overemphasis on human agency could easily result in a dehumanized world (Manemann 2014: 15–35, 99–107).

According to a further criticism, this megalomania proves Anthropocene thinking to be anthropocentric or, more generally, speciesist (Malm and Hornborg 2014; Todd 2015; LeCain 2017: 311–328; Hornborg 2017a; Bauer and Ellis 2018). Other critics complain that humanity appears to be the only agent, while the role of trees, for example, in carbon fixation and thus their impact on the greenhouse effect is overlooked (Demos 2017; Bauer and Bhan 2018: Chap. 2). The problem with the Anthropocene concept, according to Haraway, is that the Anthropocene is a humanly focused construction ("a species act"), but this concept characterizes a state of the world whose peculiarity lies precisely in the systemic entanglement of human with non-human entities: "It is not just a human species act" (Haraway, in Haraway et al. 2016: 539, similarly Manemann 2014).

From a geological point of view, however, humankind tends to be rather reduced, becoming "only" one among other geofactors. Werber believes that this "depotentiation" of humans to an actor in a network of actors is not always apparent in the Anthropocene rhetoric. After the three narcissistic insults by Copernicus, Darwin, and Freud, humans once again appear to be on top, as world builders, of the *human-made world* (Werber 2014: 243; Bajohr 2019: 66). This applies to some of its exponents (for example, Lewis and Maslin b; Ellis 2018). Ellis, for example, explicitly advocates a reversal of the grievances and sees the Anthropocene as a "second Copernican revolution" that puts humans back at the center, as Schellnhuber already noted with regard to climate change (Schellnhuber 1999).

An exaggerated "exceptionalism," which sees humans as the sole actors generating the Anthropocene, must therefore be countered by the fact that a large part of the Anthropocene effects are actually due to the interaction of a variety of things, organisms, and humans as one of these organisms. Bauer & Bhan illustrate this using the example of the materially produced anthropogenic climate change ("materializing climate change") and therefore speak of "... *sociomaterial histories*, which are conjoined products of actors that fall on both sides of the human-nonhuman-divide" (Bauer and Bhan 2018: 5. Emphasis in original).

4.5.6 Technicist and Macrosystemic Problem Framing

Accordingly, methodological approaches, especially those of the Earth System Sciences and *global change* research, are also criticized. In addition to the above-mentioned monocausal forms of argumentation and the fixation on figures, curves, and extremes, the Earth sciences are particularly criticized for always wanting to draw and stare at "the big picture." This is evident in the images used, in scales, in the forms of surveys, in certification forms, and in empirical framings. Especially when globally oriented political projects are linked to this, only the easily quantifiable variables are used instead of the more qualitative social or cultural variables. Moore identifies the guiding framing as a form of "socioecologics," namely human–environment–non-human relationships brought together as coupled systems in the explanations of Earth system dynamics and in approaches to resource planning (Moore : 38).

Moore argues that these socioecologies each construct special forms of justification in an "ecobiopolitical milieu," which can bring disciplines together, but also divide them. As examples of such effective problem framings, which also serve "Anthropocene markets" such as the trade in luxury lobsters, she cites two systemically oriented approaches, Holling's complex adaptive systems theory (Holling 1973) and the *socioecological systems theory*, for example, by Fikret Berkes and Carl Folke. However, Moore does not address the vast differences between the two approaches, such as the fact that Holling pursues a macrosystemic resilience approach, while Berkes takes a conservation approach based on local experiential knowledge (Holling and Gunderson 2010; Berkes 2003; Berkes et al. 2003).

In line with the harsh criticism of the focus on technology, there is also criticism of currently dominant ideas about biopolitics and biopower. They are regarded as anthropocentrisms, namely as sovereign anthropocentrism or "biocentric metaphysics" (Rabinow 2008; Rabinow et al. 2008; Povinelli 2014, 2016). According to Moore, such "sovereign anthropocentrism" could be addressed with a concept of *ecobiopolitics*, as it is used in medical anthropology. Such an approach is dedicated to techniques of knowledge production for the optimization of human and environmental processes. In relation to the Anthropocene, for example, the idea of "habitable space" could be analyzed at various scale levels (Messeri 2016; Moore : 37, 2019: 136–142; Chakrabarty 2020b). This could range from Amelia Moore's region of the Bahamas as a "rich source of research" to fundamental Earth System Science considerations, for example, concerning the basic conditions of human habitable and non-habitable spaces on other planets ("Exo-Earth System Science," Lenton 2018: 139).

I only partially agree with these criticisms, because side and late effects and the spontaneous emergence of change at tipping points (Barnosky et al. 2012; Lenton et al. 2019) are not only discussed very controversially with respect to the unintended effects of human action in the past, but also for human intervention considered as a countermeasure today. When reading contributions, for example, by the WBGU or by "Anthropocenists" such as Reinhold Leinfelder, one finds that the position of the feasibility ideology has not been assumed by critics, such as the theologian Jürgen Manemann (as also Paál 2016: 4). In response to Werber's criticism of the exaltation of humanity as "maker," it must be stated that such a position does not reflect the majority of scientific voices in the debate. Even Reinhold Leinfelder, whose work "Unswelt" also illustrates this position for Werber, emphasizes that humanity causes fundamental changes, but these are primarily *not intended*. The psychologist and chemical engineer David Kidner argues that classifying human environmental behavior across the board as anthropocentric underestimates the unintended effects of the industrial system. Instead of anthropocentrism, it is necessary to combat industrocentrism (Kidner 2014: 477). In this line, one can also say that the term Anthropocene also establishes a productive distance from the term "human": human in the reductive sense of the Enlightenment, as humans who have modernistically "conquered," industrialized, and state-organized nature and thus unintentionally brought about the disaster of current world dynamics (Tsing, in Haraway et al. 2016: 541).

Feasibility ideology and macrosystemic problem framing have been criticized by anthropologists and other social scientists long before the discussion of the Anthropocene. This applies especially to development policy and particularly to development projects. A classic example is the study of the systematic failure of large-scale development projects by the historically interested political scientist and anthropologist James Scott (Scott 1998). The failure of state-designed dam construction projects, for example, is typically based on thinking in terms of blueprints and large scales. The assumption is that there are general solutions that could be realized regardless of local circumstances, and that large projects would have large positive effects. The problem arises from a lack of fit between projects that are conceived at the macro level ("seeing like a state") and the local ecological and socio-cultural reality of life (Eriksen's *mismatch*, 2016: 21).

4.5.7 Actor Idealism—The Example of the Think Tank WBGU

The Anthropocene discourse notoriously departs from the original path of taking stock of scientific knowledge. Many social scientists argue that it is the central task of the social sciences to deconstruct and unsettle this very narrative. Especially from the perspective of interpretative critical approaches, it is about problematizing the assumptions behind the narrative and thus opening up the concept to a variety of interpretations (Lövbrand et al. 2015: 211).

A central criticism is that the narrative of the Anthropocene depoliticizes the development problems of our world. The narrative is a post-political narrative, namely a depoliticizing narrative about human resource use, planetary boundaries, threat to life, and the urgency of global environmental protection. This concerns not only the public debate, but also the sciences themselves. "Scientificization" entails forgetting the historically evolved and socially particular conditions of one's own knowledge, which Bourdieu refers to as the "scholastic fallacy" in relation to academics in general and which I see as a form of nostrocentrism, the unconscious universalization of one's own worldview.

A "normative overstretch" (Neckel 2020: 190) in research on the Anthropocene is as such a social fact that needs to be analyzed sociologically. In particular, the normative orientation often remains implicit in research on the Anthropocene, which is particularly evident in climate research. From a sociological perspective, biotopes and sociotopes are easily equated here. A prevailing idea is that "… political systems could analogously process the Earth system and make collectively binding decisions within the framework of an Earth system analysis" (Neckel 2020: 162). Kersten (2014) and Bruns (2019) illustrate this with WBGU's project "World in Transition—A Social Contract for Sustainability."

The panel of experts postulates a profound but inevitable transition of the economy and society that is currently lying ahead of us as a "great transformation." Based on the diagnosis of the fundamental shortcomings of the carbon-based economic model and the changes that are already taking place, a new "social contract" for a "great transformation" is called for. It would essentially require a modern Leviathan, a new *social contract*, but this time as a worldwide agreement to tackle development alternatives for "a good life within the limits of the natural environment" (WBGU 2011a, b: 1–2, 5, 27; Biermann et al. 2012; Galaz 2014; Pattberg and Zelli 2016; Nicholson and Jinnah 2016; Schneidewind and Mitarbeiter 2019; Hickman et al. 2019; Pattberg and Davis-Venn 2020).

In one year, the world burns as much oil as was produced from algae and plankton in 450,000 years (Steininger and Klose 2020). The WBGU's diagnosis of the world's problems fundamentally highlights the "fossil-nuclear metabolism" of industrial society and specifically the deficits in the implementation of global environmental policy measures, arguing that these problems could be remedied through more knowledge, more accurate data, and technological innovations relating to the Earth system. The WBGU considers it urgently necessary to intervene deeply in the organization of human coexistence. In terms of the depth of intervention, this necessary transformation is comparable to the two fundamental historical upheavals in world history, the Neolithic Revolution and the Industrial Revolution or Polanyi's *The Great Transformation*. Specifically, the optimal "navigation" within planetary guardrails and the role of politics and governance are discussed. However, the guiding objectives are fundamental, especially the global Sustainable Development Goals (*SDGs*). Critics of these say:

> In view of such a diagnostic framing, which lacks a deep engagement with social and political science insights – which, among other things, point to the causes of the multiple crises – there is still a way to go before the Anthropocene as a new conceptual figure also reaches and transforms the epistemological basis of Earth System Sciences (Bruns 2019: 58).

In policy-oriented projects and programs, human responsibility for effects of the Anthropocene is emphasized. Here, the idea of feasibility is often combined with an actor idealism and the unity concept of humanity. In the WBGU's reports, the figure of the *anthropos*, the universal human, is repeatedly invoked. Politically, humans appear there not only as a global citizen, but enters the stage as a political actor, as "global citizenship" (WBGU 2011a: 8 ff.). In contrast to the positions of classical cosmopolitanism (see Chap. 7), it is now supposed to take collective responsibility for planetary risks according to a contractual model, for example, in order to cope with climate change (critically on this Kersten 2014: 381 ff.).

This idea of a global political subject, however, is wishful thinking. The global community is not a single entity, but is rather, despite economic globalization, profoundly unequal economically, divided along religious lines, and, above all, politically fragmented. The findings of a worldwide Anthropocene threat cannot easily be used to infer the emergence of a global actor. Sociology knows that collective actors do not simply come into being as a function of their necessity. The idea of an effective global citizenship, which practically manages or controls the planet on behalf of humanity, is an actor idealism far removed from the world (Neckel 2020: 159–160).

In light of the realities of the EU or other regional organizations, such as ASEAN, or the problems of establishing human rights, it is clear that wishful thinking is being normatively overstretched here. An example is the failure of Web 2.0 as a global network culture, which promised the emergence of universally shared values, but has in fact become a battleground of contradictions (Kersten 2014: 388). Critics emphasize that this is an unrealistic idea, and it has long been abandoned by the social sciences and humanities as an idealistic fiction (Bajohr 2019; Bänzinger 2019).

In policy-oriented environmental research, there is a widespread belief that it is possible to draw conclusions about the right policy from Earth science findings. Such a naturalistic fallacy can be criticized in the argumentation of the WBGU for the necessity of a "global transformation." In the 2011 report (WBGU 2011a, b, see WBGU 2020), for example, with the goal of facilitating a joint development of the atmosphere and the anthroposphere, an alignment of the respective global social product with the global average temperature is called for. On the one hand, this is not feasible as it consists of aggregated values of the United Nations Conference on Trade and Development (UNCTAD). This means that the gross world product is not accessible for control by any authority in this world (Neckel 2020: 159). However, much more fundamental is the problem that, with reference to the Anthropocene, a scientific category is being transferred to the realm of the social (Kersten 2014: 384–386). It could also be noted that the protagonists of the "great acceleration" have brought the political explosiveness of the Anthropocene to light, albeit in an unintended direction. While the representatives want to point to the urgency of political action, the current graphs suggest that the efforts of explicit environmental and sustainability policy since the 1970s "have not been able to change a single bit" in the acceleration. Christoph Görg describes this inability of the institutions as the real scandal of the term Anthropocene (Görg 2016: 10, 11).

4.5.8 Depoliticisation and the New Geopower

Kersten (2014) and Bruns criticize the depoliticization of debates on global problems with regard to the concept of planetary boundaries (Rockström et al. 2009; Steffen et al. b) and its political implementation in the work of the WBGU. Bruns notes that in the WBGU's argumentation, the supposed singularity of the global almost inevitably leads to the demand for a "great transformation" (Bruns 2019: 58). One example of the thinking behind this is the idea of human responsibility for nature, an orientation of political control derived from it (*Earth stewardship*, Steffen et al. 2011a, b; Ogden et al. 2013; Collard et al. 2015; Folke et al. 2021; Keys et al. 2023). Critics see the Earth scientists here as the good "scientific shepherds" who are now (supposedly) taking care of the planet in continuation of earlier environmental policy. In fact, this is a form of "geopower" (Bonneuil 2015: 23; Bonneuil and Fressoz 2016; Sprenger 2019).

Here, talk of the "good Anthropocene" has particularly come under fire—an optimistic approach that seeks positive solutions for a nature-compatible civilization and often envisages technical measures to achieve this, prototypically a conscious influence on the atmosphere. The main criticism is the fact that its proponents think they can solve everything ("solutionist science"; Asayama et al. 2019). Critics, who consider this as an ideology, as well as proponents, refer to this approach as ecomodernism. Criticism is mainly directed at their naive trust in technological solutions, the homage they pay to a so-called techno-fix (Wilson 2016: 85; Sklair 2019). Haraway and Tsing consider such attitudes to be the outgrowth of narrowed

Western thinking and therefore proposed the term "Euclidocene" to highlight this hubris (Tsing, in Haraway et al. 2016: 546). Contrary to some criticisms, it should be noted that proponents of the "good Anthropocene" are by no means only found in the politically conservative spectrum (Meyer 2016). A more profound criticism states that proponents of the good Anthropocene continue to think in terms of human and technological systems, whereas it must be made clear that nature is now active in a de facto socio-ecological system (Dryzek & Pickering 2019: 10, similarly, Fischer-Kowalski and Haberl 2017; Haberl et al. 2019).

At the centre of the much-criticised human hubris is an unreflective and infinite confidence in technical solutions, including for global environmental problems. This is underpinned by a quasi-religious exaggeration of science, which turns the Enlightenment into a myth (Fladvad 2021: 144, 159). The criticism of human over-confidence also comes from deep ecologists, whose views hardly appear in most publications on the Anthropocene. Deep ecology is a normative, political philosophy that wants the best for nature *and* humans and strives for a comprehensive global ecosystem in which humans fit. A prerequisite for addressing global environmental crises would be fundamental changes in the relationship between humans and nature, as postulated by an ecological philosophy anchored on the body (Abram 1996; Naess 2013, 2016). At the level of concrete measures, deep ecologists think about systematic rewilding. What deep ecologists dislike about the idea of the Anthropocene is that the thesis is often associated with the claim that the Earth as an independent entity no longer exists without human alteration (Rothenberg 2018). This assumption is based on human hubris, the idea of human supremacy. Deep ecology, on the other hand, assumes that nature will survive human-made damage despite the current wave of extinctions because life on Earth will recover after millions of years, as with previous mass extinctions. Evolutionary ecology would argue that such recoveries in the Earth's history have each lasted many millions of years.

The cause of the depoliticizing character of the dominant narrative is seen by many critics in scientism, insofar as the Anthropocene is predominantly described in scientific terms. Accordingly, the idea is not only a powerful persuasive narrative, but a "persuasive *science* narrative of escalating human-induced environmental change" (Lövbrand et al. 2015: 212, emphasis mine). Geographers and ecologists in particular have argued that the scientific narrow-mindedness is evident in the fact that global change, within *global change* research and through proponents of the Anthropocene idea, is primarily described as environmental change rather than as social change (Swyngedouw 2007; Castree 2014c, 2015). A consequence of this is that the Anthropocene itself is not seen as a social and cultural construction and therefore the various possible responses to it are overlooked.

The dominant concepts of the Earth sciences, or more precisely of the politically effective section of these ideas, become particularly clear when they are experienced as a quasi-foreign world of thought (Horn, pers. comm.). I myself experience this time and again when I collaborate with natural scientists (Antweiler 2020): I sometimes feel alien myself, as I am an anthropologist, but have studied geology and also deal with evolutionary biology. Werner Krauß, as an ecologically active anthropologist, sums this up nicely based on his experience of a kind of culture shock when exchanging with Earth system scientists:

> The world of so-called Earth System Sciences is one of symbolic power and a more-or-less open political agenda, the world is at risk, the science is settled, and it is time for political action – this is the message in a nutshell. It is a message that is accepted uniformly (Krauß : 69).

From a cultural studies perspective, it is also a specific social situation between the years 2000 and 2010 that led to the cumulative environmental effects of human activity being perceived as a "problem" and a problem framing appearing as a global crisis. This happened particularly at conferences on global change in the context of the emerging Earth System Sciences (Uhrquist and Lövbrand 2014). As representatives of the *Science and Technology Studies* (STS) have pointed out, the specific problem framing dominated by the natural sciences does not only have consequences for the topics of further discussion. From the perspective of knowledge coproduction, it also has concrete material consequences through the environmental governance motivated by it. In short, one could say: the status of knowledge shapes states of knowledge.

This criticism of the technological orientation of natural scientists, however, overshoots the mark. A serious consideration of the possibilities of geoengineering, such as measures to artificially influence the atmosphere (*climate engineering*), does not necessarily mean that climate protection is not (any longer) necessary in the face of unstoppable global warming. In 2006, Paul Crutzen considered for the first time explicitly whether sulphate particles could be introduced into the stratosphere in order to reduce solar radiation by increasing the albedo. What is often overlooked by critics of the "technocratic Anthropocene" (such as Mathews 2020: 70) is the fact that Crutzen only saw this as a last resort, *if* we cannot get climate change under control through political and economic measures reducing consumption (Crutzen 2006a, b: 217). However, critics emphasise that Crutzen deliberately broke a taboo in his 2002 essay, where he speaks of 'managing the earth', and especially in his 2006 essay, and that his authority made him a trailblazer not only for research, but also for attempts at climate engineering (Jansen 2023: 157 f.). On the other hand, these considerations did not prevent Crutzen from advocating for political measures to reduce greenhouse gases until his death in 2021: "Concern for the Earth's atmosphere runs through Crutzen's life just as much as his commitment to environmental protection" (Schmitt 2021: 32).

The tricky thing about depoliticization is therefore the fact that Anthropocenists often propose countermeasures that can be implemented quickly due to their urgency, but forget that this would require a political decision. The determination of a global "crisis" can never be based solely on facts, findings, and diagnoses. Every therapy, to stay with the medical metaphor often used in the Anthropocene discourse, requires a diagnosis and a setting of values, namely what should be avoided as it is sick and what is considered healthy and thus desirable. In this respect, the notoriously apocalyptic representations of the Anthropocene can all too easily lead to a logic of constraints. Accordingly, one position of a critical social science would be to avoid coming up with quick solutions:

> We contend that critical social engagement with the Anthropocene does not promise any immediate solutions to contemporary environmental challenges. The research agenda advanced in this paper is more likely to unsettle the Anthropocene and to pave the way for competing understandings of the entangled relations between natural and social worlds (Lövbrand et al. 2015: 212).

On the other hand, it can of course be argued that such an attitude is itself apolitical because it remains in the ivory tower. Too strong an orientation towards solutions, for example, in the WBGU, can become a problem, but this danger of a solution orientation guided by constraints is, in my opinion, a general problem that is only intensified by the enormous force of the problems of the Anthropocene. It is by no means only the propagators of the Anthropocene idea that are subject to this danger, but also nature conservation projects (Krauß : 65) or initiatives dedicated to sustainability, the future of the planet, the "one Earth," or the "saving of the Earth," such as *Greenpeace* or *Future Earth* (Future Earth 2013). However, none of this is an objection to the concept itself. The Anthropocene as a concept is not per se technocratic or blind to questions of justice, because the urgency of these issues is precisely what brings them to the fore, as exemplified by Hurricane Katrina (Dryzek and Pickering 2019: 58–81).

4.6 Bourgeois Universalism—Generalization of Responsibility

Not all conceivable human and non-human actors are involved in a wide range of qualitatively new environmental dynamics. Here, different effects and also different temporal and spatial levels must be distinguished. Many critics argue that proponents of the Anthropocene idea lump all people on the planet together by referring to "humans" as a geological factor. When humanity is seen as a geological factor, social diversity and cultural differences are reduced to a monolithic humanity in a kind of "post-social ontology" and are treated as quasi-natural (Lövbrand et al. 2015: 213; Mann and Wainwright 2018: x). According to this criticism, the anthropocentrically colored idea of the Anthropocene overemphasizes some actors while excluding others.

> As a uniform and indiscriminate planetary process, the Anthropocene can only be understood from the perspective of a scientific aggregation of data, which becomes a scientific reality when looking at the Earth as a whole, but which does not apply as such to any single place on Earth and also does not call for the same measures everywhere (Neckel 2020: 163).

4.6.1 Overlooked Inequalities and Conflicts

Critics complain that the term Anthropocene, with its rhetorical "We," generalizes the causation and thus the responsibility for human-induced effects, "… where guilt and suffering are highly unevenly distributed" (Macfarlane 2016: 96). Central impact factors, such as the fact that average Americans consume 500 times (!) as much energy as Ethiopians, are ignored. The criticism applies in terms of socio-economic inequality, but also gender differences (Merchant 2020a: xiii, 145–146, 2020b).

The key message is that under the term Anthropocene, a problem framing is undertaken in order to iron out inequalities. This is a particular aspect of depoliticization. A depoliticized view overlooks the inequalities of causes and consequences between rich and poor, men and women, Global North and Global South, urban and rural areas, and also within increasingly fragmented cities (Ribot 2018; Bruns 2019: 53). Regarding the causation of the Anthropocene, it would be overlooked that the actions of people in the Anthropocene do not only indirectly fall back onto their own living environment, as Latour explains in his earlier idea of the Anthropocene as a composition (*loops*, Latour 2010). No, in the Anthropocene, actions notoriously affect the lives of actors who live at *other* times and in *other* places. This makes the Anthropocene a conflict (Kersten 2014). In short, the multiple causes that plunge people differently into risks and reproduction crises, would, as in the climate change debate, also be overlooked.

Many critics see the deeper background for this narrow view in undifferentiated thinking in a species unit, namely the whole of humanity, which as a collective singular becomes a quasi-personal actor. Some critics see the overlooking of inequality both in cultural anthropology and in archaeological research, which depicts the work of entire societies as undifferentiated actors. The consequence of the criticism of speciesism in Anthropocene concepts is the demand to "fractionate the anthropos" (Bauer and Bhan 2018: 4–5, 7, 18).

This criticism applies not only to the consequences, but also to the explanation of the causes of the Anthropocene. Farmers in India and marginal settlers in urban Peru contribute differently to climate change than the inhabitants of affluent countries (Baskin 2015). Just as the Anthropocene did not effectively occur synchronously, it is also causally not a product of coordinated human action; rather, it is heterogeneous, namely it is the effect of the actions of individual groups of people or specific forms of society in certain regions (Bauer and Bhan 2018; Bauer and Ellis 2018: 209). The human ecologist Hornborg points out that even during the phase of industrialization, humanity as a collective entity never acted as a clear historical agent (Hornborg 2017b). As with the climate change debate, the unequal share of the causes and corresponding justice and goals would have been overgeneralized, omitted, or even deliberately obscured (Swyngedouw 2007; Bonneuil and Fressoz 2016; Moore 2016; LeCain 2017). All human action is socially embedded, which is why this deficiency in Anthropocene research should be countered with a "socially embedded Anthropocene doctrine" (Lövbrand et al. 2015: 214).

This is where a tricky naturalization comes into play: the Anthropocene idea *de*naturalizes climate change as being human-induced, and at the same time *re*naturalizes it as the result of inherent human characteristics rather than seeing it purely as an effect of economic structures. According to the criticism, an understanding of capitalism would instead be sufficient to explain the Anthropocene (Malm and Hornborg 2014, similarly Moore 2016; Swyngedouw and Ernstson 2018; Ernstson and Swyngedouw 2019). This reduction of the Anthropocene to the effects of the capitalist economy and its specific relationship to nature has been vehemently contradicted by others, especially by some historians and geologists, prehistorians, and generally by proponents of the "early Anthropocene" (Chakrabarty 2017, 2018; Lewis and Maslin 2018). I will go into this in more detail in Sects. 7.1 and 7.2.

Overall, this criticism also argues that the Anthropocene way of thinking ignores the fact that it is essentially about political questions: what knowledge is produced, whose needs are marginalized, and who has access to decisions about countermeasures (Bruns 2019: 54–55)? This criticism is exacerbated by the accusation that science is becoming an accomplice of powerful political actors. One example of this is the German Advisory Council on Global Change (WBGU) (Altvater 2017). Some critics align this ignoring of differences and inequalities with other forms of "silencing," as can be found in numerous contributions on globalization (Trouillot 1995).

Many critics altogether reject the invocation of the broad category of humanity as well as that of the Earth because it lacks not knowledge, but political consensus. In the arguments, however, humanity and planet are again used as categories, as in the following conclusion:

> There is no way out of these multiple conflict constellations that could lead beyond the path of one humanity out of the global upheavals. To claim this would be pure ideology in the age of the Anthropocene. But the Earth itself also offers us no point of reference for how global socio-ecological conflicts of inequality can be best resolved justly, peacefully, and without catastrophic consequences *for humanity and for the planet*. (Neckel 2020:164, emphasis mine).

I generally agree with these criticisms from the early phase of the debates until 2010. I share the critics' objections with regard to the criticism of the trade of "world citizenship," which the WBGU, for example, presented as politically necessary according to scientific findings. However, it should be reiterated that the WBGU rarely explicitly refers to the Anthropocene, but it almost always does so to climate change and other selected areas of global change, such as agriculture (WBGU 2020, 2023).

The criticism that the Anthropocene would obscure or silence inequalities of any kind with its notion of a humanity universally responsible for the current crisis (Hoelle and Kawa 2021: 657; Pye 2021: 16) persists right up to the latest contributions, including in cultural anthropology. However, the assertion of humanity as a unit of causation of the Anthropocene, which has been criticized again and again since Hornborg's 2014 essay, is not found in the narrative of Earth System Sciences. This criticism is virtually a straw man, as Crutzen's first essay from 2002 clearly states that the Anthropocene can be traced back to the activities of only about a quarter of humanity (Crutzen 2002). A large part of their recent contributions is dedicated even more strongly and also quantitatively more precisely to the diversity of regional and local effects (Ellis et al. 2020). What is true, however, is the fact that most works from the Earth System Sciences still lack a social theory of the causes of Anthropocene effects (Pye 2021: 16).

The assertion that the whole of humanity is notoriously and indiscriminately being held hostage corresponds to similar insinuations about climate change research, which are repeatedly and stereotypically presented despite all evidence to the contrary, as the Canadian political scientist Ian Angus demonstrates in detail (Angus 2016: 225–231). For example, the updated IGBP curves emphasize regional differences in the effects of the Anthropocene (Steffen et al. 2011a). They show climate change not as a direct Malthusian consequence of population growth, but as an effect of capitalist economic activity. Unfortunately, even the latest criticisms do not address the AWG's responses to criticism (Zalasiewicz et al. 2017c).

The Anthropocene concept can certainly include different contributions and responsibilities, and global injustices will indeed persist if one does not think and act within the framework of the Anthropocene. This is shown by the current debate on climate justice, where people are struggling with the problem of how universal and specific perspectives on *one* big problem can be integrated (Dryzek and& Pickering 2019: 60).

Critics mainly overlook the fact that the proponents of the Anthropocene are not claiming that humanity is acting as a whole. Rather, they are saying that it is the *cumulative* effects of human activities that have the overall effect of permanently changing the geosphere. This does not mean that humanity is in any way acting as a unified actor, a point that is also often misunderstood in Chakrabarty's argumentation. One publication from the central research circle on climate change led by Will Steffen states that today, 60% of greenhouse gases are produced by only one-seventh of the world's population, that is, about one billion out of 7.6 billion people, whereas 40% are produced by 5%, that is, by three billion people (Steffen et al. 2018).

Similarly, the "anthropo" in the term Anthropocene does *not* mean that "all humans" are the cause, but that the global changes of this epoch would not have occurred without humans, that is, they are largely anthropogenic in nature. What Earth scientists, evolutionary biologists, and also some historians say, however, is that it is the specific characteristics of humans as a very special animal, and especially the effects of his cultural ability, that make this power possible. This can be shown empirically without having to evaluate it in any way. In order to understand the anthropocene impact of humans, we need to realise that humans are animals, but very special animals. We are a dominant species that has been significantly changing the geosphere for a long time, in parts since the beginning of the use of fire at least 800,000 years ago (Vitousek et al. 1997, Ehrlich and Ehrlich 2008, Smil 2019: 205–209, Desmond 2024). This can be shown empirically without having to evaluate it in any way.

Furthermore, the Anthropocene thesis states that the effects of the Anthropocene affect all of humanity, which only means that even the wealthy cannot completely escape it and that this *does not* exclude the possibility that these effects will vary greatly in quality and quantity. The criticism of the *monocular gaze* regime, that of a monolithic humanity as the cause of the Anthropocene, therefore does not apply to the basic argument. However, the objection is justified with regard to the specific concept of the "great acceleration." The host of hockey stick curves of social or sociotechnical activities, which have been increasing exponentially since the middle of the twentieth century, are generally reproduced without any regional differentiation. Some of these criticisms have been addressed, for example, by taking into account non-OECD countries. However, publications from "Anthropocene Science," that is, from *global change* research and Earth System Science, which still like to speak generically of the "human enterprise," "gender," and "capitalism," seldom explicitly appear (Clark and Szerszynski 2021: 36, 59–60).

4.6.2 Moral Claim despite Normative Indeterminacy

If we assume a multi-species approach, Anthropocene effects *by definition* are no longer completely anthropogenic. According to this approach, actions arise due to distributed agency ("distributed quality of action"). Nevertheless, as Bauer & Bhan argue, the responsibility for anthropogenic climate change lies with humans, and exclusively with humans, although differentiated according to living conditions (Bauer and Bhan 2018: 6).

This touches on another weakness of the Anthropocene concept, the issue of values. On the one hand, it is emphasized that humanity has a responsibility to find quick solutions. This is seen as a fundamental moral mission. And from this, political necessities such as the "great transformation" are derived, as in the reports of the WBGU. On the other hand, the concept remains normatively indeterminate. What normatively follows from the diagnosis of an age of humans is mostly left open (Neckel 2020: 161). In his first essay, Crutzen already deduced from the ability of humans to ruin the geosphere that they also have the possibility to repair it. From the damage to the ecosphere, he later derived the imperative to ensure the habitability of the Earth's surface through technical measures. But even this position remained normatively undefined. The underlying values are rather implicitly revealed.

This also applies to counterpositions, such as the critique of growth, for example, in contributions on post-growth (*degrowth*). These are programs or visions for a fundamentally different political and economic system in which the material and energy throughput in societal production and consumption processes is reduced. Even if post-growth is hardly ever defined, the tenor of the argument is against ruthless hominization. Economically, it is about *less*, and socially about something *different*. The aim is to create a different social system compared to the current one, especially in the fields of social organization, social institutions, and economic growth (Bonneuil 2015: 26–28; Schmelzer and Vetter 2019: 149). At its core, it is about a coexistence that aims to achieve a balance based on solidarity and also includes non-human beings (conviviality, cf. Bätzing 2023; Wallenhorst 2023: 36–44).

In normative terms, this argues against a subordination of non-human life to human purposes. However, it is usually not specifically stated which basic values should be pursued. The focus is usually on general maxims such as responsibility, for example, Hans Jonas' *The Imperative of Responsibility* (Jonas 2020), or general values, such as the value of life or the quality of life. More specific are approaches that work on concrete implementations, such as achieving a convivial coexistence through the principle of giving or reciprocity. There is a lack of determination of priorities, and it is not precisely stated on which—necessarily human—values such a position should be based. One objection within the fundamental criticisms of the Anthropocene concept is a very widespread but contradictory double statement. On the one hand, humans appear to be nothing special (anymore) or are even worth nothing anyway. In the same breath, on the other hand, man appears to be the one who is 'destroying the earth'. However, man-made environmental change can only worry him if man is worth something, not despite, but because he is Prometheus and thus has his own environment (Sakkas 2021: 159 f., cf. Wallenhorst 2023: 117–140).

4.7 Welcome to the Neologismocene—The Many Names of Resistance

> *Love it or hate it, the Anthropocene is emerging as an inescapable word for (and of) the current moment.*
> Donna Jean Haraway (2016: 535)

> *We are all struggling for an appropriate language to help us analyze the changing human and ecological condition.*
> Simon Nicholson and Sikina Jinnah (2016: 4)

When the word "Anthropocene" is used in such an inflationary way and is applied as a label to very different things, the question arises as to whether there are more accurate terms. Since the word Anthropocene is heavily promoted by opinion leaders and organizations, it is not surprising that there have been many criticisms and, above all, alternative terms motivated by this. These terms can therefore be read as "various names of resistance" (Demos 2015, 2017: 85–112). Some suggest to simply speak of the "age of destruction" instead of the Anthropocene or the "human age" (González-Ruibal 2018). Many of the new terms mirrors the manifold criticism of the Anthropocene approach, which Baskin summarises concisely:

> First, it universalises and normalises a certain portion of humanity as the human of the Anthropocene. Second, it reinserts 'man' into nature only to re-elevate 'him' within and above it. Third, its use of 'instrumental reason' generates a largely uncritical embrace of technology. And, fourth, it legitimises certain non-democratic and technophilic approaches, including planetary management and large-scale geoengineering, as necessary responses to the ecological 'state of emergency' (Baskin 2015: 11).

Many colleagues suggest terms that are linguistically similar to "Anthropocene." The word invites the creation of similar word formations with the syllable "cene" and the term's multiplication. Mentz lists 24 linguistic creations and aptly speaks of the Anthropocene as "Neologismocene" (Mentz 2019: 57–64), and a new essay documents a whopping 91 such neologisms (Chwałczyk 2020: 23–33). One form of multiplication is the hardly translatable talk of *Anthropocenes* in the plural, particularly strongly represented by the geographer and professor of inhumanities (sic!) Kathryn Yussof with a call for "a billion black Anthropocenes"... or none at all (Yusoff 2018: 87–101).

4.7.1 Inflation of Alternative '-cenes'

The multiplication of the "-cenes" came mainly during the peak phase of the debate about the Anthropocene as a cultural concept between 2013 and 2016, as a chronological analysis of the debate shows (Will 2021: 42, 207–208). In terms of content, however, they first go in various directions, and second—in contrast to the geoscientific term—focus on the causes of the Anthropocene rather than the cumulative effects. This raises the question of whether the Anthropocene actually forms a meta- or master narrative that overshadows all other narratives of environmental catastrophe (Will 2021: 207 vs. Dürbeck 2018a, b; see Testot and Wallenhorst 2023).

The most prominent alternative term is "Capitalocene" (Malm and Hornborg 2014; Moore 2014a, b, 2021a; Parenti 2016; Hubatschke 2020), which I will explain further below. In the context of digitality and artificial intelligence, one of the newest terms has emerged—the "Novacene," an epoch supposedly following the Anthropocene, proposed by the originator of "Gaia." Lovelock sees the Anthropocene as a direct consequence of the evolution of life on Earth. The Novocene will then be a promethean epoch characterised by artificial intelligence (*novacene*, Lovelock and Appleyard 2020). Machines would soon be more intelligent than us and capable of intervening directly in Earth processes. This would enable them to protect the earth (rather than humans) from overheating. Some see this as a helpful myth (Rival 2020: 9, 12). Less optimistic ideas, but also related to communication and information, which cannot be further discussed here, guide the current emergence of the term "digital Anthropocene" (WBGU 2019: 8–9; Renn 2021, see Boellstorff 2016: *Digitocene*, see Mainzer 2020).

The majority of the terms discussed here stem from the skepticism of social and cultural scientists towards an undifferentiated or apolitical conception of the "age of humans" (Adloff and Neckel 2020: 8). Many of the alternative terms are clearly used with critical intent, and a number of terms, especially those used by historians, aim to highlight the main actors, substances, or processes (Will 2021: 206). The processes lie in the past, such as in the carbon economy (*Carbocene*, LeCain), the role of combustion (*Pyrocene*, Pyne 2019, 2020), and European dominance and the Eurocentric narrowing of the concept (*Anglocene*, Fressoz), or in the future, such as in *Machinocene*, *Robotocene*, or Lovelock's *Novacene*.

Most of these new terms originate from cultural studies and the social sciences, but many have also emerged in the visual arts, in visual media, and literature. In a widely acclaimed book about the "shock" of the Anthropocene, French scientific historians Christophe Bonneuil and Jean-Baptiste Fressoz examine how issues of environmental change have been politically charged in 250 years of European and North American history. In doing so, they distinguished no fewer than seven forms following historical phases of human impact: "Thermocene," "Thanatocene," "Phagocene," "Phronocene," "Agnatocene," "Capitalocene," and finally, "Polemocene" (Bonneuil and Fressoz 2016).

This critical motive also holds true for the term "Gynocene," which opposes the male characterization of the Anthropocene as "Androcene" or "Patriarchalocene." This is about the contribution of women to political work against the unequal impacts of the Anthropocene. In some cases, the critical counter-concepts examined here only refer to certain areas or aspects of the Anthropocene or are used for specific perspectives on it (Walton 2020), such as on questions of environmental justice (for example, Davis et al. 2019). Thus, "Gynocene" is intended to promote countermeasures through the combination of eco-feminist interventionism with indigenous-inspired ideas (Demos 2015), similar to the term "Phallocene" (*faloceno*, La Danta LasCanta 2017), used by a Venezuelan eco-feminist group.

4.7.2 "-cene"—Misunderstood Geology

The alternative designations are generally modeled after the Anthropocene in terms of word formation ("-cene") (Halle and Milon 2020). The word Anthropocene and especially its ending "-cene" often encourage wordplay. This can happen in creative ways, for example, when entities that "ignore" the nature/culture dichotomy are referred to as *anthropo-not-seen* (De La Cadena 2010, 2015; Mathews 2020: 70), or when it is advocated to criticize the *Anthrobscene* or interrupt the *Anthropo-ObScene* (Parikka 2015b; Swyngedouw and Ernstson 2018). Therefore, it is important to clarify the original meaning of the word component "-cene" in nineteenth century geology.

The commonly used rendering of "Anthropocene" as "the human-made new" or as "the human new" (for example, Gibbard & Walker 2014: 19; Egner and Zeil 2019: 18) does not capture the geological meaning precisely. Charles Lyell, who introduced the use of the component "-cene" (Greek: *kainos*) for younger rock layers, named the ages based on the *ratio of extinct to not yet extinct species* in the fossil record (McPhee 1981; Burchfield 1998). Thus, "-cene" does not simply mean "new," but rather the *degree of novelty of the fossils*. Etymologically, "-cene" is a corruption of the ancient Greek *kainos* (καινός, new, present). From this, compound words were formed: Eocene (from *eos*, dawn, beginning anew), to mark the new compared to the previous Paleocene, Miocene (from *meion*, less, less new), Pliocene (from *pleion*, more, more new), and Pleistocene (from *pleistos*, most, most new), and Holocene (from *holos*, complete, completely new).

"Cene" therefore talks about the proportion of fossils from the recent past in a period under consideration. The *Mio*cene contains *few* fossils from the recent past (about 15%) compared to the *Plio*cene, which contains *more* of them (almost 50%), and the *Pleisto*cene, which contains the *most* (more than 90%). Finally, rocks from the *Holo*cene, introduced in 1885, are almost *entirely* composed of fossils from the recent past. The "cene" in "Anthropocene" refers to a time interval whose *rock layers* are or will be (since it is still ongoing) *dominated* by residues of *recent* human activity.

However, the proportion of newness can no longer be linguistically increased compared to the "completely new" (Holocene), because "cene"—contrary to popular belief—does not mean "epoch" or "age," but "new." If the Holocene already contains "completely new" fossils from the recent past, no increase is possible. From a linguistically point of view, Anthropocene is the wrong word (Visconti 2014: 382–384). More appropriate would be the term "Anthropozoic" (Markl 1986, 1998: 23, 162; on this Thober 2019). This would be a word formation suitable for the ending "-zoic" ("animal age"), which, however, originates from biological and not geological terminology and thus denotes the dominance only of animate nature by humans. Linguistically, this would correspond more to the German "Zeitalter des Menschen" (age of humankind) than "Anthropocene." It is therefore based on a misunderstanding when harsh critics of the Anthropocene as a concept partially accept the word "Anthropocene," but then linguistically criticize it and linguistically misinterpret it, namely as the "new human" in contrast to the Holocene as the "completely new" (as in Foster et al. 2010). The following objection, based on the assumption that we are probably bringing the Holocene to an inglorious end, also misses the point of the syllable "-cene":

> So no, not the Anthropocene. That name completely muddles the message. We don't name new epochs after the destructive force that ended the epoch that came before (Moore 2013).

Most of the following terms do not refer to the above geological meaning of "cene," and it is often difficult to discern where the actual empirical gain in knowledge or analytical trick lies, apart from the "a-ha" effect. Ian Angus, an eco-socialist and editor of a project on the history of socialism, believes with some justification that terms such as 'AnthropoObzön' (Anthrobscene, Parikka 2015b) can only be understood as a joke (Angus 2016). In the following discussion, it is important to realise in advance that the alternative terms to the 'Anthropocene' often—I would even say: mostly—address socially and politically important, but essentially non-geoanthropocenic topics, as Jan Zalasiewicz, a leading member of the *Anthropocene Working Group*, clearly points out:

> These terms do not, however, 'supplant' the 'geological' Anthropocene, as they represent different concepts, from different contextual backgrounds, with social science interest on the socioeconomic drivers of change rather than on resultant Earth system behavior and its petrified and strata-bound consequences (Zalasiewicz et al. 2018a: 221).

These alternative terms cannot replace the basic concept, but they were inspired by it and in some cases work creatively with it. Despite the overwhelming diversity of terms that I am discussing here, it should not be forgotten that they were inspired by Crutzen & Stoermer's short paper at the turn of the millennium, especially Crutzen's *one-pager*. This one-sided paper, but in my opinion anything but one-sided, served as a tremendous impetus for cultural studies, which is now also recognized in the history of science (Merchant 2020a: 1–3, 16; Will 2021).

4.7.3 Capitalocene

The core idea behind Capitalocene is that we live in a capitalist social formation and not simply in "society." The Capitalocene is the phase of history that has been shaped by the capitalist system..Historians have shown that the accumulation of capital increased by a factor of 134 between the years 1900 and 2008. However, historians argue against placing capital (no more than "humanity") at the center as a universal category. Instead, the historically concrete "metabolism" in the capitalist world system of the last 250 years should be taken into consideration. This would link the *history* of capitalism with the *genesis* of the Anthropocene Earth system (Bonneuil and Fressoz 2016: 222–223).

From a historical-materialist perspective, representatives of this school of thought point out that the Anthropocene represents a particular social relationship to nature that is due to the material consequences of the exploitation of nature and the use of cheap human labour. In Marxian materialism, human history is primarily a history of humankind's behavior towards nature (Herrmann 2016: 48):

> Labor is, in the first place, a process in which both man and Nature participate, and in which man of his own accord starts, regulates, and controls the material re-actions between himself and Nature. He opposes himself to Nature as one of her own forces (Marx, Capital, Vol. I, 3.: 192).

The starting point is the criticism that the term Anthropocene portrays geo-global change as the result of an anthropological variable or force. Or that it reduces the global environmental crisis to an effect of technology. In fact, global change is rather the result of a specific class formation that developed in the form of production relations as early as the "long 16th century" (Moore 2019: 596, 2019: 261–295). The cause is not the actions of humanity, but the economic system and in particular the exploitative relationship with nature that began with early capitalism. Therefore, the factually unequal distribution of responsibility for the disaster of the Anthropocene should not be obscured (Swyngedouw and Ernstson 2018; Brunnengräber 2021). Moore understands capitalism less as a socio-economic system than as a historically specific *form of organization of nature*, as a relationship with nature that he refers to as "world ecology" (Moore 2016: 6).

The term "Capitalocene," first used by Andreas Malm in a seminar in 2009, is intended to highlight the fact that the Anthropocene is not natural or represents something supra-historical. The terminology of the "Anthropocene" obscures the historical role of capitalism. The term is used in particular by authors who stand in the tradition of Marxism or are critical of the dominant political economy (*Capitalocene*, Malm and Hornborg 2014; Moore 2014a, b, 2016; Altvater 2017, 2018; Bassermann 2017; Foster 2020; Foster and Clark 2020). However, the term

Capitalocene is also used in works critical of colonialism, racism, and in works that criticize the idea of the Anthropocene from a gender-theoretical perspective (Vergès 2017; Merchant 2020a: xiii; Baer 2017 as an overview).

Some advocates of the Capitalocene concept set the beginning of the Anthropocene at the time of early capitalism. During Europe's global expansion in the "long 16th century," Europeans established a hegemony of pre-industrial capitalism (Moore 2016). Humans were able to dominate the geosphere due to mechanistic science promoted by experimentation and mathematics that emerged during the scientific revolution of the seventeenth century. This required the steam engine and resource exploration as innovative technologies. However, the decisive factor in global expansion was the use of fossil capital in the form of coal, oil, and gas, organized in a capitalist manner through socio-economic relationships, and the control of cheap labor in the form of slaves, migrants, and poorer classes. Thus, it can be said that the Anthropocene subsumes the Capitalocene, but that the Anthropocene is nevertheless implemented by the Capitalocene (Merchant 2020a: xiii: 23–24).

Jason Moore, the most publicised proponent of the Capitalocene thesis, accuses earth system scientists of using the conceptual duo of 'humans and the environment' to add a 'popular Anthropocene' to the 'geological Anthropocene' without further ado. For them, the controversial stories of modernity would become supposedly apolitical techno-demographic narratives. The two pillars of this earth system science ideology of the 'Age of Man' are James Watt's steam engine (1748) and humanity (population growth). He argues that these concepts are anything but innocent. As with Thomas Malthus and later Paul R. Ehrlich, the history of the struggles for a fairer and more democratic world is ignored. 'Man' and 'nature' are turned into an 'operational system for imperial-bourgeois hegemony' (Moore 2021a: 36).

This becomes politically relevant, as Moore says that the conceptual pairing of 'man and nature' mutates into 'ruling abstractions': practical guides for the reorganisation of human and non-human networks in the service of the pursuit of profit. In combination with the Western civilising mission, these ruling abstractions would become a political ethos of dominance: man and nature would become man above nature. According to Moore, this strategy of 'geocultural dominance' ('Prometheanism') gave rise to modern racism and sexism (Moore 2021a: 37).

All of this is framed by a 'bourgeois naturalism' and the world-historical urge to increase profitability. According to Moore, in his typically drastic choice of words, the fundamental inequalities of this world are purified by the 'baptismal fountain' of bourgeois naturalism. This is central to the climate crisis and the Anthropocene because the central means of profit accumulation is the utilisation of 'cheap' sources. These (cheaps) are above all the extraction of natural resources and the exploitation of workers. In short: according to Moore, the Capitalocene emerged as a geohistorical era that strategically brings together the strategies of domination, exploitation and 'environment-making' as an 'apparatus':

> *Cheapness* was a strategy of domination and accumulation that joined the "economic" moments of *valorisation* to an unprecedented apparatus of geocultural *devaluation* (Moore 2021a: 37, emphasis original).

Accordingly, the beginning of the phase should be set at the onset of agriculture largely carried out by slaves instead of the coal industry and the use of oil as fossil fuels from the middle of the eighteenth century (see Angus 2016; Pomeranz 2000; Zeller 2020). This is a clearly definable period of human history, "… a historically situated complex of metabolisms and assemblages" (Haraway, in Haraway et al. 2016: 555). It is the conditions of the capitalist mode of production that expanded and globalized the "metabolic rift" in societal and natural law-bound "metabolism," which has existed since the Neolithic Revolution and resulted from the colonization of non-European territories (Marx 1867: 192; Foster et al. 2010; Malm 2016: 177–185; Mahnkopf 2019: 13). The central pillars of the capitalist economy and their striking environmental consequences go back much further than the early industrial phase, namely to classical antiquity or even to the Bronze Age (so Hann 2016b: 16).

This already suggests that the term Capitalocene, when applied to the heyday of capitalism, is too presentist, as archaeologists, universal historians, and historical sociologists have objected. In relation to Asia, for example, the thesis that the rift between urban and rural areas can be traced back to specific features of modern capitalism (2020) does not fit with more recent findings of historical scholarship on China (Marks 2015). The core idea that we live under capitalism and not simply in "society" is true, but human creative abilities are not reducible to capitalism (or socialism), nor is the global environmental crisis culturally reducible to capitalism (Chakrabarty 2009, updated 2021: 23–48, esp. 35, 43). Overall, the thesis of the Capitalocene represents a confusion of categories:

> Philosophers might call this a category mistake – capitalism is a 600-year-old social and *economic system*, while the Anthropocene is a 60-year-old *Earth System* epoch. Any serious engagement with social and natural science will conclude that capitalism existed for hundreds of years before the new geological epoch began, and that the new epoch will continue long after capitalism is a distant memory. Treating them as identical can only weaken efforts to get rid of capitalism and mitigate the harm it has caused to the Earth System, so that human society can survive – and hopefully prosper – in the Anthropocene. (In passing: If our current epoch is the Capitalocene, then surely the previous epoch should be renamed Feudalocene, preceded by the Slaveryocene, preceded by … what? The Hunter-Gatherer-ocene? The fact that no one suggests such absurdities is instructive.) (Angus 2016: 233, in Orig.: 2015: 25).

From a historical perspective, it is worth considering whether it was not the imperial expansion of domination that was the decisive historical impetus for the Anthropocene. These widespread imperial expansions are historically pre-colonial, as Amitav Ghosh argues in terms of climate change:

> I differ with those who identify capitalism as the principal fault line on the landscape of climate change. It seems to me that this landscape is riven by two interconnected but equally important rifts, each of which follows a trajectory of its own: these are capitalism *and empire* (the latter being understood as an aspiration to dominance on the part of the *most important structures* of the world's most powerful states). In short, even if capitalism were to be magically transformed tomorrow, the imperatives of political and military dominance would remain a significant obstacle to progress on mitigatory action. (Ghosh 2016: 146, emphasis mine; see Thomas et al. 2016).

The term Capitalocene should also be countered by the fact that the current perpetrators are by no means only to be found in the first world, the Global North. The term Capitalocene tempts us to attribute agency only to Western elites. Historically, it is not only capitalist societies that have brought about irreversible large-scale changes and triggered human-made crises. Examples include the permanent changes to the landscape in the North American prairies in pre-European times, overuse of the environment in the Roman Empire, or the irreversible ecological change on Easter Island (see Debate Diamond vs. McAnany and Yoffee 2010; Middleton 2017).

A pure sociology of capitalism suffers from both insufficient materialism and narrow historical provincialism (Dubeau 2018; Danowski & Viveiros de Castro 2019: 103, 171). From a deep historical perspective, as with Chakrabarty, the history (and historiography) of climate change has virtually replaced the history (and historiography) of globalization, with a pure critique of capitalism running the risk of dramatically underestimating the problem of the Anthropocene. The tools of the Capitalocene cannot explain why the middle classes of the non-Western world, which were once colonized, are now so widely adopting the "Western" visions of a modernized future based on the exploitation of energy and raw materials (Chakrabarty 2021: 101).

Archaeologists, universal historians and historical sociologists agree that the concept of the Capitalocene is too presentist when applied to the heyday of capitalism. With regard to Asia, for example, the thesis that the 'rift' between urban and rural areas can be traced back to specific features of modern capitalism (2020) does not fit with more recent findings of historical scholarship on China (Marks 2015). The core idea that we live in capitalism and not simply in 'society' is true, but neither are human creative capacities reducible to capitalism (or socialism), nor is the global environmental crisis culturally reducible to capitalism (Chakrabarty 2009, updated 2021: 23–48, esp. 35, 43). Overall, the thesis of the Capitalocene represents a confusion of categories:

> Philosophers might call this a category mistake – capitalism is a 600-year-old social and economic system, while the Anthropocene is a 60-year-old Earth System epoch. Any serious engagement with social and natural science will conclude that capitalism existed for hundreds of years before the new geological epoch began, and that the new epoch will continue long after capitalism is a distant memory. Treating them as identical can only weaken efforts to get rid of capitalism and mitigate the harm it has caused to the Earth System, so that human society can survive – and hopefully prosper – in the Anthropocene. (In passing: If our current epoch is the Capitalocene, then surely the previous epoch should be renamed Feudalocene, preceded by the Slaveryocene, preceded by … what? The Hunter-Gatherer-ocene? The fact that no one suggests such absurdities is instructive. (Angus 2016: 25).

Capitalocene as a replacement term and counternarrative for Anthropocene is therefore problematic in several ways. Firstly, capitalism did not have its origins in the middle of the twentieth century, but in the world system since the fifteenth and sixteenth centuries. The Anthropocene was created by the specific global monopoly capitalism from the mid-twentieth century onwards (Foster 2022: 462–464). Secondly, the term Capitalocene sounds very critical, but it misses the really fundamental critique of human natural relations that the deep time concept of the Anthropocene conveys. The main problem is that 'Capitalocene' ignores the suddenness, totality and irreversibility of humans as a geological force. Thirdly, such a 'capitalocentric' (Foster 2022: 279) view forgets that historically there can hardly be a way back from complex societies and, above all, industrialised economies. A planetary collapse of civilisation would not be a continuation of the Anthropocene epoch, but its end:

> There can be no conceivable industrial civilisation on Earth from this time forward where humanity, if it is to continue to exist at all, is no longer the primary geological force conditioning the Earth System (Foster 2022: 463).

Nevertheless, the role of labor highlighted in critiques of capitalism can also be analyzed more focused in terms of energy use as the primary driver of the Anthropocene (Suzman 2020). Altogether, I see the Anthropocene as a result of the ubiquitous efficiency of human economic activity in general, and of the rapid increase in the use of fossil fuels under capitalism in particular (similar to Morris 2015; Angus 2016). For this reason, I find the hitherto rarely used term "Pyrocene," the age of fire use, to be an insightful way of characterizing the Anthropocene. Pyrocene refers to the enormous landscape-changing effects of fire use as well as to the combustion engines at the core of the industrial economy. The industrial economy is characterized by mostly unseen, enclosed burning fires that are baked into the industrial structure (Soentgen 2021a: 143–153, see Pyne 2019).

4.7.4 *Plantationocene*

The development of capitalism was closely linked to colonialism, which led to large-scale changes in socio-ecological processes. Against this background and in exchange with the eco-Marxist critique, Donna Haraway coined the alternative term "plantationocene" (Haraway 2016: 99–103, Haraway et al. 2016: 22), which is similar in content to the Capitalocene idea. At the centre of this proposal is an agricultural system of exploitation of workers and slaves in monocultural export-oriented agriculture that has existed since colonial times and continues to this day. Haraway defines the Plantagocene as the devastating transformation of various living things from human-managed farms, pastures and forests into ex-tractive and closed plantations based on slave labour and other forms of exploited, alienated and usually spatially transported labour (Haraway 2015: 162).

The aim of this form of cultivation is the mass production of cheap export products. From an ecological perspective, as with Ishikawa's "humanosphere", the global simplification of the landscape, the emptying of the soil and the cultivation of monocultures are emphasised here. According to Haraway, the central dimensions with which the planatage as a landscape model characterises the world (world-making) are the reduction of biodiversity, replicability and adaptability to large scales (scalability).

The system implies an abstract relationship between owner and investment area, it turns plants into investment resources and involves alienating people from the local ecosystem. In abstract terms, it is about the subjugation, exploitation and conditioning of nature and people (Folkers 2020: 3). The term "Negrocene" summarises this aspect in an even sharper way (Ferdinand 2022: 50, 58–62). The plantation stands for an imperial, externally oriented cultivation of nature based on the long-distance transport of plants, animals and labourers who—torn from their respective contexts—live and work in an alien environment (Tsing 2015: 40–44). However, more recent ethnological and environmental geographic research on contemporary plantations, such as the palm oil economy, shows that plantations are more complex, by no means isolated and not stable as a system:

> People and Plants are not quite exterminated; land is not quite emptied; labour is not quite isolated; and plantations are welcomed and opposed by different social forces and nonhuman associates that render them fragile on multiple fronts (Li and Semedi 2021:23, cf. Pye 2021).

The *Plantationocene* concept is a counter-proposal to the Anthropocene concept, which is seen here primarily as being Eurocentric. From a Plantationocene perspective, the standardization of nature through rationalized agricultural production, which is equally an exploitation of the environment and human labor, including racism, constitutes the plantation (The Plantationocene Series 2021; Chao et al. 2023; Tsing et al. 2024: 43–44). Recent studies on sugarcane plantations have emphasized the colonial relationships between humans and non-human beings (Tsing 2015: 40; Moore et al. 2021). As early as the mid-1980s, ethnohistorical studies have documented the translocal, regional, and even global effects of these new relationships to nature by means of the sugar industry (Mintz 1985). Examples of such studies inspired by the Plantationocene are soil archaeological investigations combined with art campaigns on the "memory of the soil." The occurrence of slavery on a plantation in Maryland is revealed by soil finds, which were previously missing from the public historical archives and thus remained hidden as memory anchors, for example, on postcards (Martens and Robertson 2021).

The concept makes it clear that humans had to pay for the development of the Anthropocene even before the environmental disasters of the present (Folkers 2020: 594). Such a periodization thus fits in with the thesis of authors who want to decolonize the term Anthropocene (Davis and Todd 2017; Yusoff 2018; Driscoll 2020; Taddei et al. 2022). If one sets the beginning of the Anthropocene with the colonization of the two Americas, one emphasizes the violent expropriation and marginalization of cultures (including genocides) and their anthropogenic effects on ecosystems. This also has implications for cultural anthropology, because this approach would mean working out the cumulative effects of capitalist economic activity and using indigenous voices and sources more strongly (for example, Whyte 2017, 2018).

As Anna Lowenhaupt Tsing puts it: the logic of the plantation is different from that of the commons (Tsing, in Haraway et al. 2016). These displacements of the living substance, of plants, animals, microbes, and humans formed the basis of extractive economies (Lewis and Maslin b: 171 ff., Lewis and Maslin 2018). As an early example of the causes of Anthropocene change, plantations historically precede capitalism and thus set the beginning of this era earlier than industrial capitalism. Moreover, the term appears more descriptive and less ideologically or normatively charged than that of capitalism (Haraway, Tsing, in Haraway et al. 2016: 556–557). The cited example of slave labor on the plantation makes it clear, however, that the reference to *global* and *isochronous* effects is at best loose. From a historical perspective, it is worth considering whether it was not rather the imperial expansion of domination that was the decisive historical impetus for the Anthropocene. These imperial encroachments are historically pre-colonial, as Amitav Ghosh argues on the basis of climate change:

> I disagree with those who identify capitalism as the main fault line in the landscape of climate change. It seems to me that this landscape is criss-crossed by two interconnected but equally important fissures, each following its own trajectory: These are capitalism and empire (the latter understood as a quest for dominance on the part of the major structures of the world's most powerful states). In short, even if capitalism were magically transformed tomorrow, the imperatives of political and military dominance would remain a significant obstacle to progress in mitigating the consequences. (Ghosh 2016, here from the orig. 2016: 146, own emphasis, cf. Thomas et al. 2016).

Despite all their differences, both terms, Capitalocene and *plantationocene*, aim to highlight the specific role of humans in the fabric of life and their effects in the form of permanent environmental degradation. To this end, both emphasize the notorious link between capitalism and an assemblage of exploited humans on the one hand, and an exploited and non-human nature on the other (Hoelle and Kawa 2021: 658). Currently many of these critiques referring to Capitalocene or Plantationocene are discussed in the Global South, especially on and in Latin America (Kaltmeier 2024a; cf. Fig. 4.1). Both alternative terms are best suited to an expanded Anthropocene concept that is primarily geologically, archaeologically, and historically oriented, and which differs significantly from both the geological and the geoscientific understanding of the Anthropocene. This refers to the sum of all empirically verifiable influences of the *Homo sapiens* on the geosphere at any given time (diachronous) and possibly only on a regional scale (see Anthropocene 3 in the glossary vs. Anthropocene 1 and Anthropocene 2).

4.7.5 World-Ecology

One variant of talking about human-generated global environmental processes is Jason Moore's concept of "world-ecology" (Moore 2019). Here, the word "world" initially appears as general as the word "Earth." *World-ecology* would be "Weltökologie" in German. The German translation can be confusing because in Moores usage it refers to environmental relationships as a phenomenon, not to a global science of ecology. "Environment" here includes not only farms, fields, and animals, but also class, imperial structures, and the globalizing patriarchy. In their

Fig. 4.1 Latin American perspectives; graffito in Cologne. (Photo: Author)

mutual entanglement, specific environments are socially created ("environment-making", also Moore 2021a: 37). The term is based on the idea of a world-systemic capitalism (Bonneuil 2015: 28) and the theory *of a* "world system" (*world system theory*) in the sense of Immanuel Wallerstein (Wallerstein 1988, 2004). Moore, however, greatly expands on this, recently describing world-ecology as "… a dialogue that aims to describe capitalism as an ecology of production, reproduction, and power in a web of life" (Moore 2020).

According to Moore, *world-ecology* means *oikeios*, the totality of the diverse—"cultural", sociopolitical, and economic as well as "natural"—conditions of the generation of life, species, and environments in the world context, specifically in the world capitalist framework. This goes in the direction of *oikos* as a "household of life" (Pálsson et al. 2013:11). According to Moore, the Capitalocene will soon be over, namely when the environmentally damaging effects of using supposedly cheap resources drive capitalism itself to ruin (Moore 2016: 79). Within *world-ecology*, Moore categorizes what others call the Anthropocene as "Capitalocene" (see above). Moore follows Rosa Luxemburg's thesis that capitalism only functions as long as there is an outside that can be appropriated: other countries, colonies, slaves, and women for domestic and care work and "cheap" nature": cheap labor, cheap food, cheap energy, and cheap raw materials (Moore 2019, 2020, 2021a). These resources are "cheap" only because of the ability typical of modern market economies to externalise their extraction and the subsequent environmental costs, as Karl William Kapp noted in a classic study (Kapp 1963). Moore has recently added three more categories, those of "cheap money," "cheap help," and "cheap lives" ("cheaps",

Patel and Moore 2017). However, these strong terms are rather non-analytical (see Uekötter's criticism 2021: 803).

The strongest counter-criticism to the capitalism-critical and decolonial anamnesis, such as the Capitalocene and Plantagocene theses, consists of empirical data. According to the empirically best substantiated findings, the Anthropocene begins with the "Great Acceleration", the rapid increase in human environmental effects since the Second World War. This dating suggests that domination structures such as colonialism, imperialism, capitalism, racism and sexism are by no means the only important drivers of the Anthropocene. The model of fossil industrialisation was adopted in the fight against colonialism, imperialism or even in communist countries and in countries of the Global South (Mauelshagen 2023: 96). The liberation from colonial domination and de-proletarianisation, i.e. "... also moments of social emancipation and progress in the struggle for materialised freedom (have) contributed to the planetary malaise of the present ..." (Folkers 2024: 8, addition CA).

4.7.6 Chthulucene

Although Donna Haraway herself repeatedly uses the term Anthropocene, she introduced the colourful term 'chthulucene' as an alternative term (chthulucene, Haraway 2015: 160, 2018). For Haraway, the term refers to 'collectively producing systems that do not have self-determined spatial or temporal boundaries' (Haraway 2016: 37, 2020: 103). The term is intended to show that humans are no longer in control of the world on their own. At the same time, it is intended to have a post-heroic, anti-apocalyptic or anti-cynical effect. Instead of doom, we should devote ourselves to 'ongoingness'. In short: it is not too late; despite the rapid global environmental crisis, there are still futures: something is still happening and there is still something to do. (Hoppe 2022: 7–10, 12).

Humans are part of a tightly woven web of all life. Haraway focuses on the close 'kinship' relationships between humans and other living beings. By this she does not mean biological-evolutionary kinship, but a broad (not necessarily intentional) relationality of living beings, a connectedness of 'companion species'. Whether humans or microbes, they are relatives in their all-connectedness, whereby they are to be regarded as 'cohabiters' (Haraway 1991: 181). The fitting image of today's world would be that of a compost heap, where new things arise from the fermentation of the old, in typical Haraway diction: "We are not posthuman, we are compost. We are not homo, we are humus" (Haraway 2020: 100).

Unlike the Anthropocene, Cthulucene refers less to an age than to a particular mode of thinking, being-in-the-world and storytelling and a perspective on transformation (Hoppe 2022: 2, 10). Haraway uses the part of speech that is reminiscent of the monster creature that can end all life on earth, which was invented by the writer Howard Phillips Lovecraft in 1928 in 'The Call of Cthulu'. Haraway's reference in terms of content, on the other hand, is the use of ancient Greek quel-len, which shows the creature to be less patriarchal and monstrous. It is named after a spider (*Pimoa cthulhu*), whereby the ancient Greek word khthnios refers to the chtonic powers of the earth and a close connection to the ground. Accordingly, Haraway writes the name of the spider and the terminus differently from Lovecraft, namely with an additional 'h' (Hoppe 2022: 9, 13).

As a form of political activism, Haraway accordingly demands that we should "make ourselves kin" to non-human beings (*making kin as oddkin*). Like already Lynn Margulis she speaks of sympoietic alliances between materially linked living beings. Unlike the Anthropocene, the Chthulucene is not so much an age to encourage 'tentacular' thinking in order to explore spaces of possibility for better kinship (Haraway 2016: 2, 2018: 10–14, 47–84). Haraway has thus revived a broad interest in the interrelationships between humans and other living beings, which is usually known under David Abrams' idea of the "more-than-human," especially in the circle around Donna Haraway and Anna Lowenhaupt Tsing and among some representatives of "new materialisms" (Abram 1996; Hoppe and Lemke 2021: 123–140, 158, 171, for criticism e.g. Böhme 2020; Steiner et al. 2022).

Haraway rejects the term ´Anthropocene´ as well as the term 'Capitalocene' as an era in the sense of a time segmented by humans in some places because humans do not make history alone (2016: 35, 2018: 47–48). In other places, she declares it useful as a strategic term (Haraway, in Haraway et al. 2016; Haraway 2020). Haraway advocates playing with a plethora of neologisms, which is also expressed in her saying, *Chthulucene* would resist dating and "ask for myriad names." So, it is not surprising that, in addition to the mentioned "Chthulucene" and "Plantationocene," she not only uses "Anthropocene," but also uses "Capitalocene" in passing. Taken together, however, Haraway's ideas form a concept that highlights a certain dimension of the Anthropocene, namely interspecific relationships and connections between bios and geos. Thus she follows symbiotic concepts of recent alternative biology with its eco-centred concepts, such as Margulis' macroscopic Gaia or the more microscopic holobiont as an ultimately collective whole organism (Gilbert 2017, cf. Belardinelli 2022: 25–28; Kerner 2024: 176 f.). In my opinion, however, both concepts could be more strongly grounded in the natural sciences in order to facilitate a truly analytical approach to the topic of the Anthropocene.

Haraway says that chthulucene would resist dating and 'ask for myriad names'. This is precisely where I see a problem with this term—as with many other alternative '-zän' terms—for an understanding of the Anthropocene. Despite the use of the suffix '-cene', Haraway in no way attempts to place the beginning of the Chthulucene in terms of the –cene as a period somehow geologically or historically. Haraway deals as little with geological phenomena and concepts of time as she does with historical processes. Her position is thus very far removed from the conceptual challenge that the geological framing of the Anthropocene idea poses for cultural studies. With Haraway, for example, it is striking that she deals with geological phenomena and concepts of time as little as she does with historical processes. Haraway does not attempt to somehow geologically or historically locate the beginning of the Chthulucene in the sense of the *–cene* as a period. Thus, such positions are very far removed from the conceptual challenge that the geological framing of the Anthropocene idea poses for cultural studies. An additional problem arises with regard to the causation of the Anthropocene and thus also the question of responsibility. If, according to Haraway and Morton, humans are not distinguishable as individuals and all living beings are connected by "kinship," the question arises whether causal agents can be determined at all.

The conclusion to be drawn from this is the question is of which collectives can build hopes for the prevention of human effects, who should be the actors to initiate countermeasures or to whom demands should be made: "… the addressees of all calls to action are simply not jellyfish, corals or dolphins, but humans" (Bajohr 2019: 69). This reveals a central problem of all posthumanist responses to the Anthropocene, namely how ontological questions and ethical questions can be brought together. A flat ontology, as in the particular approach of Latour, carries with it the risk of obscuring responsibilities, because the assumption of non-human things as actors or "agents" (actants) dissolves not only the distinction between nature and culture, but also that between subject and object (Angus 2016; Ribot 2018: 219).

4.7.7 Technocene and Geosocial Formations

Only the comprehensive mechanization of all areas of life has enabled the depth of human intrusion into the environment. Accordingly, techno-historical processes are of great importance for debates about the age of humankind and age dating (Trischler and Will 2020: 236). This is especially true for the Anthropocene in the version of the approach from the Second World War onwards. Alf Hornborg, a human ecologist and cultural anthropologist, proposes the term "Technocene" instead of Anthropocene, which in some respects can be seen as a refinement of the Capitalocene concept (Hornborg 2014, 2015; Cunha, Moore 2020). It is economic systems, lifestyles, and technologies that are causing the ecological crises of the present. The term focuses on the connection of the economy with technologization, especially the social and cultural genesis of technology. This corresponds to the perspective provided by recent technological history (for example, Trischler and Will 2017, 2019). Hornborg's reference to the economy is more constructivist than materialistic and thus not as narrow as that of Moore and other advocates of the term Capitalocene. Society or the social remains, in concrete practice, a realm separate from nature, so that concrete political and ethical questions of responsibility arise. Accordingly, Hornborg as an ecologist, unlike Moore, does not relate the Anthropocene exclusively to capitalist economic systems.

The main reason for the historically sudden and drastic change is seen less in the conscious politics of the elites during industrialization, but more in the objectification and fetishization of technology, which corresponds to a form of alienation (Malm and Hornborg 2014; Cunha 2015b: 263). The term Technocene claims to unsettle the dichotomy of society and nature and thus promote a less Eurocentric, "post-Cartesian" way of thinking (Hornborg 2017a, b). The term rightly focuses on the crucial role of human technology and the danger of a quasi-subjugation of humans to a regime of artifacts for global change, but it is only related to one dimension of the Anthropocene.

The term Technocene has recently been supplemented by the term "Anthrobscene." It points to the "obscene" consequences of human activity. Parikka defines this as a term that encompasses the waste of digital life (Parikka b; Swyngedouw and Ernstson 2018, see Holm and Taffel 2016). Parikka argues that laptops, tablets, smartphones, and e-readers promised us a more environmentally friendly world that is not dependent on paper and deforestation. However, he sees the actual result as our life in ubiquitous digitality, a barren land where media never dies. He thus interprets people's individual consumer desires and the needs of corporate actors as a quasi-geophysical force. In the sediments of toxic waste that we will leave behind as our geological legacy, remnants of technical media will remain. Parikka sees the material side of the Earth as essential for the existence of media and introduces the notion of an "alternative deep time" of media. I myself think, like Angus (2016: 231), that proposals such as the "Anthrobscene" can be stimulating, but are more likely to be understood as jokes. In the Technocene, humans not only produce buildings, infrastructures and waste. At the same time, we are also creating enormous cavities worldwide through the extraction of raw materials and construction measures, such as tunnels, which sometimes collapse (*sinkholes*, Bridge 2009). In the context of wordplay, which is particularly popular in art and the media, this has inspired the 'Hollowcene' (Siegler 2024: 59).

4.7.8 Urbanocene—Cities as Seismographs and Drivers of Change

In contrast, "Urbanocene," like "Technocene," is, in my opinion, a term that is to be taken very seriously (*urbanocene*, also *urbicene*, *astycene*, and *metropocene*, Heise b; Chwałczyk 2020). What it means can be summed up as follows: "The world is becoming a city," the planet is dominated by cities, and thus the fate of the geosphere is the fate of the cities… and they can be shaped by humans. Some of the strongest anthropogenic environmental problems and effects investigated by anthropologists are manifested in cities (for an overview, see Antweiler 2018). Cities, especially megalopolises, also provide the basis for most apocalyptic narratives about the Anthropocene (Brosius and Gerhard 2020).

Cities are a major driver of environmental change that is not sustainable. Cities generate around 70% of CO_2 emissions. As densely populated areas, they can only exist in the long term by building large-scale infrastructures and incorporating natural resources from the surrounding countryside or imperially appropriated areas, such as building materials or natural gas. Some urban metabolic processes, such as water and energy supply, typically disappear underground. In this way, the appropriation and commodification of nature become invisible. Added to this are the inequalities in supply and disposal that are particularly notorious in cities. These inequalities are concealed and, at the same time, environmental consumption is concealed for population groups (Kropp 2024: 323 f.).

Urban lifestyles are becoming increasingly normal worldwide, even in formally rural areas. In cities, some of the strongest anthropogenic environmental problems and effects manifest themselves at the level of everyday life (for an overview, see Antweiler 2018). The planet is dominated by cities, and thus the fate of the geosphere is the fate of cities … and they can be transformed by humans. The world goes urban and that makes 'Urbanocene', like 'Technocene', a term that should be taken very seriously in my opinion (*urbanocene*, also *urbicene*, *astycene*, *metropocene*, Heise b; Chwałczyk 2020). The term Urbanocene can also make it clear that the Anthropocene cannot simply be reduced to climate change.

Most authors understand Urbanocene as a term that is similar in meaning to Technocene. Both terms relate to a central problem for the transformation of cities. It takes a very long time and is very costly to dismantle sealed surfaces, asphalted roads and tunnel systems. Because of their materiality and scale, technical infrastructures such as supply systems are generally resistant, stubbornly resistant to change (*obduracy*, Hommels 2005: 329–341). Through sheer material persistence, infrastructures channel thinking, they tunnels environments by linking technological and social structures and they perpetuate political structures. Infrastructures, such as extensive sealing through asphalting, thus make cities inflexible in the face of a desired transformation.

Artificial sediments in cities are just one part of the by-products of industrialization, construction, resource extraction, war, and agriculture (Fig. 4.2). The term Urbanocene can also nicely illustrate that the Anthropocene cannot simply be reduced to climate change. A major problem for coastal cities like Shanghai and Shenzhen is the increase of flooding. For example, coastal districts of the Indonesian port city of Semarang are sinking up to 15 centimeters per year. The water is pushing into the city, which is literally "tipping" into the sea. Some residential areas and industrial facilities are flooded on a daily basis, and some streets are already permanently submerged. The most important factor is often not the rise of the sea level due to global warming, but subsidence due to soil compaction caused by groundwater extraction and by the sheer weight of the buildings.

The Urbanocene can be understood not only in terms of the expansion of cities, but also in terms of verticality, a phenomenon that the social sciences are only slowly opening up to (Graham 2018). This is a potential field of research where urban archaeology and urban cultural anthropology can enrich each other, and which I will address in Sect. 5.3. For some authors, the Urbanocene represents a step out of the Anthropocene, such as for Geoffrey West, who considers the Anthropocene to have already ended. He sees the Anthropocene as beginning with the increase in the human metabolic rate several millennia ago—what others call the "early Anthropocene"—and ending with the start of the exponential rise of cities that dominate the planet today (West 2017: 220–222, 268). The term Urbanocene can not only emphasize the urban genesis of core problems, but also convey the

Fig. 4.2 In-situ-preservation of human modified ground in section; station concourse, The Syntagma Metro Station Archaeological Collection, Athens, Greece. (Photo: Hoverfish 2009 CC-BY-SA 3.0); https://commons.wikimedia.org/wiki/File:Syntagma_Metro_Station.jpg

insight that the solutions to quite a few problems of the Anthropocene will be urban solutions (Müller et al. 2022, for example, for Jakarta Simarmata et al. 2020).

4.7.9 Fixation on Naming and Specificity of the "-cene"

My conclusion is that the proposed alternative terms as a social fact could be an interesting field of research for social and cultural scientists. They open a window into the many ways in which the term is understood. They offer a rich repertoire of forms in which the term Anthropocene is received, especially by critics. However, I see consistent problems with the alternative terms and also with the discussion about them. The Anthropocene is by no means old wine in new bottles, because it is based on assumptions or insights that differ from or significantly complement well-known important insights. Key terms, such as "resilience," condense the zeitgeist or the experiential signature of a period. They are supposed to open more than just a door, and this requires a certain degree of vagueness. Key terms are rarely purely descriptive, because their claim is both diagnostic and transformative in response to challenges (Bröckling 2017). The alternative terms discussed here are similarly associated with the problem that they suggest seeing only one cause where there are actually many. After a review of the alternative terms that use the component "-cene," I agree with Leinfelder:

> The alternative terms proposed mainly from the cultural or social sciences, such as Capitalocene, Pyrocene, Plastocene, or Homogenocene, also focus on important causes or findings, but are not comprehensive enough as a narrative headline (Leinfelder 2017b: 11–25).

On the one hand, many critics obviously do not know what the term Anthropocene *linguistically* means in geology and paleontology. Contrary to some explanations in the popular science texts discussed above (for example, Schwägerl 2012), "Anthropocene," from a geological perspective, does not mean the new "age of humans" or the "epoch of humankind." The "-cene" in Anthropocene does not simply mean "new," but the *degree of novelty of the fossils*. However, almost all of the alternative terms do not address the geological core of the concept, the globally synchronous boundaries and fossils. This dissolves the potential specificity of the syllable "-cene." In terms of content, most critical concepts of human-made change do not need the word, they rather misuse it. "Perhaps we should rather refrain from any further 'cene' and acknowledge that we were and are never masters in our own house" (Schüttpelz et al. 2022: 110).

Overall, I see an unproductive focus on naming debates. This is all the more true for a phenomenon that raises existential questions for humanity and the biosphere. Ian Angus expresses his similar bewilderment by saying that such debates always remind him of Douglas Adams' "The Hitchhiker's Guide to the Galaxy." There, a business consulting committee on a prehistoric planet is criticized for failing to invent the wheel. The committee chairman responds to this objection: "Well ... if you're so incredibly smart, why don't you tell us what color it should be" (Adams, trans. By Angus 2016: 231).

Taken together, the central implication of most criticisms is the demand for a "socialization" of the Anthropocene (Lövbrand et al. 2015: 213; Clark and Szerszynski 2021: 38–41). The complement, namely a "geologization of the social," the demand for a necessary geosocial thinking (Clark and Gunaratnam 2017), which I will analyze in Chaps. 6 and 7, appears much less in the critiques, primarily in those from the humanities, cultural studies, and social science. They tend to fear an appropriation of their topics by natural scientists. This surely reflects fears of marginalization and a competition for definitional power, in short: "Anthropocene power struggles" (Will 2021: 215). I see the political significance of the term Anthropocene in the fact that it is open enough to bundle different emphases in the diagnosis of the cumulative effects of human activity. On the one hand, the term inquires about effects on the climate, soils, seas, and biological diversity, for example, and about the systemic interactions between these effects. On the other hand, the term uses the strong syllable "anthropo-" as a kind of currency to draw attention to a problem affecting humanity as a whole. In this dual character, it resembles other powerful terms, such as the current buzzword "international community" (for details, see Lindhof 2019). Despite its inadequacy, the term confronts us with frightening facts, it shows both the seriousness of the consequences of our actions and the limits of our possible influence on long-term processes, and it reveals the interconnectedness of humans with the animate and inanimate world (Macfarlane 2016: 96).

I am talking here about 'alternative terms' and 'counter-narratives', but for the most part these are not simply alternative words for the same thing, but different terms, i.e. words for different things, phenomena and perspectives. Above all, two very different time scales clash here. No one has worked this out as clearly as Zoltán Boldizsár Simon (2017: 241–242). A comparison of the Anthropocene and the Capitalocene reveals the following. Proponents of the Anthropocene typically emphasise the *unprecedented* novelty from a *geological* time perspective and derive the urgency of preventive action from this. Proponents of the Capitalocene concept, on the other hand, emphasise the historical *continuity* of inequality and from this follows their call *for social justice*. This imperative of social justice has a desired future in mind and calls for proactive action. The Anthropocene concept, on the other hand, fears unwanted collapse in the future and derives from this the call for preventive and therefore reactive action. Both perspectives could complement each other, but they have different time perspectives and also different concepts of change: the concept of (geologically) unprecedented change contrasts with the processual historical change of (Western) modernity.

In my opinion, these are important insights, but Simon does not tunnels that geology is an (earth-) historically oriented science that knows sudden change as the norm. It is overstated when he considers the Anthropocene and Capitalocene concepts to be totally incompatible. He uses the Capitalocene and Technocene concepts in an undifferentiated way and does not take into account the diversity of historical and archaeological approaches. Above all, it seems highly exaggerated when Simon bluntly concludes that the use of modern historical thinking to understand the Anthropocene is '… an untenable, contradictory and self-defeating endeavour' (Simon 2017: 239).

What links the many alternative concepts discussed here to anthropology is the pervasive tunnels or between the acceptance or rejection of universal thinking. Chakrabarty, as a historian who retains the term Anthropocene, works to reconcile universal thinking about the universal history of life with postcolonial arguments about anything universal and to endure this dilemma (Chakrabarty 2009: 219–220). Most of the cultural studies or social science authors of the alternative '-cenes', on the other hand, take one side or the other (Simon 2017: 240). While the vast majority, such as proponents of the Technocene and Capitalocene, as well as the mainstream of anthropology, are particularist in orientation, a few argue in favour of a universalist perspective.

Such universalist concepts deal less with the causes of the current mega-crisis in the *past*, but rather open up a normative view of possible *futures*. One example is the alternative term 'cosmopolocene' (Delanty and Mota 2017). However, this should not be understood as concepts completely opposite to the Anthropocene; the concepts can complement each other. While retrospective perspectives, such as the term Capitalocene, emphasise the historical inequality as the cause of the current crisis, the Cosmopolocene works on how this inequality could be overcome in the future. The Cosmopolocene thus rejects the apocalyptic view of the future of the natural sciences, but retains the alarmist view of the Capitalocene with regard to the past. In addition, there is the draft of a utopian future to strive for (Simon 2017: 240–241). I conclude: we need both universalist and relativising approaches, and comparative research is needed. When humans become geological actors, we need historical sciences (Bhattacharyya and Santos 2023) and human sciences, e.g. cultural anthropology.

Chapter 5
Anthropocene Anthropology—Contributions and Opportunities

Abstract This chapter shows that cultural anthropology has always had a professional proximity to the subject of nature and has dealt with the environment in various ways. The causes of the Anthropocene may not be perceived by us as individuals, but the effects will be experienced locally and differently in the long term. The concepts and goals of countermeasures are also local. This shows that the local access and the holistic approach of cultural anthropology opens a fruitful window to understanding the Anthropocene. In this section of the book, I criticize currently dominant approaches to the Anthropocene in cultural anthropology because they avoid scientific findings, obscure the specific geological deep time, and generally provide more artistic contributions. The chapter presents very different anthropological approaches, and I also attempt to do justice to directions and positions that I do not represent based on my scientific assumptions, such as the *more-than-human* approaches or posthumanism.

> *But if we are going to come to terms with the deep histories of global transformations, we need to (…).*
> *listen to the other half of the globe, the tribalists out there and right here, talking back.*
> Jeremy Adelman (2017: 5)

> *Anthropology is the most humanistic of the sciences and the most scientific of the humanities.*
> Kroeber (2003: 144) (1956)

Anthropologists are used to closely scrutinizing their surroundings. The discipline traditionally deals with spatially small and culturally delimited units, and with temporally short-lived phases that last within the range of one generation. Anthropological research starts with everyday life, with the experiences and actions in the social microspace of human experience and action. An understanding of culture as practice is helpful for grasping how to deal with anthropocene threats in everyday life. Practice anchors social action in the material environment, in social space, connects it with social and cultural norms and focuses on routine and normality as well as on the cultural novelty and creativity of the actors. One example of this is the current external insulation of old buildings: beyond technology, it connects everyday life with social situation and social transformation (Niewöhner 2013: 47–49).

C. Antweiler, *Anthropology in the Anthropocene*, Anthropocene – Humanities and Social Sciences, https://doi.org/10.1007/978-3-031-74591-1_5

Traditionally, cultural anthropology is a science that is focused on locally limited lifestyles. As a "microscopic" cultural and social science, it is equally important for understanding and explaining human ways of life as it is necessary as a scientific basis for *realistic* political appeals for change. However, cultural anthropology also has a broader anthropological dimension. As anthropologists, we are concerned with the role of humans in environments, and we study how humans are conceived in scientific and political debates about development and future environments (Hoelle and Kawa 2021: 655). In this respect, the comparatively limited experience of cultural anthropology in working together with natural scientists is relevant (Schareika 2006; Casimir 2021).

5.1 In the Contact Zone of the Disciplines—Resonance in Cultural Anthropology

> *In anthropology – a discipline driven by fads to be chewed up, spat out, and replaced by the next "turn" – the danger is that invocation of the Anthropocene becomes almost a cliché.*
> Richard Irvine (2020: 11)

Various reactions of anthropology to the topic of the Anthropocene are conceivable. One thing is certain: the keyword is attracting attention in the field. The invitation of the world-famous knowledge researcher Bruno Latour, as a representative of another field, to give the keynote speech at the 2014 annual meeting of the *American Anthropological Association*, which was entitled "Producing Anthropology," was probably the clearest sign from the field and its most powerful organization that the topic of the Anthropocene was being taken seriously. Latour sees anthropology as one of the "lower sciences," as an "earthly" science, like soil science and ecology.

Ten years before addressing the anthropologists, Latour had already wondered how it was possible that we as humans could have been a part of natural history for so long, but realized so late what had happened. Latour then took on a positive view of the term Anthropocene, seeing the main benefit in its "narrativity," in the power of the Anthropocene concept to tell a "geostory," a story that is not simply an addition to the physical history of the Earth (Latour 2017a, b: 10). In his lecture, Latour spoke of the Anthropocene as being perhaps an undeserved "gift," a gift that could also prove to be toxic if anthropology does not accept it productively. Latour drastically emphasized the challenge posed by the Anthropocene to the often precarious state of anthropology:

> What an amazing gift! Sure it might be poisonous. But how silly it would be not to try to peek through the wrapping to take a glimpse of what is in store. Consider the situation: here is a battered scholarly discipline, always uncertain of its scientific status, constantly plagued by successive and violent 'turns' (the 'ontological turn' being only the more recent), a field which always finds itself dragged into the middle of harsh political conflicts, a discipline that runs the constant risk of being absorbed by neighboring specialties and voted out of existence by deans and administrators impatient of its methods and ideologies, a discipline that accepts being crushed under the weight of all the violence and domination suffered by

> the many populations it has decided to champion – a lost cause among all the lost causes, okay, you see the picture, and it is to this same discipline, which a few years ago, an amazing present was offered: pushed from behind by the vast extent of ecological mutations and dragged ahead by philosophers, historians, artists and activists, a sizeable group of natural scientists are describing the quandary of our time in terms that exactly match the standards, vices and virtues of that very discipline. Yes, what a gift! It is really embarrassing, especially if it is not deserved! (Latour 2017a, b: 35).

5.1.1 Between Buzzword and Radical Criticism of the Field

In an overview of the core concepts of anthropology, Lavenda & Schultz write: "Few anthropologists would challenge the vast scientific consensus supporting the view that, as a consequence of human activities, we are living in the Anthropocene ..." (Lavenda and Schultz 2020: 7). However, this leaves open the question of whether they are simply accepting another word for what is commonly referred to as global change. It could be, and this is suggested by Lavenda & Schultz's position, that most might simply be referring to the field of environmental anthropology and are thus addressing colleagues who are used to working with others who deal with the environment and global problems. However, as established in Chap. 2, at the core of the concept of the Anthropocene is a specific *geologically* substantiated assumption.

One reaction of the discipline may be to actually use the word Anthropocene only as a buzzword in order to cover other topics or specific aspects under the label. In many debates, it is noticeable that the phrase "... in the Anthropocene" is used to lend rhetorical power to any argument, which has been aptly named "Anthropocene for show" (Malhi 2017: 93; Dryzek and Pickering 2019:13, 49–50). For example, a (very good) overview article on disaster anthropology uses the word "Anthropocene" in the title ... and then never again; even the word "climate" only appears six times, and "global" only appears twice in the text (Barrios 2017). Simply labeling various topics with the word "Anthropocene" is a strategy that is at best conceivable in terms of science policy, and even there it will not be fruitful in the long term, because the half-life of buzzwords is usually short.

If we take the criticisms discussed in Chap. 4 seriously, especially those from the social sciences, a second stance could be that anthropology is explicitly maintaining distance from the Anthropocene concept and is primarily engaging in the form of accompanying criticism. Anthropologists have been very active in this area, as was already made clear in Sects. 4.2, 4.3, and 4.4. However, in my opinion, the topic is too consequential to leave it as a critique of the concept, as a blanket critique of growth-fixated capitalism or as a specific criticism of the extractive management of natural resources. All of this can also be done by other disciplines. Edgeworth, as an archaeologist, has pointed out that anthropology can contribute more than just criticism—as important as this is—to the question of the Anthropocene:

> The evolving and still fluid debate on the Anthropocene is too important for anthropologists to situate themselves outside of it. A more positive approach (...) would be to recognize that anthropological methods and perspectives – together with those of the natural sciences – are crucial to the collaborative multi-disciplinary task of reformulating the Anthropocene (...) In such a project, moving beyond critique, anthropologists surely have a vital role to play (Edgeworth 2018a: 759).

The professional relevance of the Anthropocene is understood very differently in anthropology. Some anthropologists argue that the Anthropocene, at least in the form of environmental disasters in the context of anthropogenic climate change, could lead to a "renewed project of cultural criticism" in the wake of the traditional *anthropology as cultural critique* of the 1980s. In one of the most widely read anthropological contributions of recent years, Cecil Ryan Jobson argues that the existential climate crisis requires an end to ethnographic distance as well as sentimentality and a move away from the assumption of a coherent human subject in order to create a "radical humanism."

The article, which was intended to critically examine the research contributions of anthropology in 2019, turned into a thoroughly programmatic and dramatically written text in the face of the climate crisis. The 2018 wildfires in California—coinciding with the central conference—should have shaken up the discipline, which had long been in an epistemic crisis anyway. Indeed, let anthropology "burn": "To let anthropology burn permits us to imagine a future for the discipline unmoored from its classical objects and referents" (Jobson 2020: 261). The drastic language of the text stands in stark contrast to the corresponding annual overview of anthropological archaeology, an essay that also discusses problems of categorization very critically, but does so in a much more objective manner (Kosiba 2019). In Jobson's statement, the fears of the Anthropocene in the form of climate change manifest themselves as an extensive criticism of the field and also in a violent metaphorical choice of words (see Hornborg 2017b; Malm 2021). While the Anthropocene serves here as a fundamental motif for rethinking the discipline as a whole, one of the most widely read theoretical essays in anthropology in recent years (Ortner 2018) does not mention the Anthropocene topic at all.

5.1.2 Professional Aspirations and Classic Topics

Many of the issues that have been raised by the Anthropocene have been addressed in anthropology for a long time, as Bruno Latour said in his lecture to the AAA in 2014 (see Pandian 2019). He identified three highly relevant issues in the Anthropocene: the first is human agency, the second is the connection between physical and cultural anthropology, and the third is comprised of the similarities and differences in the "ways of inhabiting the Earth" (Latour 2017a, b: 7). Similarly, although from a different theoretical perspective, Marc Augé recently noted that all the traditional subjects of empirical cultural anthropology with respect to social relations, such as lineage, kinship, intergenerational ties, and residence, have been significantly affected by global anthropogenic environmental change (Augé 2017: 66–68). They are also relevant for an understanding of the Anthropocene. Especially through their expanded contexts, these topics are even more extensive than ever, with the subject of cultural anthropology not disappearing, but shifting.

Nevertheless, a good 10 years ago, Paul Stoller stated with some justification that anthropology had not yet produced an active response to the challenges of the Anthropocene, which he sees above all in the politics of the Anthropocene (Stoller 2014). As an anthropologist, one might wonder why the discipline played no role in the formulation of the concept of the Anthropocene, even though it is about the consequences of human activity. But this is not only the case for anthropologists. From an external perspective, it is also surprising that archaeology was not consulted in the canonical essays in which the topic was formulated. The main actors in the debates in the first decade of the twenty-first century were climate scientists, geophysicists, and Earth system scientists interested in large-scale geophysical effects. Regional and local environmental change, documented by archaeologists but having no effect on the entire geosphere, was long ignored (Bauer and Ellis 2018), with Ruddiman's contributions to the early Anthropocene being the exception. Clive Hamilton has drawn from this a clear—in my opinion, premature—conclusion for archaeologists and ecologists: "butt out" (Hamilton 2014).

On the other hand, cultural anthropology sometimes makes a rather sweeping claim that anthropology, as a science that emphasizes the experiential dimension of human–environment relationships, is of course relevant to such an anthro topic. Some representatives even say that anthropology "with all its sub-disciplines" is ideal for the topic and for critical scientific "interventions" (Bauer and Bhan 2018: 7). It remains open whether the subfields of broad anthropology in the American sense are meant here, that is, cultural anthropology, linguistic anthropology, archaeological anthropology, and physical anthropology, or just archaeology (as in Bauer and Bhan 2018). If "only" the individual subfields of cultural anthropology are meant, it would not only be about the thematically relevant areas of environmental anthropology, climate anthropology, and political anthropology, but also about economic anthropology, social anthropology, and religious anthropology.

5.1.3 Contouring, Disparity, and Precariousness of the Discipline

In order to make a substantial and constructive contribution to the topic, it is crucial how cultural anthropology is organized as a discipline and whether and how it presents itself to the public. Of central importance for such a publicly relevant topic as the Anthropocene is how the discipline positions itself in society, how it presents its profile to the public, and to what extent the discipline gets involved in political processes. This raises the question of what contribution cultural anthropology can and wants to make in concert with a broad range of sciences.

Cultural Anthropology could make its contribution as a discipline or rather as an integrating part of many disciplines in a field encompassing various scales. It could be a minor accompanying science that critically examines hegemonic and universalistic forms of knowledge ("minor science," Marcus and Pisarro 2008). Conversely, it could also be expanded as an "anthropology without borders," as has been boldly postulated in the context of climate change research (*cross-scale*, Crate 2011: 183). These are just some of the conceivable positions and forms of collaboration with natural scientists (see Schareika 2006; Skinner 2021).

The recent debate on the naming of the multi-name discipline in the German-speaking world was extremely controversial and demonstrated how different the thematic and methodological orientation can be (documented in Antweiler et al.). A crucial question regarding the Anthropocene is how anthropology itself is substantively contoured in relation to the future. It could be profiled (a) as ethnic studies, (b) as research on the unfamiliar, (c) as a methodically micro-oriented part of the social sciences, (d) as part of a broad human science, as I advocate, or (e) as a moral authority for minorities and diversity, as –I perceive the currently dominant trend to be. Given the thematic breadth of the Anthropocene, one can search for an "anthropology of the future" and ask about basic orientations and future orientation as a method (Bryant and Knight 2019: 192–200). This also raises institutional questions: will anthropology remain a discipline in the future or become a kind of interdisciplinary field?

This chapter concerns anthropology, whose orientations are as diverse as its subject matter. The discipline also has a double face (as suggested by Sprenger 2021, 2022): On the one hand, anthropology is a discipline with its own scientific criteria of truth, which should be free from the norms and values of its own society and also independent of such precepts on the part of the communities being studied. On the other hand, there is a civil society side to the discipline, which typically shows itself in solidarity with minorities of all kinds: with the poor, the marginalized, with particularly vulnerable and less powerful people. Since anthropology is my discipline, this chapter is certainly more strongly characterized by my own attitude than that of others, and these I would like to briefly disclose here, as in the credo in the appendix on science in general.

5.2 Cultural Anthropology—A Profile and a Position

Le onta du respect de différances culturelles
tend à ignorer et à faire ignorer d'une part, les differences individuelles à l'interieur des cultures et d'autre part, l'existence de l'espèce humaine, du genre humain, au-delà des cultures.
Marc Augé (2017: 89)

A large percentage of anthropologists believe that the debate about the Anthropocene will move anthropology in new directions and challenge us to reflect again on the subject of the discipline and on the lives of those with and about whom we research (Moore 2016; Tsing et al. 2019: 187; Augé 2017).

Classic ethnography as the main product of the discipline was a village study about the life of a local community that came in the form of a book. To this day, anthropologists primarily study small collectives or small or manageable sections of large societies. Modern ethnographies typically describe local parts of more complex societies or sections of networks, such as a sub-network of political activists among the participants of a social movement. The local basic orientation does not preclude anthropologists from repeatedly dealing with larger units, such as whole nations, regions, or with the entire cultural world, with world culture. This is especially true for anthropologists who are not only active in the discipline, but also in the interdisciplinary field of area studies or global history, such as the classic works of Ruth Benedict on Japan, Leslie White on world culture, and Eric Wolf on the history of the Global South.

5.2.1 Scales and Units

The micro-orientation of anthropology is closely linked to its experience-based methodology. The classic method is the so-called field research. In this case, a social field—hence the name—is investigated in an experience-based manner by a researcher who lives with the people being studied for a long period of time. This is only done by a few other sciences, and this distinguishes anthropology from most research in the major social sciences. Exceptions to this are the ethnographic branch of sociology (*ethnography*), the modern mainstream of European ethnology (formerly folklore studies) with the concept of everyday life, and some geographers, such as Jonathan Rigg, who similarly encapsulates this under the term *everyday* (Rigg 2007: 1–23). Just as the spatial scope of anthropological studies is limited, so is its temporal framework. Anthropologists usually deal with the current situation and current processes in human groups. Change is usually only traced back to recent history.

> So, to take up Sartre once again, is anthropology a humanism? Yes, because it strives to emphasize the correspondence between the questions posed by different cultures of the world without alienating itself from their respective answers, because it highlights the tension between the constraints of social meaning and the aspiration for individual freedom, because it understands the boundary between cultures and the boundaries between individuals as thresholds and not as barriers, because it follows its dual vocation as a theoretical and applied discipline engaged in world history, anthropology is a humanism (Augé 2017: 90).

5.2.2 Anthropology as Experiential Cultural Anthropology

If intercultural positions of cultural anthropology are to be achieved, it is first necessary to determine what the core of general cultural anthropology should consist of. In the international and German-speaking debates, there are a multitude of positions, but little genuine consensus. In order to make my understanding of cultural anthropology clear, I will first give a primary definition and justify this. Then, I will say how cultural anthropology can be defined secondarily, in order to finally derive explicitly from these definitions how I would explicitly not define cultural anthropology—in contrast to some currently prevalent positions. How could the core of cultural anthropology be characterized, how should cultural anthropology define itself?

1. Cultural anthropology as a science, and thus not primarily as a "perspective," "viewpoint," or method. As a science, it would be determined by the focus of its content, and so primarily by its subject matter of culture. This includes, firstly, culture as cultural capability and, above all, the existential dependence on culture (cultural dependence, cultural need) of humans; secondly, this entails culture as a particular form of existence of and in groups of people (societal culture). As a science, cultural anthropology is a profession that must be learned, even if the anthropologists are *native anthropologists*. It should derive its legitimacy less from identity than from scholarship. As a science, it should seek "authority" rather than avoid it, but this authority must be methodically justified and transparently presented.
2. Cultural anthropology as a generalizing and comparative cultural anthropology, thus as an integrated part of anthropology as the science of the "whole humankind." Cultural anthropology should be situated within the framework of other anthropological disciplines. It must be part of a general anthropology that cannot be limited to a purely cultural science. It should not be isolated, as is the case in Germany or as is definitely the current trend in the USA, it should not detach itself from the other anthropological disciplines and align itself with literary studies, literary criticism, or *performance studies*. The focus should be on culture, comparison, and generalization, as used to be the case in American cultural anthropology and has become increasingly prevalent in British cultural anthropology in recent years. We do not need fewer comprehensive approaches (*metanarratives*), as is so often demanded, but more. This does not mean a blanket rejection of postmodern cultural anthropology, as it has provided important insights (problems of representation, multiple varieties of identity, constructivism, historical citation). However, the extreme cultural relativism of many postmodern anthropologists must be rejected. The issues raised by postmodern anthropologists do not preclude them from being explicitly addressed scientifically. I therefore advocate for an explicit search for generalizations and patterns as well as quantitative statements, but I oppose their totalizing use. Instead, it should be about the critical and careful interpretation of those generalizations.

3. Cultural anthropology that focuses on culture as a way of life. The holistic concept of culture, with its assumption of a coherence of thought and action, is central. However, this should not lead to the assumption of uniformity, as intra- and intercultural diversity are a fundamental characteristic of all human communities. Furthermore, culture includes not only ideational elements, such as symbols, cognition, and meanings, but also material products, such as pots and cultural landscapes. Culture is in no way reducible to either the material or the ideational; culture is materially hybrid. Despite hundreds of proposed definitions, the fundamental characteristics that remain are the high significance of the non-genetic transmission of information and the adaptive and orientational capacity of culture for human action. Within the anthropological disciplines, it should be the task of cultural anthropology in particular to construct knowledge about how actions can be described in the context of those studied and understood as coherent from a scientific perspective. A focus on cultural knowledge allows for a reduction in the fixation on difference, because knowledge is inherently transitive between people of different cultures and is assumed to be more self-evident in knowledge stocks than in culture, since they are not only distributed unequally between but also within groups of people. In the more recent discussion about indigenous knowledge (*local knowledge*), this knowledge is nonetheless often exaggeratedly presented as uniform, holistic, environmentally adapted, or as a "completely different" kind of knowledge.

5.2.3 Criticism of Other Common Definitions of the Discipline

In the literature, there are still a number of other definitions of cultural anthropology. Although many of them do indeed capture certain characteristics of cultural anthropology, in my view they are not suitable for defining the core of the discipline. Therefore, these characteristics should only be used secondarily when characterizing the discipline.

First, I see cultural anthropology only secondarily as the study of "cultures," "ethnic groups," or *shared values and ideas*, even though this is the traditional focus of cultural anthropology. Cultural anthropology should be focused less on societies ("cultures") as such, but rather on specific ways of doing the same thing or on different ways of solving the same problems. A definition of cultural anthropology as the study of cultures always runs the risk of understanding these as internally uniform and delimited and of mistakenly assuming that culture is shared among individuals. This essentialization easily leads to seeing people as mere "specimens" of cultures and thus depersonalizing them. This is particularly dangerous in times of extreme identity politics, where identity is fertile ground for political maneuvers and culture is characterized as difference in cases of doubt, which is something that often happens to us anthropologists.

Second, I see cultural anthropology only secondarily as the science of the culturally foreign per se. If cultural anthropology is defined by the foreign, however, the foreign *and* one's self should be understood in their context, because the foreign is always rational. Foreignness is not a characteristic of cultures, but of their relationship. The foreign as such should not be the object, because this speaks against comparisons as a method and against the insight of the unity of man. In particular, cultural anthropology should not be defined by the foreign as the "complete Other." Otherwise, dichotomies that hinder cognition are cemented, which can even become very dangerous politically because they unintentionally question the unity of humanity. Glorifying, mystifying, but also degrading "cultural racisms" are widespread, and they can all too easily appropriate anthropological ideas or even just anthropological terms for themselves.

Third, an epistemic definition of the discipline is logically subordinate to the thematic and theoretical definition. This would be a characterization of the discipline in terms of its specific epistemic problem (foreign culture) and its practical problem of cultural exchange (intercultural interaction) as well as its general attitude in response to this (cultural relativism). Anthropological cultural relativism, with its respect for other cultures as a basic attitude, is important for a practically engaged cultural anthropology. However, a consistent cultural relativism is equally obstructive for research, because cultures are then thought of as monoliths and thus no comparisons are possible.

Fourth, I also consider a definition of the methodology or the methods (procedures) used in the discipline to be secondary. Methods are means to an end, regardless of whether they are scientific or life-related intentions. If cultural anthropology is defined by the methodology, then this should not only be marked by stationary field research, but should also include intercultural comparison, which is neglected in Germany, as a second methodological pillar. If field research is placed at the center, then one must precisely say what is meant: participation, perspective-taking, and the second socialization of a connected social "field," because geographers, sociologists, and historians also conduct "field research." Moreover, field research as a method is too diverse and too dazzling to form the basis of a whole scientific discipline.

As a consequence of the previous considerations, cultural anthropology should not, or at least not in essence, be defined in the following ways, some of which are particularly *en vogue*:

- In a postmodernist way, as a pure representation of the foreign, the other, as an exchange of ways of speaking or reading the world, as a praise of the "cacophony," as a "discourse of diversity," or as a supplier of "alternatives to us" (instead of "for us," in the sense of Clifford Geertz (1992)), or also as pure "evocation," literary criticism, or art. All of these can be done better by artists, literary critics, or novelists (for example, Gabriel Garcia-Marquez, Salman Rushdie, and John Irving) than by anthropologists. Moreover, these goals contain a number of internal contradictions, for example, an evocation cannot forgo representation. Cultural anthropology as a cultural critique is also not possible if it does not explicitly state what it is criticizing and what it stands for. The "critique" expressed in the works presented in *Anthropology as a Cultural Critique* is in most cases directed at academics from other disciplines or at Western intellectuals. It often remains feuilletonistic, self-serving, narcissistic, and diffuse. The mentioned orientations can be criticized as "politically correct" postmodern cultural anthropology or as "moral models." However, other branches of critical cultural anthropology and representatives of humanistic anthropology also criticize the fact that the texts of postmodernist cultural anthropology and anthropological cultural criticism usually refer to other texts and little to the actual lives or experiences of real people. Classic topics of cultural anthropology are therefore also dealt with by others. The information made available to the public about other societies is almost exclusively provided by journalists, who often build on their extensive knowledge of the country, but hardly ever by anthropologists. Furthermore, postmodernist authors refer to texts that are canonical for the discipline and have a hegemonic effect, but hardly to texts in which cultural anthropology was popularized and thus became socially effective. Here, an examination of older texts, for example, by Margaret Mead and Ruth Fulton Benedict, or of more recent ones, such as those by Carlos Castaneda, Michael Harner, or Nigel Barley, would be more important than classic monographs. On their way into the university establishment, the critics have domesticated their criticism. In many ways, postmodern approaches have enriched anthropology theoretically and empirically, but in this respect, postmodern anthropology has led in the wrong direction.

- As a science that is limited to the internal or personal view of the actors (emic view), because the distinction between emic and etic is analytically useful, but in concrete contexts we always have mixed phenomena. Today, people live in exclusive identity worlds hardly anywhere. In situations of mixed, diffuse, and changing identities among those being studied as well as those doing the studying, the separation between emic and etic becomes as problematic as that between the self and "others." Anthropology should be more dialogical, reflecting the voices of those studied, but it is an illusion that it could be practiced by the subjects themselves in the sense of an indigenous anthropology, unless they are trained anthropologists. Anthropology is a profession with special concepts, methods, and validity criteria, not a hobby science.
- As a study of delimited, small, and/or culturally homogeneous ethnic groups or as an investigation of "the local" or only of "local knowledge" (indigenous knowledge). All groups of people have historically been influenced by others and are, often for a long time, networked with other groups. They are often internally diverse and culturally fractured. If this is overlooked, there is a risk of artificial "ethnicization." Small communities are less able than ever before to serve as examples (synecdoche) for larger cultural units. Even today, the concept of "society" itself is increasingly disputed, as there is always the danger of reification. The postmodern discussion has brought up a serious criticism with respect to this. The concept of the local, on the other hand, is more popular than ever, but problematic in many ways. Unintentionally, but not coincidentally, it is implied that local communities are the least developed.

- As a human (versus natural) science or as a social science applied as a qualitative or interpretative science, because there is no fundamental difference between quantitative and qualitative, and because an interpretative approach is not a peculiarity of anthropology within the social sciences.
- Through the history of the discipline or its sometimes very problematic colonial role, because this is purely backward-looking, because coincidences in the history of science also play a role, and also because much remains unclear, because there were also many anti-colonialists among anthropologists, and finally, because historians of science are more competent and detached for these questions as they have no guilty conscience.
- By means of demarcation lines to other subjects, for example, by not emphasizing the differences to sociology, because otherwise territorial fights would continue and because reality is not organized according to the historically grown and mostly not systematically laid out disciplinary boundaries.
- As a "humanities" discipline (versus "natural science"), firstly, because this separation as such is outdated, secondly, because a human being is not only spiritual, but also a living being with a body, and thirdly, because sociality, even very complex sociality, is not limited to humans and is also studied in the so-called natural sciences (biology: behavioral research, sociobiology).
- As a "cultural science," without explaining the very special anthropological concept of culture, because cultural anthropology (also as a practice-oriented project) would otherwise be lost in other so-called cultural sciences, which usually mean something completely different by "culture." There are now too many cultural sciences for "culture" to still be a symbolic resource.
- As a "moral science," because cultural anthropology as a "moral model" overlooks the complexity of the causes, tends to think in black and white ("good" and "evil"), making it all too easy to believe that good intentions lead to good results. In its rejection of objectivity, it tends not to change. The currently prevalent form of moral anthropology, which tends to see the entire discipline as a moral or humanitarian enterprise (see contributions in Fassin 2012), assumes that it is mainly the legitimization of mystifications and not material relationships and inequalities that bring about oppression. Moreover, it does not say whether there is also "good" power, and strongly tends to integrate itself into a globalist moral project (see criticism in Kapferer and Gold 2016).

What does this mean for cultural anthropology's potential contribution to the Anthropocene as a problem and discourse? What are its strengths? Using the definition of the subject matter presented in Sect. 5.2, including the specific methodological characteristics of cultural anthropology, I see the following strengths (Table 5.1):

Table 5.1 Relevant cultural anthropology: strengths of cultural anthropology in the lobal context

Subject
Societies, groups, and networks all over the world, not just "foreign cultures" Cultural characteristics and global cultural diversity, but also diversity within societies (intracultural diversity) Everyday issues that are relevant to people worldwide
Theory
Culture comprehensively conceived as the shaping of existence, holistic but not totalizing Emphasis on systemic connections, for example, between religion and the economy
Perspectives
Culturally relativistic, not judgmental Not ethnocentric Not Eurocentric or nostocentric Scientifically distanced external view and internal view(s)
Methodology
Firstly, through field research as a local and experiential approach Secondly, through a broad cultural comparison: Differences between societies, but also similarities as well as worldwide patterns (universals) Modern large-scale societies as well as other communities: *WEIRD* and *non-WEIRD*
Ethics
Emphasis on the specifics of responsibility in research on and with people Taking into account notorious asymmetries in the research situation
Researcher's relation to the subject
Inter- or cross-cultural personal experience over a long period of time Intensive local knowledge and competence in the local language
Conclusion
Cultural anthropology conducts research on and with people in collectives The particular strength of cultural anthropology is the investigation of generally important and globally relevant topics through intensive detailed studies: "Large issues in small places"

5.3 Grounded Anthropology—Nature, Climate Change, and the Anthropocene

> *How can anthropology contribute to an understanding of climate change if the object of investigation defies human perception?*
>
> Bauer and Bhan (2018: 8)

Just as geology needs to be humanized, anthropology should be grounded terrestrially.How is the Anthropocene to be thought of in cultural theory? What can cultural anthropology contribute to the topic in theory and empiricism? What could an "Anthropocene Anthropology" (Moore 2016) look like? Anthropology shares the first two syllables with the word Anthropocene, but it needs help as to how it should relate to a concept and a matter that strongly imply an interdisciplinary approach (Bubandt, in Haraway et al. 2016: 538). Although anthropologists often see the subject as inherently interdisciplinary, this interdisciplinarity usually only includes disciplines from the social sciences and humanities, and rarely includes the sciences.

Assuming that Anthropocene effects consist in the accumulation of the activities of specifically integrated place-based actors (Bauer and Bhan 2018: 13), anthropology should be relevant. Furthermore, anthropology has always strongly assumed that human communities are embedded in the natural environment. This ranges from the earlier concept of "primitive peoples" to modern medical anthropology, migration research, and environmental anthropology.

> Considering that the Anthropocene is at root a chronological designation about human activities and their relationships to the global environment, one might expect that anthropology would have had input into its formulation (Bauer and Ellis 2018: 211).

Previous anthropological (and also other cultural-scientific) contributions are only recognized by very few natural scientists, which at first glance may be due to the strong quantitative orientation (Hann 2016a: 19). Anthropology must critically ask itself why this lack of awareness exists, as there are notable exceptions of scientists who are very open to cultural studies, such as the environmental scientist Erle Ellis, who was also an advisory member of the AWG from 2009 to 2023.

5.3.1 Disparate Basic Anthropological Positions on Nature

In the current anthropological discussion, there are, to put it simply, two fundamentally different views of nature across a variety of approaches. These basic attitudes also shape the anthropological debate on the Anthropocene (Thomas et al. 2020: 95). A traditional view assumes that human communities emerge from the interaction of culture and nature. Philippe Descola has formulated it in such a way that all human societies represent "compromises" between nature and culture (Descola 2013, 2015). The task of anthropologists is to carry out research on the respective forms of this compromise. Thus, an analytical distinction is at least drawn between nature and culture. Critics call this view anthropocentric; I call this view a biocentric anthropological perspective.

The problem with this position arises mainly from the fact that both terms, nature and culture, (a) are used in everyday life as well as in the sciences, both (b) are used descriptively as well as evaluatively, and (c) the word "culture"—at least in the dominant scientific language, English—is one of the most semantically complex words, which, in its polysemy, is only surpassed by "nature" as the most complex word (Eagleton 2000: 1). As such, this is a fruitful field of research, especially because these words are so effective. Arizpe Schlosser, for example, analyzes and illustrates this extensively and polyphonically with the handling of the word "culture" based on documents and many years of participant observation at international conferences on development, sustainability, and the Anthropocene (Arizpe-Schlosser 2019). Like nature, "culture" allows for a multitude of references; as a hyper-referential term, it is open to the most diverse political use and exploitation (Kuper 2003; Antweiler 2018).

Newer approaches, on the other hand, assume that the dualism between nature and culture (*naturalism*, Descola 2013, 2015) must be dissolved. This refers to a different direction in philosophy, media, and cultural studies, which is now known as "new materialism" (Hoppe and Lemke 2021). It emphasizes the close interconnectedness of culture and the non-human environment. Some representatives of a "non-human turn" or of the "more-than-human" advocate for a fundamental "decentering" of the human sciences in order to accomplish a lot with and for the non-human. This implies, for example, that violence against humans and "non-humans" cannot be separated.

Societies are therefore parts of more complex networks (or "cosmologies"). Currently, the most important representatives of this approach are, in general, Donna Jean Haraway and Bruno Latour, and in anthropology, Eduardo Viveiros de Castro and, to some extent, Philippe Descola. In the anthropological Anthropocene discussion, it is primarily Anna Lowenhaupt Tsing. In some variants, humanity is explicitly seen as not synonymous with *Homo sapiens* (Hoelle and Kawa 2021: 659). Instead of a scientific and thus materialistic-monistic concept or biological concept, non-western models are used for theorizing within the framework of a so-called relational or "flat" ontology. This leads to the mixing of ethical and emic perspectives.

Within this framework, many anthropologists essentially represent a monistic position, but one that does not simply want to theoretically dispose of nature. Instead, they want to relativize or minimize the importance of humans as actors in the face of other "actors," which is why this approach is also referred to as "posthumanism" (Loh 2018). Under "non-human," for example, Richard Grusin includes "animals, affectivity, bodies, organic and geophysical systems, materiality, and technologies" (Grusin 2017: vii). In view of the problems of the Anthropocene, the question arises as to whether such a disparate conception can actually contribute to the frequently admonished co-production of social or cultural-scientific knowledge combined with the findings of natural science (Arizpe-Schlosser 2019: 285).

The fictions of others are a preferred research object in cultural anthropology. Myths, stories, and individual narratives are meticulously analyzed. In addition, fictional representations of facts can provide illuminating insights or sensitize people to problems, for example, through alienation (*ethnofiction*, Augé 2017: 78–79). Sometimes, however, the fictions, ideas, or worldviews of others are confused with distanced analysis. This is precisely what I observe in writings on the "ontological turn." There, the worldviews of indigenous people are sometimes presented in such a way that they are seen either as role models for ecologically meaningful action in the Anthropocene or even as an alternative to so-called "Western" science (Kohn 2013; Holbraad and Pedersen 2017; see the criticism in Bessire and Bond 2014).

This is partly reflected in more recent debates on the pluriverse. Following Zapatista conceptions, this *umbrella term* refers to a "world of many worlds," a compilation of concepts, worldviews, and practices from all over the world that focuses on diversity instead of modernist universalism (Escobar 2018; De La Cadena and Blaser 2018; Canaparo 2021). The now globalized activist project of the pluriverse or the "multiple (one) world" aims at a shift in human civilization towards diversity through experimental alternatives and a development concept aligned with socio-ecological environmental justice (Blaser and De La Cadena 2018; Kothari et al. 2019: xvii; Omura et al. 2019). This sometimes leads to exaggerated statements, such that the insight into the global environmental crisis is new only for Western societies. Indigenous communities in the Americas, for example, have already known this for 500 years and have processed it in their dystopias or their *science (fiction)* (Hoelle and Kawa 2021: 660; Whyte 2018: 228–234).

In contrast, it must be noted that the experiential background of these apocalyptic perspectives is local or at best regional in nature. What remains unnoticed in this, in my opinion, reduced, view of Western science is that traditions of relational ontologies also existed in so-called Western sciences in pre-colonial times and still exist today (Clark and Szerszynski 2021: 165). Many works suffer from aligning themselves with the supposedly correct side (see Ellen 2021: 3–5). I have fundamental concerns about this, similar to how Marc Augé formulated his in relation to cultural anthropology in general:

> It even went so far that anthropologists, seduced by the fictions of others, became more like their precursors and, by breaking with the imposed distance of the anthropologist's role, simply transformed themselves into more or less learned guarantors who spoke in the name of the culture they had set out to study (Augé 2017: 78).

A third view is a decidedly monistic view that emphasizes the uniqueness of humans, but is clearly materialistic. Generally, this position, despite all the differences in detail, is held by the developmental psychologist Michael Tomasello, the entomologist and evolutionary biologist Edward O. Wilson, and the primatologist Carel van Schaik. In the Anthropocene debate, it is mainly represented by Chakrabarty, the environmental historian John McNeill, and—in some of his many contributions—by the geographer Erle Ellis. Critics see it as an epistemology that is "... characterized by uniformity, formality, soullessness, and indisputability" (Danowski and de Castro 2017: 170, see Weber 2019). Among anthropologists, this approach is represented by comparatively few colleagues, most prominently by Thomas Hylland Eriksen. I myself tend towards this position, with an even stronger evolutionary emphasis, which I will explain in the following chapter.

Overall, one can ask critically why anthropological contributions on the subject notoriously argue against a dualism between culture and nature, but then still cling to what I believe is ultimately an idealistic monism under the banner of "flat ontology", which retains the dualism, albeit subcutaneously. In a nutshell, the question is: How can the genuinely geological category be "socialized", and how can sociology and ethnology be geologised (Delanty and Mota 2017; Clark and Szerszynski 2021: 38–41,46–49; Leggewie and Hanusch 2020; Hanusch et al. 2021)? What could a cross-disciplinary synthesis of a "geoanthropology" (Barker and Barker 1988; Renn 2020: 375–376) look like?

Behind these different perspectives on nature and humans are different views of what cultural anthropology is all about. They imply visions of how cultural anthropology could or should develop in the future. At present, the most common contributions of cultural anthropology include: first, more or less harsh criticisms of the term Anthropocene, and second, ethnographic case studies on local impacts or local perceptions and on ways of dealing with Anthropocene effects. The Anthropocene, as I will often note, challenges cultural anthropology in many ways. For example, what do ecological relationships between far-flung communities and between *several* species mean for the subject and methods of cultural anthropology?

5.3.2 Nature and Environment in Cultural Anthropology—Milestones of a Long History

Cultural anthropology has a historically grown proximity to environmental issues and yet also a distanced relationship to nature. Unlike sociology, which deals with modern societies, early cultural anthropology had a proximity to the topic of nature in the division of labor between the humanities and the social sciences. This was because cultural anthropology was institutionalized as "ethnography" from around the middle of the nineteenth century, as a study of the so-called "primitive peoples," groups of people who are characterized by a supposedly special proximity to the natural environment. On the other hand, there was also a tendency in early cultural anthropology of the early twentieth century to avoid both nature as a topic and biology. This is often associated with Alfred Kroeber's concept of the "superorganic." In fact, Kroeber did not deny the biotic basis of culture and the universals, but only wanted to treat culture autonomously for methodological reasons (Kroeber 1917; Jackson Jr. 2010: 134–140; see Halvaksz 2021: 247).

There is also a clear reference to the environment in the field of social theory through Émile Durkheim, one of the founding figures of cultural anthropology and sociology. Durkheim addressed the relationship between culture as a "microcosm" and nature as a "macrocosm." By means of a specific assumption that cultures project social categories onto the environment, he argued that this relationship formed the foundation of human thought as well as cultural social organization, which was codified in religion. These topics were taken up by anthropologists forming important schools of thought, especially in functionalism, and were applied in particular by Edward Evan Evans-Pritchard in ethnographic works on environmental use and meteorological or "ecological time" (Evans-Pritchard 1940).

The combination of environmental research with anthropology as a science of human existence was etymologically established at around the same time as the first use of the word "ecology" by the German zoologist Ernst Haeckel. In his "General Morphology of Organisms" from 1886 (Haeckel 2019), he emphasized the significance of *oikos* as pertaining to household, city of residence, and family. At the center of human ecology as well as cultural anthropology are connections between human systems and their environments (*coupled human and natural systems*, Orr et al. 2015: 156–157; pioneering Greenwood and Stini 1977). Ecology and cultural anthropology also resemble each other in their scales of investigation. Just as cultural anthropology focuses on social processes, groups, and networks, ecology within the life sciences is primarily concerned with symbiotic communities and ecosystems at the meso and micro level.

Against this background, it is hardly surprising that cultural anthropology, as compared to other human sciences, addressed the human–environment relationships or rather the culture-environment relationships comparatively early in the history of the discipline (Pálsson et al. 2013: 5). One of the few syntheses on the subject already cites a variety of topics, approaches, and methods under the umbrella term "ecological anthropology" (Orlove 1980). I will mention only some milestones below. Since the 1950s, cultural ecology has been concerned with the interrelationship of current human lifestyles with the natural environment (McNetting 1977). However, older approaches to the exploration of long-term cultural change (*cultural evolution*) and also archaeology were included (Steward 1955). Approaches to a broader environmental anthropology emerged as early as the 1960s (Vayda and McCay 1975; Ellen 1982).

Since the 1980s, many research directions have been added, as well as many collaborations with numerous other disciplines which have focused on "coupled human and natural systems" and, above all, on political aspects and supralocal relationships. This can be seen in the six more specific research overviews on ecological topics in cultural anthropology, which appeared in the *Annual Reviews of Anthropology* between 1991 and 2012; the overviews were on the ecology of materials, behavioral ecology, indigenous peoples and environmental politics, historical ecology, complex adaptive systems, and on climate and culture. The next truly broad overview on the anthropology of the environment, which appeared only 35 years after Orlove's synthesis, shows this enormous breadth and signals this through the long-standing umbrella term of "environmental anthropology" (Orr et al. 2015).

One specific environmental school of thought that emerged in the 1980swas ethnoecology, which was particularly concerned with local environmental knowledge, was strongly influenced by cognitive cultural anthropology (for example, Conklin 1955), and partly argued from a culturally deterministic perspective. A second direction was historical ecology (for example, Ballée contributions in Descola and Pálsson 1996a, b), which highlighted how long supposed natural landscapes have mostly been shaped by humans. I will discuss both directions in Sect. 5.3.

Beginning in the mid-1990s, there was much discussion about the concept of nature and what contribution cultural anthropology as a theoretical science could make to the topic of the environment, especially through anthropological cultural theory and cultural relativism. This gave rise to important critiques, for example, of ecosystem models from which culture was notoriously excluded, partly by anthropologists themselves (Milton 1996: 55–59). In this context, modern *environmentalism* as a discourse, ideology, or political program was also comparatively examined. This resulted in empirical studies that debunked as myths some of the widely held assumptions about the environmental orientation of indigenous groups in various schools of ecologism (for example, Ellen 1982, 1999; Milton 1996, see 2002). This explicitly cultural-theoretical discussion has unfortunately somewhat subsided, although some of the topics, if not these theories, re-emerged within the newer approaches of *multispecies* anthropology.

In the 2000s, a more politically engaged critical approach emerged (*critical ecological anthropology*), which I will discuss below. The tenor of many recent contributions to environmental anthropology is one of dismay at the realization of the extensive and sometimes radical effects of human activity on ecosystems and, above all, the impression that these could be irreversible and harmful to all. Orr and co-authors describe this as pathos reminiscent of the tragedies of Sophocles. Where the "end of history" was recently celebrated due to the success of democratic regimes, human history now appears to be nearing its end due to environmental impact, with the main geological trace being the mass extinction of living beings induced by humans. Because of this break in perception and the rapid increase in knowledge about human–environment relationships, anthropologists have since become more concerned with the material and energy flows between societies and their environment, with the relationships between individuals and their collectives in the use of natural resources, and with the personal experience of these flows and interactions (Orr et al. 2015: 155; Haenn and Wilk 2006; cf. also Lockyer and Veteto 2013).

Anthropologists have therefore long been concerned with the interrelationships between cultures and their natural environments. In Germany, for example, there was a large DFG project on humans and the environment as early as the 1980s, in which human–environment relationships were examined using New Guinea as an example (Schiefenhövel and Heeschen 1983). A number of anthropologists were significantly involved in this project, working together with linguists and natural scientists from various disciplines. Anthropologists have adopted the Anthropocene concept from this perspective (Orr et al. 2015; Bauer and Ellis 2018: 209). This also applies, albeit to a much lesser extent, to long-term human–environment relationships and to spatially expansive processes, and through to global environmental change. In newer overviews on environmental anthropology, the topic of the Anthropocene appears several times (for example, Kopnina and Shoreman-Oumet 2017).

Given these topics and this concern for the environment, it is understandable that cultural anthropology took less time to engage with the initially foreign topic of the Anthropocene, which emerged in 2000, than with the topic of globalization, which was "new" to the discipline in the 1990s. This happened in very different ways, but some contours or anthropological discourses can be distinguished. What issues of global environmental change were anthropologists mainly concerned with before the concept of the Anthropocene emerged (Moore 2015a: 34–35; Orr et al. 2015: 155–162)? Historically, the earliest topic was environmental degradation, which was examined by John Bodley in many publications (Bodley 2016, first published in 1983). Here, the embedding of local communities in global environmental changes was central. Furthermore, the focus was on the local population's specific perception of changes as natural or artificial. A more recent direction can also be classified here, which has been known as the "anthropology of sustainability" since 2015 (Brightman and Lewis 2017). This orientation critically responds to development projects and also critically deals with the concept of development. Thus, it converges with critical orientations in development anthropology (for overviews, see Gardner and Lewis 2015; Antweiler 2019a).

The second major topic was the worldwide loss of biodiversity, which, following the *Rio Convention on Biological Diversity* in 1992, led to many anthropological studies. In anthropological studies, the topic of biodiversity has been examined since the 1980s, particularly in connection with local or indigenous knowledge. Here, special attention has been paid to the social construction of local knowledge as a concept with political, economic, and legal implications, for example, as a driver of indigenous communities' self-awareness or as a legal category for the implementation of property and knowledge claims (Nazarea 1999, for an overview, see Antweiler 1988).

Despite all the danger of romanticizing indigenous people as quasi eco-saints, which can mainly be found in globalized popular culture, there is still one empirically verifiable finding that can be important for a response to the Anthropocene. As is often emphasized by non-anthropologists, members of indigenous communities often see themselves not only as protectors of nature, but as "guardians of a world that is larger and older than they are" (Christian 2018: 330). Unfortunately, however, there is little reliable anthropological data on the prevalence of such truly global ecological orientations. Indigenous experiences are not only relevant for sustainability-oriented values, for which they are often invoked. They can also provide concretely useful knowledge, such as for the conservation of biodiversity.

One particular direction that focuses on species extinction is known as the "anthropology of extinction" (Sodikoff 2012). This is primarily concerned with how the reduction in biodiversity is perceived and how ideas about ecocide are socially constructed. However, extinction or the word component "-cide" is used here as a general analytical concept. It is not only about the extinction of other species, but also of languages, non-western knowledge systems, and ways of thinking: ecocide and ethnocide are brought together on several levels. The "sixth extinction crisis" of species is thus only one example of the effects of humans as *apex predators.* This could be combined with ethnoprimatological research that examines the human nexus in which a large part of today's great apes exist. Recent regional comparative studies show, for example, that the closer primates live to humans, the less diverse their behavior is Kühl.

These anthropological approaches, which brought together politics and the environment, fit well with two eminently interdisciplinary research fields: Science and Technology Studies (STS) and political ecology. Political ecology in particular focuses on the actors with their different interests and their notoriously unequal power (Lit.). However, it is precisely these convergences that raise the question: "But what about specifically anthropological engagements with the Anthropocene?" (Moore 2015a: 34). A key focus of this research has been on the diversity of actors, who, in addition to scientists, are involved in the generation of knowledge about climate change, such as journalists, science managers, administrative staff, politicians, and environmental activists in the Global North and Global South, as well as the leaders of marginalized groups of indigenous movements. By combining political ecology with research on ontology and multispecies relationships, new concepts have also emerged. One example is the concept of "plural ecologies" (Sprenger and Großmann 2018: xii; Duile et al. 2023). The term stands for cultural environmental settings in which seemingly incompatible concepts and relationships coexist in the sense that different people, such as those in a rural community, have different attitudes towards nature or carry out different environmental practices. Thus, in areas of Southeast Asia, for example, animistic or nature conservation ideas can exist in one and the same community alongside usage, efficiency, or growth-oriented environmental concepts. This leads to parallels, but also to tensions, negotiations, and, sometimes or temporarily, to hegemonies.

Both in political ecology and in STS, the aspect of knowledge access, knowledge control, and more generally of power differentials is central. As a result, there was a renewed attention to non-Western concepts of animals, plants, the environment, and ... humans. Local concepts of dealing with nature were examined and in some cases discussed as a model for sustainable interaction with the environment. In particular, the Latin American concept of "good living" (*buen vivir*) has been the subject of many research contributions, which have also aroused strong interest outside the field.

An example of an area of conflict where different concepts of various actors collide and power differences are significant is the phenomenon of invasive species, especially the strategies for limiting or eradicating non-native species. Not only do the interests of different *stakeholders* collide here, but also concepts of living beings and space that are used as conflict resources, such as "protectable" species or "threatened spaces," which—as social facts—are anthropologically questioned. Other fields of such environment-related conflict situations, where language means power, and where terms therefore have concrete consequences, are carbon trade, forest management, and the palm oil industry.

One specific research field that is relevant to questions about the Anthropocene is the interdisciplinary study of disasters, which go beyond catastrophic weather events such as Katrina (*disaster studies*, Oliver-Smith and Hoffmann 2020). This involves the examination of the impacts, perceptions, and reactions to volcanic eruptions, earthquakes, *tsunamis*, and floods. In my opinion, this research area could be paradigmatic in several ways for an anthropology of the Anthropocene, because such disasters represent Anthropocene phenomena on a small scale and raise the question of nature and culture in an exemplary manner. At first glance, the causes of natural disasters are clearly geological or geophysical in nature. However, hazard research and investigations on vulnerability carried out in the fields of social science and cultural studies show that so-called natural disasters often prove to be culturally co-determined.

This applies above all to the form of the effects and, in some cases, also to the culturally influenced etiology, such as the vulnerability of certain groups of people in the case of avalanches, landslides and shallow earthquakes. In short: natural disasters are also social disasters (Felgentreff and Glade 2008; Niewöhner 2013). This insight of disaster research (hazard research) coincides with aforementioned insights of political ecology and historical ecology, which revealed the anthropogenic character and the politico-economic shaping of many spaces that had been understood as natural landscapes until now. Furthermore, this fits with the prehistoric and historical research into the collapse of complex societies, for example, due to indigenous overuse of natural resources or due to colonial structures and exogenous influences, such as the controversial discussion in relation to the Easter Islands (Tainter 1988; McAnany and Yoffee 2010 vs. Diamond 2005).

Another closely related field is moral anthropology, in the sense that it primarily deals with moral tensions and debates that are significant beyond local communities, such as international development, global health, social justice, human rights, and security. All these topics involve divergent interests and controversial positions. Many have always seen a moral mission, such as opposing oppression and rejecting cooperation with the military. Especially in the global context, anthropologists cannot avoid this, and the topic of the Anthropocene calls like no other for collaboration between a wide range of actors (Tsing 2005a, b). An example of possible methodological approaches to such topics is to make experts on globally popular concepts themselves, for example in geopolitical scenarios (e.g. Hannerz 2016) and scientific models (e.g. Roscoe 2016), explicitly the topic of anthropological investigation.

5.3.3 Flexibility and Novelty: Enter Gregory Bateson

In addition to the major approaches to dealing with certain topics or aspects of the environmental relationships that groups of people have, there are theoretical approaches that have been largely forgotten in today's mainstream cultural anthropology. These include neo-evolutionary concepts such as Julian Haynes Steward's *cultural core* and Leslie White's insights into the relationship between energy consumption and cultural complexity (White 1945). I will discuss this in Sect. 5.9. A thinker who is at least as important for the topic of the Anthropocene is the great anthropologist and cyberneticist Gregory Bateson (1904–1980). Bateson was a genuinely ecological-minded scientist, probably the anthropologist who first fundamentally considered the relationship of humans and societies to their environment at all levels (psychological, social, economic, political).

Bateson was strongly influenced by the dynamic cybernetics of the 1970s, the science of the regulation and control of systems. His thinking was fundamentally ecological, focusing on systems, relationships, and processes. In doing so, he discovered fundamental processes of complex and dynamic systems using very different examples. He showed that mutually reinforcing growth processes can lead to the collapse of systems unless a third party intervenes and changes the relationship. Bateson described this as "schismogenesis," for example, in the arms race between the USA and the USSR during the Cold War and in interethnic processes of ethnicization and increasing dissimilation (Bateson 1985: 99–113). In general, such processes can be referred to as "runaway processes" (Eriksen 2016: 21–23, 27, 131). They are important for understanding the Anthropocene, as they are typical of anthropogenic effects, where the convergence of initially unproblematic growth processes leads to such runaway processes.

In addition to the networking of biological systems around equilibria, Bateson emphasized the breaks, the fundamental vulnerability of complex ecosystems to disturbances. In systems heavily influenced by humans, "metapatterns" and self-destructive repercussions can occur (Bateson 1985). The central conclusion from this is that cultural adaptation to environments can never mean fixation on a pattern. Adaptation can precisely consist in adapting to fluctuating environments instead of adapting to a certain dynamic at a certain time. A systemically built-in flexibility is required. This is even more true in an Anthropocene world: We need flexibility, "... to get out of the groves of fatal destiny in which our civilization is now caught" (Bateson 1985 (1972:74).

A recent key insight is that complex systems can change quickly due to small causes, or that real systems have the potential to switch between different basic modes. In the Anthropocene, this affects large systems such as parts of the geosphere (Clark and Szerszynski 2021: 47). In ecology today, we speak of "fundamental novelty" in ecosystems (Kueffer 2013, see Van der Leeuw 2019). Unintended effects of human activity are only recognized at a late stage. The classic example from cultural ecology is a study on the pig cycle of the Maring in New Guinea (Rappaport 1968). In a cycle that is initially in equilibrium, the number of pigs, the economically and symbolically most important animal of the described society, can suddenly reach a level where the pigs destroy the fields, the basis of agriculture. Such system dynamics lack a limiting mechanism, a thermostat, so to speak, that slows down the overheating of the system. Eriksen considers the Anthropocene as a whole to be an example of the increase in unintended effects due to such runaway processes and metaphorically terms the phenomenon as "overheating" (Eriksen 2016: 22, 152–153, 2018), which does not refer to the specific atmospheric warming that I will now turn to.

5.3.4 Climate Change Anthropology in the 2000s

A third and central theme in environmental anthropology is anthropogenic climate change, which, since the 2000s, has attracted even more attention than species extinction and environmental degradation, and natural disasters (Dove 2014). Relatively early on in the first decade of the twenty-first century, this led to the demand for a separate sub-discipline, the anthropology of climate change (Crate 2009; Hastrup 2016; Baer and Singer 2018). This is usually associated with a call for the active involvement of cultural anthropology, which particularly consists of documenting the local effects of climate change and making the views of the local population visible. This can fill conceptual gaps and also data gaps in the dominant, macro-scale, and quantitative aspects of research (Moore 2015a: 35). Within contemporary environmental anthropology, direction that consider climate anthropology as critical and also applied cultural anthropology play a dominant role, as can be seen in conferences and edited volumes. These research approaches are known as *anthropology of climate change, climate ethnography*, or explicitly as *critical environmental anthropology* (Strauss and Orlove 2003; Crumley 2007, similarly Bauer and Bhan 2018).

One of the more recent topics is the study of communities that are particularly vulnerable or susceptible to the consequences of climate change. Such climate ethnographies have highlighted political questions that extend beyond the local living environment. This, in conjunction with the climate discourses and global institutions that have emerged worldwide since the 2000s, has led to a fourth topic, namely political decision-making structures. Studies such as those on the "politics of global warming" have focused on the central actors of power, such as governments and institutions. In this context, the centers of climate change science, where the dominant concepts are devised and implemented, are examined with a critical eye (for example, Lahsen 2005; Worster 2008). Here, the central contribution of the discipline is seen in interlinked detailed studies, such as in "critical, collaborative, and multisited climate ethnography" (Crate 2011; Gibson and Venkateswar 2015).

Anthropogenic climate change is only one area of Anthropocene change, but an important one, and the anthropology of climate change is also theoretically relevant for the anthropology of the Anthropocene. In my opinion, Bauer & Bahn built a substantive bridge between the anthropology of climate change and Anthropocene research through their consideration that the climate itself can be seen as an assemblage, in the sense of Deleuze & Guattari and in the use of DeLanda. In the case of climate, very different factors of the most diverse materiality play a causal role, which interact with each other and whose respective effectiveness depends on each other. Climate arises from the interaction of different materials, physicochemical processes, energy transformations, material objects and organisms, including humans. These active forces also exist on different scale levels, ranging from the gravitational effect on the parameters of the Earth's orbit on the macro scale, to the respiratory activity of bacteria at the micro level (Bauer and Bhan 2018: 3–4).

When Bauer and Bhan conceive of climate as an assemblage and as a "dynamic organism" and also regard it as living due to its scope and its concomitant temporality, this seems at first glance to be similar to the concept of *Gaia.* Bauer and Bhan consider *Gaia* to be a more accurate characterization of climate than the term hyperobject. Hyperobjects in the sense of the literary critic Timothy Morton (Morton 2013: 1; Morton 2009) are *gargantuan* objects, which not only entail an extreme distribution in time and space in relation to humans, but also exhibit a continuous logic or order. This is not in the case with the climate. I would point out that *Gaia*, in the sense of Lovelock & Margulis, focuses on the coevolution of air components and organisms and is therefore not a spatially, but certainly and objectively a much more limited concept.

The concept of assemblage is mostly used to study micro-relationships, such as those between different materials in a landfill or in a craftsman's workshop. Since the atmosphere, where the climate occurs, is only one sphere of several Anthropocene-relevant Earth spheres, one could—taking Bauer and Bhan's argument further—understand the Anthropocene-shaped parts of the geosphere in this sense as a geodynamic macro-assemblage, which would go beyond the idea of a "global assemblage" as a transnational constellation (Ogden et al. 2013: 342).

With such terminology, however, it is always necessary to ask what the scientific added value is, for example, as compared to the descriptive terminology developed in the scientific fields of geography and ecology for corresponding processes. A comparable example where such a question arises is the term "scapes," according to Appadurai, which is widely used in programmatic texts and is now partially being replaced by the term "assemblage" (Orr et al. 2015: 161). The term *scapes*, in my opinion, is stimulating, but nothing more. It is theoretically much less fruitful and, above all, less empirically realizable than that of the network (Antweiler 2020).

Beyond environmental anthropology and the anthropology of climate change (Table 5.2), there are works in various subfields of cultural anthropology that are thematically or conceptually related to the Anthropocene without using the word, as some of them were written long before the "discovery" of the Anthropocene. Examples are the historical-anthropological works of Sydney Mintz on transnational economic relations and their lasting effects on local ecosystems (Mintz 1985) and the works of Erik Wolf on the influence of non-Western societies on the global history of the world (Wolf 1982), which will be discussed below.

5.3.5 Current Reception and Positions on the Anthropocene

In international cultural anthropology, the concept has been widely received since around 2010. The week-long annual conference of the *American Anthropological Association* (AAA) in 2014 featured several competing panels on the topic each day, and the scientific researcher Bruno Latour gave a much-noted plenary lecture (Latour 2017a, b). In contrast to climate change, however, the Anthropocene has hardly made it into introductory works, textbooks, and reference works in cultural anthropology as well as sociology, except for mentions (as an exception, for example, Robbins and Dowty 2019; Robbins and Dowty Beech 2021; Palmer 2020). This applies even to extensive reference works on globalization or globality (for example, Kreff et al. 2011; Niederberger and Schink 2011; Kühnhardt and Meyer 2017).

To this day, the topic of the Anthropocene in cultural anthropology has been overshadowed by the more narrowly focused engagement with human-induced climate change. In one overview paper about anthropological work on climate change, the word Anthropocene does not appear (Crate 2009). Ten years later, this has changed somewhat. Now, the Anthropocene is featured in both the content and subtitles of anthropological books on climate change (Bauer and Bhan 2018). However, the Anthropocene in the context of environmental issues is addressed rather marginally in other discussions, especially with respect to human-animal relationships (*multispecies anthropology*) and non-Western concepts of the world and the environment as well as the *ontological turn* stemming from them (Viveiros de Castro 2012; Pickering 2017; Tsing 2019b; Grana-Behrens i.V.: 10–17). This applies even to more recent anthropological overviews on the Anthropocene (for example, Mathews 2018). I will go into these debates in detail in Sects. 5.3 and 5.5.

Table 5.2 Directions of environmental anthropology in broad terms and further anthropological directions and their potential for anthropological Anthropocene research

Direction	Representative	Leading Theories, Sciences	Methods	Anthropocene-relevant Aspects
Anthropology of climate change (*climate ethnography*)	Crate and Nuttall (2016), Barnes and Dove (2015), Dietzsch (2017)	Various; climate cultures	Observation, participation, critique	Anthropogenic Climate change as a paradigmatic area of the Anthropocene
Research on environmental degradation; cultural anthropology of waste	Tsing (2015), Bodley (2016), Resnick (2021)	World system theory, capitalism	Ethnography, comparison	Soil sealing, species loss, waste, assemblages
Disaster anthropology	Fortun (2001), Barrios (2017)	Historical genesis of crises	Ethnography, comparison, partial observation	Perception, vulnerability, coping, politicization
Cultural ecology, ecological anthropology	Steward (1955), McNetting (1977)	Neo-evolutionism, adaptation, cultural Core	Cultural comparison, long-term field research	Infrastructure, Technofossils
Political ecology (*new ecology*)	Leach (1970), Blaikie (1985), Swingedouw, Scoones (1999), Maffi (2001)	Marxism, postcolonial theory, dependency theory, Arena	Actor analysis, discourse analysis. Intersectional analysis	Interest groups, discourses on the Anthropocene, landscape protection
Extraction of material resources	D'Angelo and Pijpers (2022)	Extractivism	Field research and Documen tary analysis	Interests, political context
Systemic *anthropology*	Vayda (1983, 2008, 2013), Walters and Vayda (2018), Lansing and Cox (2019)	Regulation, feedback, complexity theories, coupling, niche construction	Quantitative modeling, simulation, Progressive contextualization	Micro-, Meso-, macro-relations, non-linear dynamics
Ethnobiology (*ethnoscience*)	Frake (1962), Werner (1972), Medin et al. (2002), Ellen (1982, 1999, 2021)	Structuralism, cognitive science, laypeople as scientists, logic, Ethnotheories	Domain analysis, triad tests, analysis of taxonomies and causal thinking	Local environmental knowledge, species protection, universals in nature concepts,

(continued)

Table 5.2 (continued)

Direction	Representative	Leading Theories, Sciences	Methods	Anthropocene-relevant Aspects
Anthropology of trouble	Rappaport (1993)	Systems theory,	Comparison, observation	Crisis also for theory generation
Explicitly anthropological-ecological cultural theory	Milton (1996)	Structuralism	Theory analysis and theory construction	*Environmentalism* as ideology or political program
Exploration of alternative ontologies	Descola (2013), Kohn (2013, 2015), Viveiros de Castro (2019)	Philosophy, Poststructuralism Metaphor theory	Ethnography,Critique of Exceptionalism and eurocentrism	Indigenous world models, end time concepts, Future concepts
Multi-species studies (more-than-human studies)	Haraway (2016), Tsing et al. (2020), Bubandt (2018, 2019)	Cyber studies, literary studies	Theory, Metaphor formation, Critique of capitalism	Species interaction, interaction with material objects
Phenomenology of interaction with environment	Ingold (2011b), Vetlesen (2019)	Philosophy, Phenomenology, *embodiment*	Ethnography, landscape morphology	Perception and experience of change
Cultural evolution (social evolution)	Steward (1955), Sahlins and Service (1960), White (1949), Hodder (2020)	Energy, physics, entropy, *constraints*	Diachronic comparison, natural experiments	Energy use, cultural Core

The Anthropocene has definitely arrived in cultural anthropology. However, the topic is still not part of the established canon, and often, the tendency has been to adopt the word rather than delve into the conceptual core. There is a huge amount of articles from cultural anthropology on the subject and many chapters in anthologies on the anthropology of climate, sustainability, or in contributions to research on the "more-than-human." In keeping with the intensive engagement with human–environment relationships, the first sentences of the latest handbook on environmental anthropology emphasize the subject's proximity to the topic of the Anthropocene:

> As so many of us are all too aware, we have entered the Anthropocene, marked by the large and active imprint that humans have made on the global environment. Environmental social sciences, and environmental anthropology in particular, have long focused on the interaction between human societies, cultures, and complex environments – both in physical and symbolic terms (Kopnina and Shoreman-Ouimet 2017: 3).

Unlike anthropogenic climate change, the topic of the Anthropocene has not yet reached the canon of education (Table 5.3). The topic of the Anthropocene has been, for the most part, either touched on in chapters on globalization or the environment, or dealt with en passant under the keyword "anthropogenic" on less than one page. Even in the indexes of introductory works, the topic of the Anthropocene was rarely mentioned until around 2015. This also applies to anthologies on theories and even to anthologies on practice-oriented or rather engaged cultural anthropology, or *public anthropology*, or even development anthropology. Most of the new encyclopedias of cultural anthropology and anthropology still do not include a keyword on the topic, unlike the *Cambridge Encyclopedia of Anthropology* (Chua and Fair 2019), which is available online. The more recent English-language handbooks of cultural anthropology do not mention the Anthropocene at all (Fardon et al. 2012; Carrier and Gewertz 2013), and even the recently published handbook only cursorily addresses the topic in five places and otherwise as a sub-aspect of another topic—legal anthropology (Pedersen and Cliggett 2021: 358–361).

The first US college textbooks to devote more than a few lines to the topic appeared in 2020 (Bodley 2020: 406–417; Palmer 2020; Kottak 2021). In the German-speaking world, there is a lot of interest, but still few explicitly anthropological publications on the topic. The Anthropocene is hardly mentioned in established German introductory books and textbooks. A master's thesis from Cologne in 2016 is a pioneer (Krämer 2016), and the German Society for Social and Cultural Anthropology made it the theme of their 2021 conference (DGSKA 2021).

The possible contributions of anthropology to the Anthropocene depend strongly (a) on how deep the Anthropocene is set historically and (b) on whether cultural anthropology is narrowly conceived as relevant to the present, focused on local communities, and oriented towards field research, or as part of a broad anthropology as a science of humankind (Hann 2016a: 2). When the Anthropocene's beginning is set with the advent of the atomic bomb in 1945, many contributions are conceivable—and there will certainly be many approaches to this. In contrast, if one assumes an early Anthropocene, then the contributions on the currently dominant topics and theories of cultural anthropology are rather limited. However, this limitation would only concern the Anthropocene as a geological phenomenon, not the

Table 5.3 Milestones in the reception of the term Anthropocene in cultural anthropology

Start ca.	First or early occurrence	Examples
2012	Conferences	Increasing mentions
2014	Institutions, conference focus	American Anthropological Association: AAA Meeting (2014), Latour (2017a, b) (2014)
2018	Introductions, textbooks	Brown et al. (2020) (orig. 2018), Ingold (2018: 11)
2019	Special issue of a journal	*Current Anthropology* 2019
2019	Lemma in reference works	*Cambridge Encyclopedia of Anthropology*
2020	Textbooks: comprehensive consideration	Bodley (2020: 406–417), Palmer (2020), Kottak (2021)
2020	Theory overviews	Lavenda and Schultz (2020: 197–201)

concept as a current *social* fact. What approaches can be found in cultural anthropology on the subject of the Anthropocene? Inspired by Chua & Fair (2019: 4–12), I will attempt a systematization by asking about the key concepts and guiding theories, the empirical methodology, and the normative or ethical implications (Table 5.4).

Table 5.4 Cultural anthropology's approaches to the topic of the Anthropocene

Approach	Key concepts, theories, theorists	Methodology	Normative or ethical implications	Example
Empirical research in regions and life-worlds directly affected by the Anthropocene	Locality, local knowledge, indigeneity, vulnerability, heterogeneity	Ethnography, field research, collaborative research, island ethnography	*Advocacy* for the locally affected	Rudiak-Gould (2012), Lazrus (2012), Tsing et al. (2024)
Critical analysis of the concept as a problem space	Postcolonial theory, power critique, deconstruction, Narratives, imagology, *truth claims*, *Anthropocene spaces*	Discourse analysis, image analysis, ontains critique, dichotomy critique,alienation of the familiar	*Empowerment* of local concepts	Moore (2015a), Haraway et al. (2016)
Constructive future-oriented reinterpretation	Non-dualistic nature-culture theories, non-Western nature and culture concepts, conviviality, *livability*	*Transgression*: *multi-species ethnography*, collaborative projects, experimental construction, new forms of representation	Solidarity with non-human beings, expansion of the spectrum of alternatives	Tsing et al. (2019), Jobson (2020), Bryant and Knight (2019)
Repoliticization of the concept	Political economy, political ecology, Marxism, inequality, world system theory, colonial theory, "production of nature"	Inequality analysis, historicization, capitalism analysis, interest analysis (stakeholder), raw material extraction	Emphasizing the responsibility of the powerful, solidarity with precarious people and vulnerable collectives, criticism of teleology	Tsing et al. (2019)
Institutionalization of Anthropocene thinking	Cultural theory, international transactions, development anthropology	Institutional analysis, anthropologization of discourse	Goal: enable worldwide sustainability awareness and cosmopolitanism	Arizpe-Schlosser (2019)

Amelia Moore has outlined the vision of an *Anthropocene Anthropology*. She envisions an open engagement with the concept of the Anthropocene, though not as a new sub-discipline. Moore sees this primarily as a critical engagement with the topic and advocates an anthropological awareness "of" the Anthropocene instead of an unreflective acceptance and work "in" the Anthropocene (Moore 2015a: 27–28, 36–40). At the core is the examination of ideas, concepts, and models, especially notions characterizing life, change, and "the environment."

As anthropologists, we can clarify how these ideas, shaped by conceptions of the Anthropocene, manifest themselves in the everyday lives of groups of people, wherever in the world they live. In addition to the areas of environmental anthropology mentioned above, other subfields of cultural anthropology can also contribute to this. An example is the anthropology of tourism, where islands in particular concretize the phenomenon from the problem site and island tourism itself is motivated by the Anthropocene. Here, specific notions about the Anthropocene converge with scientific institutions, political interests, and normative ideas about environmental education, as Moore presents in a monograph on the Bahamas (Moore 2015, see examples in Gren and Huijbens 2015).

5.4 Localization—Cultural Anthropology as an Advocate of Small Scales in the Anthropocene

> *Smaller scales do not simply vanish as larger ones emerge.*
> Mary Stiner et al. (2011: 247)

> *(…) analysis of local contexts can have global significance.*
> Stuart Kirsch (2018: 136)

The way the sciences deal with the Anthropocene varies. Historical evidencing is characterized by narrative access, by narrativization (Will 2021: 225). The anthropological approach is characterized by localization and contextualization, which is obvious due to the field research orientation and the holistic concept of culture. The contributions that cultural anthropology can make complement well with environmental history approaches, because:

> Nature and history are increasingly becoming intertwined. We need the history of nature in order to write history. We need the big drama and the Earth sciences of the sciences. And we can't leave these stories to scientists alone, because their stories lack details of the human (Sörlin 2018: 23).

Research on the Earth system, complex systems, and the findings of *Big History* research have all shown that complex adaptive systems can only survive in very specific environments. To do this, they must be able to read and decipher not only universal conditions and rules but also local information (Christian 2018: 91). This ties in well with a subject conception of cultural anthropology, which I think Thomas Hylland Eriksen has summed up wonderfully: anthropologists research "large issues in small places" (Eriksen 2023).

> Anthropologists are also concerned with accounting for the interrelationships between different aspects of human existence, and usually investigate these interrelationships taking as their point of departure a detailed study of local life in a particular society or a more or less clearly delineated social environment. One may therefore say that anthropology asks large questions, while at the same time it draws its most important insights from small places (Eriksen 2023: 3).

I take this to mean an engagement with the profound problems and questions of humanity, hence with anthropological questions in a broad sense. The methodological approach of cultural anthropology is carried out by studying local ways of life based on detailed ethnographic case studies.

5.4.1 Local Contemporary Perspective and Time Continuity

Cultural Anthropology provides empirical access to how people perceive environmental phenomena and produce environmental change, and thus environmental history and environmental knowledge (Reuter 2010a, 2010b, 2015; Bauer and Bhan 2018: 7, examples in Hornidge and Antweiler 2012). The explicit access to Anthropocene phenomena through local case studies makes cultural anthropology unique among the social sciences:

> (...) the starting place for anthropologists is almost always a richly detailed ethnographic case study, whereas other social scientists are likely to begin with a formal model and look for ways to test it (Orr et al. 2015: 156).

Overall, however, the Anthropocene calls for delocalization, for a decentering of local views and anthropocentric cosmogonies: "The ideology of place, of the anchor point of meaning and relationship, is shattered by the decentering that the truths of macrophysics impose on us" (Augé 2017: 231). Voltaire, on the other hand, recommended that we must cultivate our garden, that is, that we remain within the scale of our own history. The contribution that cultural anthropology can make as an "advocate" of the local perspective is particularly important because the temptation of scaling up is especially threatening when solutions for major problems are sought, and the Anthropocene represents a mega-macro problem.

Despite all global effects, there are communities that are not fundamentally connected with the rest of the world in ecological terms. Analytically, groups that operate in local environments can be distinguished from others, the industrialized majority, who are integrated into the global technological system through the use of external resources ("ecosystem people" vs. "biosphere people," Dasmann 1989: 304). Above all, however, the effects of the Anthropocene are experienced and lived locally (see examples in Hoffman et al. 2022). In addition to de-localization, we urgently need localization, as indicated by one tenor of cultural studies contributions; we need it in order to tame, to domesticate the dominant macro-oriented research on the Anthropocene:

> (...) there are serious discrepancies between the dominant processes of theorizing planetary multiplicity, and the multiplicity or alterity of which social thinkers speak – and which speaks for itself (Clark and Szerszynski 2021: 53).

This discrepancy is evident in research on global change. In the documents of the *Intergovernmental Panel on Climate Change* (*IPCC*), for example, the atmosphere is necessarily systematically globalized (IPCC 2021). Regarding climate change, Sheila Jasanoff asks: "... will scientist's impersonal knowledge of the climate be synchronized with the mundane rhythms of lived lives and the specificities of human experience?" (Jasanoff 2010: 238). The dominant framing in international bodies and institutions sees climate change exclusively as a risk to the global atmosphere and thus excludes climate change as being *long-term* change in the weather at *individual* locations.

Globally operating institutions have long been an important subject of studies in development and organizational anthropology, and which are often critical (for example, Arizpe-Schlosser 2019). However, contrary to what many critics, especially from cultural anthropology, suggest (for example, Tsing et al. 2019), a narrowed macro view is not a privilege of natural scientists or neoliberal economic thinkers. Just as some representatives of the *Earth sciences* prefer to stick to the macro view, some environmental activists, who only keep an eye on the whole world or living world and the future, also lose sight of the reality of local communities in the here and now (Eriksen 2016: 137).

Apart from the present orientation, which separates cultural anthropology from geology, classical cultural anthropology involves a bias towards time that it shares with classical geology and evolutionary studies: the emphasis on continuity. This entails a tension with the concept of the Anthropocene, which emphasizes a break or rupture. Even though the basic assumption in modern cultural anthropology is that it is normal for cultures to change, this change was seen as continuous. This continuity consists in regular rhythms such as the annual cycle of agriculture and the ritual calendar, which have also determined the prototypical field research duration of one year. Other rhythms that imply continuity are the biographical course of people's lives and the succession of generations. Thinking in terms of continuities is virtually part of the "deep structure" of anthropological theory formation. This is evident in the classic titles of monographs such as "Tradition and Change of X." Until recently, the focus of the discipline has been on *persistent* forms of action and meaning. The prerequisite for the emphasis on persistence was a concept of time in which regular rhythms provided a medium where continuity unfolded (Robbins 2007: 9, 13).

The concept of continuity is strongly linked to the traditional social science dualism of non-modern vs. modern cultures. In cultural anthropology, this is perpetuated by the idea of "history-less" peoples or "peoples of eternal return." In the twentieth century, continuity in theory was emphasized by anthropological and sociological functionalism, which was primarily interested in the conditions for the persistence of society. Breaks or sudden changes are thus easily overlooked or seen as an exception. In modern cultural anthropology, abrupt change, such as that resulting from sudden global economic effects, natural disasters, conflicts, or wars, plays an important role, which is also reflected in theoretical concepts such as vulnerability.

5.4.2 *The Challenge—Large Issues in* Large *Places?*

How can the scales of geological relevance be linked with the socially and culturally relevant scales that are important for individuals—"connecting the Earth system with tonight's avocado salad" (Thomas et al. 2020: 7)? In Sect. 6.7, I will discuss the challenging general question of whether scales can be uniformly calibrated at all (*nested scales*). The challenge for cultural anthropology lies in the fact that cultural anthropology not only investigates localized—albeit sometimes multi-local—communities or networks, but also focuses on the experiences of the people studied.

The challenge here in terms of experience is that some global consequences of human action cannot *in principle* be perceived by humans and that their complexity is also exorbitant. This applies not only to long-term effects such as climate change or species extinction. For the perception of many current or recent parameters of the Anthropocene, we need, for example, satellites with global measurement capabilities, *remote sensing* technologies, sensors in remote locations, and computers that can process everything digitally or prepare it visually (Jasanoff 2004b). Now, when humans "make" Earth history, they are, metaphorically speaking, the new authors of the book of nature, or more precisely, co-authors. However, we humans are losing our role as sovereign interpreters of this book due to our dependence on measurement technologies. We therefore need new scientific structures or rather knowledge infrastructures (Edwards 2015; Folkers 2020: 592; Renn 2020).

At the level of human experience, the Anthropocene is primarily manifested in crises of reproduction (Eriksen 2016: 27). People are experiencing disruptions in central aspects of their lives or their living environments, which are generated by an *accelerated* change imposed on them from *outside*. For example, they have problems keeping their children healthy and equipping them with future-proof knowledge. People are experiencing disruptions in their economic autonomy and their right to self-determination, they are experiencing increasing vulnerability, and they are developing a growing awareness of risks. This can be easily combined with a perspective that is exemplified in Amerindian communities, where an awareness of past and dynamic stabilization instead of the acceleration of change is promoted:

> Indigenous practice emphasizes the *regulated production of transformations* capable of *reproducing* the 'ethnographic present' (...) and thus preventing its regressive and chaotic proliferation (Danowski and Viveiros de Castro 2017: 76).

In the social reality that arises through the interaction among people, communities, and networks, these crises manifest themselves locally in different ways. The concept of vulnerability is important here: societal conditions are not simply characterized by poverty, but by susceptibility, insecurity, and defenselessness in externally determined local worlds. Tourism is an example of how vulnerability is linked to the increased energy use typical of the Anthropocene and to external decisions that may later lead to permanent local dependence or loss of autonomy. One case is Greece during the economic and refugee crises of 2008–2009 and 2015 (Eriksen 2016: 69). As with economic globalization, the local responses to these Anthropocene reproduction crises vary accordingly: adaptation, protest, or attempts at avoidance or isolation. Such Anthropocene-induced crises of reproduction and the reactions to them must therefore be considered at the local level.

However, a large part of the effects of the Anthropocene are spatially distributed and also temporally extended. Depending on the interpretation, they represent a widely distributed assemblage or a hyperobject. Morton has shown this using the example of climate change. They are only limitedly perceptible to humans because they have non-local and enveloping qualities (Morton 2013, example in Vince 2014). Massive changes in the geosphere, such as the spread of radionuclides or the dispersion of plastic particles in the oceans, cannot be directly perceived due to their distribution (Bauer and Bhan 2018: 8). The most obvious challenge for cultural anthropology therefore concerns the scales:

> The Anthropocene is a planetary-scale phenomenon, whereas the methods of anthropologists and systems ecologists are suited to studies of local, community-scale processes (Orr et al. 2015: 156).

Even the local effects of human-made global environmental change are not always directly perceptible, except through measuring instruments. They can only be determined as a global fact through the collaboration of people worldwide, and require data, diagrams, and visualizations (Bajohr 2019: 71; Hüpkes 2019). This is especially true for geochange, which is rapid according to geological understanding, but slow according to human time scales (Morton 2009). Research on climate change has shown that the quantitative scope of the Anthropocene can only be grasped by means of a colossal data collection and the use of computer models. The anthropogenic change of the planet is definitely a *large issue*, and this applies to space, a large part of the geosphere, and time, namely large sections of the Holocene. In addition, Anthropocene change is based on a large number and variety of nonhuman contributions that "collectively produce environmental histories on different scales of space and time" (Bauer and Bhan 2018: 7).

The historian Chakrabarty in particular has worked out that this not only raises the question of who the subjects and the objects of action are, who is to be understood as a social actor, but also which novel conceptions of history, which historical imaginations, are formed (Chakrabarty 2009, 2012; Chakrabarty 2020a, b). The central challenge of cultural anthropology posed by the phenomenon of the Anthropocene therefore consists in a combination of different scales in space and time on the one hand, and a diversity of factors of anthropogenic causes and effects on the other. As with the topic of globalization, cultural anthropology, which is based on ethnographic research and examines local settings of culture, is challenged by large-scale, long-term, and materially diverse phenomena. Hornborg puts it in a nutshell:

> What could anthropology contribute to human survival and sustainability in the Anthropocene? Most anthropologists are preoccupied with understanding local experiences rather than global processes. The *sine qua non* of their research is ethnography (Hornborg 2020a, b:1).

This implies a renewed reflection on ontological assumptions (concept of culture, holism), epistemic assumptions (understanding, immersion), and the resulting question of methodological approach (field research, cultural comparison). Extending Eriksen's dictum, we could say that it is no longer only about large (cultural) issues in small places, but also about large anthropic questions in superhumanly large spaces and over long periods of time. However, there are also completely different attitudes towards this:

> By abandoning the universal liberal subject as a stable foil for a renewed project of cultural critique, the field of anthropology cannot presume a coherent human subject as its point of departure but must adopt a radical humanism as its political horizon (Jobson 2020: 259).

As illustrated in Sect. 5.3, recent cultural anthropology has provided a wealth of contributions on the handling of anthropogenic climate change. Significantly fewer studies deal with other anthropogenic impacts, such as soil and coastal change, which are of central importance for the Anthropocene. This is also reflected in the fact that in anthropological texts that explicitly deal with the Anthropocene, the vast majority of examples concern anthropogenic climate change (for example, Gibson and Venkateswar 2015: 10–11; Bauer and Bhan 2018). Similarly, only a few studies are dedicated to the *social and cultural* preconditions for the emergence of Anthropocene social systems in human history (Hann 2016a: 19). Both point to research gaps in the area of conceptualization as well as in social and cultural theory.

Chua & Fair argue that the classic anthropological approach, namely the critical juxtaposition of the foreign against the individual, the known, and the familiar, and classic methods such as the participatory observation of small life-worlds, are well suited to providing empirical depth and nuance to this multidisciplinary field (Chua and Fair 2019: 2). Here, the distanced external view (*etic view*) and the internal perspective (*emic view*) can be introduced. However, it is clear that this macro topic is empirically challenging for cultural anthropology, which continues to be oriented towards particularities, and that it also has important political and ethical implications. Anthropologists must deal not only with material culture and novel types of art at once, but also with stones, microplastics, and microorganisms. But as I see it, this coincides with the holistic concept of culture: cultures need material carriers. Even if meanings and sense are formed linguistically, societal relationships are shaped by discourses, and there is also virtual reality—none of these processes can do without a material substrate, such as food, ores, and computers, as it also follows from the perspective of "societal relations to nature" (Becker et al. 2006; Görg 2003a, b, 2019: 172; Brand 2018). In addition, political actors are both the objects and actors of research, and anthropologists also have to deal with foreign beings, such as natural scientists.

5.4.3 Anthropocene Everyday and Life-Worlds

A central contribution of the discipline lies in the local perspective of cultural anthropology. What does the Anthropocene change mean for respective local life-worlds? And conversely: What brings about the cumulative effects of local lifestyles and their transformation due to globalization in larger spatial units and longer periods of time? Many of these effects of human activity are almost impossible to quantify, making qualitative data important. For anthropologists, local life and local knowledge are primarily relevant as causal factors of experience and action (Horn and Bergthaller 2022:111).

Since anthropological research is based on everyday experiences and actions in the social micro-space of experience and action, it can be equally important for explanations and for appeals for changed action that are realistic because they are socially grounded in the local communities or networks concerned. This might refer to the everyday consumption, e.g. to non-utilization of quite practical plastic blister packages (Fig. 5.1). Thus, anthropological perspectives can also critically examine a central fallacy in the thinking of modernity, namely blueprint universalism in the sense of "one size fits all." In his first book, Pierre Bourdieu used the example of Kabyle farmers in Algeria to show that if their *agency* falls below a certain minimum, they are simply not in a position to practice cultivation methods that would conserve their own resources and thus secure the future (Bourdieu 2000: 87–103).

Fig. 5.1 Wordwide consumer culture: a blister package of candies. (Photo: author)

Eva Horn beautifully illustrates the problems of an abstract ecological universalism, its decontextualized ideas of equality, and quantitative benchmarks by using the example of the "ecological footprint" in rural Sulawesi in Indonesia:

> Perhaps it is simply absurd to address a rice farmer in Sulawesi and a student from Stuttgart as a common 'humanity' or to calculate their ecological footprint, which in the case of the rice farmer is surprisingly high due to the methane emissions of rice fields, and in the case of the much wealthier, cycling, and organic-vegan-eating student is surprisingly low. Perhaps the rice farmer, his tools, the climate-dependent water management of the village, the debts of his family, the health of his water buffalo, the snail infestation in the rice field, his seeds, the number and future prospects of his children, and the geography of Sulawesi should instead be considered as a specific sustainability problem, while the student's bicycle, her enlightened ecological lifestyle, her choice of studies, the German climate policy, the robust social system, and her therefore presumably manageable number of children should be understood as another (Horn 2017: 16).

The methodological benefit of cultural anthropology's contribution could lie in complementing macro approaches that come top-down from large spatial and temporal units, with approaches that empirically start from smaller spaces and short time spans. Anthropologists see an opportunity here to take local interests into account through bottom-up research and to bring in local or traditional knowledge for solutions and to integrate both forms of knowledge (Gibson and Venkateswar 2015: 10–11). This can include locally limited sections, such as parts of landscapes and place-based climate narratives (Krauß 2020: 5–6; Bremer et al. 2021) or medium-sized to very large units, such as deltas (Krause 2018; Bollig and Krause 2024: 110–127). Around 12% of the world's population lives in river deltas. Deltas are often fertile, which is why intensive agriculture is it here. Many mega-cities are located in the area of deltas, e.g. Kolkata (Calcutta). Around 50 million people live in the Bengal delta of the Ganges and Brahmaputra in India and Bangladesh alone.

Many of these people, for example in the Sundurbans, have experienced flooding for generations and therefore have a traditional knowledge of where flooding and drought are imminent. They think of water and land together instead of separating them in law and management like the post-colonial state. They know which areas of this extremely dynamic habitat are more habitable than others, where only gods can dwell safely. What does environmental change mean for people and how do they deal with it? Which dynamics do they not it? What resettlements are necessary and what human costs will they entail? Which colonial concrete dams can be dismantled and replaced by green dams and mangroves? Are there at least limited generalisable solutions (Bhattacharyya 2018; Bhattacharyya and Santos 2023, as a living narrative see Ghosh 2005)? Deltas thus invite collaboration between, for example, earth system scientists, geologists, environmental historians and economists with anthropologists.

Based on the experience in climate change research, social scientists know that globally oriented research guided by global-scale models notoriously tends to ignore these local and regional levels. Tsing sarcastically notes: "... the scale is global because the models are global," which, as will become clear, falls short as a criticism, or more precisely: falls too small. As anthropologists, we emphasize that the Anthropocene, even more comprehensively than climate change, affects all of humanity, but manifests itself in local life-worlds:

> The view from anthropology suggests that global environmental problems still happen locally in the Anthropocene, and that it is from here that we have to compose our research agendas *and the idea of a common world to be taken care of* (Krauß 2015a: 61, emphasis mine).

There is a large conceptual and empirical gap between the "safe operating space for humanity" of Earth System Sciences, which looks from a space perspective, so to speak, and long-term viability on a local level (Pálsson et al. 2013:7). This is accompanied by the fact that internationally organized decisions on global climate change policy that are based on large-scale models restrict the possibilities for local decision-making. This is rather evident in areas where the population is particularly vulnerable due to Anthropocene effects, especially sea-level rise (Lazrus 2012). This is precisely what anthropologists can contribute to in local studies, for example, research on the Pacific islands of Tuvalu, Kiribati, and the Marshall Islands, and on the Maldives in the Indian Ocean. The inhabitants of these particularly hard-hit places often find it difficult to adapt to change due to poverty, infrastructure deficiencies, and geographical location (Crate and Nuttall 2016; Hastrup 2016, contributions Hornidge and Antweiler 2012).

Bauer & Bhan illustrate the relevance of local studies using the phenomena of climate and weather as an example of how a microscopic research approach can also contribute to macroscopic phenomena (Bauer and Bhan 2018: 18–19). Weather usually refers to the local state or local dynamics of the atmosphere over a short period of a few days. Climate, on the other hand, refers to long-term patterns of weather in a given region, such as over weeks and months. Thus, climate change *per definitionem* cannot be perceived individually. This also applies more generally to the accumulated effects of human activity, which are spatially dispersed and spread over time. Chakrabarty says that we humans cannot experience ourselves as a species, that there can be no phenomenology of the species *Homo sapiens* (Chakrabarty 2009:220, similarly Morton 2013: 5).

The central subject of the Anthropocene is hardly accessible to the senses, and our brains are not evolutionarily designed in such a way that it is easy for us to understand change (Marshall 2014). Therefore, it is not to be expected that support for political measures against global warming can be built on the basis of facts and climate narratives alone. This has also been shown by the experiences with Covid-19, which is not visible as a virus and can only be directly experienced individually. As with Anthropocene change, the cause of Covid-19 is not visible, but its consequences can be directly experienced by the individual or by other people, unlike climate effects, for example. In this respect, Covid-19 is not the Anthropocene (contra Schüttpelz 2022).

In this sense, an anthropology of the Anthropocene should also focus on the experience of weather, instead of always just talking about large-scale or global climate or other worldwide dynamics. An ethnographic focus on weather would allow us to explore tangible phenomena that represent Anthropocene effects in their impact on everyday life. While *global* climate change is not directly experienced by people, climate change certainly is (Bauer and Bhan 2018: 19). Weather is more than just a faint reflection of the climate if we see climate as the sum—or the cross-section—of all weather over a given period, be it for a locality, a region, or the planet. Weather is concretely experienced and is of existential importance.

Many groups of people are certainly aware of long-term weather changes. Farmers are noticing earlier planting dates, parents are lamenting that they can no longer go fishing with their children on vacation, forestry officials are considering how they can still organize a year-round hunting season. For a local community of people and the other organisms living there, global climate change is thus fundamentally local *in its consequences* (examples in Vince 2014). Climate change is therefore anything but "posthuman," but is at least partially perceptible locally through its material effects, as exemplified by the fact that traditionally established practices of land management no longer work (Bauer and Bhan 2018: 21). Climate effects manifest themselves in the form of local effects on specific ecologically situated collectives. We need insights into local dynamics and place-based perceptions, dialogues, and local knowledge. Bauer & Bahn speak of climate as a "broadly distributed material assemblage that transcends human perception," but whose constituents can be experienced. In short: climate is also to be understood as cultural (Krauß 2015a).

The line of reasoning here was presented in relation to the effects of climate change, the sinks in the cycles, but the same applies to the factors, local sources as contributors to accumulated warming. What was been argued here using the example of weather in relation to climate applies analogously to the localization of Anthropocene phenomena in general. Their emergence is also embedded in local and at least partially perceptible actions and assemblages.

With regard to the effects, all this is particularly true for the effects of the Anthropocene beyond the climate, that is, for "more material" and lasting effects, such as the increasing sealing of soils by roads and buildings. Here, cultural anthropology oriented towards the local and present forms of life could methodologically work together with historians of Big History, who are oriented towards the very other end of the spectrum of temporal and spatial scales. Methodologically, field research informed by the geosciences and deep time could cooperate with *Big Historians*, who are increasingly writing "little big histories," that is, researching the extreme deep-time histories *of localities* (Quaedackers 2020). One example is the analysis of the development of a river in Asturias in northern Spain, from the first local traces of humans from the Lower Paleolithic over 200,000 years ago to the end of industrialization in the 1980s. Here, the climate and elements such as carbon and gold were shown to have established a deep-time continuity (García-Moreno et al. 2020).

5.4.4 *Anthropocene Spaces, Places, Non-places, and the* One *World-Place*

An anthropological benefit of the concept of the Anthropocene also, or even primarily, lies in establishing an iterative process of conceptualizing anthropogenic change. Amelia Moore has an Anthropocene-aware anthropology in mind (Moore 2015a), which amalgamates diverse forms in *one* argument about the special character of the contemporary world. Through cycles of "positive feedback," ideas about planetary change would emerge and thus inspire new alliances between different disciplines and a multitude of other partners. This would be suitable for the creation of productive research frameworks in order to explore planetary change in "Anthropocene spaces," such as the Bahamas.

> I see the Anthropocene as the most recent iteration of the positive feedback cycle producing ideas about planetary change: the more researchers and policy-makers promote anthropogenesis as a global issue with political stakes and the more transnational action takes place in its name, the more we will see shifts in understandings of global transformation, sociality, ecology, and landscape (or marinescape) formations on multiple levels. These will in turn inspire new alliances and materializations (Moore 2015a: 36).

In these Anthropocene spaces, the political subjects are scalable; in the case of the island space of the Bahamas, between the island itself, the island as a nation, the Caribbean as a cultural region, and even the entire planet as a "fragile Earth island" (Moore 2015a: 37). She emphasizes the diversity of the materialities involved, and therefore such an Anthropocene-aware anthropology should also include the "politics and poetics" of material measures (Moore 2015a: 36–37, similarly, Jobson 2020: 262). With this, Moore is thus alluding to the representation debate in cultural anthropology, which had strong critiques of science. In her argumentation, a characteristic feature of many new anthropological texts on the Anthropocene is that the references to geology or the life sciences are marginal. I wonder whether Moore's version or vision of an *Anthropocene anthropology* should instead be translated as "Anthropocene cultural anthropology" rather than "Anthropocene anthropology."

The Anthropocene also raises new questions with regard to smaller sections of space, that is, concrete places of life, because there is the looming possibility of seeing the planet as a world-place. In 1994, Marc Augé distinguished between "places" and "non-places" (*non-lieux*, Augé 2014) based on urban anthropological studies and using the example of transportation. Places are therefore spaces where social organization and social cohesion can be read, because social relationships as well as their historical development are geographically reflected in permanent symbols. In contrast, non-places, such as supermarkets, airports, or other traffic spaces, lack symbolized locations, behavioral cues, and readable social relationships in a given space.

According to Augé, in the age of globalization, the places represent the texts and the non-places the contexts in which every place is located. For example, every metropolis, with its inequalities, has the fragmented city-world as its context. Due to the expansion of the scale of vision in the age of the Anthropocene, completely new spaces of encounter or thought are now emerging beyond urbanization, communication, and the Internet, because the non-places will form the context of every possible place in the future. The concept of non-places can be useful for the Anthropocene because consumption is the central driver of Anthropocene effects. The common feature of the various non-places is consumption, from which individuals can hardly or do not want to escape, whether it is in tourism, in communication, or on their mobile phones (Augé 2017: 32–34, 46).

The unprecedented shift in dimensions in a planetary age is iconically reflected in the project of space tourism: the goal here is not an unobstructed view of the mountains or the sea, but of the entire Earth itself. Such a concrete planetary view makes a world society conceivable. Augé asks whether the world, from the perspective of these tourists, will appear as a place or as a non-place. The wealthy tourists will see themselves as travelers in a non-place, but the planet itself could appear as a world-place. In view of the inequalities, the question immediately arises as to how these wealthy tourists will appear to the excluded (Augé 2017: 33).

Outer space or other planets do not seem very inviting as a home, even if they are at least conceivable by means of *terraforming* (Marsiske 2005; Messeri 2016). However, due to the perceptible distinctiveness and wholeness of the Earth when viewed from space, combined with the crisis-ridden consciousness of the Anthropocene, the Earth could also appear as a "home planet." Thus, the unity of humanity *could* be perceived as reality, as the result of evolution, and as problem, and this could be politically emphasized in a metaphorical manner as "home human" (*Heimat Mensch*, Antweiler 2009). In Chap. 7, I will pursue the tension, well illustrated by Augé, between the anthropologically well-founded need for places (Augé 2019: 45–49) and the new non-local contextualization that is obviously necessary in view of the Anthropocene.

5.4.5 *Field Research Beyond the Purely Human Dimension*

The increased diversity of objects and subjects allows for new methodological approaches to empirical research. Just as the orientation of cultural anthropology as cultural criticism has led to experimental forums of anthropological representation since the mid-1980s (Marcus and Fischer 1999), so has the reality of the Anthropocene entailed experimental approaches to field research (e.g.. Tsing et al. 2024). Gibson and Venkateswar point out that by taking the manifold entanglements of humans seriously, other "subjects" also become participants, such as horses or elements of the cultural landscape (Candea 2010: 4).

> Although traditionally focused on human societies, many anthropologists now approach fieldwork as a site that is full of dynamic relationships, pathologies, and interactions that move beyond the concept of the human altogether, to examine the multifarious composition of life forms that make up our world (Gibson and Venkateswar 2015: 6).

If the former field settings now turn out to be permeable ecological archives, then this means nothing less than an "epistemic crisis" for the discipline (Jobson 2020: 261). Jobson illustrates this using the debates and his own experiences during and after the *AAA* meeting in San José in 2018, as severe forest fires were raging. With the dissolution of small and clearly defined field units and the official melting away of the academic distance from the research object, the clearly defined collectives with which one feels solidarity become indistinct, and the question arises as to which part of the culture one criticizes as an anthropologist:

> In lieu of a facile resolution to this epistemological crisis, anthropologists were called to dwell with the contradictions of San Jose to dismantle a comforting register of ethnographic sentimentalism and cultural critique. To let anthropology burn permits us to imagine a future for the discipline unmoored from its classical objects and referents (Jobson 2020: 261).

The combative expression, indeed the violent vocabulary, is understandable based on activist intent and in view of the extreme experiences, as in some recent contributions to the environmental crisis (Tsing 2019c; Malm 2021). The strong language, however, easily obscures the fact that old wine is sometimes poured into new wineskins. When Jobson, in line with Tsing et al. 2019, refers to the "permeable ecological archives" as "permeable patches" and contrasts this with "hermetic field sites" (Jobson 2020: 263), this disregards the debates that have been going on for years about alternatives to localized field research, such as those in urban cultural anthropology or to *multi-sited ethnography* in anthropological migration research and translocal climate research (for example, the contributions in Coleman and Hellerman 2011).

With regard to the important cooperation with the natural sciences in the face of the Anthropocene, it should be noted that there have been field research-based directions in biology since the nineteenth century. These include not only naturalistic ecology per se and evolutionary research. Research on biological socialized (taxonomy), which is central to biodiversity research, is also field-based. Their work is typically experiential and involves physical experiences—exemplified by excursions and expeditions. In contrast to biological laboratory work, such working methods show a strong affinity to ethnographic research. In postcolonial regions, materially heterogeneous settings, "foreign nature" and dealing with a wide variety of people and groups play a role in this natural biological research, and all of this invites collaboration with anthropology (Bogusz 2022: 13–19 using the example of taxonomic research in Papua New Guinea, cf. West 2016).

5.4.6 Diversity, Inequality, and Mutual Relations

This leads to another argument for the relevance of cultural anthropology: the traditionally strong engagement with diversity and inequality. The social sciences as a whole can illustrate (a) the different effects of human action in the global system and (b) the socially uneven distribution of these effects on human and non-human collectives (Pálsson et al. 2013:7). Anthropologists can testify to Anthropocene effects on individual societies or communities, and they can recount how individual communities in specific sociocultural settings and geographical contexts deal with such effects: "bear witness and contribute stories" (Gibson and Venkateswar 2015: 6). Cultural Anthropology has the advantage of having its eye on the entire range of human collectives, that is, including the "majority world" of Asia, Africa, and Latin America instead of just *WEIRD* people (*Western, Educated, Industrial, Rich, Democratic*, Henrich et al. 2010). From a comparative anthropological, socio-psychological, and sociological perspective, the western influenced social systems appear to be very special and, above all, particularly consequential (Henrich 2020, see Stichweh 2018).

In the Anthropocene, however, it is about much more and about a *much wider range of* diversity. This is where anthropologists can introduce their strength in paying attention to nuances (Gibson and Venkateswar 2015: 9). In addition to cultural diversity, there is also the diversity in the biosphere and the geosphere. Here, cultural anthropology can contribute knowledge on differentiating the effects of human action (Leggewie and Hanusch 2020: N4). With their view on diversity and variability as well as their attention to relations and details, anthropologists can contribute to making hidden aspects and overlooked people visible, thus working towards a "sociology of the absent" and also to a development of possibilities, a "sociology of emergence" (Escobar 2016: 15). In this sense, the question is whether cultural anthropology can determine the extent to which modern society describes itself by means of the idea of the Anthropocene (Werber 2014).

A central point, which is mentioned in almost all opinions on the strengths that cultural anthropology can contribute to the topic, is the insight into the diversity of relationships between people and other things, living beings, or the environment as a whole. This is actually the topic of ecology. In recent cultural anthropology, this diversity of content is reflected in a wealth of words or paraphrases, especially in programmatic anthropological texts. As an example, I would like to take a text by two New Zealand anthropologists, Gibson and Venkatsvar, who have dealt with horse-assisted therapy, political ecology, and indigeneity in earlier works. They speak of "connectedness," "inextricable connections," "multifarious associations," and "relationality," that is, a network of complex relationships. As seen from our all-encompassing *sense of connection*, the environment appears as "our surroundings." Species interdependence is emphasized. In this view, the other living beings, which have been largely ignored until now, do not simply appear as organisms, but as "Earth-others." In keeping with Haraway, they "inhabit" the world together with humans (*coinhabit, cohabiters, companions;* Tsing 2010a, b: 192, 2012b; Gibson and Venkateswar 2015: 11, 12, 14, 15; Tsing et al. 2019: 187).

All of this is suitable for combating the separation, the "rift," or the splitting off of humans from nature and the disruption of ecological systems. The authors summarize their research stance with their continuous trope of "entanglements." Consistent with the poet D. H. Lawrence, they invoke a "metaphor of connectedness and implicit oneness with the world" (Gibson and Venkateswar 2015: 5). By *entanglement*, they do not mean the entanglement between different societies (Randeria 1999) or the entanglements of objects and human action (Hodder 2012) that is referred to in globalization research, but rather what Karen Barad calls specific material relationships in the ongoing differentiation of the world (Barad 2010: 265) and what they call "embodied entanglement." They write that it is the task of anthropologists to face the reality of "multispecies communities" and to "write from a place of occupied connectivity" (Gibson and Venkateswar 2015: 18).

Other authors say that anthropologists should write narratives from the "interior of the Anthropocene," the goal being to "write from the place of occupied connectivity" (Gibson and Venkateswar 2015: 18). This would be a means against the attitude of looking at Anthropocene situations and dynamics with a purely external objectivity, in other words, a means to avoid an objectivist gaze (Rose 2013). Gibson & Venkatesvar see the central contribution of cultural anthropology to the exploration of the Anthropocene in "critical engagement." If one reads texts such as Gibson & Venkatesvar's carefully and repeatedly in the sense of a *close reading* and applies the critical attitude propagated by the authors, the strong normative undertone becomes abundantly clear. The choice of words also occasionally reveals a romanticizing attitude, an attitude that can also be described as "anthropological sentimentalism" (Jobson 2020: 259, see comments: American Anthropologist 2019).

5.4.7 Multiscale Sustainability Instead of New Localism

The title of Sect. 5.4 was intended to be a question. On the one hand, although cultural anthropology has a definitive strength in its local access to concrete realities of life, it does not have to limit itself to local contributions. On the other hand, from a comparative perspective, there is another argument with regard to the scales of human communities and societies. If we juxtapose human societies throughout history and from different, recent cultural settings, comparatively systematic correlations will emerge between the extent of the collectives (population, area, complexity) and their typical systemic problems. This simultaneously raises the empirical question of the level at which the most appropriate scale-oriented solutions can be found that could make it possible to contain the Anthropocene (Bodley 2020: 440, Table 14.2). In a world marked by political and economic inequality, the central question here concerns the level at which social power and the associated growth can best be limited in the sense of the Pareto optimum or the *summum bonum* (Bodley 2020: xii, 35, 414, 437).

A fundamental finding is that many social problems increase with the size of the units (population, area, complexity (Table 5.5). Societies that grow economically and expand territorially typically run into problems. The most successful countries in the world in terms of sustainability are microstates with fewer than 10 million inhabitants, such as some states of the *Nordic Council and Nordic Council of Ministers* and micronations, or small island states such as Cuba and Dominica in the

Table 5.5 Societal scale and the *summum bonum*

Goals	Organization, Institution	Population size (in persons)	Authors
Local cosmopolitanism, development of "moral conversations"	Small group	2–5	Appiah (2006, 2008)
Reproduction (preservation, humanization)	Household (extended family, kinship)	5–25	Kohr (1957)
Collective identity, stability, social cohesion, social intelligence	Medium-sized collectives (*Dunbar's number*)	150	Dunbar (1993)
Sociability (community)	Village ("clubs")	500	Bodley (2020), Scott (2017)
Prosperity (wealth, time autonomy)	City (markets, factories)	5000	Kohr (1957)
Security (justice, peace, defense)	Large city (courts, town halls, armories)	10,000–20,000, up to max. 15 million	Kohr (1957)
Expressive culture	Metropolis, state, ontains, churches, museums, universities	100,000, up to max. 15 million	Kohr (1957)
Limitation of power, sustainability	Small states	Max. Ten million	Bodley (2016: 302–303)
Limiting the Anthropocene, preserving diversity, cosmopolitanism	Planetary society	Approx. 9 billion	Appiah (2006, 2008)

Caribbean (Bodley 2020: 441–446, 456–465). At the project level, medium-sized but long-term development projects are successful, such as the famous *Vicos* project of a planned and anthropologically accompanied cultural change and other community-based projects (Seithel 2000, for example, Wergin 2018).

This raises the general question of whether smaller socioeconomic units, especially units below nation-states, tend to make more humane living situations likely, that is, lifestyles that are more secure, healthier, and more environmentally friendly (Bodley 2020: 456–460; Scott 2017). However, the question always arises—especially among environmental researchers and anthropologists—as to whether this does not also involve romanticizations. Historically, there were indeed larger power structures that were not characterized by strong economic inequality and political hierarchy (Parzinger 2019). If these findings by Bodley can be confirmed, this would suggest a general orientation towards the principle of subsidiarity: larger units or higher levels of organization should only be established when smaller units or lower levels cannot solve a problem.

However, the Anthropocene presents a fundamental political problem. While about two-thirds of all politically independent countries are small states, they only make up just under one percent of the world's population. They may have developed practical and sustainable solutions over a long period of time, but they are up against large nations, powerful decision-makers, and the one percent of the extremely wealthy. The unresolved fundamental question in the Anthropocene, as a problem that affects all of humanity, is how the cooperation and coordination of smaller units could lead to tackling the Anthropocene as a global problem. This involves political questions that I will not pursue further here due to a lack of expertise, but it also raises the question of common—that is, cross-cultural—norms and values, which I will address in Sect. 7.4 on the topic of cosmopolitanism.

I have touched on these aspects here because anthropologists can ask themselves at what levels their primarily local approach can be helpful for addressing large-scale systemic issues of the Anthropocene as a mega-macro epoch. Here, the breadth of anthropological expertise should be emphasized. Cultural Anthropology is the science that has an overview of all societies of the world. A broad cultural comparison can point to realistic solutions for Anthropocene problems through *proven* examples of concrete implementation. This would be a task for an applied or practical cultural anthropology that focuses on global human problems or the dark sides of progress instead of limited applications (Bodley 2016; Antweiler 2019a, b).

A collaboration between anthropologists, primatologists, and historians could, on an empirical, comparative cultural, and diachronic-historical basis, address the question of which human problems can only be solved through social cooperation, which value basis promotes social cohesion, and which counterforces, such as non-cooperative or free-riding behavior, are to be expected (Antweiler et al. 2019; Pries 2021). In addition to individual ethnographic descriptions of sustainable cultures, the typology of different system models can also be useful, such as Eric Wolf's classification into kinship-based, tributary, and capitalist production systems (or "economies", Wolf 1982) or John Bodley's tripartite division into tribal, imperial, and commercial "worlds." The ecological and material footprints are in fact extremely different between the first two and the capitalist forms, and this reflects the "great acceleration." The use of 1 hectare of land and 1–4 tons of material turnover per capita and 1–4 percent of the global biocapacity in the tribal world is contrasted with 2.2 hectares of land use and 10 tons of material per capita and 122 percent of the biocapacity in the commercial world (Bodley 2020: 414; Fischer-Kowalski and Haberl 1993, 2017).

5.5 *Patchy Anthropocene*—A Programme in the Mode of Demarcation

> *The Anthropocene lacks the debate about "the Anthropos," or more precisely, about the social dynamics underlying the* great acceleration.
>
> Christoph Görg (2016: 11)

Anna Lowenhaupt Tsing, who, along with Donna Haraway, is one of the central figures (so as not to say stars) in the anthropological debates on the Anthropocene, has, together with Andrew Mathews and Nils Bubandt, designed an anthropological program on the Anthropocene under the title *Patchy Anthropocene* (Tsing et al. 2019).

In 2019, the three authors published a special issue of the leading specialist journal *Current Anthropology* (Rutherford 2019). They dedicated themselves to the topic 60 years after a famous symposium of the Wenner-Gren Foundation on the "Man's Role in Changing the Face of the Earth," which took place at Princeton University in 1955 and led to a 1236-page publication (Thomas Jr. 1956). This conference provided the groundwork for the establishment of an interdisciplinary study of the global environment. Until then, the focus had mostly been on individual communities or societies embedded in their respective environments under the title "Man and His Habitat" (Arizpe-Schlosser 2019: 268).

In their programmatic introduction, however, a reorientation of the discipline is now called for in response to the Anthropocene (Tsing et al. 2019: 186). Tsing et al. explicitly developed a quasi de-disciplining of cultural anthropology in relation to the Anthropocene, unlike other current demands for a repositioning of cultural anthropology, which are more closely oriented to current environmental disasters, the consequences of the current capitalist economy, and authoritarian politics (such as Jobson 2020). I will first present the approach of Tsing and co-authors here and will only critically comment on it in the following section, thus representing a very selective reading of the matter and also of the term Anthropocene.

5.5.1 *"Patchy Anthropocene" and a* Retooling *of Cultural Anthropology*

The authors begin with the observation that the Anthropocene requires a spatial as well as a temporal analysis. Instead of celebrating or rejecting the Anthropocene as a concept, it is about engaging with it critically and curiously. For Tsing et al., this demand is accompanied by a critique of anthropology, which is still strongly deconstructivist and particularist in orientation:

> In the face of the challenges of the Anthropocene, anthropology must dare to be *more than the voice of parochial alterity* , dare to allow anthropological stories of the 'otherwise' into *concrete transdisciplinary conversations* about planetary structures that 'change everything' (...). We need to *reclaim, in a new register, anthropology's heritage of daring to make big claims* about humans and about the worlds that humans cohabit with others *instead of being content to deconstruct* such claims. But we have to make such claims with all the circumspection that also is the trademark of anthropology (Tsing et al. 2019: 187, emphasis mine).

Tsing and colleagues point out that the deconstruction and critique of modernist or universalist approaches are very important, but in the face of the realities of the Anthropocene, they are not sufficient as a contribution of the discipline. They argue that locally limited studies of particular life-worlds, à la "here-where-I-did-my--fieldwork-everything-is-completely-different," are also not sufficient. Tsing et al. also distance themselves from anthropology that focuses heavily on meanings and symbols, emphasizing that the "webs" in which people are entangled go far beyond the significations in Geertz's webs of meaning (Geertz 1977:5). In the Anthropocene entanglements of living beings, many weavers are involved in addition to the humans "weaving meanings."

Tsing and co-authors also distance themselves in many other cases, namely from transhumanism, green capitalism, and eco-modernist concepts of a "good Anthropocene." Together with Buck-Morss (2002) and Hamilton et al. (2015), they characterize ideas of the "good Anthropocene" or even a "great Anthropocene" as dreams of a large, universal, and scalable world that have brought us to this catastrophic situation in the first place. They regard all the aforementioned directions as "hopeful politics of technological transcendence" (Tsing et al. 2019: 192).

Then again, what is their own program for an anthropology of the Anthropocene? Under the "patchy Anthropocene," the authors understand the thematization of the materially unequal conditions under which humans and other living beings live in landscapes that are increasingly shaped by industry (Tsing et al. 2019: 186, 2024). The authors take the terms "patch" and "patchy" from a branch of landscape ecology, in which landscapes are seen as inherently heterogeneous structures characterized by patches. These patches at any scale. The background to this is Tsing's interest in a conception of capitalism that does not assume progress. Tsing had previously argued that a capitalism without growth would be conceivable, in which a concentration of wealth is possible: value is produced on unplanned patches that capital can appropriate (Tsing 2015: 5).

The central research object in the *patchy Anthropocene* approach should be the connections between humans and other living beings, the "more-than-human relations," "more-than-human space." In terms of change, it is about a history of relationships between many species: "more-than-human histories" or also "multi-species history" (Tsing et al. 2019: 186, see Tsing 2015 "holobiont," Haraway 2016). The authors understand the term *patchy Anthropocene* as a conceptual tool for investigating landscape structures characterized by multiple inequalities. Tsing et al. examine these processes using "landscape structures," understood as historically emerging patterns of human and non-human assemblages.

As examples of such morphological landscape structures, they cite a forest, a plantation, a city, or a region of intensified agriculture, as described by Clifford Geertz in Java (Geertz 1963). Tsing et al. advocate a comprehensive literacy for landscapes. Increased *landscape literacy* could enable anthropologists to see landscapes for themselves again. At least implicitly, the authors thus distance themselves from Geertz's later goal of understanding the emic perspective by strongly emphasizing observations. They say that landscape literacy could free anthropologists from their imprisonment in the concepts and cosmologies of their interlocutors and encourage them to pay more attention to their own observations again (Tsing et al. 2019: 188).

Particular attention should first be paid to industrially induced simplifications of environmental conditions, as in monocultures (*modular simplifications*). Simplification is a key element of modern agriculture (Fig. 5.2). A second focus should be on uncontrolled proliferations triggered by industrial activities, such as diseases. Anthropocene patches are places that reveal intersectional inequalities *among humans*, because despite all emphasis on multispecies relationships, an

Fig. 5.2 Simplification of the landscape in Zülpich, Germany. (Photo: author)

Anthropocene cultural anthropology in the sense of the authors should be focused on humans (or human groups) (Tsing et al. 2019: 188). The argument here is that human deprivation and inequality have been shaped throughout history *by non-human agency* as a reaction to human action. The Anthropocene patches emerge within the relationship between simplifications and proliferations.

The prime example of an industrially induced simplification of environmental conditions (*modular simplifications*) and their corresponding untamed spread (*feral proliferations*) are plantations, otherwise known as simplified or standardized landscapes (see Tsing 2019a). As a form of industrial agriculture, tropical plantations combine a drastic and permanent change in the environment with certain forms of work, such as forced labor and slavery, to produce world market products. They combine regimes of human existence with the determination of non-human life, which the authors collectively refer to as *regimentation* (Tsing et al. 2019: 189). The simplification of the environment facilitates the spread of tropical plant diseases, also referred to as proliferation. Historically, plantations are associated with colonial subjugation and specific colonial forms of labor organization. Indigenous communities and local ecosystems were eradicated to transform the landscapes so that the cattle could be protected and controlled, although this was not always successful (Ficek 2019).

The modular simplifications of the environment consist in reducing the number of cultivated crops in plantations to ensure a densification of the individuals of the desired species. In extreme cases, the number of cultivated plants on plantations is reduced to a single crop (monoculture, *monocrops*). The goal is to eliminate everything that is not necessary for the reproduction of the economic product, and this applies to plants as well as animals. In line with Appel (Appel 2012), modularity means that the work is temporally disciplined and spatially contoured in a modular manner, with concepts such as efficiency and risk serving as guiding principles. This includes work norms, time concepts, alienation, and forced labor. Very similar structures of simplification can be found in animal production, such as the standardization of husbandry and nutrition in the mass rearing of chickens.

In the wake of large-scale environmental simplification, there is often a "wild" or an "untamed" increase (*feral proliferation*) of organisms, for example, of salmon bred in hatcheries as opposed to wild salmon on the US Pacific coast. Such environmental simplifications facilitate the spread of diseases, such as coffee rust (*Hemeileia vastatrix*). This coffee pest, a fungus, grows on coffee plant leaves and reduces the productivity of the plants. It first appeared in plantations in East Africa, probably spread across the Atlantic through high winds in the stratosphere, and is now found in most plantations in the Americas. The fungus has a particular affinity for such industrial monoculture coffee plantations, where the plants grow without shade. There, the fungus then also spread to smallholder plantations and, despite the mixture with shade-providing and thus wind-reducing plants, it spread rapidly across entire landscapes from 2012 onwards (Perfecto et al. 2019: 234). In a similar way, viruses that emerge on chicken farms can also spread to humans. The "wild" nature of *feral proliferations* here refers not only to life forms, but also, for example, to pesticides, toxins, or toxic substances escaping from industrial infrastructures and forming radioactive landscapes (Tsing et al. 2024).

By structures, Tsing et al. mean "phenomenological markers of increased field sensitivity," forms of the world that attract our gaze, call for attention, and refer to a longer history. Metaphorically, these structures resemble the morphology of trees that reveal the history of growth. Unlike in structuralism, the authors do not use the term "structure" to refer to binary thought structures, and unlike Radcliffe-Brown (1940), they do not distinguish between structure and change. The authors also use the term "system" but want to clearly distinguish it from the systems of the Earth sciences. They therefore advocate an open concept of system, which understands systems as "thought experiments with the aim of understanding structures."

5.5.2 Models and Cosmologies

This broad concept of systems can include very different understandings of systems, and the authors support an approach that Viveiros de Castro (2019) calls "ontological anarchy" in the same publication. Tsing et al. distinguish three models: (a) ecological models, which mainly come from the Earth sciences, especially Earth system models and climate models, (b) non-secular cosmologies, that is, religious ideas, prototypically in concepts about spirits in animisms, and (c) concepts of political economy, for which the Green Revolution is the prime example (Tsing et al. 2019: 190–193). Tsing et al. refer to these concepts as kinds of system thinking, but also speak of "kinds of system building." All three are systemic ways of thinking about the world, but also ways of acting that construct the concrete material world in a systemic form. In Geertz's words, they are also "models for" actions, not just "models of" something (Viveiros de Castro 2019). Tsing et al. consider the comparative analysis of the kinds of system thinking to be a central contribution of cultural anthropology to the Anthropocene:

> We argue that a key contribution that anthropologists can make to Anthropocene studies is to juxtapose these alternate kinds of systems-thinking – thus opening attention to multiple ways of gaining traction on observations of landscape structure (Tsing et al. 2019: 189).

Some examples of ecological models are the models of the Earth system, which are based on the massive aggregation of data. According to Tsing et al., these are models that aim to model the planetary totality of the entire world, thereby obscuring both diversity and other perspectives. This also applies to totalizing models on a smaller scale, such as models by hydrologists of river deltas. Here, the heterogeneity of landscapes is simplified and local experiences are omitted or silenced. According to Tsing et al., this category of ecological models also includes concepts of carrying capacity, and they consider such models to be examples of hyperobjects (Morton 2013). Hyperobjects, in the sense of Timothy Morton, are phenomena that cannot be directly experienced by humans due to their distribution, such as climate and climate change as opposed to weather. According to Tsing et al., however, these hyperobjects are presented by science or institutions in such a way that they are measurable and can thus be used to model future developments.

The second form of the models of thought and action are non-secular cosmologies, or simply put, religious ideas. According to Viveiros de Castro, they resemble ecological models in that spirits, for example, also anticipate developments and results ("models of") and want to change the world with a certain intention and are therefore normative ("models for," Viveiros de Castro 2019: 296, 300). Tsing et al. see geology as a form of cosmology, as imaginative extrapolations in which observations of landscapes are amalgamated with ideas of deep time. This also reveals the romantic roots of geology as a science (Tsing et al. 2019: 191; Khan 2019). Similarly, Szerzynski sees the concept of "risk" in Earth science, which he interprets as the comingling of a secular category and risk, with quasi-religious undertones (Szerszynski 2017, 2019).

Tsing and colleagues thus see a potential in the fact that ideas of ghosts, spirits, monsters, and the like can unsettle or undermine the predictions from ecological system models ("to unpredict"). Ghosts can typically leave the systems from which they emerge. The "uncanniness" of the Anthropocene could complement the worldview that has grown from model-made realities with a new verisimilitude. This fits with the project of Tsing and other authors to design "arts of living" that respond to the Anthropocene and thematize ghosts, as well as further "others" in addition to animals, such as trees (Tsing et al. 2017; see Waterton and Saul 2020: "arboreal others"). It offers the chance to escape from the "enslavement" of rationalist and reductive thinking and models for analyses. Felix Lussem relates Tsing's monsters to the current pandemic:

> With Tsing et al., SARS-CoV-2 can therefore be understood as a monster of the Anthropocene that frustrates modern fantasies of endless growth – based on the separation of humans from their environment, on the technical transcendence of nature – through its all too risky proximity to humans and thwarts the material networks of capitalist circulation. In this context, monsters are thus figures that irreducibly demonstrate their presence and transgress boundaries believed to be safe – just as the virus breaks through the immune barrier of humans and hijacks their cellular metabolism for its own monstrous multiplication (Lussem 2020).

For Tsing et al., non-secular ideas ("cosmologies") are of central importance precisely because they open up the hope for an "unruly nonsecular Anthropocene," a model that can be set against the dominant concept of a unified and homogeneous Anthropocene (Tsing 2012b, similarly Szerszynski 2017; Daston 2021). The authors also see this as an opportunity for cultural anthropology to return "home," namely back to the world of invisible beings, unpredictability, and doubt (Tsing et al. 2019: 191–192, 194):

> Ontological anarchy is therefore not exterior to systems thinking. It is the ghost that stirs within and between systems. In this sense, anthropologists need to take natural science models as seriously as spirits (Tsing et al. 2019: p 191).

Tsing et al. thus have a nuanced position on the *ontological turn*. They advocate a "critical description" of Anthropocene inequalities rather than demanding to deal with what they call "philosophies of the Anthropocene." They thus oppose both the Geertzian tradition of emic-oriented cultural anthropology and—but only partially—approaches in the sense of the *ontological turn*. They oppose the projection of radical alterity onto idealized ethnographic others, but use insights from ontological studies, such as ghosts, as one of the three system models or "thought experiments," as they also call them.

The third form of systemic thinking for Tsing et al. is composed of "political economies." By this they mean environmental models that are guided by ideas of dominant economics. One case is the intensification of tropical agriculture in the form of the "Green Revolution." This program is based on concepts of rationality, efficiency, and risk in order to increase yields, especially in wet rice cultivation in South, Southeast, and East Asia. Critically speaking, however, this is accompanied by specific normative ideas of social organization, such as the concept of the family as a unit of agricultural production, and the farm as a production unit. Within the known diversity of family types, this means a modularization and a normative setting that goes undiscussed in the "charismatic package" of the Green Revolution (Tsing et al. 2019: 192).

The authors argue that this also obscures local forms of knowledge, such as the knowledge of small farmers. A specific cosmology, namely the Anthropocene ontology of one world, would level out the diversity of forms of knowledge through a claim to exclusivity, which Law dramatically characterized as an "epistemicide" (Law 2015). Tsing et al. conclude that "... an anthropology of a patchy Anthropocene requires attention to a diversity of modes of knowing as well as modes of living" (Tsing et al. 2019: 192). The authors conclude from their considerations, in a programmatic manner and in their characteristic choice of words, that a renewed cultural anthropology is necessary:

> To describe a patchy Anthropocene that is politically made, ecologically remade, and uncannily unreal, we need both observations and thought experiments. This is bold research on planetary conditions of livability that addresses the challenges of the Anthropocene by retooling anthropology's critical attention to specificity, context, and difference (Tsing et al. 2019: 196).

5.5.3 A Non-academic Project in Scientific Garb?

Having outlined the argumentation of this important contribution, I will now critically examine the article by Tsing et al. In terms of topic and presentation, the contribution is an illustrative example of several recent anthropological contributions on the Anthropocene. I chose the text because it is explicitly programmatic, and two of the authors, Tsing and Bubandt, are among the most active and most widely received anthropologists on the subject of the Anthropocene. The article is relatively new and introduces a special issue of *Current Anthropology*, the leading anthropological journal, which, in contrast to what is usual in *Current Anthropology*, does not include any discussion papers in relation to the article in the supplementary issues. I will examine the article as a scholarly contribution, not as an artistic one.

Since I will strongly criticize the article by Tsing and co-authors here, I would like to emphasize in advance that I greatly appreciate other works by the authors, especially the studies by Anna Lowenhaupt Tsing—despite her often exuberantly meandering prose. This is especially true for her studies on the human-made, partly chaotic transformation of tropical landscapes and livelihoods in Kalimantan in Indonesia driven by globally organized resource frontiers, which she presented from the perspective of the affected people (see Tsing 1993, 2005a, b). I would also like to mention her innovative recent works on the culture-nature nexus and cultural encounters using the example of the global supply chains of the *matsutake* mushroom (Tsing 2015, 2020; Choy et al. 2009; Matsutake Worlds Research Group 2009). The other two authors, Nils Bubandt from the Aarhus working group (*AURA*) and Mathews, have also made important contributions; Bubandt, for example, on non-secular environmental concepts (Bubandt 2018, 2019) and Mathews with an overview of the topic in the prestigious *Annual Reviews of Anthropology* (Mathews 2020).

The article by Tsing et al. analyzed here is in the context of projects on the art of living on a "destroyed planet in the more-than-human Anthropocene" (Tsing et al. 2017; Tsing 2020). This also includes the "Feral Atlas," a *digital humanities* project for the scientific and artistic representation of human–animal relationships and for the establishment of a broad collaboration between anthropologists, other scientists, artists, and architects, for example. There, the contributors were invited "... to explore the ecological worlds created when nonhuman entities become tangled up with human infrastructure projects" (Tsing et al. 2020), especially by means of large-scale industrialized infrastructures such as dams or oil platforms, Tsing et al. 2024: 249–256.

The article by Tsing et al. introduces a special issue and claims the *Patchy Anthropocene* to be a new approach to the topic. However, the authors do not address other recent and equally programmatic approaches from their own discipline of cultural anthropology, namely the works of Nathan Sayre (2012), Pálsson et al. (2013), and Gibson and Venkateswar (2015). It is even more surprising that works by colleagues who explicitly use landscape concepts do not appear, although they are published in English and easily accessible. I am thinking of the studies by Werner Krauß (e.g. 2015) on thing politics and the localization of climate policy driven by nature park concepts on the German North Sea coast, or the conceptual approach of Amelia Moore, who also presented a spatial concept with her "Anthropocene spaces" (Moore 2016, see Sects. 4.2 and 5.2).

5.5.4 Landscapes and Patchiness—Missed Connections

Tsing et al. emphasize in their article how important landscapes are to them (Tsing et al. 2019: 187–189). In their extensive argumentation, however, they do not address the long and still intensive empirical exploration of landscapes outside of cultural anthropology and—what is especially surprising—the long discussion or even the current state of discussion on landscape concepts. There is not a single reference to landscape geography, to the archaeology of landscapes (*landscape archaeology*), or to landscapes as part of environmental psychology. The interdisciplinary research field of *landscape studies* (Benediktsson and Lund 2010; Küster 2012; Leibenath et al. 2012; Krauß 2019; Howard et al. 2019), where anthropologists are also involved, is likewise not engaged with here.

The concept of patches wants to emphasise heterogeneity in every respect, but it is all too open. In a new "Field Guide to the Patchy Anthropocene, Tsing et al. emphasise that these patches can be of any scale, from a small patch on a leaf to the Black Sea or an entire ocean. Patches can be man-made, like a shopping mall, or human-influenced but unintentional, like the giant plastic vortex in the Pacific; they can be areas or connecting corridors (Tsing et al. 2024: 15, 35).

Concerning these Questions of heterogeneity of landscapes it would be useful to have an exchange with concepts of those cultural geographers, landscape ecologists, and representatives of a comprehensive "landscape science" (Küster 2012), who assume that the "invention" of landscape was probably the most consequential discovery of humanity and that, because of the close interweaving of nature and culture in every landscape, one should not differentiate between natural and cultural landscapes (Küster 2012: 38, 307). Accordingly, humans also did not begin to "destroy nature" at a certain point in time. Especially in relation to landscape, the Anthropocene raises anthropological questions about spatial and temporal aspects of relationships to land, which are not even mentioned by Tsing et al. and for which diachronic models exist (for example, Stoffle et al. 2003). A central aspect is the “normal” presentism of humans in dealing with landscape. Here, geological temporality and cultural temporality are often intertwined in disparate ways (Crumley et al. 2017; Irvine 2020; see Kühne 2018). I will pursue this topic in Sect. 6.6.

The plea by Tsing and co-authors for a *landscape literacy*, with their call for anthropologists to bring in more of their own observations, seems like a rejection of the representatives of flat or alternative ontologies, but contradicts other statements in the article and also other contributions in the special issue edited by the authors. It is astonishing that the entire body of research on the readability of landscapes for actors is completely ignored. There are many studies on this in environmental psychology, environmental sociology, anthropogeography, and *design studies*. Tsing and co-authors mainly work with the concept of *landscape structure* and use the word “structures” in an unusual way, namely in the above-explained phenomenological sense as sensually perceptible forms of the world. They differentiate this from Radcliffe-Brown’s scientistic concept of structure, but without referring to any phenomenologically influenced anthropological literature, such as that by Ingold or Jackson.

The idea of the patchiness of landscapes is central and also gives the article its title. They take the term from landscape ecology, where patches are understood as units of different scales in landscapes, whereby landscapes are understood as units of heterogeneity. Tsing et al. say that they have modified the term *patch* (in what way?) and emphasize that *patches* reveal landscape structures, that is, morphological patterns of the arrangement of humans and non-human beings, with the mentioned plantations being an example (Tsing et al. 2019: 188; 2024: 15, 35–46). It remains open whether the abiotic components, such as skeletons or soil minerals, are also included also among the non-human beings here.

For their central concept of *patchiness*, the authors refer to only two older publications without discussing them in any way (White and Pickett 1985; Wu and Loucks 1995). Tsing et al. do not deal with the specific framework of these texts, namely natural disasters in the first case and ecological dynamics in the second case. Tsing and co-authors do not address the other contributions to the anthology, the first text of which serves as the introduction. Both texts contain precise definitions and typologies of the dynamics of landscape patches, which Tsing et al. do not discuss at all, even though their text mentions much of it several times, so there would have been room for it. *Patchiness* is an established field of research in ecology, especially in *community ecology* (Mittelbach and McGill 2019: 250–264; Begon and Townsend 2021: 339–349, 599–602), but this is left out. The entire research on ecological diversity and the dynamics of landscapes in the study of landscapes and also in the research field of resilience is not even mentioned by Tsing and co-authors.

The concept of the model is not clarified at all, and corresponding debates in the philosophy of science, scientific theory, and, for example, the technical sciences, are not used. The concept of “simplification” is concretized using plantations and other examples and is also sufficiently explained conceptually, even though existing studies on plantation work are not used (McKittrick's objection already in 2013 to Haraway) and no reference is made to conceptual literature on monocultures or monostructures. However, the term "modular" is then used by the authors 21 times, without it being clear what this concept of modularity, which they take from Appel (2012), exactly means. Both terms could also have resulted in wonderfully conceptual references to cultural studies on forms of standardizing simplification in business and management, such as McDonaldization in the sense of George Ritzer (Ritzer 2021: 33–49).

5.5.5 Self-Legitimation and Defensive Attitude Towards Natural Science

Sources from the natural sciences, such as ecology, Earth System Science, and geology, are hardly referred to at all in the article. If so, very few older titles are mentioned rather than discussed (for example, Rockström et al. 2009), although the field is developing extremely rapidly. Despite all the emphasis on the importance of interdisciplinarity and the litany-like "beyond xy," the natural sciences appear to be dissociated in a Manichean way from cultural studies approaches. However, the scientific debate is far more heterogeneous than is commonly perceived (Görg 2016: 10). This attitude also gives rise to contradictions in the argument. For example, Tsing et al. refer to Swanson's argument in the same issue, saying that anthropologists who have prejudices against quantitative analyses should take numbers seriously. Tsing et al. do not see this as an invitation to open themselves up to scientific questions or even findings. Instead, they reduce the use of numbers to their significance in the criticism of models, such as on the carrying capacity of ecosystems (Tsing et al. 2019: 191).

For the authors, the natural sciences, prototypically the Earth System Sciences, are only one of many forms of knowledge or belief systems (similarly Watson-Verran and Turnbull 1995). With respect to geology, reference is only made to its romantic side in its period of origin; otherwise, it appears as one cosmology among others, such as animism (Tsing et al. 2019: 191). The above-mentioned infatuation with language is matched by the pervasive celebration of the disordered, the untamed, the anarchic, and the uncanny, which is why words like *uncanny*, *unruly*, and *anarchic* characterize the text (cf. Tsing 2012b). Despite the celebration of "ontological anarchy," discussions on science and animism within the so-called *ontological turns* are only casually alluded to through several sweeping references to Viveiros de Castro (2004a, b).

Tsing et al. offer some interesting ideas, but the text breathes a strong defensive attitude towards the natural sciences. At the beginning of their text, Tsing et al. advocate for concrete transdisciplinary conversations about "planetary structures that change everything," and they emphasize that anthropologists should pay attention to peculiarities and diversity without becoming "parochial" (Tsing et al. 2019: 186).

Despite this, however, it seems to me that the authors themselves want to establish a church. The emphasis on the more-than-organic beings seems like an updated version of Kroeber's concept of the "superorganic" (see Danowski and de Castro 2017: 84), which, however, is to be understood quite differently, namely as a scientific-pragmatic "divorce" (see Jackson Jr. 2010: 137–144).

Current Anthropology is a journal that is explicitly dedicated to the entire breadth of anthropology. The authors argue that today, the terms "man," "Earth," and "change" are no longer satisfactory terms as they were in the phase of the great anthropological designs of the 1950s (Tsing et al. 2019: 194–195). The Earth System Sciences are consistently criticized and (in my opinion, historically correctly) classified as a child of the Cold War and American hegemony in climate-related technologies. However, in terms of content, the authors do not really acknowledge Earth scientists, but rather reduce them to technophile model builders, which they only partially are, and dismiss them by *othering* them.

Fabienne Will argues that Tsing's approach is less radical than Haraway's. Her concern is "… *not primarily* about the exclusion of knowledge and approaches of disciplines outside the field and the adoption of a counter-position to the debate and the Anthropocene as a geological concept" (Will 2021: 245, emphasis mine). However, in the programmatic article discussed here, Tsing and co-authors pursue a mission against the supposedly all-consuming, universalist-Western Earth System Sciences. Throughout the article, they do not address the macro-dimensions of time and space in the Anthropocene at all, but celebrate the small and the diversely speckled. They seem to want a translation, a bridge between different disciplines, but have no sense that geology is a special natural science, namely a largely non-experimental science of natural history. Thus, Tsing et al. do not recognize the difference between geological working methods, time concepts and model representations, and the models and simulation methods used in Earth science. This reveals a danger, which the science and environmental historian Fabienne Will has formulated as follows:

> Translation, however, also carries the risk of using the Anthropocene without a thorough examination of the concept, because the significance of one's own contribution supposedly increases (...) (Will 2021: 205).

When I read the text not as an anthropologist, but as a geoscientist, I realize that the connection is cut: "The number you have dialed is unavailable." In this article on the Anthropocene, with a bibliography of no fewer than 94 titles, three Earth System Science titles and not a single geological work are cited. As an anthropologist, I am always surprised by the limited number of the same authors that are cited again and again, which borders on a citation cartel. Given the topic, it is surprising that Tsing and co-authors strongly emphasize the concept of systems, but do not mention the latest synthesis on the topic of environmental anthropology by Orr et al. (2015), in which the colleagues explicitly emphasize systemic perspectives and discuss many facets.

As a reader of the article, when I imagine the representatives of the other anthropologies, such as those from archaeological anthropology or physical anthropology, which are part of the comprehensive *four-field* anthropology, I am struck by their complete absence in Tsing et al. I wonder how an invocation of "ontological anarchy," without clearly defined concepts and with an explicit rejection of any generalization, is supposed to initiate a discussion with analytically oriented disciplines—or even just a conversation (sic!):

> As anthropologists come into conversation with Earth systems modelers, then, it seems important to hang on to our capacity to remember what Viveiros de Castro calls 'ontological anarchy'. Models can coexist with other modes of systems-thinking, including cosmological alternatives (Tsing et al. 2019: 191).

5.5.6 Commitment to Vagueness and Sprawling Analogies

Opaque word formations combined with what I believe to be exaggerated criticism, also lead to vagueness and contradictions. On the one hand, Tsing and co-authors, as quoted above, call for the re-engagement of cultural anthropology with big questions: "We need to reclaim, in a new register, anthropology's heritage of daring to make big claims about humans and about the worlds that humans co-inhabit with others ..." (Tsing et al. 2019: 187). At the same time, they only positively highlight deconstructivist theoretical directions in cultural anthropology from around the time of the Second World War and knocking down generalizing approaches labeling them as hegemonic and "universalistic," as "world-engulfing 'big anthropology'" (Tsing et al. 2019: 193–195).

Wishful thinking also plays a role in these arguments, which repeatedly leads to exaggerations, such as when it is said that representatives of the Global North have been "reduced to one among many global players" in cultural anthropology since the 1990s, even in theory production. This by no means corresponds to the actual image in publications and the actual power relations in institutions, but is rather wishful thinking. Mexican, Brazilian, and Indian cultural anthropology may now be more important than their German counterpart in a global comparison, but as a general finding, this is also wishful thinking. Such a thesis can surely not be derived from the anthology on *World Anthropology* cited by Tsing et al. (Ribeiro and Escobar 2006). In their general reckoning with recent cultural anthropology, Tsing et al. are as extreme as Jobson's total criticism of cultural anthropology due to the behavior of colleagues in view of the forest fires near a major anthropological conference in 2019 (Jobson 2020), but far away from the differentiation of a fundamental category critique, such as that found in Marilyn Strathern (Strathern 2020: Chap. 2).

Some terms used by Tsing and co-authors raise the question of whether they are not just new words for known phenomena. One example is modular simplifications. Inspired by his interest in the epistemic crisis of cultural anthropology in the face of the Anthropocene, Jobson applies the concept of modular simplifications from Tsing et al. to cultural anthropology as an organized discipline, an original idea. He refers to Tsing et al., who say that the approach of the *patchy Anthropocene* invites greater attention to the "wild proliferation of anthropologies beyond the established canon and the curated paths of anthropological knowledge production" (Tsing et al. 2019: 189). Jobson advocates a "decolonial turn 2.0" for cultural anthropology because:

> (...) our own modular simplifications – disciplinary associations, academic departments, tenure and promotion committees, and peer-reviewed journals – need to be dispensed or significantly revised in favor of new measures and values of intellectual work (Jobson 2020: 263).

I wonder what the added value of this terminology is compared to conceptualizations in science studies or the sociology of knowledge, for example, in Knorr-Cetina, Woolgar, or Latour. The text by Tsing et al. contains not only formulations such as the one stating that non-established anthropologies should "spread wildly," but also many rapturous statements: ghosts, according to the authors, would not care about boundaries, whether they are spatial, conceptual, or physical (Tsing et al. 2019: 191). Some formulations, in my opinion, are more suited to an artistic text than a leading scientific journal of anthropology:

> As Ficek shows, sometimes cattle destroyed the ecological relations that were meant to nurture them. Sometimes *feral cattle supported rebellious human subjects* as they spread beyond and before advancing empires (Tsing et al. 2019: 189, emphasis mine).

> Models enable a kind of literary imagination, a sensing of new social collectives and relations, even as they may silence attention to landscape structures (Tsing et al. 2019: 190).

> The modular and the feral are messily entangled in landscape structures (Tsing et al. 2019: 190).

> In the uncanny valleys of the Anthropocene, spirits, ghosts, and monsters of many kinds proliferate (Bubandt 2018; Tsing et al. 2019: 191).

Without this being explicitly stated as a goal, the article seems to have rather non-scientific claims, which also aligns with Tsing's undertaking, *Arts of Living on a Damaged Planet* (Tsing et al. 2017). The volume offers many artistic contributions and stands out for its creative peculiarities, such as by inviting the reader to turn the book around, and even carries different subtitles depending on the reading direction, in which the ghosts of the Anthropocene are invoked on one occasion, and its monsters on another. In my opinion, such artistically motivated approaches also involve the call, within so-called "realistic genres" such as ethnography and journalism, "(...) not to re-exclude supernatural beings as 'belief'" (Mathews 2020: 71). One example is a study on Taiwanese agriculture in which ghosts and plants and animals, in line with Haraway, appear as "odd kin," occurring alongside rice farmers as actors and analyzed as an example of non-simplified rural culture (Tsai 2019: 343).

Like many recent contributions, the article by Tsing and co-authors is heavily characterized by jargon and is abundant with wordplay (Table 5.6). As examples of jargon, I see creations like "to unpredict" or "unprediction" as well as "epistemicide," "co-species toleration" and "co-species collaboration," and "para-sites" (Tsing et al. 2019: 191–193). The authors almost repetitively speak of "more-than-human

Table 5.6 Examples of jargon and neologisms in programmatic texts on the Anthropocene

Term or linguistic expression	Possible meaning	Source
To undo	Avoid, prevent	Mathews (2020: 70)
To unpredict, unprediction	Predict, unsettle, emphasize uncertainty and unintended consequences	Tsing et al. (2019: 191–193), Mathews (2020: 70)
Speculative technologies	Technologies, the consequences of which are uncertain, e.g., CO_2 storage	Mathews (2020: 70)
More-than-human	Humans in nexus with other living beings (and sometimes things)	Tsing et al. (2019: 13 places); Mathews (2020: 6 places)
Para-sites	Situations or settings where different beings and things meet	Tsing et al. (2019: 191–193)
Epistemicide	Loss of culture-specific or local knowledge	Tsing et al. (2019: 191–193)
Temporalities	Time levels, micro-, meso-, macro-frames	Mathews (2020: 71)
Alienation	Alienation as separation of living beings from their life processes	Tsing (2015: 290)
Anthropo-not-seen	Entities that "ignore" the nature/culture dichotomy	Mathews (2020: 70)
Anthro-unseen	Overlooked entities because they "ignore" the nature/culture dichotomy	Mathews (2020: 72)
Unfolding ontological unruliness	Emerging entanglement of humans with other living beings and things	Mathews (2020: 77), see Tsing (2012a, b)
Anthropocene ethnographies	Ethnographies of Anthropocene dynamics *or* Anthropocene-theory-oriented ethnographies (?)	Mathews (2020: 74–76)
Analytics	Approach, methodical, theoretical, or both	Mathews (2020: 75)
Noticing	Attentive observation, as opposed to analysis	Tsing (2015: 17–25), Tsing et al. (2019: 188–189), Mathews (2020: 75)
Regimentation	Connection between human and non-human life, e.g., plantations	Tsing et al. (2019: 189)
Cosmopolitical registers	Unclear: Cosmopolitan political forms or related language forms	Mathews (2020: 77)

relations" when linking humans with other living beings, without even mentioning similar concepts such as "coupled human and natural systems" (Orr et al. 2015: 156–157). Critics say that differences in the exploitation and alienation of humans and non-humans are leveled by such a vocabulary (Davis et al. 2019). The authors prefer to use their hyphenated word creation and their love of wordplay in a politico-scientific way to quickly demand a "more-than-American anthropology" in contrast to the earlier volume of the Wenner-Gren Foundation on the subject (Thomas Jr. 1956; Tsing et al. 2019: 194–195).

Pascal Goeke criticizes Tsing's Matsutake book (Tsing 2015) for its manipulative use of personal pronouns such as "we" and "us" and notes an "observation all of its own" (Goeke 2022: 90). Something similar can be seen in this essay. In order to show that this article is no exception—not only in its extremely selective reading of the phenomenon, but also in linguistic terms—from among the publications on multi-species cultural anthropology and especially the influential group around Anna Tsing, I have examined an article by the second author, Mathews (Mathews 2020). It is presented as an overview and was published in the leading journal *Annual Reviews of Anthropology* (*ARA*). Nevertheless, parts of it are shaped by the same program, and there are sentences such as "... sandy soils become social through histories of mining that produce collapses" (Mathews 2020: 73). Such language stands in stark contrast to texts on similar topics penned by more scientifically oriented anthropologists, as the overview on climate anthropology in the same issue of the *ARA* shows (O'Reilly et al. 2020).

A content analysis of newer studies by Haraway would lead to a similar result, although I would like to state in advance that this play with language is evaluated quite differently by some, namely positively (Hoppe 2021). How can one explain this form of texts? I can only make assumptions here. An artistic aspiration might explain the heavy use of many words with unusual meanings, the high proportion of fashionable vocabulary and cryptic formulations. A more realistically oriented colleague describes such texts as full of rhetorical charm that goes hand in hand with metaphorical "excesses" (De Pina-Cabral 2017: 5). Another explanation is simpler, but also seems likely to me: this heavily jargon-laden language is motivated by both content and aesthetics, forming a reflex against what is perceived as the technicist language (and also the numbers and graphics) of the Earth System Sciences. The authors' explanation would probably be that a new reality, namely the Anthropocene, also requires a new language:

> From the feral and unexpected encounters between bovines and humans in Panama, as from those between marabou storks and salvagers in Uganda, the illegitimate hope of a kind of co-species collaboration in ruins emerges for which we do not have a language (Tsing et al. 2020: 193).

Marilyn Strathern's more recent publications (Strathern 2020) make it particularly clear to me what gains in stimulation and productive uncertainty, but also what losses undefined words and the multiplication of open metaphors bring to the conceptual understanding of the Anthropocene. The two quotes from a review by Tatjana Thelen get to the heart of this tension:

> Instead of merely focusing on overcoming human-centered approaches in the social sciences – Strathern quotes here the authors of neo- and postmaterialism, multispecies ethnography as well as the so-called ontological turn – a more radical reflection on the conceptual inventory of political criticism is needed. (...) More fundamental to Strathern than the extensively discussed dichotomies of nature vs. culture, body vs. consciousness, human vs. animal seems to be the concept of relation (which has *diffusely positive connotations* in both scientific and colloquial terms). (...) In *deliberately undermining scientific 'clarity'* and convention, Strathern ultimately offers no solution to the epistemological problems she has uncovered. At the same time, it is this *self-commitment to vagueness and rampant analogy* with which she continually provides new impulses and challenges to the all too often epistemologically naive social sciences (Thelen 2021: 4, emphasis mine).

In my opinion, the approach of Tsing et al. reveals an example of how inter- and transdisciplinary arrangements lead to openings towards the evidence of other disciplines. This is to be welcomed, but these openings in turn entail processes of closure (Wenninger et al. 2019, regarding the Anthropocene, see Will 2021: 25). Narrative practices, such as localization, contrast formation, and, in the case of Tsing et al. in particular, translation, entanglement, and delimitation, themselves generate new inclusions and exclusions. Tsing and co-authors are very open to artistic approaches, emotional arguments, and creative forms of writing. However, this leads to them using geoscientific words but largely excluding geoscientific arguments in the factual dimension.

To what extent this also leads to the selective inclusion and exclusion of individuals as knowledge producers in the social dimension of evidence building would first have to be clarified. Mechanisms of the academic market would need to be examined: attention economy, neologisms, and citation circles. Even well-meaning connoisseurs of the positions of new materialism note a frequently "exaggerated rhetoric of innovation" that obscures historical lines of tradition and conditions of origin in order to claim theoretical originality (Hoppe and Lemke 2021: 163, similarly Ellen 2021).

Tsing et al. work with a narrativization in the form of a translation of the Anthropocene concept and, for example, the concept of landscape, in which localization (in time and space) and differentiation (by subject areas) are prominent. With the evidence practice of entanglement and translation, knowledge from outside the discipline, such as systems science, is mobilized to underpin the importance of their own discipline. This corresponds to Haraway's seemingly paradoxical mobilization of geoscientific knowledge, especially geological chronology, as an argumentative resource, although her argument for the *Chthulucene* is delimiting (Will 2021: 253). As an *umbrella concept*, the Anthropocene can be used well... or also misused, as I believe is the case with Tsing and co-authors. If this happens superficially and on the basis of a legitimation motive, there is the danger of knowledge from outside the discipline being effectively excluded, which in this case comprises scientific conceptions and insights.

5.6 Concepts—Meaning, Embodiment, and Assessment Procedures

Alf Hornborg points to another central contribution of cultural anthropology to Anthropocene research, namely the study of meanings (Hornborg 2020a: 2, 2020b). This is the approach that has dominated large parts of cultural anthropology since Clifford Geertz under the label "interpretive cultural anthropology" or "semiotic approach." At first, one might think that this could hardly make a substantial contribution to the understanding or even to the explanation of Anthropocene change, since the Anthropocene is primarily determined by material flows.

However, Hornborg argues that these material flows can be traced back to economic market activity and the use of technology.

5.6.1 Concepts of Resources

This economic activity is largely based on specific values and norms. Contrary to the understanding of the dominant currents in economics as material-related research, the core concepts are heavily laden with meaning. At its core, it is about money as a symbol or a sign of money (*money signs*) and the assumption that everything and anything can in principle be exchanged for anything else. Economic analysis is therefore an analysis of signs based on "cultural" concepts of money and similar assumptions about "utility," "preference," and "willingness to pay":

> It is ironic that we tend to think of economists and economic concerns as 'materialist', when, on the contrary, mainstream economics, ever since Victorian times, has been couched in a highly *idealistic* focus on human desires and subjectivities. (…) We need to realize how peculiar the cultural idea is that anything can be exchanged for anything else – whether fresh resources for resources already dissipated, or local produce for products transported halfway around the world. (Hornborg 2020a, b: 2, emphasis in original).

Modern economics therefore has just as much to do with values and symbols as it does with industrial ecology and material flow analysis, and even more so, according to Hornborg. Accordingly, it is difficult for both economists and anthropologists to think about concepts such as "social metabolism." On the other hand, one could argue that cultural anthropology can contribute to a rethinking of neoclassical economics, which is related to it in its idealistic axioms. Marcus and Fisher had already seen the treatment of material processes as the core of their now classic project *Anthropology as a Cultural Critique* (Marcus and Fischer 1999: 141 ff.).

A central contribution of cultural anthropology is the description, analysis, and possibly critiques of units, limits, etc., which are used in natural science research on the Anthropocene. Here, cultural anthropology can apply its comparative and critical approach to a part of its "own" ("Western") culture. Pálsson et al. see in this the central role of the social sciences in quasi auto-sociological research with political implications:

> One step would be to massively research the roots of the social measurements of success, the roles and functions of reward systems, to investigate and critique their methodologies and tacit and explicit assumptions, and to analyze their implementation politics (…) our work should be guided by the possibility of social sciences and humanities research as the intellectual input for a New Political Economics that can discard the most dysfunctional elements of the present (Pálsson et al. 2013: 10).

This highlights another strength of cultural anthropology, which it shares with other fields of *cultural studies*, namely the critical view of concepts, terms, and narratives. It can enrich the repeatedly called for "micro-social reflection" (Leggewie and Hanusch 2020: N4) on the topic of the Anthropocene. In the climate change discourse, there are studies on language, such as "ecological debts," on metrics used, and on scientific visualizations (for example, Schneider 2018). Finally, anthropologists, together with humanities scholars and social scientists, could contribute to social movements in the Global South that arise in response to the experience of the Anthropocene. So far, there has been much work on environmental movements.

5.6.2 Environmental Cognition and Indigenous Allegories

In cultural anthropology, there is no term that encompasses individual and collective knowledge, perception, and experience of the natural environment (Orr et al. 2015: 157). There are only terms that denote different aspects or facets of environmental cognition or phenomenological experience, such as traditional knowledge, local knowledge, *folk biology*, *ethnoscience*, and *cognized environment* and *symbolic ecology* (as examples, see Schareika 2004; Antweiler 2008). In cultural ecology, the importance of knowledge for the adaptation of communities to their environment is emphasized (Steward 1955; McNetting 1977). This aspect is gaining new significance in the face of climate change and the demand that humans should take *stewardship,* responsible management (Orr et al. 2015: 158).

Particularly in research based on cognitive cultural anthropology, it has been shown that there are often detailed and observation-based conceptions of nature and living beings (Werner 1972; Orr et al. 2015: 158). Such conceptions have been made especially in the area of classifying environmental units, processes, and perceptions. Classic examples are studies on the indigenous classification of flora and fauna (Berlin 1992). Recent works have shown that taxonomic conceptions, in which several species of living beings are conceived as belonging together, "guide" certain deductive conclusions about functional similarities among the living beings (Atran and Medin 2008, according to Orr et al. 158). Such environmental cognition is usually strongly linked with concepts of causality as well as with religious concepts and cultural values (for example, Bennardo 2019). Taxonomic information is thus not only an ordering of the environment or world, but a way of making landscapes "actionable" (Orr et al. 2015: 158). Recent studies have examined the varying distribution of environmental knowledge in communities or networks. To date, a certain weakness of research on environmental cognition is the lack of research on the change in local knowledge through events and in a transgenerational context. Transgenerational learning processes based on environmental experiences have been demonstrated, for example, by Roy Ellen in Indonesia (Ellen 1999).

The scientific character of these conceptions has been recognized since the 1980s, which is reflected in terms such as *ethnoscience*, *folk science*, and *folk biology*. People have an "empirical-objective orientation" (Rudolph 1973), which can be developed precisely through experiences with their material environment (Atran 1998; Antweiler 2007). These studies also found universal ideas about nature, living beings, and the environment (Frake 1962; Atran 1998; Medin et al. 2002; Atran and Medin 2008). Especially in the field of environmental knowledge, there seem to be more cognitive universals than in other fields of culture, which is obvious since the human psyche develops in certain environments (evolutionary psychology). These research works can be brought into dialogue with studies on Anthropocene allegories that are theoretically quite differently informed by religious anthropological studies (DeLoughrey 2019).

One topic to which cultural anthropology could contribute a great deal is that of the personal or collective experience of local effects of the Anthropocene. How does the Anthropocene manifest itself not only in cognition, concepts, and classifications, but also in physical experiences and non-linguistic forms? Here, one could build on a cultural anthropology that is interested in individual but culturally embedded experiences and emotions, and study universal as well as local-specific affects (Antweiler 2017; Röttger-Rössler and Slaby 2018; Röttger-Rössler 2020). This applies, for example, to deep grief at the loss of habitual environments or feelings of guilt in the face of man-made permanent environmental change, which is experienced as "ecocide".There are initial research approaches to negative and positive emotional reactions to climate change and psychological pressure caused by rapid changes to the Earth (Cunsolo and Landman 2017).. In addition to "Earth emotions" such as "environmental anxiety" and "ecophobia," there is a specific emotion known as "solastalgia," a kind of homesickness experienced while at home. Similar to nostalgia, a feeling of loss is experienced in the face of significant change or destruction of one's own living space (Albrecht 2019, on the Anthropocene, 70–72). In view of anthropogenic change, an emotional nexus that can be described as "guilty grieving" is particularly significant; Craps 2023). While we know a more and more about human reactions to drastic environmental changes, the perception and possibly fearful experiences of animals have hardly been researched (Soentgen 2018).

There are a number of studies on lived experiences with the natural environment by Tim Ingold, who approaches the topic from a phenomenological perspective. Ingold ethnographically investigated behaviors and especially sensory and bodily experiences, for example, during hunting in arctic environments and among people who live as pastoral herders and interact with their animals and elements of the landscape and at the same time partly produce them. He argues that culture, both individually and collectively, is formed particularly through environmental experiences ("sentient ecology"). These experiences with the environment stem from lived interaction with the material environment and are largely manifested in non-linguistic form.

People act in landscapes, they move physically in these landscapes, and these landscapes contain traces and also objects of material culture, such as human buildings (Ingold 2011a, b). With this research approach, everyday experience is assigned a more important role than in cultural anthropology, which is oriented more towards language and non-secular aspects of culture. In this respect, one question posed by the Anthropocene would concern how people of different cultures not only conceptualize the increasingly asphalted or concreted environment, but also move in it and deal with it in a way that shapes their experience (Fig. 5.3).

Fig. 5.3 Graffiti as an urban expression of human and environmental diversity in Cologne, Germany. (Photo: author)

5.6.3 "Ontologies"

Many anthropological studies on the relationship between humans and nature are based less on symbols or cognition than on the worldview of the cultures studied, on so-called local or indigenous "cosmologies". These ideas are also taken up by natural philosophers in view of the Anthropocene (for example, Vetlesen 2019, contributions in Kirchhoff 2020).This includes most approaches of the "anthropology of ontology" (also "anthropology of ontologies," *multiple ontologies approach*). Four anthropological approaches can be distinguished here: "symmetrical anthropology" (Bruno Latour), "monistic anthropology" (for example, Philippe Descola), "anthropology beyond the human" or "anthropology of life" (Eduardo Kohn), and "artifact-oriented anthropology" (see Grana-Behrens in prep.). Many of these approaches are influenced by *Science and Technology Studies (STS)* and currents of posthumanism. If one wants to relate these to the phenomenon or problem of the Anthropocene, a problem will arise. These approaches largely pursue an attitude that is diametrically opposed to other "ontologies" or "cosmologies" of the so-called Western worldview, which is why they are also referred to as *alternative ontologies*. In the studies, which contrary to the designations are actually mostly concerned with epistemics rather than ontics, the local system of cognition or knowledge is sometimes not presented in a detached manner, but is rather interpreted as being superior to universal science, as an alternative to effects of the techno-centric science of "the West", which causes disastrous environmental effects.

These assumptions make collaboration with the natural sciences that goes beyond collegial tolerance difficult, because the aforementioned assumptions necessarily come with ontic and epistemic assumptions that are hardly compatible with the materialistic assumptions of natural sciences. They also contradict historical-materialist, Marxist, and conventional anthropological concepts of culture. This means that the natural sciences are overlooked—fields such as evolutionary biology and human ecology, which are of central importance for the Anthropocene (Belardinelli 2022). I had already noted this in Sect. 5.5 regarding the works of Tsing et al. Eduardo Kohn follows the special approach of an *anthropology of life* (Kohn 2013, cf. Grana-Behrens in prep.). When he follows heterogeneous assemblages in the rainforest of the upper Amazon in Ecuador, he does use concepts from the natural sciences. However, these are not definitely concepts from evolutionary ecology or the ecology of complex systems, such as those of Lansing, but ones that are difficult to implement in empirical research, such as Deacon's "biosemiotics."

A similar divide exists between biological animal behavior research (ethology) and *human-animal studies*, which was inspired by actor–network theory (ANT, e.g. Wirth et al. 2016; Krebber 2019). An even greater disparity exists between biological evolutionary ecology and multispecies ethnography (Kirksey 2015; Ameli 2021). Although both fields actually study similar topics, such as the co-construction of landscapes, the terminology and the guiding ontic premises as well as epistemic assumptions of multispecies ethnography differ vastly from approaches in ethnoprimatology and cultural primatology, which study the interaction relationships between humans and other primates from a behavioral science (ethological) perspective (for example, Fuentes 2006; Fuentes and Baynes-Rock 2017; Riley 2020). An ethologically informed ethnography is only slowly developing (Hartigan Jr. 2021: 850–853).

Fortunately, primatologists interested in anthropology rarely write sentences of the following kind: "... multispecies *entanglements* become central aspects of anthropogenic ecolog*ies*" (Malone and Ovenden 2017: 8, according to Gibson and Venkateswar 2015: 14, emphasis mine). For a research program that wants to proceed empirically, such diffuse and all-pluralizing language is too ambiguous and therefore makes it almost impossible to operationalize research questions. If the authors boldly conclude that "anthropologists are simply expanding their horizons" (Gibson and Venkateswar 2015: 9), then it must be clarified what constitutes this continuum of anthropology, for example, with respect to cultural anthropology per se or to the range of the four-field approach of North American anthropology. Given the disparate knowledge cultures of these research approaches, it is at least premature or overly optimistic when Gibson & Venkateswar present conventional ecological and "ontological" research fields as complementary:

> Yet both trajectories of research and those based on more radical new ontological considerations cannot easily be separated, particularly because within the discursive space of the Anthropocene they complement one another. Indeed, they reside within a continuum of anthropological work that illuminates how worlds are made, reshaped, and understood (Gibson and Venkateswar 2015: 9).

Here, a fruitful connection to cultural anthropology-oriented primatology would be possible. Ethnoprimatologists study interactions between humans and other primates, the mutual, coevolutionary development and shaping of their respective environments. The focus is on environments in which humans and other primates permanently share habitats and therefore, for example, have coevolutionary medical histories (Riley 2020: 84–88). To this end, the researchers also use observation methods similar to those employed in cultural anthropology and apply them to non-human primates. They use precise methods of non-participatory observation or highly participatory procedures. This is especially true for the study of cultural competencies in non-human primates in so-called cultural primatology, where such investigations complement laboratory studies. In this case, cultural primatologists and ethnoprimatologists study a common niche for non-human and human primates (*hominoid niche*, Malone 2022: 160–180).

5.7 Culture—Anthropological Holism *Revisited*

Even though there are well over a hundred definitions in the anthropological literature, almost all of them deal with innovation and the non-genetic transmission of information between generations (transmission) in the context of social communities. A conception of culture that is particularly helpful for the Anthropocene is to understand culture as a quasi-experimental method as well as a mechanism for *striking a balance between inherited and newly created knowledge* (Arizpe-Schlosser 2019: 280). The task of anthropologists is then to precisely explore this interface. In relation to the global environment, it would be the task of cultural anthropology to secure experiences that allow for the consideration and choice of various actions to promote a transition towards sustainability and less imperial relations to nature. In this work, a deep discussion of the role of cultural theory in environmental issues cannot be conducted (groundbreaking Milton 1996), but only some aspects specifically relevant to the problem of the Anthropocene can be addressed.

5.7.1 Concept of Culture—Janus-Faced Reception in Global Environmental Institutions

One approach to making the Anthropocene fruitful for cultural anthropology, and *vice versa*, is an analysis of the use of the concept of culture in institutionalized debates on global change. This is because in such debates, it has causal effects on decisions consequential to the Anthropocene. This approach is taken by the Mexican anthropologist Lourdes Arizpe Schlosser. She had previously traced in development debates how the concept of culture was used in international documents on global change until 1990 (Arizpe 1996). In a more recent study, she expands on this with an investigation into the use and impact of the term Anthropocene in international debates. It turns out that culture has become more important internationally than many other terms since the Second World War, and that this has served as a "dreamcatcher" and at the same time as a "golden account" of one's own civilizational achievements (Arizpe-Schlosser 2019: 1).

Arizpe Schlosser comes to the interesting conclusion that the Anthropocene as a term emerged, in a parallel development, simultaneously with the worldwide success of the expanded concept of culture, both in science and in the public sphere (Arizpe-Schlosser 2019). Such studies on the development of the term also open up interesting questions for the history of science. Haraway, for example, asks why the concept of the Anthropocene emerged just when concepts of autopoiesis, symbiogenesis, and endosymbiosis became popular in the life sciences (Haraway 2016: 61).

Studies in the anthropology of science are also methodologically progressive. Arizpe Schlosser does not only evaluate conference documents, but also informal discussions and experiences gained from observing participation in many such conferences. As a basis, she used her many years of experience in international institutions, especially her position as president of the *International Union of Anthropological and Ethnological Sciences* (*IUAES*) and as a member of the executive committee of the *International Social Science Council* (*ISSC*), where she herself participated in the council's "International Human Dimensions Programme." Among other things, documents, her own records, and participant observations at over 40 international conferences during the period between 1990 and 2016 were used. Taken together, this goes well beyond a purely organizational anthropological approach or a cultural anthropology of public policy: Here, the heuristic approach via an "ethnography of transactions" is combined with elements of *multi-sited ethnography* and autoethnography. In such a study, it also becomes possible to clarify the role of the anthropologist as a co-producer of internationally effective knowledge on sustainability and also on the Anthropocene (Arizpe-Schlosser 2019: 3–5, 271–272, for an example, see Antweiler 2021b from an EU project).

5.7.2 Culture—Mode of Adaptation and Innovation

The aforementioned use of culture as a mode for political organization or *governance* of nations, ethnic communities, and religious groups at the often contested micro- and meso-scale level is referred to by Arizpe Schlosser as "laoculture" (Arizpe-Schlosser 2019: 3, 259–261, 311). The following is about what she calls "ontoculture," the way in which we become humans and develop our unique abilities, with creativity generating certain freedoms in the face of organismic limitations. Even from the perspective of evolutionary biology, culture in humans as organisms that are "wired for culture" is the means of adaptation (Pagel 2012; Henrich 2016; Laland and Chiu 2020; Sterelny 2021). Humans can only survive as individuals by using socially transmitted information. This means that there are no humans who do not, not yet, or no longer have culture. Even for collectives of *Homines sapientes*, some information that is not transmitted genetically is vital for survival.

In cultural anthropology, culture usually stands for the way of life of a collective, which differs from the way of life of other (national, religious, linguistic, spatial) collectives. According to Marc Augé, culture can be understood as "... a relatively coherent totality of representations and principles that guide the organization of relationships between individuals in a community, which is thus constituted as a society" (Augé 2017: 62). Central to this are the relationships of the self to the other, that is, social relationships, whereby I emphasize the material ones in context more strongly than Augé. The everyday English expression "way of life" hits it quite accurately. People learn culture from other people (socialization, enculturation), and they generally adopt more from social partners than they innovate themselves. Central to this is the respective *shaping* of human existence: people reshape what already exists (Rudolph 1973). This fits with the Marxist approach of "social metabolism" and "natural relations," where the central statement is that it is the shaping of social natural relations through historically specific forms of organization that makes them different from other animals (Görg 2019: 172).

Both cultural anthropology and archaeology focus on enduring habits formed through tradition and the resulting collective habits in action, experience, and material culture. This can be subsumed under the term standardizations (Hansen 2011: 7). If it is to be reduced to intentional action and its effects, culture can be defined as the sum of the effects of innovations (Rudolph and Tschohl 1977: 13). These can be new innovations or previous innovations that have been passed on. Such an understanding of culture must be explicitly distinguished from the bourgeois-normative concept of culture, which highlights certain ways of life as valuable and denies culture to other people, collectives, or phases of history.

5.7.3 Holism—Culture Crosses Substance Categories

A prototype for a broad understanding is arguably the most consequential of all definitions for theory formation, that of Edward Tylor (who equates culture and civilization): "Culture or civilization taken in its widest ethnographic sense is that complex whole which includes knowledge, belief, art, morals, law, custom, and any other capabilities and habits acquired by man as a member *of society*" (Tylor 2005, 1, emphasis mine). Tylor's definition includes the entirety of human lifestyle: it is a holistic (from the word "whole") or totalizing definition. This implies that culture is in no way physically reducible, neither to material (objects, tools, writing) nor to the psychological (thoughts, emotions) or social (relationships, networks). Culture is human-made, such as artifacts in the broadest sense. In addition to artifacts in the narrow sense, there are socio-facts and menti-facts. In short, culture is materially hybrid. This holistic concept of culture, however, does not coincide with some concepts of culture in cultural anthropology.

A narrower understanding is shown by the semiotic or symbolic and meaning-oriented concept of culture (Geertz 1992). This interpretative concept understands culture as a dimension of meaning in societal life. Collective systems of meaning are articulated in socially binding worldviews, cultural knowledge, and symbols. Both on the part of the actors and methodologically on the part of science, it is about understanding meaning. The dominant understanding of culture in North American *cultural anthropology* is significantly narrower than the Tylorian concept. Culture is defined there as cognition or as cultural knowledge (culture as cognition, culture as knowledge). This refers to action-guiding worldviews or collective representations, such as cosmologies, cosmogonies, classifications, and routine knowledge. It is usually reduced to an even narrower area, namely to the knowledge, worldview, or symbolism shared by all individuals in a collective (*culture as shared knowledge*).

While culture is mostly meant in a comprehensive way in cultural anthropology, with the social being one area among others (such as religion and economy), it is usually the other way around in sociology. There, culture is understood as one of several subsystems of the social system. Parsons, for example, sees culture as an area for the production and dissemination of ideas, but more recent cultural sociology sees itself more as a perspective of general sociology (Moebius et al. 2019). It also focuses on the ideational aspects (knowledge systems in discourses, habitus, and lifestyles), but emphasizes, in the sense of a *cultural turn*, that culture is a dimension that permeates all areas of society.

When emphasizing a holistic concept of culture in the political context, it should never be about merely emphasizing interrelationships. No. It is about showing the systemic links between different elements and dimensions in a social system, which is now a geosocial and geocultural system. Especially in the political context, anthropologists can contribute—locally or supralocally—to solving with the holistic concept of culture one of the biggest problems in working against the Anthropocene: the sectoral political action.

The metaphors described in Chap. 3 are instructive as manifestations of thinking and feeling about the Anthropocene crisis. They can be considered as a social fact, and they have effects that ultimately also manifest themselves in the extra-psychic world. However, the metaphors, especially the alarmistic ones, also have a problematic side, especially if we are looking for societal solutions for the Anthropocene as a problem. They can easily make the Anthropocene appear as an amorphous danger. But what we need from the perspective of social theory are solutions that address the complexity of today's functionally differentiated societies. This implies recognizing that the different entities of society (economy, politics, media, sciences) function according to different conditions for success.

For realistic solutions, it must also be clear that the actual functions and expectations of these systems are notoriously divergent, as the climate change debate has made clear in recent times. Any hope for a quick "one-size-fits-all" solution fails to recognize the social complexity (Nassehi 2020: 908). In the Anthropocene, in addition to scientific knowledge, cultural and social theory and the empirical evidence derived from them are needed more than ever to address problems in cooperation. Only in this way can we identify where conflicts of interest exist and where cultural milieus, institutions, and organizations intersect, at which point solutions can be applied (Blok and Jensen 2019). A current example of such an approach is the negotiation over a common CO_2 pricing. In view of the otherwise stalled climate summits due to free-riding or withdrawal, this seems to be emerging as a solution for achieving genuine cooperation in climate change negotiations (Ockenfels 2020: 907). To systematically uncover such solutions, a revival of systemic or holistic cultural concepts and corresponding empirical methods would be useful, but these would need to be linked with sociological social theory.

5.7.4 Culture Is Fundamentally Collective—Social Learning and Enculturation

However, Tylor's definition also emphasizes a second aspect, namely the fundamental sociality involved in the acquisition of a respective culture. Individuals acquire culture in the context of social relationships, that is, as members of collectives. Contrary to what can often be read, however, Tylor writes "as a member of society," not "as a member of *a* society." This is a big difference, about as big as the difference between the statement "all people speak a specific language" and the statement "all people have the ability to speak" (Klass 2003: 14, 24). The difference points to two levels of universals; on the one hand, biotic universals, which apply to all (healthy) individuals (and represent species characteristics), and on the other hand, "cultural universals," which can be found in all cultures, but not in all persons (pancultural patterns, Antweiler 2018: 97–155).

Because cultural content is transmitted through non-genetic communication, people must be members of social associations in order to grow into the culture (socialization, enculturation). This is a universal fact. More precisely: to acquire (any) culture, people must live in (any) social organization. Even if we as individuals are sometimes alone, it still applies. People generally live in clusters of several people. These are groups or groupings within which and with whom one spends one's life. Despite all personal differences, people are connected through shared patterns of action and thinking (standardizations) that apply to a particular collective. Formally speaking, these are collectives, but in terms of content, these collectives form cultures through their *consistently average identical action*. Culture and collective are thus "inseparably" linked, but they can be analytically separated into the form aspect (collective) and the content aspect (culture).

These carriers are collectives and therewith people and their products. Biologically speaking, people need other people to survive. This is especially true after birth, when the individual depends on others for care. However, these others do not have to be the father or mother, they can be any other person. Human individuals typically belong to various collectives. This multiple membership is particularly pronounced in modern (large and complex) societies. However, this characteristic, which Hansen refers to as polycollectivity (Hansen 2011, 120), is a basic feature of human societies. In principle, it also occurs regularly in smaller and simpler structured cultures. The normality (or salience or virulence) of polycollectivity is one of the characteristics in which collectives of human primates differ from collectives of other primates.

Recent works on the history of cultural anthropology show that central features of the "totalistic" concept were still missing in Tylor's work. These are primarily the historical dimension, the systemic integration of cultural aspects in a collective, and the aspect of cultural boundaries. This was first achieved by Franz Boas and his students (including Ruth Benedict and Margaret Mead). They emphasized the particularities of individual lifestyles (particularism) and spoke not only of culture, but explicitly of cultures in the plural (Vann 2013: 34, 40). Culture is what people in groups make of what they find: the sum of *artifacts* in the broad sense. It is contingent in the sense that it could be different locally according to our knowledge, which can be empirically demonstrated, among other things, by the fact that it is actually different elsewhere. The documentation of cultural diversity, especially beyond the *WEIRD* societies, is of central importance for the human sciences. If we take the number of languages as a simple indicator, there are around 7000 cultures! Only the full range shows the diversity of human possibilities. For this, extreme individual cases are also important, because: "what is real is also possible."

The classical approaches to anthropological cultural theory of the 19th and 20th centuries mentioned above left two aspects underexposed: (a) the differences within collectives (intra-cultural diversity) and thus individuality, and (b) the role of materiality, be it as objective culture or in the form of the material environment (nature). Both concurrently lead to overlooking one motor of cultural change: people learn not only from other people, but also through their *individual* interaction with objects. In social groups, learning also occurs by means of the material objects of their predecessors. The emphasis on diversity and interactions is also important because the functional systems of the economy and politics react very differently to the "great acceleration."

> The anthropos should be taken to include domesticated animals and plants as well as people in all their differentiated and unequal heterogeneity, along with other entangled entities and forces held to be part of broader assemblages producing climate change and other transformations (Edgeworth 2018a: 759).

5.7.5 *Material Culture and Verticality—Geosocial Formations and Urban Anthropology*

The Anthropocene is often associated with the horizontal expansion of cities. However, cities are also characterized by verticality (Graham 2018). One effect of the naturalizing everyday perception of the ground on which we live is to see it as an inherently *horizontal* phenomenon. Thus, human-made verticality, which not only extends to a height of 1 kilometer (*Kingdom Tower*, Jeddah, 1008 m), but also to a maximum depth of 3900 meters with the *Tautona Mine* in South Africa, does not come into view. Cities in particular are the main sites of a geology of the now or of "geology-making" (Ellsworth and Kruse 2013; Irvine 2020). This becomes apparent, for example, when one considers the totality of the world's underground railways. Here, anthroturbation, the human stirring up of the ground through mining and drilling, similar to the digging by other organisms in the ground (bioturbation), comes together with human objects such as pipes and tunnels (archaeosphere) and the upper part of the deeper layers, the "deep geology." Thus, underground railways can be quantitatively represented and analyzed as a permanent proxy for population growth and energy consumption (Williams et al. 2020: 437, Fig. 20.3). This verticality is especially evident in cities and particularly when one considers where cities get their supplies from and where city dwellers dispose of their products, that is, the entire urban metabolism:

> From 10,000 meters under the sea to 35,000 kilometers in orbit above the Earth's surface, the infrastructure that supports urban life has reached unimaginable dimensions underground, in water, and in space (Bélanger 2021, trans. CA).

This calls for a "vertical urbanism" (Harris 2016), and here lies a possible field of future collaboration between urban cultural anthropology, cultural anthropology of material culture, and urban archaeology. In old cities in particular, the human footprint extends in the depths across the entire width of the cities, as they de facto stand on *artificial* ground of considerable thickness (Redman et al. 2004; Graham 2018: 282–284; Goudie 2019: 187–231). Classic examples are the tells, mounds created by repeated settlement, for example, in the Middle East (Edgeworth 2014). But today's citizens of New York City also live on at least 8 meters of human-made geology, those of Manchester on 10 meters, and those of Rome on about 15 meters. The metaphorical concentration of power, the "city on the hill," often manifests itself in an artificially elevated location. Pits, trenches, cellars, backfills, foundations, tunnels, and other subterranean traffic infrastructure create unusually diverse structures. The complex stratigraphies are particularly an issue in old and continuously inhabited cities, where urban archaeology can address Anthropocene topics (Price et al. 2011; Denizen 2013: 40; Edgeworth 2014).

As a city with cold winters, New York generates a kind of urban glacier every year by collecting tons of dirty and salty snow along with tire debris, brake pads, and exhaust pipes. In summer, these deposits then appear as melting "terminal moraine" (MilNeil 2013). Given the scale, it is foreseeable that the constant material flows will have an impact on geological time periods. The geology of the city has long been intensively shaped by anthropogenic activity, and one can watch it happening today. It is therefore only logical that Smudge Studio in New York published "A Field Guide to the GeoArchitechture of New York" (Kruse and Ellsworth 2011). In addition to sedimentation, the intervention in existing layers in cities through the extraction of existing rock and relocation is extremely intensive. By removing hills and filling in hollows and valleys, the landscape is flattened. Mexico City was built on a lake near the Aztec capital Tenochtitlan and sank by 10 meters in the twentieth century. Chicago was built on an artificially elevated level above swamps in the nineteenth century. The notoriously flat Manhattan of today was built on land that the Lenni Lenape indigenous people called *Mannahatta*, "the island of many hills" (Sanderson 2013).

These examples of urban verticality show how earlier cultures shaped the material substrate on which people live today. More relevant for a cultural anthropology of the Anthropocene, however, is that we can anthropologically track how geological strata are being created *in situ* in cities today. We produce assemblages of old tram tracks, heating pipes, electricity and telephone cables, forgotten gas networks, ash as well as petrochemical residues, and bones contain unused raw materials. These "hibernating infrastructures" are already being used as "mines" in some cities (Wallsten 2015). They could become significant, as only half of all metals ever extracted have been used so far. In the Anthropocene, human societies are not only "geosocial formations" in the sociological sense.

The deposits in the context of urban verticality can be understood as geosocial formations *in the literal sense* and thus constitute a focus *material culture studies* related to the Anthropocene. Methodologically, this means that we need more vertically informed ethnographies ("3-D ethnographies," Harris 2016: 608–612). Landscape architects focused on verticality, for example, can show the extremely complex network of pedestrian pathways in present-day Hong Kong or Toronto (Bélanger 2007). Methodologically speaking, cultural anthropology can make many contributions here, from a three-dimensional architectural-anthropological documentation to citizen-oriented cognitive maps and methods of critical mapping (*countermapping*; Radjawali et al. 2017). The research topics under a three-dimensional urban cultural anthropology informed by the Anthropocene can range from the power embodied in infrastructure, the urban power vocabulary strongly characterized by verticality, to theoretical connections between anthropogenic verticality and local cultural responses.

5.8 Local and Present-Oriented—Opportunities for Field Anthropology

> *Our present is a disinterested present.It draws on resources long in formation paying its debts into the future fossil record.*
> Richard Irvine (2020: 106)

With its local and present-oriented approach, cultural anthropology could offer a chance to turn the Anthropocene into a more ambiguous, but nevertheless truly fruitful technical term, a "boundary object." Boundary objects respond to the fact that there are objects that are not fully naturalized in the social and scientific worlds of the actors involved and that classifications routinely collide (Star and Griesemer 1989: 297). Boundary objects should be malleable enough to be applicable in the local context and robust enough to bring together a wide variety of actors, such as humanities scholars and cultural scientists (Star and Griesemer 1989: 393).

If cultural anthropology is conceived firstly as being interested in local life-worlds, secondly as present-oriented research, and thirdly as methodologically oriented towards experiential approaches, then the phenomenon of the Anthropocene also poses a problem, or to put it positively, a triple challenge. As with anthropogenic climate change, most Anthropocene effects of human activity are (1) large-scale, (2) long-term from a human perspective, and (3) difficult to individually experience in everyday life:

> Much of the biological and chemical traces of existence left by humans are difficult to comprehend as a day-to-day concern and thus only an intangible and vague worry for many (Gibson and Venkateswar 2015: 7).

5.8.1 Localism

The majority of anthropogenic effects, such as climate change, soil acidification, and species extinction, can hardly be sensually perceived by actors in their local life-worlds. Many of the effects can only be detected with the use of measuring devices, which are often not available or accessible locally. In order to detect changes in long-term or even geological time scales, long series of measurements or mathematical modeling is required, both of which are not accessible in many localities.

In addition, many of the effects go far beyond the boundaries of the local life-world; they are classic *transboundary* phenomena. All these problems can definitely not be understood as local phenomena (Wapner 2012). The effects of one's own actions are also only directly perceptible to a limited extent. This is especially true for the effects of one's own activities in regions outside of one's own residential area, one's own roaming area (in the case of nomads), one's own commuter routes or migration routes. At best, only through mass media can an indirect experience be gained. This is where the anthropological approach via field research on local life-worlds, the emphasis on local contexts, and culture-specific embedding clashes with the differentiated local effects of human action, which are emphasized by anthropologists:

> (...) we cannot develop prescriptions that hold true across contexts, i.e., we cannot decontextualize or disembed. We will be working out methods and strategies for achieving the committed life within our contexts and amongst those who are near to us (Rose 2013: 5).

Global impacts, in the narrower sense of worldwide effects, such as global warming, are *by definition also* local and can indeed be perceived locally. People in localized communities on coasts or on islands in Southeast Asia, for example, are certainly aware of the changes in sea level, often in a critical form (Hornidge and Antweiler 2012). People in the Mekong Delta in Vietnam experience the effects of deforestation in the distant Himalayas in their everyday lives, which come in the form of changes in water level, a change in the flow regime over the years, or silting. However, information from non-local sources is required to be able to see or interpret these impacts as the result of the actions of other people or other political regimes in the Himalayas, such as deforestation in the headwaters of the Mekong. Interpretation plays a major role here, because the locally perceived effects could also be *partly* due to large-scale climate change in the Himalayas or indirectly to non-human climate change in the Himalayas that is melting the glaciers.

Above all, the perception of local effects *as part of global change* requires knowledge about the effects elsewhere on the planet. Such knowledge can only be obtained indirectly, that is, not based on personal experience, through mass media, blogs, or through personal contacts on the internet, for example, contacts within one's own diaspora. Multi-local fieldwork provides a certain, but only limited, remedy here. This can reveal networks and entanglements, but only between a few local settings.

Naturally, a local orientation provided by cultural anthropology always carries the risk of becoming narrowly localistic. Thus, some critics argue against studies on anthropogenic climate change, which notoriously highlight local effects, local adaptation potentials, or the vulnerability of local communities, claiming that they undervalue the larger political-economic drivers of climate change (as reported by Orr et al. 2015: 161). Anthropologists conduct research on social relationships within groups and networks that are so small that they can be studied by individuals. These relationships are then embedded by anthropologists in contexts, and since globalization emerged, these contexts have often been of a worldwide nature, which has led to field research being carried out in several places (*multi-sited ethnography*). In the Anthropocene, the contexts are—at least potentially—always and everywhere global. With the shift in contexts and the expansion of the scales of the actors, the object of cultural anthropology shifts.

Thus, the social relationships to be determined in social networks themselves belong to the surrounding context and help to define it, including the social embedding of the cultural anthropologists themselves (Augé 2017: 67). A current application-oriented example is the transdisciplinary use of knowledge on local narratives in conjunction with narrative theory to practically address the effects of anthropogenic climate change in coastal management in a combined development effort:

> Local narratives represent these changes, expand the problem definition of climate change and express the multiple entanglements of weather, climate and society (...) Narratives of change serve as a localization device and as a starting point for the co-development of climate services for action. Collaborations between science and humanities on the one hand, and between researchers and local actors on the other are an open-ended process (Krauß 2020: 1).

Despite all its limitations, cultural anthropology can contribute a great deal to the topic of the Anthropocene. One of the fundamental problems in the Anthropocene is the dissociation from material resource genesis, production, transport, and consumption. This separation exists on many spatial as well as several temporal levels. Worldwide, but also predominantly in the local economy, people consume resources in a short period of time, whose genesis required millions of years. In human time frames, these resources are not renewable. In spatial terms, especially people or societies living in affluence can also temporally and spatially separate the material effects of their lifestyle from their perception à la "out of sight, out of mind." Waste is made invisible or exported, an option that poorer people do not have.

The consumption of goods and services is cut off from resource genesis, from economic production, as well as from disposal. These separations and relocations can be documented and analyzed ethnographically, for example, through historically oriented anthropological research (Irvine 2020: 94) and through multi-local research. The particularly problematic narrowing of perspectives in the Anthropocene, the reduction of space and time, leads to a focus on the present, both for the actors and for cultural anthropology as a discipline.

5.8.2 *The "Presentist Present"*

There are also time-related objections to the promise of cultural anthropology's great contribution to the topic of the Anthropocene. Christopher Hann mentions the temporal orientation towards the present or the synchronous rather than diachronic orientation of the discipline. This is especially a problem if cultural anthropology is understood as a strictly present-oriented science. Anthropologists can provide local case studies, but the local cases cannot be considered as universal microcosms without further ado. This scaling problem also applies to the time segments and theorized periods represented by cultural anthropology (Hann 2016a: 14):

> It is far from obvious how, if at all, research into geological time can be advanced by such ethnographic methods. Ever since Bronislaw Malinowski and Franz Boas distanced themselves from 19th century evolutionism, the main currents of anthropology (at least in the Anglophone world) have swirled in the directions of synchronic (or at best ‚processual') microsociology and cultural studies (Hann 2016a: 1, 2).

Cultural Anthropology itself has a problematic relationship with the present, which is manifested in the phrase "ethnographic present." This synchronous bias can be seen as a pragmatic limitation: anthropologists carry out research on people they can talk to. This can be referred to as "analytical presentism," as a temporal perspective of anthropological description and analysis. Anthropologists examine ways of life as interactions between elements of the present; anthropologists do not usually travel through time: "To observe human action is to observe things as they are done" (Irvine 2020: 81). Cultural Anthropology privileges the here and now: the current culture is distilled as a snapshot of a longer history.

The ethnographic present can be criticized as a vague formulation that allows longer-term social change and history to be masked, as is objected to by representatives of historical cultural anthropology and history of science. From a geological, deep-time perspective, the ethnographic represents an anthropocentric abstraction: a present-oriented chronocentrism. This also applies to the few anthropologists, such as Erik Wolf, who have dealt with longer periods of history. The major approaches that define "modernity" based on cultural change that originated from the North Atlantic between the 16th and the 19th centuries remain partial and Eurocentric:

> When it comes to historical periodization, however, ethnography is obviously insufficient and proposals privileging the last half century, or just the last quarter of a century, seem inadequate (Hann 2016a: 1).

Cultural anthropology can contribute to the Anthropocene, especially in terms of the spatial and temporal aspects of the relationship of people and cultures to the land. Particularly in relation to landscape, the Anthropocene raises questions that are accessible with anthropological methodology. A central aspect is this "normal" short termism (presentism) for humans in the temporal perspective of people's interaction with the landscape in which they live. Here, geological temporality and cultural temporality are often intertwined in disparate ways (Irvine 2020; Yusoff 2013, 2015). Although current ways of life, especially agriculture, are causally rooted in geological deep-time conditions, the people who live in them are usually trapped in present-time thinking and experience.

Irvine demonstrates this using the example of peatlands in England, which have been in a state of flux for a long time due to geomorphological dynamics as well as earlier "ecological engineering" by humans. Today's economic pressures and the currently increasing vulnerability, however, are now leading to efforts to establish a sharp boundary between dry and wet areas. The current balance between land and water is seen as spatially fixed, and the boundary is secured through surveillance and management. Today's agricultural use, as well as environmental protection measures and green urban development in England, for example, are based—despite conflicting interests—on the universally shared idea of a fixation of change.

Presentism dramatically reduces the depth of time horizons: the long-term effects of environmental changes are increasingly becoming "unthinkable" (*temporal lock-in, fixing flux*, Irvine 2020: 66–69). Such presentism among people whose lifestyles anthropologists study represents a social fact that is central to the problem of the Anthropocene. As such, it can be described anthropologically or sociologically as time orientation in cultures. In the context of the Anthropocene, it is particularly important that a present orientation that can be observed today is not simply "established," but is itself historicized. We must ask: "What does it mean to map out geological chronology as a common home?" (Irvine 2020: 106, see 91–93).

Even classical field research based on current research in local settings can contribute to histories of local presentism by analyzing how people fit themselves into larger spatial and deeper temporal frameworks. For example, I could imagine a systematic cognitive anthropological study of local concepts of landscape history. Or possibly an analysis of references to an imagined past or nostalgic images of the future(Irvine 2020: 92). Long-term anthropological field research or repeat studies in one place could make a special contribution.

An example of repeated research in one and the same place is Chris Hann's study of a settlement in the Hungarian lowlands. He documented how social transformations brought about by the Anthropocene were experienced, imagined, and also processed in practice in a rural setting from the 1960s to the 1980s, in conjunction with state-sponsored acceleration ("socialist civilizing process") (Hann 2016a: 6–14). The intergenerational changes in Hungary, for example, can only be properly explained in the context of the prolonged pre-socialist immigration to the Carpathian region, or even further back, from "thousands of years of social evolution in Eurasia." The often overheated globalization in the twenty-first century is a consequence of earlier historical processes: "This is a pan-Eurasian story which begins in the late Bronze Age" (Hann 2016a: 19, see Hann 2015). Hann adopts Eriksen's phrase of "overheating" and emphasizes that it occurs unevenly in local spaces and can be delayed:

> (...) even in places where the term itself remains unfamiliar, the familiar method of ethnographic research can shed light on uneven processes of acceleration *and deceleration*, overheating *and cooling*, taking place in the world today (Hann 2016a: 18, emphasis mine).

One example of long-term regional processes that ultimately culminated in extremely large-scale effects is an archaeologically documented group called the Yamnaya, also known as the Pit Grave "culture." Between the late 4th and after the middle of the 3rd millennium BC, these people lived as farmers and cattle breeders in the Pontic steppes of Central Eurasia, today's southern Russia (Anthony 2007: 121–122, 412–457). They spoke a precursor of the Indo-European languages and thus shaped the languages of those civilizations that would later dominate the world. These nomads moved on horseback and in wagons and permanently changed the flora and fauna of these steppe areas through the combination of horses, horse-drawn wagons, transport of materials, cultural diffusion, and their own spread as a population. The link between culture and environment was so intense that large parts of Eurasia became a regional interaction system and transcontinental communication corridor. Although this did not yet constitute an Anthropocene, in a "planetary" reading, the dynamic Yamnaya, together with other Bronze Age cultures, formed an "Earth event" that shaped later mobility revolutions (Clark and Szerszynski 2021: 123–126, 132–133).

A deep-time-informed cultural anthropology can pay more attention to the contribution of indigenous societies to human history than established histories of civilization or teleologically oriented evolutionary narratives (Scott 2017; Graeber and Wengrow 2021). It would enrich an archaeology that, from an explicitly long-term perspective, deals with directed change, materially tangible entanglements with the environment, and the notorious self-reinforcing processes and irreversibilities involved (Hodder 2020: 389–396, 404–408). This kind of cultural anthropology opposes specialist programs that seek to limit cultural anthropology to the present world or to current problem situations. It is incompatible, for example, with the program of a cultural anthropology of the contemporary, as proposed by Paul Rabinow (1944–2021) (Rabinow 1996). Rabinow argues that some phenomena cannot be understood as a continuum of the past because emergent phenomena result from a variety of elements that come together in an assemblage. Such phenomena can therefore only be researched in real space and real time during their emergence. A historicizing approach would run the risk of distracting us from the relevant practices of everyday life. An anthropology of the contemporary should therefore operate on a small scale within a short time horizon (Rabinow 2008: 4–7).

I would argue against Rabinow that an investigation of the current emergence of a phenomenon by no means excludes a clarification of its historical genesis. Rather, it should be substantiated by it in order to know from what a phenomenon emerges. This would be the model according to which an evolutionary explanation of emergent complexity would proceed, for example, through a step-by-step reconstruction of the genesis of an evolved system back to the earliest causes (*reverse engineering*, Driscoll 2015; Queloz 2021). With regard to ways of thinking in the Anthropocene, it would, as Irvine rightly says, just be a matter of analyzing the temporal distancing from the past and future caused by presentism as a social fact, instead of replicating it by narrowing the anthropological approach. It is precisely the conditions of the presentist present that require a historicization (Irvine 2020: 98). If today's behavior has a delayed effect in the distant future, as "slow violence" (Nixon 2011), then we need an analysis of its current causes, but also an understanding of the duration, disruption, recovery, and reconsolidation of geo- and biosystems; in short, the temporality of nature. This understanding requires an environmental-historical perspective combined with an Earth historical approach:

> The fascination with deep history lies in the fact that it reveals the many contingencies that came together to bring about, for example, the industrial revolution, the last maximum glacial expansion, or the Qin Dynasty (Scott 2017: 16).

A central insight of scale considerations guided by the perspective of deep history is that small-scale systems and processes do not simply disappear in the face of emerging large-scale phenomena. Given the importance of this reflected view of scales, I will return to this in Sects. 6.7 and 7.3.

5.9 Cultural Change and Cultural Revolution—Forgotten Specialist Literature

> *... we remain an overwhelmingly fossil-fueled civilization that has been recently running vigorously into fossil carbion rather than moving away from it*
> Vaclav Smil 2021: 273, emphasis orig.

The period after the Second World War was a heyday for culture change studies in cultural anthropology. This is precisely the phase in which the Anthropocene is currently considered to have begun. The focus was on rapid, accelerated, and sudden change. This was often researched in situations of culture contact between societies that had previously been foreign to each other. These societies were typically not only foreign to each other, but also of different political and economic powerful. Prototypical are studies on so-called situations of first contact in the Pacific.

5.9.1 Adaptation and Sociality as the Basis of Anthropocene Action Effects

A central theme was the adaptation of the less powerful partner to the structures of the more powerful partner (acculturation, assimilation). Only later did the insight arise that societies influence each other through contact (*transculturation*). Another focus of culture change studies included studies on the migration of farmers who migrated to cities and adapted to urban life there. A classic example is the work of Paul Stirling from the late 1940s on peasant urbanities in Anatolia (Stirling 1965, see Hann 1994). These were local studies, but they showed such a change in the economy, politics, religion, and thinking that pointed to a break in epochs in their simultaneity.

From the perspective of research on accelerated cultural change, Thomas Hylland Eriksen identifies a more recent turning point in planetary history than the often cited years of 1945 and 1964—namely the end of the Cold War. At the beginning of the 1990s, globalization suddenly became more spatially comprehensive and significantly more dynamic. The combined gross national product increased dramatically, as did global mobility and the extinction of species. It is a matter of accelerating change on multiple scale levels (*multiscalar accelerating change*, Eriksen 2016: viii, 131–156). Thus, in a historical phase, we are simultaneously observing markets that are "overheated" and an explosion of identity politics. Eriksen speaks of a phase of intensified "heating" of the planet (*overheating*, Eriksen 2016: 22, 152 f.). Within and between these scales, tensions and conflicts arise. The decisive factor here is that various processes that were once functional come together, thereby leading to unintended side effects and *runaway* processes, some of which are partly ubiquitous.

Cultural Anthropology could also contribute to the social conditions of Anthropocene societies in history. One contribution of cultural anthropology that has so far hardly been attempted could be that of carrying out research on the local ways of life and social relationships that are causally relevant to Anthropocene effects. This could be linked with a tradition of cultural ecology, which has remained unnoticed in the German-speaking world—the explicitly cause-seeking "event ecology" (Vayda 2008, 2013; Walters et al. 2008; Walters and Vayda 2018). Based on pragmatist philosophy, an attempt is made here to avoid premature assumptions about the causes of a phenomenon. This is sensible given the many conceivable triggers of Anthropocene effects. Methodologically, we start from an event or effect and gradually deduce possible causes instead of the other way around. This involves proceeding backwards in time and inferring spatially inwards, or gradually moving outwards from micro to meso and macro contexts (progressive contextualization). In relation to historical explanations, the approach of event ecology resembles the procedure of *reverse engineering* in evolutionary research.

What are the social prerequisites, the special social relationships, and what are the psychological causes ("subjective" factors, Hann 2016a, b: 3) that make Anthropocene transformation possible? These questions require not only an understanding of human–environment relationships at the micro level, but also a macro approach. I believe that cultural anthropology can make an important contribution that has not been addressed at all by the largely established authors mentioned above.

To this end, it would be important to use a state of discussion that is no longer among the standard references of today's anthropologists, at least outside the US, namely works from comparative historical anthropology (Hann 2016a: 1–3) and studies on cultural evolution (Antweiler 1990, 2008). The question concerning the social and cultural prerequisites for the emergence of Anthropocene societies requires us to go beyond the current dominant emphasis in the discipline on synchronous or contemporary historical studies as well as the focus on local particularities, cultural uniqueness, and diversity (diversity, plurality, difference).

> Ever since Bronislaw Malinowski and Franz Boas distanced themselves from 19th century evolutionism, the main currents of anthropology (at least in the Anglophone world) have swirled in the directions of synchronic (or at best 'processual') microsociology and cultural studies. To the extent that these schools have emphasized the diversity of humanity and carved it up into 'societies' and/or 'cultures', *the emergence of the Anthropocene arguably sounds the death knell of the discipline*, since Anthropocene advocates tend to argue that the previous bases of sociocultural difference are everywhere eroding before our eyes (Hann 2016a: 2, emphasis mine).

In other words, we must not overcome, but supplement the almost exclusive concentration on current forms of society, on small social units, and the methodological focus on ethnographic procedures. According to Hann, if we as anthropologists want to understand the social preconditions of Anthropocene effects and use the holistic approach, we can bring into play authors who, *as anthropologists*, deal with long-term change, such as Goody (Hann 2016a:1) or David Graeber (for example, Graeber and Wengrow 2021).

A prerequisite for this is to realize that today's anthropology primarily deals with current social forms that are threatened by modernity, while its central social-theoretical basis is a child of European modernity. Marx, Weber, Durckheim, Giddens, Habermas, and Foucault—despite all their differences and the concept of multiple modernities (Eisenstadt 2002)—take Western modernity as a *template* (Hann 2016a: 5, 14). In some recent positions, it is argued that the Anthropocene, as an epochal change, is symptomatic of the permanent epistemic crisis of anthropology. If the "decline of the West as an intellectual and political project" is then diagnosed, and even celebrated in some cases (Jobson 2020: 260), it should be clear that this would put the discipline at risk.

Jack Goody and a few other authors have dealt with the emergence of agriculture and cities and examined the new forms of economy, new forms of political communities (*polities*) and the associated cosmological ideas (Goody 2010). Here, a historically informed anthropology can tie in with the thesis of an early Anthropocene, which has recently been postulated by some geologists and archaeologists (Ruddiman 2010, 2016a, b; Lewis and Maslin 2018). A central question, of course, pertains to which cultural preconditions formed the basis of societies that, through their economic practices, had the first regional or even worldwide impacts on the geosphere. Eric Wolf places the beginning of his history of world relations with the emergence of European overseas territories (Wolf 1982), as does Wallerstein's world-systems theory. From the perspective of a historically oriented cultural ecology, the timescales need to be extended further, because from an environmental science perspective, the human ecosystem can largely be equated with what we call "culture" (Herrmann 2019: 28–38):

> The emergence of a differentiated human culture is to be seen as a consequence of strategies of appropriation and exploitation of natural resources, from which arises an appropriation of the world that progresses from ideas to their implementation and in this process produces its own systems of meaning (Herrmann et al. 2021: 39–40).

"The 'Neolithic revolutions' mark, with the sedentarization of humans and the prospective creation of animal husbandry, the beginning of the agro-urban-industrial complex and thus the precarious state of the global ecosystem today" (Herrmann et al. 2021: 41).

5.9.2 Evolutionism **Revisited—*Cultural Evolution and Directed Change***

The Anthropocene implies the renewed use and reflection of prematurely buried theories on the increase of social and cultural complexity from a cross-generational perspective, such as in social evolutionism. Directed change (cultural evolution, social evolution) is central to the topic of the Anthropocene. Contrary to popular belief, directed change was not the focus of Darwin's theory formation. For this, such as for the emergence of ultra-social communities and states, we also need complexity theories (Lansing and Cox 2019; Thurner et al. 2018: 224–312). For long-term and directed change *in the environmental macro-area*, namely geohistorical change in the Anthropocene, I believe we need a combination of evolutionary theory, complexity theory, social theory, and cultural theory (Chakrabarty 2018; Ellis 2015: 290–300, 2018: 107–145). Among the individual theories, a genuinely evolutionary approach represented by the more recent complex synthesis discussed in Sect. 7.2 forms, in my opinion, the most fruitful theoretical approach (Zeder 2018).

Today, more than ever before, I believe that the central questions raised by classical evolutionism—albeit inadequately answered—still require clarification: Why did the majority of human societies develop in certain directions at the macro level, why do long-term trends emerge at the local level of individual collectives? There are fruitful approaches to this in anthropological cultural evolutionism, such as *general evolution* and *multilinear evolution* (White 1949; Steward 1955; Sahlins and Service 1960; Morris 2015: 35). How do we generally explain the not necessarily, but typically or moderately increasing complexity in the trajectories of cultures? These are central questions that need to be addressed in order to understand what has enabled humans to produce Anthropocene effects. What are the *enabling conditions* of Anthropocene-effective societies?

Directed long-term change is a problem area that has particularly allowed for fruitful exchange between anthropologists, archaeologists, and historians, especially since neo-evolutionism (Adams 1978; Antweiler 1990, 2012a, b; Turchin et al. 2018; Morris 2015; Tang 2020). It is about several levels of directionality (*general directionality* and *specific directionality*, Hodder 2018: 16, 126). Archaeologists ask themselves how the increase in human-altered materials, the increase in the material growth of object orientation, and the increase in the entanglement of people and cultures with material things can be explained. At the micro level, the question of direction also arises, namely about the specific path dependency between successive artifact innovations that presuppose each other cumulatively. A good example of the use of cultural evolutionary knowledge to understand the Anthropocene is James Scott's reading of the cultural revolution, "Against the Grain" (Scott 2017). Scott is interested in the central theme of cultural evolutionist cultural anthropology from the 19th century onwards, namely in how complex societies emerged. He critically deals with directed change, types of societies, and seeks the central myelin sheaths of long-term human history, such as the Neolithic Revolution and state formation.

To this end, anthropologists could establish a connection to newer concepts of cultural evolution as biocultural coevolution. Particularly useful are insights into knowledge economies as the societal totality of knowledge-related institutions and processes, as well as insights into the history of knowledge with respect to epistemic evolution (Renn 2020: 377–407, 2021: 4). Today's dependence on knowledge to solve world problems was made apparent by Covid-19, probably the most global event in human history to date. The Anthropocene offers the opportunity, indeed the necessity, to bring the knowledge economy together with the existing consequences of the gradual historical expansion of human knowledge, that is, a history of knowledge economies:

> Without sedentary agriculture and animal husbandry, the early advanced civilizations would not have been conceivable, without these advanced civilizations there would probably have been no modern science, and without the scientific revolution of modern times, neither colonialism nor the industrial revolution would probably have been possible. This long-term co-development shows how our knowledge and thus our creative power have increased and at the same time the associated, unintended, or consciously accepted consequences have been potentiated (Renn 2021: 3).

The classic approaches of anthropological cultural evolutionism (along with the criticisms) can be very well linked with the largely independent approach of deep history (see Sect. 2.4). By considering the phenomena of *upscaling* and *downscaling*, long-term history appears as a sequence of such J-curves, which, metaphorically speaking, stand on the shoulders of earlier J-curves (Stiner et al. 2011: 247). The global population increase that culminated in the Anthropocene, for example, was historically based on political innovations and traditions, such as the organization of information flows and urban development. Such an in-depth perspective shows that human history is "punctuated" by numerous sudden jumps in population size, energy flow, efficiency, complexity of political organization, and degree of connectivity.

Thus, the perspective provided by deep history as the "architecture of the present" can once again raise questions of classical cultural evolutionism on directed cultural change, which have not been answered to this day. In this way, its speculative assumptions can be transformed into empirical questions and corrected based on concrete data. One example is the ability to extract matter and energy from the environment, for example, through the enormously consequential use of fire (Glikson 2013; Soentgen 2021a). It has been shown, for example, that the height of "scalar leaps"" is not decisive for their long-term effects relative to later leaps (Stiner et al. 2011: 259–264). In relation to scalar leaps in the political power structures (*polities*) of human history, for example, the leap from communities of a few dozen to 100 people to societies of thousands was no less important than the leap from millions to hundreds of millions. Similarly, the historical leap in the extraction of biomass per unit area from 0.1 percent to 1 percent had no less revolutionary an impact than that from 10 percent to today's level of over 90 percent. The same applies to the ability to socially organize energy flows in complex resource-dependent societies (White 1945; Adams 1978; Turchin et al. 2018).

The supposedly radical transformations of modernity are based on nested hierarchies; they contain both innovations and modifications of existing forms and patterns (Stiner et al. 2011: 246–247; Turner 2021: 100–122). When newer large-scale units are added, the smaller ones nevertheless remain relevant and create continuities. Even in today's globalized world, we continue to live in families, clans, patron-client networks, and other small entities that have been described by political anthropology worldwide and are also central to evolutionary biology ("Dunbar's number," Dunbar 1993). People experience the anthropogenic transformations of their living environment and react in families, small groups, and networks—offline as well as online.

By means of a perspective that explicitly focuses on nested scales, *upscaling* and *downscaling*, we can set aside the "telescopic" models of ontogeny, ascent, and decline for other imaginations of history and other theoretical frameworks. Where conventional models privilege the moment of beginning ("birth," "origin," "development") and continuous trajectories, it is now, metaphorically speaking, more about the complex process of emergent becoming (Stiner et al. 2011: 269). The focus is on nested scales and the historical processes of expansion and contraction of the scales of sociality, including rapid systemic change.

Taken together, a deep historical view of cultural anthropology offers the chance to revive the largely forgotten research activities in classical evolutionism through a current long-term perspective of history that is both globalist and informed by evolutionary biology as well as geology (*global history*, *deep history*). This would also enable a reconnection with classical anthropological research on cultural change (*cultural change studies*) and archaeology, which has been partially discontinued in American cultural anthropology and almost completely lost in cultural anthropology in Europe since neo-evolutionism:Movement across scales of analysis can help us to navigate this complex terrain of human adaptation: specifically, it can help us to draw the lines between natural and cultural change that so effectively thwart deep historical analysis (Stiner et al. 2011: 266).

Chapter 6
Conditio Humana—The Geologization of Culture

Abstract This chapter asks what the Anthropocene means for a new science of man. The idea of the Anthropocene clearly goes far beyond the widespread concept of "anthropogenic," which is elucidated by the clarification of the difference between geological and historical or archaeological periodization concepts. Only by connecting the materially open concept of culture in cultural anthropology with the concept of planetarity will the fundamental material entanglement of man and nature become clear. For a causal explanation of transgenerational change and especially the dynamics of the Anthropocene, I connect the idea of the Anthropocene with the concept of niche construction from evolutionary ecology. This reveals a threefold heritage that is crucial for understanding the Anthropocene: the *genetic* heritage, the *cultural* heritage, and the anthropogenic *geospheric* heritage.

> *Humans live in and with nature.*
> Oliver Haardt, (2022:109)

> *All social formations are „geosocial formations".*
> Nigel Clark & Karen Yussof (2017)

> *For a social theory in the Anthropocene, the question arises (…) as to whether and in what way the Earth system should be included in the analysis.*
> Frank Adloff & Sighard Neckel (2020: 10)

The current use of geological and paleontological findings and ways of thinking in the human sciences is not without precedent. Geological thinking particularly shaped conceptions of world history and time concepts between the Enlightenment and the emerging modern age. As a recent historical study on the emergence of the historical sciences shows, geology played a constitutive role in the emergence and development of modern historical thinking. The founding fathers of historical scholarship and the philosophy of history, such as Leibniz, Schlözer, and von Müller, as well as Voltaire, Herder, and Meiners, had engaged with geological writings. They used geological terms and their meaning, especially "deep time" and the concept of "revolutions of nature" for the development of a historical concept of revolution. (Schulz 2020: 31–51, 184–251, 281–298).

C. Antweiler, *Anthropology in the Anthropocene*, Anthropocene – Humanities and Social Sciences, https://doi.org/10.1007/978-3-031-74591-1_6

> In a defiant act against the very idea of dualisms, we embrace the paradox the concept of the Anthropocene offers, namely that while reinforcing and reflecting the self – appointed dominant place humans arrogate, there is potential to break down divisions prevalent since the Enlightenment (Gibson and Venkateswar 2015: 9).

The term Anthropocene itself suggests that under the idea behind it, nothing less than anthropological foundations are to be negotiated. We humans no longer live in small worlds on a large planet, but are major actors on a small planet (Rockström 2021: 103). But we must also turn our view around: our Earth and its geology play a major role in the history of humanity, for example, in the form of earthquakes, volcanic eruptions, and plate tectonics; the Earth has co—created us (Dartnell 2018; Headrick 2021; Haardt 2022). The concept of the Anthropocene is potentially revolutionary, in that it raises the question of the extent to which established boundaries, especially the boundary between nature and culture that is constitutive of modernity, can be blurred (Trischler and Will 2020: 237). If the spheres mix and human societies are part of a large assemblage of very different components, if nature is rapidly *anthroposized* (Crutzen and Schwägerl 2011), then it can no longer be kept at a distance. Long droughts and storms in the 1930s turned the American Geat Plains into a "Dust Bowl" that could hardly be farmed and exacerbated the global economic crisis. The 1906 earthquake in San Francisco plunged the global insurance industry into a crisis. Environmental historians have been able to show that the 1815 eruption of the Tambora volcano in Indonesia played a major role in the drying up of the first long wave of innovation.

> One must consider the historical development of the relationship between man and nature from both a human and a geological perspective, i.e. examine how the cultural reality of man and the planetary reality of the earth have developed in relation to each other (Haardt 2022: 115).

6.1 "Anthropogenic"—Humans *in* Nature

> *We live in and with nature.*
> Oliver Haardt (2022: 109)

> *The environment is changing faster than society, the near future is proving to be not only increasingly unpredictable, but perhaps impossible.*
> Danowski & Viveiros de Castro (2019: 101)

These challenges posed by the Anthropocene's forced torpedoing of the nature/culture dichotomy can be well illustrated by the term "anthropogenic" (Fig. 6.1). As a descriptive term, "anthropogenic" stands for human—made changes in the natural environment. In geography and the Earth sciences, this is usually related to landscapes. In conventional understanding, "human – made" change is seen as the antithesis of untouched nature, which is regarded as timeless and in cyclical equilibrium (Sayre 2012: 57). However, the realization of the causal role of humans in climate change has made it clear that it is difficult to work empirically with such a dichotomy. As archaeological, ethnohistorical, and anthropological studies on the

Fig. 6.1 Overgrowth on a multi-storey car park in Cologne, Germany. (Photo: author)

use of fire show, for example, human collectives *act* through the Earth: ("...speaking and acting through the Earth," Clark and Szerszynski 2021: 55).

The term "anthropogenic" was first used in the 1920s by Tansley (1923: 48) in his work "Practical Plant Ecology." It stood for ecological factors beyond climate and soil and was created in response to the problem of understanding ecological change in plants. Clements had shortly before introduced the concept of ecological succession into plant sociology. According to this concept, climate and soils determine the temporal sequence in plant communities, all the way up to a stable "climax community." Tansley insisted that such successions are often only understandable if human influence is taken into account, because this stops succession processes and shortens climax stages (*anthropogenic climaxes*). Today, it is recognized that there is nothing comparable to the climax stages in cultural landscapes: here, continuous work and maintenance must be performed in order to keep the landscape stable, which it is not by nature (Bätzing 2020: 39, 2023).

In this case, it mattered that Clements worked on North American landscapes that were considered pristine, while Tansley studied English landscapes that were obviously heavily influenced by human activity (Hall 2005). Carl Sauer, as a historically oriented cultural geographer, supported Tansley's criticism of the lack of consideration given to humans in the descriptions of ecosystems. He pointed out that humans have been strongly shaping the environment for millennia and described human activity in drastic terms:

> The second great agent of disturbance (next to the climate) has been man, an aggressive animal of perilous social habits, insufficiently appreciated as an *ecologic force* and as *modifier of the course of evolution* (Sauer 1950: 18, additions and emphasis mine).

The dualistic conception contained in the term "anthropogenic" goes back at least as far as the pre—Socratics. As Glacken showed (1967, after Sayre 2012: 58), there are three related ideas in it: (a) humans shape the environment, (b) the environment influences humans, and (c) the environment and humans or culture fit together (*matching*). They are adapted in a way that is oriented towards harmony or points to a divine plan. The idealizing notions of pristine nature as the antithesis of human civilization intensified during the time of the early Industrial Revolution through experiences with the density and pollution in cities, such as those in England. These notions were particularly widespread among the bourgeois elites, who retreated to their country houses (Williams 1980).

Sayre argues that today's debates perpetuate the dualistic concept of that time, even if the conclusions are often different. The climate, soils, and humans are (a) seen as distinct entities, even if they are seen as interacting. Nature is (b) seen as static or cyclical. Humans (c) appear as a factor of change, and (d) humans who do not or did not have such an effect on their environment are considered indistinguishable from animals. Even today's criticisms of the assumptions that indigenous groups lived harmoniously in and with their environments often still think in this way (Sayre 2012: 60).

> The conceptual scaffolding put in place to define the anthropogenic some 80 years ago has proved remarkably tenacious. Many scholars have criticized the human – nature dichotomy embedded in the concept, but even in so doing, they have had to employ it and, seemingly, to grant it continued de facto conceptual purchase (Sayre 2012: 61–62).

In the natural sciences, however, it became clear even before the term Anthropocene emerged that most aspects of the functioning of the world's ecosystems can no longer be understood without considering the strong, indeed *often dominant* influence of humanity, as an early article in *Science* made abundantly clear in its title: "Human domination of Earth's ecosystems" (Vitousek et al. 1997: 494). In the face of environmental crises, the supposed harmony collapses, and the two poles of nature and culture merge into something that can hardly be grasped with the usual conceptual tools. Tansley's supposed untouched deserts, mountains, tundras, and oceans now form the "anthropogenic," as Sayre calls it (Sayre 2012: 62). While the term "anthropogenic" created a contrast between nature and culture, the term Anthropocene explicitly includes the human—influenced climate and anthropic soils.

6.2 Cultural History Is Grounded in Earth History—The Geosphere as Palimpsest

> *The human cannot be divorced from the planet that supports it, and so our existence needs to be understood in relation to the deep time of Earth history.*
> Richard Irvine (2020: 15)

From a social constructivist perspective, it is commonly said that humans and societies as actors "construct" their life-world. This construction does not only concern meanings and knowledge, as the mainstream of social constructivism in the humanities and cultural studies repeatedly emphasizes, but humans also create the phenomena of nature or the geosphere, e.g. anthropocenic future technofossils (Fig. 6.2). Nature without construction can no longer be had, just as culture is unthinkable without nature. This undermines conceptions of an environment external to culture or society, that is, an environment without human actors (Latour 2014: 6, 2018: passim). This entails contradicting the problem framing that is still dominant in the natural and technical sciences, where nature is seen as an object insofar as it is postulated that there should be a political reaction to nature, which thus appears to be external to society (Wapner 2014: 38, Bauer and Bhan 2018: 3).

As humans make raw materials accessible, extract them, and use them in work processes, geological forces and properties become a part of us. We must therefore realize that we are not only capable of acting as geomorphic agents, as Crutzen & Stoermer's classic argument states. No, we are also geological subjects, because we have some things in common with the geological forces in that we use them and incorporate them through use (Yusoff 2013: 781). This is a general characteristic of living systems, which in humans assumes larger dimensions and faster dynamics.

The explicitly science-oriented "Anthropocene science" has been criticized above for obscuring the social consequences and sometimes traumatic effects of the use of coal or the extraction of raw materials, for example. Conventional historical accounts also tend to attribute the Industrial Revolution to certain individuals and times. Analyses from the Earth sciences are larger in scale, but focus on the transition from solar energy to fossil fuels, and also on the present or the recent past. From a geo—social perspective, the conventional representations are not too

Fig. 6.2 Candidate for a future technofossil, Kuta beach, Bali, Indonesia. (Photo: author)

scientific, but *not geological enough.* This again shows the difference between geology, with its concrete field orientation and at the same time deep historical view ("hard rock geologists") on the one hand, and the globally oriented and—from a geological point of view!—present—oriented Earth System Sciences on the other: two perspectives that can learn from each other (Clark and Szerszynski 2021: 24, 64–65, see Davies 2016: 69–111).

It is not only the social inequalities that need to be more strongly emphasized, but also the consequences that human interaction with geo—resources has had on the dynamics of geological forces. In short, analysis must not only become more politically sensitive, it must also become more geological. If human activities now mix so extensively with biotic and also non—organic processes, we are dealing with socio—ecological systems in the narrow sense of the term (Folke 2006: 287 ff.). If we are dealing with socio—*planetary* couplings on a large scale, we can no longer speak of a "non – human" nature that is separate from humans and is beyond human agency. In the Anthropocene, all human collectives are to be understood as *geo—social formations* (Clark and Yusoff 2017; Clark and Szerszynski 2021: 28, 48).

One example is Jane Bennett's analysis of the mutual effects of metals, heat control, and metalworkers and the formation of the "metallurgical," which depends on a specific geological resource: iron ore (Bennett 2020). From this perspective, metallic alloys are not just amalgams of different metals, but alloys formed from geological and human forces. If we want to "socialize the Anthropocene" from a cultural and social science perspective, then the findings on extractive environmental use, the dynamics of complex systems, and non—linear dynamics call for us to conversely "geologize the social," as some sociologists or geographers demand. This can be even more theoretically grounded if the theorization of the Anthropocene is informed by the anthropological cultural theories just discussed, by cultural concepts that explicitly *do not* reduce culture to specific materialities. In view of the Anthropocene, we can thus ask about the *anthropogenesis* of social formations (Yusoff 2016).

Core ideas of the supposedly outdated nature—vs.—culture concept persisted and continue to persist in non—religious areas. A current example is the idea of supposedly untouched wilderness areas in rural spaces, which has been widespread in the field of ecological criticism and also among the population since the 1970s and is considered to have a higher ecological value as compared to cities. One example from the field of social sciences is anthropological and sociological functionalism. There, an analogy was used, which states that the relationship of the organ to the survival of the organism is like the relation of the social subsystem to the functioning of society (Bruck 1985). This way of thinking assumes that organisms and societies are *fundamentally different* things, but that there is a *specific* similarity between them. It is then no wonder that there is still no certainty as to what consequences the Anthropocene will have for environmental science and environmental policy:

> Curiously, almost everyone asserts that the nature – humans distinction is fundamentally flawed and must be transcended, but even saying so seems to make doing so more difficult.

> For ecologists and environmentalists, the politics of the anthropogenic are fraught with both practical and ideological complications (Sayre 2012: 58).

Some say that the idea of the Anthropocene would dissolve the great divide between nature and society, or even speak of the end of nature ("after nature," "post – nature," Wapner 2012; Morton 2013; Lorimer 2015; Purdy 2015: 3). Long before the Anthropocene discourse, the environmental journalist Bill McKibben wrote what was to become the first non—fiction bestseller on the climate catastrophe and gave it the title "The End of Nature." The second nature or commodified nature had subsumed the first pre—human nature, and since then McKibben referred to the drastically changed Earth with the use of the art term "Eearth" (McKibben 1989, 2010). Others argue that "thinking beyond nature" has not only been possible since the actual spread of consumerism and capitalism, but that it is historically much older (Bauer and Bhan 2018: 13, vs. Morton).

Still others argue, contrary to all of this, that the Anthropocene precisely emphasizes the difference between humans and (the remaining) nature due to the unique power of human activity. Humans have now become the "dominant force among the great forces of nature" or have "replaced" nature as the dominant environmental force on Earth (Crutzen 2002; Steffen et al. 2007: 614, Ruddiman et al. 2016a, b: 38). Another formulation states that humanity has gone from being a factor in the *regional ecological or biological* history to being a factor in a *geophysical* history that changes the climate and life on the planet by modifying the Earth system as a whole. In turn, critics see this as evidence of an outdated notion that there has ever been an Earth system external to humans (Bauer and Bhan 2018: 5, Bauer and Ellis 2018: 215). However, the critique overlooks the fact that Steffen and Ruddiman are not saying that nature has been replaced by humans as a factor, but only as the main geophysical force, as the dominant force.

For a long time, the humanities and the social sciences were the powerful monopolies that determined the *Conditio Humana* (Leggewie and Hanusch 2020: N4). This was epochally challenged by Darwin and is now being intensified by approaches such as political ecology, biosociology, and the theses on the Anthropocene. As the metaphors and narratives have shown us, the Anthropocene is very often perceived as a tragedy, especially when the focus is on the loss of diversity of life (for example, Wilson 2016). Upon closer inspection, however, the diagnosis of the Anthropocene contains a twofold message, precisely because it also fundamentally questions the established division of the sciences. It implies the possibility of a new science and a radically changed politics… amidst a ruinous dynamic. In a quasi—schizophrenic way, this tragedy also opens up a promise, namely a new beginning for the human sciences:

> For in the Anthropocene, nature is no longer what conventional science imagined it to be. And if the notion of a pure nature – *in* – *itself* has died in the Anthropocene and been replaced by natural worlds that are inextricable from the worlds of humans, then humans themselves can no longer be what classical anthropology and human sciences thought they were (Haraway et al. 2016: 535, emphasis in the original).

When it is argued that humans significantly generate global change that permeates the geosphere (*geofactor*, *driver*), the nexus of two major categories is at play: the connection between the Earth and humanity. Conventional discourse and thinking refer to this as the human–environment relationship, with humans considered as cultural and the environment as natural. The Anthropocene understands the human–nature relationship differently from the way it was seen in classical ecology, namely as much more profound than fundamental. The context in which we live today is globally the result of human action, and this has given rise to words such as "biosocial" (Ingold and Pálsson 2013) or "biocultural." As words, they are still dualistic, but they signal the importance of cultural theory and social theory for understanding Anthropocene change. What is commonly referred to as the "environment" (or the "ecological" as a phenomenon, *the ecology of...*) is not only something other than the social, but is itself also socially interwoven. Leinfelder and Schwägerl therefore speak of the human "*Us* world" instead of "environment" (in German: "*Uns*welt" instead of "Umwelt"; Leinfelder 2012), which is reminiscent of the concept of the "co – world":

> The core of the concept is the insight that humans are not isolated and surrounded by an alien environment that is disturbed by them, but that they are increasingly becoming part of this environment, which could perhaps better be called the 'us world' (Schwägerl and Leinfelder 2014: 237).

The disaster emphasized by the discourse of the Anthropocene can be a gift for the human sciences and especially for cultural anthropology, because the topic raises fundamental questions, and does so on the basis of concrete and current problems. It could even prove to be a "poisonous gift" (Latour 2014: 15, 2017a), namely when it leads to the dissolution of the category of the human being or—perhaps even more dangerously—turns it into a fetish. In a programmatic essay, Gísli Pálsson and eleven co—authors summarize the goals and challenges of the humanities and social sciences with regard to the image of humankind in the face of the Anthropocene as follows:

> The larger conceptual task remains to reframe Anthropos for the modern context. This will mean reorganizing our own house in a radical sense, expanding our tools and visions beyond 'business – as – usual' – a task that has just begun in several fields of scholarship. (Pálsson et al. 2013: 4).

> We suggest that it is essential to *fundamentally rethink* the environment – humanity relationship. To characterize the Anthropocene by means of quantitative data is one thing, to describe and understand how it perceives human interaction, culture, institutions and societies – indeed, the meaning of being human – is truly another and a major challenge for the scholarly, literary, artistic, practitioner, and policy communities (Pálsson et al. 2013: 10, emphasis mine).

The collapse of a simple difference between humans and nature does not yet provide a clear guiding scheme. It can be read as an empowerment of humans or as their removal from the center of events (*decentering*, Clark 2019). Accordingly, there are completely divergent consequences that are drawn from this in the interpretative struggle for a discursive restoration of the human being. They move between the extreme poles of neo—humanism, which seeks to conceptually grasp humans on the one hand, and post—humanism, which conceptually bids farewell to humans on the other hand (Bajohr 2019: 64).

6.3 Anthropos *and* Prometheus—Homo *and* Anthropos

> *Anthropology is less the study of culture as an object of understanding, than the culture or cultivation of humanity as a method of change.*
> Anand Pandian (2019: 11)

The discussions about the Anthropocene notoriously unsettle conceptions about humans and humanity. With the name "Anthropocene," the idea of the Anthropocene constitutively presupposes an *anthropos* when humanity or "the human" is depicted as an actor on a planetary scale. As a technical subject, the anthropos plays a pivotal role in the Anthropocene. It connects humans as cultural subjects with humans as geological actors (Trischler and Will 2020: 240). At its core, it is about two things, about the "... relationalization of human existence in the universe and the relativization of the anthropocentric perspective" (Hanusch et al. 2021):

> The anthropos should be taken to include domesticated animals and plants as well as people in all their differentiated and unequal heterogeneity, along with other entangled entities and forces held to be part of broader assemblages producing climate change and other transformations (Edgeworth 2018a: 759).

This can mean a relativization or decentering of the human being or—precisely through the relationalization—a recentering: a renewed emphasis on the special position of humans, as found in variants of Neohumanism. For proponents of a technocratically oriented variant, the goal is to acknowledge the reality of the Anthropocene, to meet the challenges of humanity optimistically and proactively, for example, by artificially influencing the climate. It is about seeing opportunities in the Anthropocene in order to make it a "good Anthropocene." This attitude is known as ecomodernism, and for its proponents as well as for some others, a reversal of the modern decentering is necessary (for example, Asafu-Adjaye et al. 2015; Lewis and Maslin 2015a, b, Ellis 2018). The pessimistic variant of Neohumanism not only considers a reversal to be impossible, but also sees the situation as completely hopeless. What remains as a task for humanity is to learn a humane way of dying. For this, we need a vision of what we are and want to be as humanity, and this requires the humanities and cultural studies (Scranton 2015).

A middle variant of Neohumanism argues that the special position of humans has been proven by the Anthropocene and thus advocates for a "new Anthropocentrism." This is not a speciesist anthropocentrism focused on self—interest, nor is it technocratically oriented. Instead, it is based on a human—centeredness that sees human freedom as meaningful and its effectiveness as a duty of responsibility (Hamilton 2017: 36–75). This justification of responsibility is reminiscent of Hans Jonas' principle of responsibility (Bajohr 2019: 68), with the important difference that he rejects all forms of anthropocentrism. For Hamilton, the most important practical approach corresponding to the new anthropocentrism is a worldwide change in consumer habits. For this "new anthropocentrism," or this "negative" teleological Anthropocentrism in view of the destructive power of humans, the former anthropocentrism was not centric enough.

These three neohumanist positions, which put humans back at the center of attention, contradict the mainstream in the humanities and cultural studies, an anti—humanism with a rejection of any essential determinations of what constitutes a human being. After the human as a meaningful category seemed to be done with Foucault's dictum of "death of man," the human being or humanity is returning to the debates with a vengeance with the topic of the Anthropocene. Hannes Bajohr aptly speaks of the "return of man at the moment of his supposedly final farewell" (Bajohr 2019: 64, 2020: Subtitle).

This explicit expatriation of the human being from cultural studies and philosophy—with the exception of philosophical anthropology—goes back at least to Foucault's thesis on the death of every subject, in particular his dictum thesis on the end of the human being as the central figure of cognition in the modern human sciences. As he succinctly put it, one could "very well bet that man will disappear like a face in the sand on the seashore" (Foucault 1974: 462). As a consequence, however, this farewell can turn out differently—as a determination by negation or as a negation of determination (Bajohr 2019: 72). It can occur as a minimal anthropology, as a negative determination of being due to the destructive power of humans, such as in Günther Anders' work, where it is described as "futurelessness" in the wake of the atomic bomb (Anders 2018). Or as a rejection of any determination of a core humanity, as a veto against any anthropology in the form of a rejection of any determination of being, as in Ulrich Sonnemann's critique of any total theories (Sonnemann 2011).

In more recent discussions, these two positions have been synthesized in the form of critical posthumanism and a new negative anthropology, both of which debunk the supposedly universal human being. Timothy Morton argues for a philosophical and cautiously realistic response to the Anthropocene because we must face the reality of Anthropocene effects (Morton 2017). There is compelling evidence for the human—made contribution to climate change, and such clear evidence makes any consistently constructivist and anti—human argument obsolete. Seen in Foucault's image of the beach, human beings return to a deeper level beneath the sand… climate change is not caused by jellyfish or dolphins, but by humans. One can speak of humans as such, namely as a species determined by its destructive power. Colebrook, as a cultural theorist, also emphasizes the undeniable fact of humans via their effects (Colebrook 2017).

However, Morton advocates a redefinition of humans, which is greatly expanded under the explicit title *humankind*, first by including non—human species, and second the Earth itself. Morton wants to drive out Marxism's anthropocentric fixation on human labor and property. Thus, a planetary species unity with many "earthlings" is, for Morton, also the unity in which a Marxist politics based on extended solidarity should be located, a "left holism" (Morton 2017: 1, 18–27). The concept of humanity in a post—Anthropocene view then includes all animals and all things. Thus, the concept of humanity becomes a reductive variable in "weird essentialism" (Morton 2017) and in "implosive" holism (Colebrook 2017) in the sense of a flat ontology. This is seen as a counter—concept to concepts of humanity based on ownership, property, demarcation, and sedentarization, taken collectively as "ontology" (Bajohr 2019: 69). However, this reduces the concept, as Bajohr rightly points out, almost to the concept of person, as is sometimes the case in animal ethics (Bajohr 2019: 69).

Overall, these positions approach an anthropocentrism that sees a broader and not only human—related center as compared to previous anthropocentrisms. This largely corresponds to the biocentrism discussed in environmental ethics, which sees living beings as moral objects, or even to ecocentrism and holism, which see ecosystems or even all of nature as a moral object (Widdau 2021: 105–122). Nevertheless, in political terms, the addressees are humans. In contrast to these positions and posthumanism à la Braidotti, the more recent negative anthropology in philosophy holds the determination more in abeyance:

> Systematically, negative anthropology denotes an approach that refuses to define humans in terms of characteristics that are essential to them, *but nevertheless makes them the center of its interest*. This distinguishes it from critical posthumanism, because where the latter wants to keep humans completely at bay, negative anthropology still holds them as a variable that cannot be dissolved, but also cannot be removed from the equation (Bajohr 2019: 72–73, emphasis mine, see Dries 2018; Bajohr and Edinger 2021).

6.4 Environment and Culture—Biocultural No Man's Land and Social Theory

> *The most important epistemological fact of our present for the common fate of mankind is the rediscovery of the nature—culture continuum.*
> Marina Garcés (2019: 116)

In the Anthropocene, nature has, so to speak, collapsed into culture in people's perception. While nature used to be seen as a threatening force, it has been regarded as fragile and weak since the middle of the twentieth century. It is now itself considered to be threatened by humans and worth protecting. There are calls for nature conservation, for humans to take care of nature out of a perceived responsibility. The decisive historical marker for this shift in the perception of nature was, after many precursors, Rachel Carson's consequential non—fiction book ´Silent Spring' from 1962, in which she addressed global environmental damage based on the effects of DDT on insects and birds (Carson 1962, cf. Kerner 2024: 15–81 on Carson's visions).

In the face of the Anthropocene, it is postulated in much more far—reaching proposals that non—human components of the environment be perceived as companions, as in Haraway's concept of *companionship*. More narrowly defined, but comparable, are bio—centric or patho—centric ideas of granting basic rights to non—human beings, such as higher primates, which can be assumed to be capable of suffering, as demanded by the Giordano Bruno Foundation, for example. On the other hand, in the Anthropocene, nature is again being perceived as a force that strikes back. Especially in the context of anthropogenic climate change, events that make parts of the planet uninhabitable such as drought or flooding, are seen as a response from nature, which is often personalized (Eriksen 2016: 17).

Ecology, particularly human ecology and cultural ecology, have seen the relationship as a *human–environment* relationship (human ecology, for example, Odum and Reichholf 1980) or as a *culture–environment* relationship (cultural ecology, for example, Moran 2010, 2016). In the first case, it is primarily about relationships between humans as living beings and their natural or human environment (Glaeser 1996). For example, what physiological adaptations do the bodies of people who perform long—term industrial work show? In the second case, the focus is on relationships between human collectives, such as cultures, ethnic groups, societies, and their respective natural environments. How do human groups in the high Himalayas, for example, culturally adapt to their cold, high mountain environment? In these models, humans as individuals or collectives form an additional external factor, complementary to internal factors that are otherwise inherent to the system. The assumption here is that this factor often exists, but is sometimes missing and often weak, thus only representing a "human influence" on the system. The concept of social ecology comes much closer to the idea of the Anthropocene (Glaser et al. 2012; Fischer-Kowalski and Haberl 2017).

Röckström et al. introduced the concept of planetary boundaries (Rockström et al. 2009; Steffen et al. 2015a, b). These are parameters that allow for the continued existence of humanity, a "safe operating space for humanity." Planetary boundaries are evaluated in terms of nine dimensions (for example, land use), which together are indispensable for the stability of the geosphere and biosphere. At first glance, this is reminiscent of the concept of limited resources, which has been discussed in relation to the planet since the *Club of Rome* (Meadows et al. 1972, 2012). At its core, however, it is a completely different concept that focuses on the coupling of human societies and the Earth system. Much more than in all these concepts, the idea of the Anthropocene demands that *every* description of the status of our planet give prominence to humans, cultures, and societies. Especially with regard to the unequal distribution of environmental degradation in poor countries and poor regions of wealthy countries, the social sciences have a lot to say. This is more than the unequal side effects of environmental stress, but rather "a political geography of equity and justice" (Pálsson et al. 2013: 7).

6.4.1 *Are We all Natural Peoples Now?—Ecological Nexus and* **Biophilia**

The geosphere consists of hybrid systems that link humans with society and the environment. The ozone hole, for example, is a hybrid object because it was created by natural forces and cultural factors (Bodley 2012: 5, 66–69). In the Anthropocene world, we have appropriated nature, but also lost it. Cultures largely follow a technological paradigm where we have developed a dispositional knowledge *of nature*, but the orientational knowledge *in nature* seems to have been lost (Mittelstraß 1992).

One anchor for such orientational knowledge could be the human proximity to the world of organisms. "Biophilia" is generally understood as a human need for a close connection with the natural environment, especially with plants and animals. Wilson introduced the concept mainly on the basis of personal experiences and defined biophilia as "… the inner tendency to focus on life and lifelike processes" (Wilson 1984: 1, similar to Dubos 1968: 39–42). Proponents of the thesis assume that biophilia represents a genetically conditioned affection for living things ("inclination to affiliate with life," Kellert 1993: 21). An inclination towards nature conservation has often been derived from this, and biophilia has been seen as an aid to promoting more conscious action to preserve biodiversity. Some literary scholars go further and see geophilia, an affection for stones, not only in geologists; they see it more generally as a tendency of matter to affiliate with other matter, more specifically as an affection for matter, regardless of whether it is alive or not (Cohen 2015: 19, 27–66).

This is where critics of the biophilia hypothesis come in. The metaphors of biophilia and biodiversity implicitly convey moral views. The concept contains a naturalistic fallacy), thus confusing genesis with normative validity. The concept is a secularized claim to consolation for non—religious people who feel a sense of responsibility for creation (Potthast 1999: 163 f.). More importantly, however, it should be noted that there is hardly any empirical evidence for biophilia. The biophilia thesis is more of a hermeneutic interpretation of the human—nature understanding than hard evolutionary biology. The environmental preferences that have been documented so far are in fact much more specific: the preference for open, uncluttered terrain, proximity to rivers or springs, etc., and they do not require the biophilia thesis (Potthast 1999: 165, see Milton 2002).

Any discussion about the Anthropocene affects both the concept of nature as well as the concept of culture. It concerns the dualism of nature and culture and also the evaluations behind it, such as the polarization into good nature and evil culture or bad humans. We live in an earthly epoch in which we are essentially only confronted with ourselves. With the Anthropocene, humanity has lost "… nature as the other of itself" (Schiemann 2020: 250). Many discussants see the most important contribution of the idea of the Anthropocene in the insight that humanity is not an external force and that culture becoming visible would prove it to be a normal part of the biosphere, thus enabling us to overcome an economic system that ignores nature or externalizes it.

Even if one only refers to Western literature, one will find very different concepts of nature and a multitude of cultural concepts. A consistent feature, however, is the definition of nature as relationship, as relation to another. In most cases, nature is defined in relation to culture and *vice versa*, less often in relation to the social. Nature is one half of a pair of concepts, this being thought of as polarity, as duality, or as (mutually exclusive) dichotomy (*naturalism*, Descola 2015). It should be noted that the clear separation between nature and culture is historically recent and only is represented in some societies in space and time:

> (…) (the) idea that nature is a universal reality that cannot be influenced by our social and moral actions (is) a relatively recent idea in Western thinking (…) (and is) (…) at best an idea (…) that is specific only to a limited number of societies, and this only (…) for a relatively restricted period of time (Platenkamp 1999: 6, Erg. CA).

The Earth System Sciences and the environmental humanities, which emerged in the wake of the discovery of the Anthropocene, aim to resolve this polarization. When humans become a constitutive element of socio—ecological systems, they are no longer an external factor, as in conventional ecology and geomorphology. They are effectively taken from the cultural sciences and humanities and brought into a comprehensive Earth System Science (Bruns 2019: 56).

6.4.2 *Beyond Dualisms?*

As a prominent representative of posthumanist orientations, Haraway coined the term "NatureCulture" (Haraway 2003: 2) early on to capture the inseparability of nature and culture due to their close causal interdependence. For her, the ideal type is the indissoluble relationship between humans and domestic dogs, which, as domesticated animals, are to be understood as part of humans themselves. Haraway wants to counter this with the concept of a *sympoiesis* of self—regulation (*autopoiesis*). "NatureCulture" also includes relationships between humans and microbes, bacteria and the genetic as well as the ecotonic levels of connection. Kirksey means something similar when he speaks of "emergent ecologies" (Kirksey 2015, see Kirksey and Helmreich 2010).

The vocabulary used in the often programmatic writings, however, underlines the fundamentally dualistic view: nature and culture should be "connected," "coupled," or "linked." Often, alternative terms that overcome the dualism are sought. Some terms are intended to confuse, for example, through a missing typographic space between two word components and the absence of a hyphen, as in "NatureCultures" ("NaturenKulturen," Latour 2015; Gesing et al. 2019: 7). Linguistically, however, the proposals remain notoriously binary, such as "biosocial," the Spanish *socioambiental*, or the word I use, "biocultural" (examples in Table 6.1). A new attempt to break free from the notoriously binary framework is the four-word term "PeopleAnimalsNaturesCultures" (inGerman "MenschenTiereNaturenKulturen",

Table 6.1 Dualistic terms for the linked phenomena of nature and culture

Term English, Spanish	Fields of use and examples from authors
Human—Environmental—Relations, human—Environment systems	Common in institutions and interdisciplinary fields, e.g., in *environmental humanities*
Man—Environment—System, socio—Ecological system	In institutions and in earth system sciences
Bio—Social	Ingold and Pálsson 2013
Bio—Cultural	Evolutionary biology, rare in cultural anthropology: Greenwood and Stini 1977
Bio—History	Harper et al. 2016
Geo—Cultural	Latour 2017a, b
Geohistory	Chakrabarty 2000
Material culture	Common in cultural anthropology
Cyborg	Haraway 1991
Hybrid networks	Latour 2015
Geo—Social	Latour 2015
Nature cultures, multiple nature—Cultures, nature—Culture borderlands	Haraway 2003: 2, Kirksey and Helmreich 2010, Gesing et al. 2019, Fenske and Peselmann 2020
—	Ameli 2021: 7, 69–77
Socio—Natures, socionatures	Castree 2017
Socioecological transitions, span.: *Socioambiental*	Fischer-Kowalski et al. 1993, Danowski and de Castro 2019
Sociomaterial histories	Bauer and Bhan 2018: 5, Maasen et al. 2018
Sociometabolic	Fischer-Kowalski et al. 2017, Haberl et al. 2019
Sociocultural evolution	Sociology, Archaeology, e.g. Trigger 1998, Blute 2010, Turner 2021
Cosmopolis	Danowski and de Castro 2019: 88
ResourceCultures	Knopf 2017
Anthropo—Not—Seen	De La Cadena 2015
Societal nature—Relations	Görg et al. 2020
Extended phenotype	Dawkins 2016 (1999)

Ameli 2021: 7, 69–77). This seemslinguistically too unconventional and uncomfortable; moreover, this term perpetuates the separations, just in a multiplied way.

The Anthropocene is characterized by the increasing presence of networks that consist of various actors. The social is no longer only related to humans. There is not only the "social question" of inequality, but also the "geosocial question" (Latour). In addition, technologization is increasingly affecting the bodies of living beings, including human bodies, as emphasized in Posthumanism. Planetary change is long—term and includes *geos*, *bios*, and human *techne*. Therefore, the Anthropocene discussion promotes the search for explicitly non—dualistic terms more than previous debates or discourses. This is based on theoretical considerations, but also concrete empirical geoscientific findings. For example, some volcanic eruptions in the Arctic and Iceland seem to be a result of glacier retreat caused by human activities (Pagli and Sigmundsson 2008). In relation to the planet, synthetic approaches have already become firmly established and institutionalized. The geosciences or Earth System Sciences have established themselves alongside classical geology, geophysics, and geography. The problems of late modernity are marked by rapid population growth, expanding raw material extraction, species extinction, and the collapse of some ecosystems.

In response to these findings or diagnoses, some propose a conceptual zooming out. "Our world of concern has been vastly expanded" (Pálsson et al. 2013: 99). Within the social science approaches to the Anthropocene, Latour in particular takes a position against the actor idealism criticized above, which sees humanity as an acting agent. In his critique of the modern distinction between humans and nature, he highlights the links between humans and non—human actors (*actants*). This entanglement of a multitude of actors, humans, animals, plants, and things with each other, and ultimately the entire ecosystem, make up the hybrid character of what Latour, in one of his many neologisms, describes as a "post – natural composite" and designates with the unitary term "Gaia" (Latour 2010, 2018).

Latour, Haraway, and other posthumanist authors want to dissolve the distinction between acting humans and a supposedly passive nature. They attribute agency to nature as they do to humans, thereby deliberately leveling out ontic differences between action or intentionality and behavior, between subject and object, and also between the living and the non—living. To overcome polarizations, Latour recently worked with the concept of the "terrestrial" ("atterrir," 2018), which, however, is then specifically related to humans, unlike older approaches such as that of the rhizome (Ingold 2011a, b; Ingold and Pálsson 2013). In addition, in the "confusion of nature" (Latour 2010: 19) or in the "multiverse," not only is the distinction between human and non—human actors, such as animals, plants, and fungi (Sheldrake 2020), leveled out, but also the difference between actors and artifacts.

All these approaches raise eminently challenging questions, which Pálsson et al. summarize as follows: "Could we benefit from applying the same theoretical frameworks to both the nano – world of bodies, cells, and genes and the giga – world of the globe?" (2013: 9). From my point of view, despite all the differences in detail—for example, between Haraway, Tsing, and Latour—it remains unclear how one should work with these terminologies in terms of social theory, what theory of society can be formed from them, and above all, how this approach can be used for the empirical investigation of people and collectives in the Anthropocene. How can "compositionist thinking" be made ready for theory without taking a metaphorical flight to *Gaia*? As Haraway's and Tsing's works have already extensively demonstrated, working or playing with freely floating concepts entails the risk of impoverishing social science analysis.

In addition, there is the threat of ethical overload due to the demand for full responsibility for every aspect of living nature. When responsibility is applied to the whole of nature, making it appear completely dependent on humans, we end up with what the Italian philosopher Agostino Cera criticises as "petification". This petification of nature results from a supposedly absolute (i.e. not just total, as with Hans Jonas) responsibility of humans for nature, a full responsibility (omni-responsibility). This would go hand in hand with a reification of nature and making the difference between the animality of animals and the special human animal disappear and—albeit unintentionally—virtually ending up with planetary management or comprehensive geoengineering (Cera 2020: 34–35, 2022, 2023). Haraway's levelling, which is actually a monistic conception of nature, when combined with the attribution of absolute responsibility, leads to a practical-political approach to earth management that Haraway could not have intended.

A few cultural scientists and sociologists work empirically with Latour's theory, for example, on the socialization of CO_2 (Laux 2020: 130–143). Above all, the expansion of his classic model of modernity by means of differentiation theory, which Latour accomplished in his main sociological work "Modes of Existence" (Latour 2016), invites analysis. There, he suspends the term "society" in favor of the "collective" and distinguishes 15 special modes of existence that produce specific world relationships in modern collectives. He also introduces a multitude of abbreviations to illustrate the diversity of networks (Latour 2018). Once again, however, the problem lies in an exuberant theoretical language, which makes connections to established approaches in social science or cultural theory difficult and is partially replaced by new terms in his last, rather programmatic work, the "Terrestrial Manifesto" (Latour 2018). The majority of sociologists seem thoroughly disillusioned, at least in view of the few concrete analytical as well as normative statements with respect to the phenomenon of the Anthropocene. Some consider Latour's grand farewell of modernity to be premature and the "flight to Gaia" to be little more than an empty phrase:

> But how can one contribute anything substantial to the sociological analysis of global ecological disruptions and conflicts when one's own judgment is exhausted in the metaphorical talk of 'spheres', 'hoops', 'cycles' and so – called 'loops'? This is where the loss of order – creating categories in a conception of the Anthropocene, which believes it can transfer the analytical discernment of social theory into hybrid composites in order to ultimately take refuge in obscure unitary concepts such as 'Gaia', takes its revenge (Neckel 2020: 166).

From an economic and materialistic perspective, the problem is that the specific causes of extreme changes in nature, such as the capitalist—driven extraction of natural resources, cannot be adequately identified as causes, as Hornborg especially argues (Hornborg 2017a, b, see Kersten 2014: 393 ff., 397 ff.). In the "Terrestrial Manifesto," Latour reasons that the progressive warming of the Earth's atmosphere is due to a plot by global elites (Latour 2018: 25–30). This comes close to a conspiracy theory, as critics rightly object (Neckel 2020: 166).

Marxist and other materialistic approaches, on the other hand, admit that there is no simple opposition between humans and nature, but maintain the distinction between humans as natural beings, who are indeed a special part of nature, and the rest of nature (for example, Foster and Angus 2016). They also make another analytical distinction other than that between humans and nature, namely that between *society* and nature. This analytical distinction is important for the neo—Marxist argument that capitalism is based on a particular, namely extractive, relationship with nature. The basis of capitalism thus is a relationship with nature in which nature appears as something external, as something that can be externalized from the economic balance sheet; in short: as a cheap raw material.

Capitalist societies have a special relationship with nature (Görg 2003a, b). In particular, today's variant of globalized capitalism can only reproduce itself through nature, the exploitation of raw materials as well as the bodies of people. Nature is thus devalued, as is labor, money, food, energy, care, and of course, living beings. Due to the inherent relationship with nature, neo—Marxist approaches therefore insist that capitalism does not deeply change nature as a side effect, but does so inevitably (Altvater 2018; Patel and Moore 2017; Moore 2020: 9–13, 51).

6.5 Beyond Sustainability?—Ecological Breaks Versus Holocene Thinking

We cannot rewind ecosystems back to a state untouched by humans.
John Dryzek & Jonathan Pickering (2019: 9)

Humans have evolved culturally and perhaps genetically to be unsustainable. We exibit a deep and consistent pattern of short-term resource exploitation behaviours and institutions.
Peter Richerson et al. (2023: 1)

Sustainability refers to future problems and permanently secured opportunities for humans and other living beings. It is about future-proof action today, and the topic is acquiring great significance in the face of the Anthropocene. This is also one of the strongest arguments for declaring the Anthropocene as a new geological epoch (for example, Mittelstaedt 2020: 137–140, Jørgensen et al. 2023). The debate on sustainability, which emerged in its current form in the context of global environmental awareness in the mid—twentieth century and the UN environmental conferences from 1972 onwards, has a number of structural parallels with the Anthropocene debate. The proximity of "sustainability" and the "Anthropocene" exists not only thematically in dealing with the relationship of current dynamics to the future, but also in terms of debate system. Both can be seen positively as *container terms* that allow for a productive handling of many meanings, or negatively as *catch—all terms* or *buzzwords* (Will 2021: 259).

The idea of continuity or permanence, which is central to sustainability thinking, and also the continuous time concept of classical cultural anthropology (see Sect. 5.4), fit in with the idea of uniform and slow change in *classical* geology, paleontology, and evolutionary research, which these disciplines refer to as gradualism. In geology, this goes back to Hutton and Lyell, who worked out that the processes that took place in the past, such as erosion and uplift, were of the same kind as those that can be observed today This applies to both laws and the rate of change. This principle is known in geology and paleontology as actualism or uniformitarianism (Henningsen 2009; Bjornerud 2018: 78–79, 120–122, 134–135).

In geology, this principle was dichotomously opposed by the idea of abrupt change, by catastrophism. However, with these stances, which are repeatedly discussed as polar positions, one should not overlook the that there were already various geological and theological variants of catastrophism in the nineteenth century

(Irvine 2020: 163, Sepkoski 2021). From a geological point of view, the Anthropocene should, for one thing, be about re-emphasizing the uniformity of geological change, but at the same time to expect abrupt phases of change as 'normal', both with and without human influence.

6.5.1 Continuity and Fundamental Ecological Novelty

In the geosciences and biosciences, there are ongoing debates to this day about the extent to which the basic assumption of continuous change, that is, evolution instead of revolution, underestimates the significance of sudden change. A position that considers abrupt geological and biological change to be normal is referred to there as punctualism (Gould 1987; Turner 2011: 37–57). This became more prominent in the 1980s, when it became clear in paleontology that catastrophes such as mass extinctions, that is, species extinction across many systematic categories of plants and animals, occurred repeatedly in the Earth's history, such as the five major extinction events (Raup 1992; Raup and Sepkoski 1982; MacLeod 2013). Today, it is known that these major catastrophes were not just interruptions in the history of the Earth and life, but played a transformative role in the further development of the biosphere and geosphere. This insight is known as "neo—catastrophism," a perspective that identifies deep time itself as eminently historical (Irvine 2020: 162, Sepkoski 2021).

What does the Anthropocene mean for notions and ideals of sustainability? The findings on the Anthropocene imply a revision or radicalization of the balance—oriented concept of sustainability. The goal of sustainability comes from forestry and was later elaborated in the development context and then popularized. Today, it is an extremely overused word with different meanings. At its core, however, it is always about the use and consumption of natural resources and the consideration of the interests of future generations of humans, thus always involving normative issues. Scientists play an important role in dealing with these issues, the normative implications of which have become clear in climate change research (Krauß 2015a: 61–62).

Sustainable economic activity consists in shaping the way of life of the present society in such a way that the life opportunities of future generations, that is, our children and grandchildren, are secured. It is about securing the same opportunities that people of the current generation have and thus (a) about the near future, (b) about controlled or planned continuity of current conditions, and (c) primarily about the interests of people or collectives. The idea of sustainability essentially states that humanity should only take from nature what it needs, give back what can be recycled, and leave non—renewable resources in the ground as much as possible, especially fossil fuels (Merchant 2020a, b: 145). The goal of current sustainable solutions is a balancing of interests, a kind of "succession of wealth." The normative guideline is the imperative not to satisfy current needs at the expense of the needs of people in the future.

Among the many notions of sustainability, two broad directions can currently be distinguished. The classic conception since the Brundtland Report is "sustainable development," a concept that is tied to development and growth and is macro—oriented. The concept of "sustainable livelihoods," on the other hand, focuses on local situations, basic needs, economic security, and solutions such as locally sustainable cultivation methods, bioregionalism, and indigenous environmental concepts (Merchant 2020b). Despite all differences, ideas of sustainability consistently contain the notion that humanity can "get a grip" on unwanted global change through certain measures or framework conditions, such as market guidelines (cf.. Eriksen and Schober 2016). This formulation alone shows the general idea of conscious human agency behind it. These ideas form the core of the development paradigm and many other modernist concepts. This is demonstrated by the global Sustainable Development Goals (SDG). However, the diagnosis of the Anthropocene is heading in a different direction:

> The complex and somewhat chaotic implications of anthropogenic environmental change undermine our capacity to respond along the lines of the modernist 'management' of the past, emphasizing human mastery and control (Pálsson et al. 2013: 9).

The findings on the Anthropocene imply a revision or radicalization of the sustainability concept, which is geared towards stability or balance. With regard to long—term stable systems, we can learn something from previous environmental disasters and ecosystem tipping points (Fagan and Durrani 2021; Frankopan 2023). About where the Achilles' heels of the Earth system actually lie (Haller 2007; Jahn et al. 2016; SONA 2021; Block 2018, 2021; Bubenzer et al. 2021: 26, Wendt 2021). However, if abrupt and systemic change has occurred in the Anthropocene, we cannot be sure if these lessons from past transformations will simply apply analogously to the future, as the main proponents of the Anthropocene idea state:

> In summary, we conclude that Earth is currently operating in a *no – analogue state*. In terms of key environmental parameters, the Earth System has recently moved well outside the range of natural variability exhibited over at least the last half million years (Crutzen and Steffen 2003: 253).

The same applies to the concept of resilience, which has already been mentioned several times. Resilience is usually understood as the ability of a system to absorb disturbances and withstand them while maintaining the basic structure. This can affect systems without humans or with humans (Folke 2016: 256, *social resilience*, Edwards 2015; Resnick 2021). How does a rice field regenerate after a plague of insects, how does a tropical rainforest react to deforestation or fire, how does an island economy in the Pacific regenerate after a typhoon, a tsunami, or a volcanic eruption? Behind this is the idea of the continuity of an ecosystem, and it is about returning to balance, albeit a dynamic balance. This is drawn from the complexity sciences and their insights into non—linear processes since the time of Prigogine and Stengers.

The complexity of the phenomena also has consequences for the role of the natural sciences and for the public dissemination of scientific knowledge in the age of social media (Renn 2020). In the case of nuclear energy, experts disagree on the non—technical consequences of using an innovative technology, which has been known as the "expert dilemma" since the 1960s. In the Anthropocene, however, there is a new problem in the matter due to non—linear dynamics, which can often only be grasped with stochastics, as has become clear in the case of climate change:

> What is new now is that where one has traditionally expected exact natural science, one finds room for interpretations and sees a plurality of opinions (Renn 2021: 26).

In view of the Anthropocene, there are no clear and thus desirable equilibrium states. Politically, this calls for an Anthropocene construction of new *dynamics*, an adaptability or transformability (*transformative resilience*), rather than a Holocene "conservation" (Dryzek and Pickering 2019: 37–39). Any form of true resilience is dependent on our knowledge of the role of humans in the geosphere, because:

> In the Anthropocene, humans do not act against the backdrop of an unchangeable nature, but are deeply intertwined with its structure, shaping both its immediate and its distant future (Renn 2020: 2).

6.5.2 **Double Bind** *in the Anthropocene—Contested "Nature Conservation"*

However, if we assume that the geosphere is undergoing directed change in the time horizons of the Earth system anyway, and that the entire Earth system is changing even more strongly and abruptly in the Anthropocene, then the conservative, that is, conserving, core of the sustainability idea can hardly be maintained. Sustainability can therefore no longer be thoughtlessly considered as a desirable guiding principle *per se*. Sustainability as a concept and guideline involves problems, paradoxes, and probably also hardly resolvable conflicts of objectives: dilemmas (Haller 2007; Benson and Craig 2017; Adloff and Neckel 2020: 12, Adloff et al. 2020; SONA 2021). This also applies to the political dimension of indigeneity, for example. The idea of sustainability, which emerged strongly in the 1980s, is a fundamentally normative concept that aims to ensure that future people have the same opportunities as we do today. In the face of the Anthropocene, the idea of sustainability—along with ideas, boundaries, vulnerability, disenfranchisement, and entanglement—can also lead to restricting the freedom of indigenous communities (Chandler and Reid 2019).

Sustainability in the Anthropocene is thus a good example of double bind situations. Gregory Bateson, whose important idea about runaway processes (schismogenesis) was explained above, contributed a term that is particularly helpful in relation to sustainability: the *double bind*. A double bind is a contradictory message that leaves the listener with no choice and is also socially dysfunctional, such as: "Be spontaneous!" Bateson described this using individual psychological processes in pathological family relationships and as a cause of schizophrenia. However, a double bind can also exist for human collectives or even humanity: "Grow green!" As Erikson aptly notes, we are currently in a *double bind* worldwide, knowing, on the one hand, that the current neoliberally underpinned growth in human history, fueled by non—renewable energy, is effectively already in a "runaway process"; on the other hand, we are also striving for sustainability.

A *double bind* is not simply a dilemma. A dilemma, such as the one between class—oriented politics, which strives for material improvements, and green politics, which de facto aims for a reduced material standard of living, can theoretically be solved by means of principles of equality and social justice. In contrast, there is a fundamental contradiction between growth and sustainability. Combined as a goal or vision, this entails a double message. We do not know whether we are evolving—in terms of evolutionary allegories—into a peacock, whose overgrown tail is selectively functional or at least harmless, or—more likely—into an Irish elk, which became extinct because of its oversized antlers. Eriksen has clearly stated that he finds such allegories more helpful than "stories about long—term global sustainability" (Eriksen 2016: 7, 22–24, 152). Concepts of sustainability have historically been strongly influenced by ideas of untouched nature.

> The search for untouched nature has always been the main driving force behind journeys of discovery and exploration as well as adventure and self – awareness vacations. (…) Not only is nature sought out, but it is also construed through these journeys and, to a certain extent, brought forth in the first place (Hupke 2020: 6).

A conservation of nature is hardly conceivable in light of the finding that we are already in the Anthropocene. We cannot redevelop ecosystems into systems where there are no human influences. Fire is one example, which is an important factor promoting biodiversity in ecosystems. On the other hand, humans have probably been using fire to change their landscapes for almost 100,000 years. In the Anthropocene, the effects of fire are changing due to interactions with anthropogenic drivers, such as land use and invasive species. Thus, effective biodiversity conservation today requires the combination of natural fire regimes with targeted human activities (Kelly et al. 2020: 1, 7–9). The idea of a possible return to a historical starting point or a stable *baseline* is unrealistic, and thus all concepts of "conservation" are systematically being challenged. Simple ideas of greening the modern world do not take these facts seriously when they aim for slow adaptation. Hence, the cultural and social sciences are also being questioned (Shoreman-Ouimet and Kopnina 2016). Today's conservationists are aware that they are protecting landscapes that have long been anthropogenically shaped or at least influenced and are therefore at best "near – natural" (Baur 2021). Nevertheless, many ideas on conservation, (*preservation*, *restoration*) still indicate "Holocene thinking" (Dryzek and Pickering 2019: 9) instead of truly striving for convivial and thus non–dualistic nature conservation (Büscher and Fletcher 2020).

All in all, an "Anthropocene sustainability" appears to be a *contradictio in adjecto*, because a basic insight into the Anthropocene is precisely that of the suddenness and irreversibility of human impacts (Hernes 2012). A meaningful adaptation to current changes can mean a maladaptation in the future. Contrary to continuity, the idea of the Anthropocene emphasizes breaks within temporal changes and fundamental "ecological novelty" within ecological systems (Kueffer 2013: 23). The human causes of such sudden upheavals have been documented primarily by glaciologists. Despite the experience of such sudden changes, the cultural idea of pure nature does not fade.

In the idea of sustainability and the concepts of nature conservation influenced by it, there are further central tensions (Ziegler 2019: 273–274, Büscher and Fletcher 2020). The idea of sustainability tends to view "… nature as an exogenous variable that should not be completely exploited for the sake of humanity in order to allow it to continue to access natural resources." (Adloff and Neckel 2020: 12–13). However, the insight into the Anthropocene as an issue means that it is now not only about sustainability, but also about habitability, that is, about permanent foundations of "livability." This habitability on Earth no longer exclusively concerns human interests (Chakrabarty 2021: 83). Accordingly, it is an open question, for example, whether biodiversity should be promoted out of human interests, as the idea of sustainability would suggest, or out of the "interest" of specific configurations of humans and other living beings ("more – than – human") or of an entire biome or even nature as a whole. What we need in any case is a genuine geo—anthropology, because:

> We do not live in a stable environment that merely serves as a stage and resource for our actions; we belong to a dynamic in which humans and the non – human world play an equal role (Renn 2021: 2).

In the current political discussion, for example, it is specifically about whether an ecosystem should be accorded not only an ethical, but even a legal status (Adloff and Busse 2021). A sustainability—oriented position is necessarily anthropocentric, whereas the idea of habitability is biocentric, but it also needs humans to claim an intrinsic value for nature. In the face of impoverished floras and faunas and the interspersion of invasive species, some conservationists emphasize that we are already living on such a "tamed" planet through growth and consumption that wilderness has almost completely disappeared in favor of entirely "novel ecosystems" (Crist 2013, 2020). This seems exaggerated and shows an insistence on the distinction from untouched nature. However, it seems especially extreme to me when it is concluded without hesitation that biodiversity as a goal is already a lost cause (see Thomas 2017). The often associated thesis that the interests of humans are inadequately considered in conventional nature conservation (*new conservation*) is also simply not true (Wilson 2016: 84–85, 93). All these are possibly consequences of an insecurity, which has interestingly been described once again using a disease metaphor, namely as an "ecological anxiety disorder" (Robbins and Moore 2013).

Sustainability concepts assuming that enhanced technologies and changed environmental awareness would improve the material cycles ignore the systemically disruptive character and temporally discontinuous progression of these changes (Pálsson et al. 2013: 8). They underestimate the importance of global power relations as well as the effectiveness of fundamental human behavioral tendencies (short—termism and meso—scale perception, Morrison and Morrison 1992) and cultural value orientations (for example, prosperity as a positive norm). For this reason, the classic goal of sustainability is no longer strictly tenable.

Sustainability concepts applied everywhere must acknowledge that much in the universe happens *in*dependently of human influence (Leggewie and Hanusch 2020: N4, Hanusch et al. 2021). A realistic near—term goal could be that grandchildren have the same number of opportunities or options, but *qualitatively* different ones. The talk of "future sustainable Anthropocene societies" (for example, in Pálsson et al. 2013: 10) can therefore, in my opinion, only refer to the more distant future. In an even more radicalized form, a changed concept of sustainability would mean keeping the future open for other experimental structures instead of striving for continuity and controlled "management" (Horn 2020).

A concrete, optimistic, but controversial proposal by environmental journalist Emma Marris is to develop and preserve "novel ecosystems," that is, anthropogenically influenced spaces of mixed semi—wilderness, instead of "untouched nature," which in any case hardly exists anymore: ruderal formations or "rambunctious gardens" (Marris 2013; Stoetzer 2018). However, in the sense of principle of prevention, this cheerful vision must be countered by the fact that we currently know very little about possible irreversible systemic effects, such as those caused by invasive species. With current methods, hardly any prediction can be made about the long—term development of biodiversity in such hybrid systems (Wilson 2016: 88, 95, 100, 117).

6.5.3 Conflicting Time Horizons

Another further tension between sustainability goals and the value of habitability as implied by the Anthropocene arises from the different time horizons or different temporalities (Adloff and Neckel 2020: 13). While sustainability as a development goal has the next or the coming generation (of humans) in mind, an Anthropocene—informed position suggests that much longer periods, namely a time in the future, should be considered. The Anthropocene emphasizes the long—term history of nature, evolution as "natural history." In terms of the long time perspective, it is not only about grandchild—compatible solutions, but also about the interests of the "children of the children's children."

The importance of thinking about such a distant future and the consequences of current decisions for it has become clear through anthropogenic climate change. Differently chosen time horizons raise far—reaching ethical and theological questions regarding those responsible for causing it as well as the sphere of those affected by its effects (Ott 2016, 2019). What is once again evident here is the fundamental divergence between a human—centered perspective, which is focused on short periods of time and small spaces, and a biocentric and geocentric perspective that takes into account long temporalities and large geospaces or life systems. Climate change has also shown that a conception of distant futures requires concrete knowledge of the deep time of the past, because without this knowledge, no long—term trends can be identified. Linked to this is the problematic fact that the change in such systems can hardly be planned.

Finally, climate change has also shown that a normative claim to distant futures can lead to objections in the present being delegitimized by a reference to higher goals (Adloff and Neckel 2020: 13). In my opinion, this is a fundamental political problem that does not only arise in the Anthropocene. As a Southeast Asian scholar, I cannot help but think of the almost classic arguments of the Singapore government, that democratic rights, such as full press freedom, must momentarily be postponed in order to ensure a secure economic future for the constantly precarious island state. However, in view of anthropogenic climate change, this danger, which is exacerbated in the broader Anthropocene, is now becoming relevant on a macro scale. The argument of "interests" can be used to argue on a much larger spatial scale as well as on a social scale—and for a much longer temporality. This extends to a cosmopolitan level: cosmopolitics. A current example is probably the decision to abruptly change Germany's nuclear energy policy under Angela Merkel with the argument of future dangers.

These considerations coincide with concepts about the fundamental importance of technology in current and increasingly in future societies. Here, concepts of the "risk society" (Beck et al. 1994) and ideas about ontains or societies are relevant. Anthropocene societies could essentially be such societies, much more than today, and for which disruptive change is fundamentally "normal." People will live in societies that consider conscious interventions in the geosphere with long—term consequences and also in the living world—including human bodies—to be the norm.

> It is now time for us to *articulate the culture of emerging Anthropocene societies* by drawing upon natural scientists, humanities scholars, and social scientists, emphasizing the new fusion of the natural and the ideational. In regard to the transition to fully Anthropocene societies, adapted to the new human condition, how can the contemporary syndromes of anxiety, drift, and self – delusion be transformed into a more positive task of building a culture of sustainability? (Pálsson et al. 2013: 8, emphasis mine).

This involves normative questions, but also the anthropological question of what human needs are and how they can be satisfied for specifically situated people and communities, a question discussed in classical cultural anthropology (Malinowski 1944), which has unfortunately long been relegated to the background in postmodern anthropological cultural anthropology. A sensible alternative to the concept of sustainability, which maintains the core contents but remains realistic, is the concept of reproduction explained above. Reproduction can be understood as the ability of a person, a system, or a social field to continue to exist without constantly having to adapt to *exogenous* change. As organisms, humans reproduce biologically and socially. Social systems must reproduce their carriers physically and their social arrangements transgenerationally. At the level of human experience, the central problems of the Anthropocene appear as the reproduction problems of cultures. Drawing from her long experience, among other things as a senior UNESCO official and president of the International Union of Anthropological and Ethnological Sciences (IUAES), Lourdes Arizpe Schlosser concludes:

> (…) what I see happening at present in the development of the concept of the Anthropocene: a lack of political and ethical ideas that self – organizing groups could latch on to (Arizpe-Schlosser 2019: 286).

A central problem is the lack of a clear direction in international policy and leadership, for example, in the sense of the concepts of *degrowth* (Schmelzer and Vetter 2019). As a result, there is a lack of compellingly plausible ideas and values that local initiatives could orient themselves to.

6.6 Deep Time and Social Times—Paleontology of the Present

Having a history is not a human privilege.
Edward O. Wilson (2016: 169)

The last thing a fish would ever notice would be water.
Attributed to Ralph Linton

In societal dealings with the Anthropocene, contrasting time scales collide. With the title formulation "Geology of Mankind" (Crutzen 2002), Crutzen initiated a debate about seemingly incompatible conceptions of temporality right from the outset (Will 2021: 34). This problem was generally put on the historical agenda by Reinhart Koselleck (Kosseleck 2013). In the face of the Anthropocene, on the one hand, the consequences of human action are discussed in geological time dimensions. On the other hand, urgent action is invoked in the face of existentially significant problems. If the planetary boundaries are not to be exceeded, tomorrow is already today. The topicality of political and economic interests meets long—term, socio—ecological consequences and geological time depths that exceed the time scales of human history by several orders of magnitude. This has already been made clear in superficial topical debates about radioactive waste, "final disposal," and "eternal burdens," a problem that is hardly mentioned in debates about the Anthropocene (Görg 2016: 12, see Brunnengräber 2019, 2021).

The vast majority of people in the world experience the Earth beneath them as largely static. While almost everything else in life is constantly changing, the rock beneath us appears as stable, as "the ground on which we stand." Sudden events, such as landslides, earthquakes, a volcanic eruption, a tornado, or a tsunami, are easily noticed. Such events are part of the normal experience of people in some parts of the world, such as Indonesia or the Philippines, who over the course of time have developed their way of life into *cultures* of *disaster* (Krüger 2012), minimizing risks through local knowledge. In contrast to such "natural disasters," geological crises run infinitely slowly from a human perspective due to the presentist scale of everyday life. They do not appear to people as disasters because they are hardly perceptible to humans without measuring instruments. From a geological perspective, however, even the quiet terrain on which we live, contrary to the dictum that nature does not make leaps, is anything but static (Ghosh 2016, Gordon 2021).

The postcolonial theorist Gayatri Chakravorty Spivak noted that there is a "planetarity" that is simply inaccessible to human conceptions of time (Spivak 2003: 88). However, I believe that we can and must visualize this based on existing knowledge, because the Anthropocene has abruptly revealed the interweaving of time levels. The human ecologists Andreas Malm and Alf Hornborg called for a deepening of social history in the face of the Anthropocene (Malm and Hornborg 2014: 66). In this chapter, my aim is to show that a deep—time perspective is indeed compatible with an anthropological, location—based, and realistic focus (see for sociology Clark and Szerszynski 2021: 53). I argue that "deep time" does not merely represent an abstract concept, but is a part of the phenomenal world that has an impact on everyday life, not just in resource extraction. The Anthropocene as a fact requires the merging of the short and long time scales of crises and, unlike other earlier diagnoses of the times or of society, such as modernity, postmodernity, late modernity, hypermodernity, or the multi—option society, the Anthropocene concept also makes this possible with its planetary long—term perspective:

> The concept of the Anthropocene suddenly created a bridge between geological and historical time. It became clear that the time scale of human history is inextricably linked with the geological time scale (Renn 2021: 3).

> The Anthropocene demands a retrospective of the current situation, a kind of 'paleontology of the present' in which we ourselves have become sediment (…) (Macfarlane 2019: 97, cf. Mitchell 2005: 325).

6.6.1 Planetary Transformation, Individual Biographies, and Missing Epochal Terms for the Future

> *It requires a different sense of time to adequately perceive the impact of slow—motion disasters as they are happening.*
>
> Stuart Kirsch (2014: 28)

The transformation of the planet in the Anthropocene and also the new image of the eminent human role in it not only question our shared responsibility, but also our own positioning as a person in time (Augé 2017: 50). Firstly, we live, on average, significantly longer than previous humans, which makes us perceive the current acceleration of history more clearly. Secondly, this means that several generations live together for longer. The extended lifespan will soon lead to a systematic simultaneity of four generations. If anthropologists have always said that the relationships between alternating generations, grandparents and grandchildren, are more relaxed than between parents and children, other forms of connectedness can now be explored (Augé 2017: 56).

With these consequences of a longer lifespan, the accelerated transformation of the planet is becoming more biographically tangible for everyone. This is currently evident worldwide in confrontations or also in the estrangement of generations. As Augé emphasizes, rapid changes in technology increase the objective connectedness *within* a generation because it has been socialized with the same technologies. A clear example of this is the experience of using electronic devices and media. As Margaret Mead pointed out in the face of rapid cultural change *avant la lettre*, grandparents can and must now learn from grandchildren, and parents from children, which complements the classic vertical direction of socialization and a horizontal one. However, rapid change also brings with it the problem of youth without a role model (Mead 1971). Therefore, I consider the term "acceleration"—not only for the Earth system (*Great Acceleration*), but also in relation to the experiential level of everyday life—to be anthropologically helpful in capturing the environmentally related escalation logic of late modern societies (Eriksen 2016: 10–15, see Rosa 2021: 181–200).

Among students of geology and paleontology, it is known that one needs simplifying tools to somehow elucidate the length of geological time. It is not only in popular scientific writings that we find depictions where the entire history of the planet is portrayed as 1 day, and man, as is well known, only appears a few seconds before midnight. However, every student of geology and even seasoned practitioners are always aware that this is only a pragmatic tool that does not really solve the perception problem (Gould 1987). Geologists experience this challenge concretely when they work in outcrops. When they walk or drive through landscapes, they are interested in those places where one can directly look into the history of the Earth. McPhee illustrates this with geologists working on road construction projects for Interstate Highways in the USA. A central argument of a cultural anthropology informed by geological deep time is the insight that the material living conditions of groups of people are the product of a long development over deep time.

We work with epoch terms and abbreviations for the past, such as "BP" (before present) and MYA (*million years ago*), but we have no corresponding terms for the geological future. No one yet speaks formally of the "after present" (AP) or "in millions of years" (Macfarlane 2019: 97). The rhythms of geological time and those of social time initially appear to be independent of each other. Only the findings on the Anthropocene have made it clear that there are not only connections here, but that there can also be tensions between our short—term everyday orientation and the unspeakably long periods of physical processes on which the present rests. As environmental historians emphasize, with the "Great Departure," the exit from the organic economy of the "old biological regime" (Braudel 1981), the natural limits that had previously existed in the form of locally available energy were not only exceeded.

No, with the creation of the anthroposphere, the natural rhythms were also drastically changed. Instead of using regularly occurring local wood as an energy source, the energy sources that originated in the geological past are those that are now being used. Wherever animal urine had previously served as manure and soils were quickly depleted of phosphorus—and still are today in poor areas of the world—and wherever external sources of phosphorus (guano) were only available to a limited extent, synthetic fertilizers were introduced with the Anthropocene (Marks 2020: 177). Because of these time tensions that are only being experienced now in the Anthropocene, it is important to deal with societal time horizons. From an anthropological point of view, it is a matter of lived everyday life being set within deep time instead of being abstracted from it (Irvine 2020: 1, 18, 174).

> (…) if we conceive of Earth processes as effectively static, too gradual to be relevant to human inquiry, then this evades the changes instigated through human *geological* agency – changes that can only be fully understood as *transformational mark in* deep time (Irvine 2020: 11, emphasis mine).

In cultural anthropology, time is an understudied but nevertheless classic theme, especially in the cultural anthropology of religion, the cultural anthropology of art, and in economic cultural anthropology. Anthropologists have studied how time is perceived as well as how actions are taken in time. They have investigated the rhythms around which everyday life is organized, how time is understood in cultures (working hours, leisure time) and how it is measured (for example, calendars). In particular, ideas about the direction of time, from the past to the present and into the future (linear, cyclical, oscillating time), have been examined. Further questions have been asked as to what extent local concepts of time are preformed by the respective language and how far they are determined by social circumstances (Whorf 1956; Evans-Pritchard 1940; Gell 1992; Munn 1992).

While these investigations mainly dealt with concepts, symbols, and worldviews with respect to time, there are also quantitative studies on the use of time. Economic anthropological, cultural ecological, and evolutionary ecological studies have also investigated how much time people use for which activities in their everyday lives (*time allocation studies*). An important finding here is that the self—perception of time use often differs significantly from the allocation of time according to quantitative data.

Evans—Pritchard made a distinction that is central to anthropological time research and, in my opinion, helpful for understanding how people deal with the Anthropocene in his classic study of the Nuer's time concepts. In a precursor to *multi—species anthropology*, he depicted ecological time and structural time among these herders in Sudan. Ecological time is the time regime characterized by the interactions of humans with domesticated animals, water, and the rest of the natural environment, in this case, for example, in the form of the dynamics of the land surface. Various rhythms come together here, such as physical (rain, seasonality), biological (grasses, fish), and social, particularly the milking cycle in the course of the day and the relocation of settlements between village and field camps throughout the year. These rhythms pertain less to temporal references to external time frames than to interconnected activities, and they are also symbolically linked (Evans-Pritchard 1940: 36, 59, 66, 72, 95 f.).

Structural time in the sense of Evans—Pritchard, on the other hand, is the time that indicates when one event took place in relation to another, such as the initiation into an age category. This time is often used when it comes to social distance, especially to people from other subgroups, clans, lineages, or ethnicities. While ecological time is strongly related to materials and material flows, structural time is more abstract. Cultural values, models, and behavioral expectations play a central role here, and in the case of the Nuer, for example, norms and ideals about age groups and kinship groups. At first glance, structural time appears to be directed or cumulative, but it moves within culturally predetermined structures. Thus, structural time becomes static due to the actors' assumption that the time span from the beginning of the world to the present is unchangeable (Evans-Pritchard 1940: 95, 105–108).

A more specific question that has been less investigated so far, and which is central to the topic of the Anthropocene, is that of the extent to which time depth is experienced and conceptualized and how far this influences action (Shryock and Smail 2011). What is considered "very far back in the past" and what is not in a community under study? What is "distant future" in relation to inheritance of land, and what is it in relation to risks in agriculture? Which time registers are used, and where are the respective time horizons in the pursuit of short—term goals, for instance, economic goals, or in the course of one's life, in genealogical topics, and even time registers in which landscape change is conceptualized?

Who thinks in terms of deep time or controls deep time processes, and how are the boundaries between time registers culturally established? While the recognition of the importance of deep time is indeed geologically factually based, talk of *deep time* is anything but apolitical, as is always the case with the globalization of time concepts, as global historical studies on modernity and particularly multiple modernities have shown (Conrad 2016). The geographer and professor of *in*humanity (!) Kathryn Yusoff notes that "(the) grammar of geology is foundational to establishing the extractive economies of subjective life and the Earth under colonialism and slavery" (Yusoff 2015). In general, we can ask: "Whose time is deep time?" (Irvine 2020: 5). Historically, the idea of geological deep time first had to be established in eighteenth and nineteenth century England, where it was not only about findings and interpretations, but also about institutions and power, as history of science studies have shown, for example, on James Hutton (1726–1796) and Adam Sedgwick (1785–1873) (Rudwick 2005; Hüpkes 2020b; Gamble 2021).

Therefore, this is less about foreign or non—Western concepts of time, but rather about the cultural handling of time depth (Irvine 2020: 3). This can involve questions about the respective rhetoric, about resource management, and about the relationship of the actors to the land beneath their feet. A central theme regarding the Anthropocene is the relationship between human rhythms (social time, cultural time) and rhythms in the non—human sphere. This is not only about the rhythms of pets and domesticated plants, but also about climatic rhythms and, for example, erosion rates. Is there a harmony of vibrations, or are there notoriously occurring frictions, such as asynchronies? Furthermore, the extent to which people mentally distance themselves from such dynamics or even deny them socially must be addressed. Accordingly, Irvine programmatically says: "The task for anthropology, I argue, is to analyse the conditions of this extraction from deep time and not to replicate it" (*disjuncture*, Irvine 2020: 5).

In relation to the Anthropocene, the interesting thing about the Nuer is that ecological time is not solely determined by social guidelines à la Durkheim. Durkheim argued that the world is only accessible through categories that have become social, and that all categories of the understanding of time are thus exclusively socially created (Gell 1992: 4). Using the Nuer as an example, Evans—Pritchard demonstrated that categories of time arise through interactions that do not exclusively represent human sociality. Elements of the environment, such as animals and landscape, play a major role in the concepts of time. Even Durkheim himself, however, makes reference to the tension between ecological determination and socially constructed categories of time (Irvine 2020: 25). Regarding deep time, it should be noted that environmental time among the Nuer is integrated into structural time, with the effect that the environment is rather seen as *emic* than static in a long—term perspective. The Nuer do see cyclical rhythms, but only the course of the year has a direction. Otherwise, the changes in nature are conceived emically as change within a constant abstract structure (Irvine 2020: 23).

A deep—time consciousness in cultural studies would mean that time is not only thought of in relation to humans (Ellsworth and Kruse 2013: 24). A cultural anthropology informed by both the matter and the concept of the Anthropocene would have to assume that deep time represents a reality that is initially independent of human existence. Humans are existentially integrated into long—term flows of matter and energy. This means that the existence of deep time is independent of the respective cultural representations (Gell 1992: 55, 141, Bloch 2012; Irvine 2020: 106). The deep—time entanglement of human ways of life limits the range of possible culture—specific time concepts. While many time concepts are conceivable in principle, only those that are also viable within the real material framework will endure.

However, such a naturalization of deep time does not mean that social and cultural factors play no role in the current effects of deep—time processes. Analytically, we should distinguish between the level of time consciousness, especially in terms of the past, present, and future ("A – Series," Gell 1992), and the level of human existence in time from earlier to later ("B – Series"). Culture—specific time consciousness is particularly important for the division of time and also for all types of time—related social boundaries (*demarcations*, Irvine 2020: 107).

The socio—political framework of the demarcation of time boundaries begins with the fact that there were historically specific circumstances that led nineteenth century England to deal with it and recognize it as a social and political problem. In this context, geology was institutionalized, geological chronologies were introduced, and quite practical geological mappings were established. For social systems and cultures that are significantly differently situated (sic!) than the regional context in which the term originated in England, this can seem like an imposition. Studies on the politicization of time in the early phase of globalization would point to this aspect, especially in the form of calendar time (Postill 2002; Ogle 2013). What role did the discovery of deep time play in the spread of industrialism and imperialism? If a decolonization of the Anthropocene is demanded (Todd 2015; Davis and Todd 2017; Taddei et al. 2022), one can consistently go further and ask to what extent the geological timescale as such was constituted colonially. With Irvine, we can critically ask: "Is the export of geological chronology another dimension of this (general) incorporation into 'global time'?" (Irvine 2020: 107; own addition).

A concept that fundamentally includes multiple time frames, but analytically distinguishes them, can prevent certain contemporary societies from being assigned their own time. With such a hierarchizing operation, they would be separated from the overall flow of human history. The segregation or "alteration" of other cultures (*othering*) as "timeless," "ahistorical," or "prehistoric" has been a classic problem of social evolutionary—based cultural anthropology since the nineteenth century. They are thus denied simultaneity (*coevalness*, Fabian 2014: 37–70). Similarly, however, it is also a tradition in cultural anthropology, as in other cultural and social sciences, for human cultures to be rationally "cast out" of geological time. As Irvine plausibly argues, from a deep—time perspective, placing societies *outside of geological time* is also a hierarchical approach (Irvine 2020: 126).

6.6.2 Political Geology—The Long Present

Deep time also continues into the future and can lead to a call for a "political geology" (Bobbette and Donovan 2018; Bobbette 2023 on Java, Indonesia). Deep time is mostly addressed as a time of the past. In the "face of the Anthropocene" (Angus 2016; title), however, it is of central importance to look deep into the future, and further than just to the next generation. The accumulation of human activity, for example, leads to large parts of the Earth's surface being sealed: they are being covered with asphalt or concrete (Højrup and Swanson 2018). From a geomorphological perspective, the constant change in the Earth's surface is the norm.

Humans *slow down* landscape development because the natural flow of matter, such as through erosion, denudation, or sedimentation, is restricted by asphalt or concrete (Zalasiewicz et al. 2014b). In geological terms, concreting is an intervention that fixes the landscape, while anthroturbation, which far exceeds bioturbation, agitates it. From a deep—time perspective, the short biographical time is thereby oktroyed on the planetary history of the geosphere through a special form of presentism. The current situation is thus quasi "fixed" in Earth history (Irvine 2020: 139). However, it remains open as to how long it will be perpetuated in this way.

From an anthropological point of view, it is therefore of interest how the depth of time—or how deep time—is perceived socially and culturally and, above all, at what depth of time one's own environment is perceived. What we need is to examine the *relationship* between different *simultaneously* existing materials temporalities *and* time perspectives. From an anthropological point of view, it is particularly interesting that it is not only people in Western cultures who usually ignore a deep—time perspective or even explicitly close themselves off or resist it. People in other life situations can also resist it, sometimes for very specific local reasons. A short—term—oriented attitude can even be a method of distancing oneself from the concept of *deep time*, which is perceived as Western or colonialist and comes from England.

This concerns resource extraction in today's Global South or in postcolonial settings. European settlers brought geology into the country and with it deep—time concepts, and they imported methods of geological mapping on which resource extraction is based. The extraction of raw materials can be interpreted as "mineral presentism." Geological time is "contracted," so to speak, by employing deep—time generated resources practically as a service, for use in the shallow time horizon of the present (Irvine 2020: 128).

With the present exploitation of raw materials in parts of the Global South, various time concepts are consequently intertwined, as Povinelli shows using the example of Australia (Povinelli 2016). The import of deep time clashes with culture—specific concepts of materiality and time, and also specifically, to some extent, with concepts of deep time. Elizabeth Povinelli concretely illustrates this using the example of the updating of fossils by indigenous Australians:

> A fossil, a bone, a set of living now recently deceased people – for my old friends, all are in the same time and same space of signifying material mutuality (Povinelli 2016: 69).

A privileging of deep time would insist on seeing the "fossils" as part of the past. This, however, would remove them from the Aborigines' field of action. This raises questions as to what extent the Aborigines are not only structurally and spatially, but also conceptually displaced in this process of resource extraction. The temporal entanglements and possible time conflicts in this local case of resource extraction can stand here for a microcosm of global resource extraction, a central marker of the Anthropocene as such.

All this calls for a new, literally deepened, perception of temporalities in *field settings* in cultural anthropology. Field research sites can indeed be seen as a manifestation of processes that have evolved over long geological periods, with cultural life representing just *a part* of these processes. This can counter an exaggerated presentism in cultural anthropology, which—unlike archaeology—tends to underestimate the long—term development of the material environmental basis of actions. The (materialist) thesis of the geological foundation of human cultures in deep time therefore by no means implies that social and cultural conditions should be neglected. But it does mean limiting the Durkheimian privileging of social construction, especially the thesis that time is exclusively or primarily created through a collective representation of time.

> It is in ontains or these entangled, contested, and often distorted relationships between the biographical and the geological, *rather than attempting to transcend them*, that we might come to an understanding of what it means to be coeval in deep time (Irvine 2020: 128, emphasis mine).

6.7 Planetarity—Scale Clashes and Two Sides of Human Agency

> *It is a world of our making, but not of our choice.*
> George Monbiot (2014)

> *If we recognize the Earth materially, epistemologically, and ethically as a planet and explain human (co –)existence through it, this has consequences for the way we think about society in general.*
> Claus Leggewie & Frederic Hanusch (2020: N4)

Until recently, humans lived in the "normal" of hunting and gathering small groups that tended to be egalitarian and rather stationary. This way of life in small communities for almost the entire history of mankind has shaped our genomes and cultures. Life as small group herd animals shaped our psychological abilities and cognitive and emotional limitations. We are descendants of creatures that were evolutionarily successful as meso-scale orientated beings. Humans are ontains towards local worlds, recent perception and proximate cause exploration. Humans are good at perceiving rapid changes in medium-sized objects in our immediate environment and we think in short time frames (Jamieson 2014; *cognitive myopia*, Weber 2017).

The Anthropocene, on the other hand, confronts us with long-term change, networked trends and spatially widely to globally distributed objects. Many phenomena prove to be "hyperobjects" that are difficult for individuals to perceive, cannot be grasped intuitively and are therefore difficult to think about (Morton 2013: 1). The problem is that we are dealing with barely conceivable phenomena and we have to think about connections "… that are not—or not without contradiction—the subject of our experience (Horn 2016: 87, Hartung 2022: 18, Hübner 2022). We find it difficult to ontain the effects of our actions on people living far away or on people of later generations (Meneganzin et al. 2020:12–14). These individual limitations and many other social and political limitations at a collective level prevent effective action, as is particularly evident in the ignorance and inactivity with regard to climate change (Jamieson 2014: 61–104). To put it bluntly, despite all the individual activities, all this leads to a *societal* denial of the Anthropocene.

All members of such small-scale societies had approximately the same non—genetic information. In evolutionary terms, humans have been living in the aforementioned "abnormal" of a highly populated world with global communication and supra—local migration for a very short time. We live in "ultrasocial" large—scale societies with strong inequality and cultural diversity. And we all live on an earth's surface that is being rapidly changed by humans and their technologies. Unlike storms and earthquakes, we do not see many processes of destruction and disintegration. Species loss and climate change are sudden and rapid catastrophes in geological terms, but gradual on a human time scale: "disasters without an event". Today, we need technological aids to ontains key aspects of man—made change, such as the increase in greenhouse gases. There is a mismatch between evolutionary psychological tendencies and the challenges of the current situation ("mismatch"; Ehrlich and Ehrlich 2022: 778–780; cf. Ehrlich and Ehrlich 2008).

The anthropologist Thomas Hylland Eriksen notes that in the face of the Anthropocene, the clash of scales (Eriksen 2016: 132) is more culturally relevant than Huntington's clash of cultures. The attention to different, especially large scales in space and time mainly comes from geology, paleontology, evolutionary research, and the directions of *deep history* and *Big History*. What is the benefit of no longer distinguishing between history and Earth history, of allowing world history to merge into Earth history, as Chakrabarty proposed (Bänzinger 2019)? What orientation can Earth history as a science of the planet bring, and what additional insight does historical knowledge offer when it is understood as knowledge about the history of the Earth system? These questions arise because there is no lack of knowledge about environmental damage, but rather of unifying concepts and the common political will to decide on the consequences to be drawn from it (Neckel 2020: 165).

Scales are central to the topic. Ignoring them can lead to hasty conclusions, for which I will give an example using cultivated landscapes. From a macro perspective, the world is characterized by anthromes (Ellis and Ramankutty 2008; Ellis et al. 2020); on a smaller scale, however, things look different. One might assume that agriculture is leading to a significant decrease in landscape diversity worldwide because almost all areas with vegetation are used for agriculture in some way. In large parts of the world, natural landscapes are quantitatively and qualitatively being transformed into cultivated landscapes, mainly through deforestation, drainage, irrigation, and terracing, which leads to reduced diversity of life and cultures (Maffi 2001; Blackbourn 2006; Bätzing 2020: 37, 198, Bätzing 2023). However, this does not simply lead to a leveling out of differences. First of all, some areas, such as parts of mountains, volcanoes, karst areas, and peatlands, are resistant to almost all forms of use and therefore continue to be elements of the cultivated landscape. More importantly, however, the diversity of the forms of use of the cultivated landscape reflects the mosaic of the natural landscape:

> The countless natural differences between wetter and drier, steeper and flatter, deep and shallow, sun – exposed and shady locations are usually not leveled because the amount of work involved would be too great. Instead, these areas are used differently, so that cultivated landscapes consist of a mosaic of different forms of use (Bätzing 2020: 37).

6.7.1 Scales, Scalings, and "Battlefields of Knowledge"

Different scales lead to different perspectives. During the last interglacial phase about 125,000 years ago, the atmosphere was slightly warmer than it is today, and the sea level was about five meters higher than today's. From a geological perspective, this is a comparatively low value when you consider that the global sea level during the last ice age was about 160 meters higher than today. If the sea level rises slowly, from a human perspective, to five meters higher than today, which is sudden from a geological perspective, this will have massive consequences for human communities, especially for those living near coasts and deltas (Edgworth 2018a, b, c). As any look at a world population map will show, humanity largely lives in fertile, low—lying flatlands and along coastal areas.

This is most evident in the beautiful composite images of the world at night, where the cities of the world light up. This is the result of the human response to the slow adaptation of the coasts to the hitherto comparatively stable climate in the Holocene. In the next century or two, many large cities will have to contend with flooding. This is already evident today in large coastal metropolises that are simultaneously struggling with land subsidence. The clearest example is probably the Indonesian capital of Jakarta on the island of Java, which is aptly referred to as the "city of the Anthropocene" (Chandler 2017) mainly due to its combined anthropogenic environmental problems, and which, according to current plans, will be relocated as the capital to the province of Kalimantan on Borneo, that is, to another island, in a long process beginning in 2024. It is crucial that the Anthropocene is perceived and responded to by humans locally, mostly in the form of reproduction crises (for example, Simarmata et al. 2020 for Jakarta).

The extinction of species, one of the clearest signals of the Anthropocene, is an example of the problematic nature of the Anthropocene's spatial and temporal scales, which are difficult for humans to grasp. When a feared current or imminent extinction of many species is referred to as the "Sixth Great Extinction" (Kolbert 2014, see Martin and Wright 1967; Barnosky et al. 2012), it is thereby considered to be one of the major mass extinction events of the entire history of the Earth. During these previous geological crises, not just 75–95% of the species died out, but a large part of all animal and plant families. The previous five extinctions thus represent the peaks of extinction before the usual background rates (Wignall 2019: 19–20). Only with a geohistorical classification does it become clear that these extinction crises represent something completely different than a change in the environment that causes some species to go extinct. There are only a few individual species capable of producing such Earth system effects. From a geological perspective, it is only humans… and the aforementioned cyanobacteria. Edward O. Wilson, the doyen of biodiversity research, has pointed out that the current extinction is primarily affecting large fauna, and he sarcastically named the Anthropocene the "Eremocene" or the "Eremozoic," the period in which humans become lonely (Wilson 2016: 27). The term "sixth great extinction" makes the drastic effects of humans in the biosphere clear, but it can also lead to people seeing the current species extinction as another "natural" mass extinction, and thus assuming that the living world would recover afterwards. This would be forgetting that the current species extinction is slow from a human perspective, but conceived in geological terms, unlike the five preceding defaunations, it is anything but a *slow—motion catastrophe* (Sepkoski 2021: 227, 298–301, Hannah 2021: 193–194;).

However, the current truly catastrophic loss of species among large terrestrial vertebrates with small populations should not lead to an exaggeration of the overall extinction of species. In the six extinction events to date, many plankton populations have also collapsed, thus breaking down the oceanic food chain, and these extinctions affected almost the entire spectrum of habitats. The current biodiversity crisis has not (yet) shown any of the characteristics of the five major mass extinction events. The extinction rates are lower, and there is still no extinction crisis among marine invertebrates and in the deep sea (Wignall 2019: 18–19). In descriptive terms, different levels of scales, not only with regard to the loss of biodiversity, can be distinguished, all of which can be discerned from the grain of sand up to the entire planet (Eriksen 2016: 28–29), and all imply political—ecological tensions (see Marston et al. 2005; Jessop et al. 2008).

The first (1) level is about the physical extent of a human collective and its infrastructure. Here, for example, small societies can be distinguished from populous societies, and spatially extensive societies from small ones. Spatially local, regional, and global units and processes can be distinguished as well as temporally short, medium, and long—term developments. When considering the Anthropocene and the role of humans in the long—term change of the geosphere, it is often about a combination of population size or spatial extent and social or cultural complexity, such as the role of colonial empires in the transportation of species in the early Anthropocene.

Therefore, we can (2) distinguish a social scale or a complexity scale. Tendentially, populous societies are more complex than societies with a small population, for example, in terms of the number of professions or status positions. A spatially or demographically larger society requires more integration and coordination due to the larger number of actors. This is considered in the classic contrast between "community" and "society" and is a central theme of neo—evolutionary theory. Collectives of equal population size can, however, be differently complex.

On a third level, the time scale is particularly relevant in the context of the Anthropocene (3). Regarding the past and the future, we are talking about geologically long periods of time, deep time. As humans, however, we have a daily and a biographical time frame, even if the horizon of the future is always present. Likewise, the political and economic worlds tend to be rather short—term—oriented. However, there are also clear exceptions, such as state investment in infrastructure, the development of a drug, investment in a raw material mine, or the framework of environmental movements that advocate for sustainability. Such long—term processes are difficult to represent in economic concepts.

Fourthly (4), it is about the mental representation of the three scales. Do the individual actors perceive their community, society, or their social networks as small or large, as simple or complex, as autonomous or dependent? What is the general cultural, that is, supra—individual shared image (*emic view*) of their own collective? The image of the social scales can deviate from the real scales. A society can be perceived or portrayed as less interconnected with larger systems than it actually is. A clear example is the strong networking of actors and thus high mutual accessibility worldwide, as demonstrated by s*mall world studies,* which is significantly underestimated by almost all respondents. On the other hand, in communities that practice little division of labor and are only embedded in small networks, an awareness of previous networking and long history and current globality can still be widespread, as Fredrik Barth demonstrated (Eriksen 2015).

In relation to the Anthropocene, different ideas about the size, unity, and diversity of the Earth and humanity play a role, especially when it comes to measures. Is humanity perceived only as the sum of individual persons or as a unit? The perception of time is also crucial both for the recognition and diagnosis of the Anthropocene and for political responses. What is the temporal horizon that one imagines back into the past or forward into the future when making decisions? All of this has political implications.

The importance of conceptual ideas about scales in the Anthropocene is demonstrated by the fact that proposed solutions to the crisis of the Anthropocene typically call for upscaling or downscaling. Most of the time, *scaling up* and *scaling down* are proposed simultaneously, in accordance with the motto "act locally, think globally" (Eriksen 2016: 29, Clark and Szerszynski 2021: 74–76, 85–87). The physical or physiological possibilities of scaling up or down certain solutions are often misjudged because they are not intuitive (West 2017; Sommer 2021, and Jon 2021: 8–32 using the example of cities). We are supposed to or want to think about the whole world, about all of humanity and futures long after our death and strive for worldwide political coordination. On the other hand, we want or are supposed to scale down socially and spatially: produce locally, consume regionally, post—tourism: "small is beautiful." Between these levels of scale, tensions, frictions, and clashes are notoriously occurring in the Anthropocene (*clash of scales*, Eriksen 2016: 29, 131–156).

These problems exacerbate dilemmas that were already at the center of the globalization debate. Large—scale and long—term phenomena are typically standardized, while the local level is specific, often unique. The demands of proponents for upscaling are opposed by the demands of globalization critics for downscaling. Within the debates on globalization, this became particularly clear in the question of the relevance of local knowledge for development (Antweiler 1988). Representatives of different forms and regimes of knowledge—on the one hand, local such as indigenous knowledge, and on the other hand, global, external, or universal knowledge—fight for consideration.

Similarly in debates on the Anthropocene, local and thus context—rich experiential knowledge meets abstract and thus scalable "expert knowledge." Even within these poles, there are tensions between different directions, for example, between scientific and cultural knowledge. In the "battlefields of knowledge" (Long and Long 1992), knowledge regimes of different scales are used to substantiate claims or to legitimize certain policies. In the Anthropocene, these conflicts of scale are much more extensive and pronounced than in development and globalization issues. This is one of the theoretical challenges, namely to *analytically* distinguish the dynamics of the Anthropocene from globalization processes, although they are closely linked to them.

Scales also open up an approach to the formation of a unified consciousness of humanity, and here psychologically effective universal metaphors such as "home" can be of significance. From the perspective of evolutionary biology and deep history, "kinship" is probably the metaphor with the greatest mobilization potential, although it is actually a concept for small social units, or one that is central to non—modern societies. Contrary to earlier models of social evolution, however, large—scale units do not historically replace the earlier smaller ones, but complement them. Thus, the cultural invention of "kinship" and "family" as concepts is not just relevant in small social systems or on a small scale. No, it is also of central importance in complex large—scale societies, for example, to motivate national allegiance.

Kinship can be found as a conceptual domain in all present and in historically known human societies. Kinship functions as a motivating and solidarity—generating idea. Kinship is also an ecologically consequential idea, as has been shown historically in its role in the formation of large political entities. In the exploration of the history of discourses, it manifests itself as a "genealogical" approach. Kinship is a concept that promotes resonance and social cohesion (Antweiler 2021a). The special Anthropocene potential of the concept lies in the fact that kinship can be scaled down or up. The concept of kinship allows for the inclusion of non—human actors in considerations of law and justice, as in Haraway's call for humans to become kin with animals or to "make kin" (Haraway 2016; Wolfstone 2019; Mattheis 2022, making *kin*, see 4.7). However, "kinship" also allows for metaphorical extensions that Haraway does not address, but which may be even more important for a response to the Anthropocene (*deep kinship*, Trautmann et al. 2011: 160–162). At various levels, we can create close connections through gifts, sharing, marriage, and hospitality: "kinshipping." Spatially, kinship also allows for extensions that go beyond the neighborhood, the region, and the nation to concepts of a "human family" (*scalar play*, Stiner et al. 2011: 253, 267–268). The guiding model behind this does not necessarily have to be the bourgeois family.

6.7.2 Planetarity in Space and Time

The idea of the Anthropocene emphasizes a large space with the geosphere, but it is primarily an idea related to time and concepts of time, such as time periods (sic!). As an epochal term, the Anthropocene is a deep—time alternative to terms like modernity, late modernity, postmodernity, and multiple modernity. All of these are primarily period concepts, even if they sometimes go beyond that, especially the concept of postmodernism, which undermines thinking in epochs. Bruno Latour sees the Anthropocene primarily as an alternative to the talk of modernity as an epochal concept and modernity as a characterization of societies; he positions the Anthropocene as a *quasi*—stratigraphic marker cheekily alongside the *golden spikes* of stratigraphy:

> (…) what makes the Anthropocene a clearly detectable golden spike way beyond the boundary of stratigraphy is that it is the most decisive philosophical, religious, anthropological and, as we shall see, political concept yet produced as an alternative to the very notions of 'modern' and 'modernity' (Latour 2010: 77).

In the idea of the Anthropocene, geological time horizons and historical time dimensions are interconnected. Both temporalities, the geological ("vertical") and the historical ("horizontal"), come together. Because of the significantly extended temporal depth compared to historical and also prehistoric time scales, this is often referred to as "deep time." Historically, deep time was a discovery of geology in the nineteenth century (Gould 1987: 13–38).

It was the discovery and, above all, the correct interpretation of fossils that led scholars such as Nicolas Steno (1638–1686), Robert Hooke (1635–1703), Georges—Louis Leclerc Compte de Buffon, James Hutton, and Giovanni Arduino (1714–1795) to reject the biblical short chronology of the Earth's history of 6000 years. These central figures in the discovery of geological deep time were natural philosophers who investigated physical, physiological, and philosophical questions in equal measure, which also applies to Darwin and Alexander von Humboldt. The aforementioned divide between the natural sciences and other sciences has only existed since the 19th century. The study of deep time was only scientifically established with the emergence of geology, paleontology, and the establishment of Darwin's theory of evolution (Gould 1987: 15, Davies 2016: 15–40).

The Anthropocene has raised the question of whether geological time scales and other orders of time can be reconciled. The classic concept of scale assumes that a uniform scale can be established from the micro level right through to the macro level: second, minute, hour, day, week, etc. Small sections are sub—units of larger units, which is similar to the principle of the Chinese box, the Matryoshka doll, or also the Windows principle (*nested scales*, Thomas et al. 2020: 8). This is the concept of consistent proportional equivalence, which is also suggested by cosmology and the Earth sciences. The geological time scale (*GTS*) is one such scale, a pragmatic way of dividing geological time into smaller units. Even if the units at a given level, such as epochs, are of different lengths and represent different degrees of upheaval, epochs within the Quaternary, the period in which we live, for instance, are comparable to epochs as sub—units in the Devonian period.

Such a common standard of proportionality is intended to enable communication between very different sections of the phenomenal world and to use different informal and formal measuring instruments for this purpose. This concept of scale is suitable for making comparisons as well as finding patterns and correlations. It allows causal relationships to be traced and syntheses to be determined. In geological stratigraphy, the layers are distinguished from each other by means of various phenomena, such as index fossils, magnetic or chemical traces, physical markers in ice cores, or other "proxy data." The dating methods used are relative time determination, for example, by means of stratigraphic layering or tree—ring dating, or absolute dating methods, for example, by means of radiocarbon dating. The problems that arise in this context are similar to those that historians have with periodization.

Proponents of the idea of the Anthropocene usually automatically assume that such a unified concept of scale for the Anthropocene is sensible. Schwägerl emphasizes that the Anthropocene effortlessly combines short—term, even ephemeral, processes with long—term geological change in an integrated time scheme. Thus, the Anthropocene does not only overcome the dualism of nature and humanity and connect all disciplines, but also the time scales, which he encapsulates in the term "neurogeological" (Schwägerl 2014: 30). Regarding historical change, concepts have developed that emphasize the unity of historical change, such as "geohistory" (Chakrabarty 2012, 2018, 2019), an approach that links natural history with cultural history.

Some go even further in terms of the intended synthesis scope and temporal range. The most extensive implementation of this scale concept can be found in the concepts of *Big History*. There, the temporal scale is maintained from the here—and—now to the cosmic dimensions of the *Big Bang*. The approaches of *Big History* and *deep history* connect human history not only with prehistory and early history, but also with the history of the Earth's and trace it back to cosmic history (Christian 2018; Shryock et al. 2011; Spier 2010). In time scales on the Internet, you can find temporal data at the macro level at will or go to micro levels at any level and view details there or zoom back to the macro levels (*Big History Project*).

Tsing, on the other hand, argues that such a time scale, which she refers to as *precision—nested scaling*, (Tsing 2012a; Tsing et al. 2024: 54–58), does not make sense for the Anthropocene. It is hardly possible to develop a common scale when it comes to ethical values, art, fashion, or rituals. Such phenomena "resist" being included within a time scale. This also applies spatially. Local life—worlds cannot be mass—produced without difficulty (Eriksen 2016: 151). If this is attempted, it inevitably ends in a simplification beyond recognition or in the supposed universalization being thwarted from below or from within (localization, glocalization). Tsing thus brings in an argument known from the philosophy of knowledge. She argues that one should rather develop an archipelago of interrelated but inconsistent formulas for objects, experiences, meanings, and material effects of human action. This implies working within different frameworks, where each of these frameworks brings not only different quantities, but also *incommensurable* differences of type. This fits with Tsing's concept of friction between experiences, values, and perspectives that differ in scale and quality (*friction*, Tsing 2005). Working with sprawling scales (*sprawling scales*) opens up the possibility that the image of reality that emerges from one time frame is different from the reality that emerges from other time scales. Thomas et al. capture this in a beautiful image such as this: "It's like opening up a Russian doll to find a tiny golden frag inside who spits a helicopter in our faces" (Thomas et al. 2020: 9).

Anthropologists have already made similar arguments outside of the Anthropocene debate. The most prominent position here is probably Marilyn Strathern, who has repeatedly argued that phenomena that are discussed under one term in cultural anthropology should not be regarded as one phenomenon, but as a family of only partially similar phenomena. The best—known example is initiation, where massive differences are shown (Strathern 2020). Strathern has most recently clarified this in the above—mentioned book on the phenomenon of relationships (Strathern 2020). Similar considerations can be found in the concept of family resemblance according to Wittgenstein, where similar phenomena, such as prototypical games, show the same characteristics, but no single member exhibits all characteristics. Comparable approaches can also be found in empirical work, especially in those from medical cultural anthropology and from medicine itself, where families of similar symptoms are grouped under "syndrome."

I consider scales as tools, as a method to see patterns in the diversity and to create order. Scales open up certain questions and close off others, while different scales open up different qualities. Scales are heuristic tools and follow pragmatic restrictions and are constantly to be revised, and the scale should be changed again and again. An early publication that, by means of photos taken from a porch in Manhattan, gradually zoomed into the cosmos and back, shows the eye—opening effect that a change of scale can have (Morrison and Morrison 1992). However, in my opinion, Tsing throws out the baby with the bathwater when she dismisses nested scales as generally unhelpful for questions of the Anthropocene. The reasoning seems to me to be strongly guided by the motive of differentiating the humanities and cultural sciences from the geosciences or, more generally, the natural sciences (see Sect. 5.5).

As Thomas et al. point out, the clash of different scales is a central obstacle to multidisciplinary research (Thomas et al. 2020: 11). But with the reference that the authors then make to Strathern's example of initiation, the question of time scales is all too quickly equated with the problem of anthropological category formation or any formation of general terms. The fact that larger scales obscure the view of small worlds is trivial. From an Anthropocene perspective and for cultural anthropology, it is less trivial that, conversely, small scales obscure larger contexts. It is also about more than just scales, namely about causality and path dependency.

The Anthropocene primarily concerns questions of life and survival on a wide range of spatial and temporal levels. The choice of respective scales implies epistemological, but also political and ethical consequences, as is particularly evident in examples from Africa (Hecht 2018). Even beyond ethical and political settings, the level of threat is extremely dependent on the chosen time scale. From a geological perspective, the Anthropocene may hardly be a threat because *Homo sapiens*, like every species, will eventually become extinct, just as our planet will disappear in an expanding sun in at most five billion years. However, when it comes to the next thousand years, things look very different. Humanity will predominantly be living in cooler regions with less space than today' and will probably suffer from massive pollution and environmental conflicts.

Based on current trends, the geographical niche will change more in the next 50 years, that is, by 2070, than it has in the last 8000 years. An extremely dystopian forecast gives humanity a 50–50 chance of surviving until the year 2100 (Rees, 2003, according to Thomas et al. 2020: 12). That would be 80 years from now. According to *business—as—usual* scenarios from risk research, it could come to the point where 3.5 billion people will have to live in areas that are too warm for them. It is not only humanity, but also such inequalities that are now being revealed by the geological view of human history.

The Anthropocene challenges historians, for instance, historians of technology, to supplement conventional linear temporalities with non—linear concepts of time and to additionally locate history on the vertical time domain, that is, on the geological time scale. Even within a conventional concept of scale, the problem of scalability arises (West 2017).

6.7.3 *Geology of the Household and Geological* Agency

Stratigraphers, geologists, and evolutionary biologists are concerned not only with slow, continuous change but also with discontinuous change, abrupt change, and planetary crises. Crises in the history of life are one of the central research areas in paleontology and biological evolutionary research. The more recent history of technology accordingly demands that historians of technology also conceive "layer models of historical temporality" (Trischler and Will 2020: 240).

The Anthropocene is both a period concept and, secondarily, a spatial concept. Temporally, it is a geologic—stratigraphic extended view of a *longue durée* approaching the present. This concept of time is not only relevant in terms of natural history or the history of civilization, but also has immediate consequences for questions of evaluation. Claims regarding boundaries and limitations must always be contextualized and historicized. For example, actions that are considered positive today, because they promote jobs and growth, may suddenly appear problematic from the long—term perspective of the Anthropocene. What is only controversial in the short term appears to be criminal, as murder or suicide, in the long term. Nixon named the example of the disastrous effects of unbridled growth on the global sphere as "slow violence" (Nixon 2011; Folkers 2020).

In response to Chakrabarty's call to do justice to environmental change and deep time on a global scale, there is also a search for possibilities of representation. Mathews then brings together the macro—historian Fernand Braudel, John McPhee, a science fiction author, and the ethnologist Evan Evans—Pritchard as positive examples of how to move deftly across the time and space scales of human history and natural history (Mathews 2020: 71). So far, I have mainly highlighted the questions that a cultural—cum—planetary perspective raises for cultural social and ecological issues. Is it conceivable, for example, that people "... in their thinking, feeling, and acting could be capable of disregarding the human in order to understand themselves instead as a planetary force among others" (Neckel 2020: 165)?

The problems of the planet that humanity is becoming aware of in the Anthropocene can also conceptually revitalize geology itself. In a profound "planetary social thinking," our planet can now be considered in a broader field of time scales, future visions, and trajectories than was previously customary in geology. The crisis—like nature of the Anthropocene raises the question of the extent to which sudden changes are part of the normality of the geosphere. Gradualism and the principle of actualism in geology are once again and more intensely being questioned in the face of a socially and culturally created catastrophe.

Analogous to cultural diversity, a concept of "planetary multiplicity" (Clark and Szerszynski 2021: 3, 8) would ask what capacities the Earth has—at all scale levels—to become different from what it is, to differentiate itself. An approach that fully considers the interrelationships between humans and the geosphere combines planetary multiplicity with earthly diversity (*earthly multitudes*, Clark and Szerszynski 2021: 9). Earthly diversity consists in the various ways in which human collectives respond to this planetary multiplicity, that is, how the planet—more precisely, of course, only its surface—is imagined, what is known about it, and how it is dealt with. In doing so, humans must deal with the specific materiality of the Earth; indeed, they must, metaphorically speaking, "think" dynamically like the Earth (Yusoff 2013: 780, Clark and Szerszynski 2021: 10), and humans as well as cultures will use their experiences with the Earth to do so:

> Viewed both as a species and as a heterogeneous ensemble of lineages and collectives, humankind has a vast amount of experience of living with and through the *dynamism* of the Earth (Clark and Szerszynski 2021: 10, emphasis mine).

> In this—distanced—view, the way humans interact is not only to be understood as an imprint on the Earth or as an inscription of human action in landscapes, but as an active interaction of human and geological forces from different places of the geosphere, whereby humans produce themselves and the landscape. Geological *agency* is not simply the manifestation of human agency, but a co—production of effects together with geological forces. The authors encapsulate this well in one of their less personalizing formulations of the Earth: "(…) modes of human creativity voiced through the materials of the Earth and forms of planetary generativity embodied in the medium of human activity" (Clark and Szerszynski 2021: 74).

In relation to the experiential dimensions of everyday life, we need a kind of domestic geology (*domestic geology*: 66–71). Clark & Szerszynski demonstrate this using the example of the iron, whereby the geology—related aspects of gender and social inequality also become exemplarily clear (Clark and Szerszynski 2021: 55–74). Braudel illustrates similar connections using the example of the loom (Braudel 1981: 337). Thus, the often—critical social science postulate that the Anthropocene should be "socialized" (for example, Malm 2016) must be supplemented by a complementary call for a "geologization of the social" (Clark and Szerszynski 2021: 11, 46–49, 54). Such geologisation is politically relevant. It calls for different temporalities relevant to action to be related to each other:

> While we desperately need to build a sustainable culture (culture of long-termism), we are in a situation that requires immediate action, in which everything drives us to act in the short term (short-termism). Therein lies the paradox (Read and Alexander 2019: 120).

6.7.4 Power of Action vs. Effects of Action—Two Sides of Human Agency

A central challenge for cultural anthropology and the human sciences as anthropology in the broadest sense concerns agency and thus the concept of action. In the Anthropocene, there is a divergence between actions and their effects (Horn and Bergthaller 2022: 99). Actions are local and usually follow an intention. However, their consequences are often very different, and the effects are usually in many places and we do not know to what extent causes and cultural consequences are also linked over great distances (*telecoupling*), such as in the case of biodiversity (Carrasco et al. 2017; Liu et al. 2018). Moreover, the consequences are largely unintended ("tragedy" in the narrow sense of the word, Horn and Bergthaller 2022: 106). This disjunction of action and effect (Horn and Bergthaller 2022: 99) is anthropologically relevant and could constitute a focus of research.

There is often generalized talk of agency or of problems that would exceed human agency. In contrast, it is essential to distinguish between two forms of agency: Is it (a) agency as intentional action, intentional behavior or (b) agency through non—intentional behavior or unwanted side effects through accumulated intentional actions (Horn and Bergthaller 2022: 100 ff., 105, 222). The first form of agency is related to the idea of culture as a “second nature”. Culture has an adaptive function according to a materialistic view of humanity. It is intended to tame imponderables that stem from the first nature, to compensate for risks, and generally to ontains if not eliminate the contingency of life. In today’s societies in particular, the ultimate goal is to ontains control over first nature, primarily through knowledge and technology:

> The chance of genetic origin is to be eliminated through a targeted selection of genetic material, the risk of disease is to be reduced through this same pre-selection and targeted diagnosis, the process of ageing is to be halted and even unpredictable death, the greatest threat from the point of view of modern man, is to be made controllable. (Hartung 2022: 15).

With increasingly artificial, technologically ontains nature, there would ultimately be a decoupling from the first nature. However, this ontains of the difference between the two natures must remain an illusion, because fundamental living conditions are still not controllable or controllable by humans, such as earthquakes, volcanism and the main Earth-systemic dimensions of the climate, such as the inclination of the Earth’s axis relative to the orbit around the sun. Contrary to transhumanist visions, we as organisms are irrevocably dependent on the first nature.

In risk societies, particularly in the Anthropocene, however, there is much that can no longer be controlled, especially the unintended effects of people’s actions. This second face of agency, the increasingly escalating agency of humans, puts the assumption of the sovereignty of individual and collective actors to the test. Societies can influence the system, but not control it. Humanity steps back as the principal, nature steps “into action”: “People have intervened deeply in the planet and have become a geological force themselves, but they find no way out of the environmental destruction they have caused” (Leggewie and Hanusch 2020: N4). Following Conolly, the group of authors around Gísli Pálsson summarizes the problem as follows:

> After all, there are good grounds for speaking of *distributive agency*, which emphasizes that the Anthropocene is not the result of *Homo sapiens* acting in isolation, but is only made possible through a diverse network of technological, cultural, organic, and geological entities (Pálsson et al. 2013: 7, emphasis in the original).

Among the many actors around the Anthropocene, it makes sense to distinguish “formative actors” from other actors, as Dryzek does, and to examine their *formative agency* accordingly (Dryzek and Pickering 2019: 105). Formative actors can be considered those who determine which basic principles are applied for the diagnosis of a case or for the solution of a problem. Such formative actors are involved in small cases, such as the establishment of a small sluice gate, or in large questions, such as the handling of the term “global justice.” They define the form in which sustainability, justice, or similar general concepts should be understood in practice. The main function of formative actors is to offer or politically establish meanings. This must be distinguished from other actors who apply the principles.

A relevant example for the Anthropocene involves the central actors in the field of biodiversity and the related concept of ecosystem services (Dryzek and Pickering 2019: 106–109). Formative actors also exist among scientists, particularly natural scientists with regard to the Anthropocene. An analytical distinction can be made here between those who see themselves as "policy – relevant" and others who want to be "policy – makers." The former see themselves as experts and the latter more as policy advisors ("mapmakers" vs. "navigators," Dryzek and Pickering 2019: 121). Both types of actors play an important role in the human shaping of econiches.

Chapter 7
Human Niche Construction and Niche Heritage- Building Blocks for Synthesis

Abstract This chapter brings together what the culturalization of the planet and the geologization of culture mean for societies and for us as individuals. In view of the diverse debates, only building blocks can be provided here, which will not be enough to produce a conceptually inhabitable building. Hence, the controversial discussion about scales is combined with concepts of world citizenship in the Anthropocene. I have then formulated the guiding question as a cosmopolitan-informed question of world ecology: *How can intensively networked cultures peacefully coexist on a planet that is heavily influenced by humans but limited in size, without all having to become indistinguishable?* Using Southeast Asia as an example, I finally show what a "provincialization" and a localization of the Anthropocene could look like on site.

> *If we consider ourselves and our societies as integral parts of the Earth system, and we take serious the new properties that humans bring to the Earth system, then this requires a new kind of Earth System Science.*
>
> Tim Lenton (2016: 123)

> *... we remain an overwhelmingly fossil-fueled civilization that has been recently running vigorously into fossil carbon rather than moving away from it.*
>
> Smil (2021: 273, emphasis orig.)

Humans actively and permanently change their environment. The environmental changes in one generation become partly permanent and form the basis of the supposedly "natural" living conditions of the next generation t + 1. From the perspective of evolutionary biology, they even determine the conditions of natural (sic!) selection. This is the core of niche construction theory, which for humans, in contrast to other animals, is of fundamental importance for long-term change (*human niche construction*, Laland et al. 2000; Laland and Chiu 2020). In view of current commercial interests and dynamics, there will be no rapid decarbonisation, for example (Smil 2021: 289). This means that, in addition to genetic and cultural heritage, future generations will find environmental conditions that have been

C. Antweiler, *Anthropology in the Anthropocene*, Anthropocene – Humanities and Social Sciences, https://doi.org/10.1007/978-3-031-74591-1_7

significantly shaped by their ancestors, i.e. the lifestyles of people living today: Environmental conditions as an Anthropocene niche inheritance (Meneganzin et al. 2020: 10).

Such niche construction has had very diverse effects in the history of humankind, on humans as well as on other organisms and the inanimate world. Broad anthropogenic changes in biodiversity, for example, testify archaeologically to human expansion in the late Pleistocene, the Neolithic spread of agriculture, the era of island colonization in the Mediterranean and the Pacific, as well as the emergence of early urbanized societies and commercial networks (Boivin et al. 2016: 6389–6393). As recent biological and archaeological data show, the long-term evolutionary effects of human action go beyond the extinction of species. A central reason is that humans are regulators in food pyramids through their subsistence activities and cause the decline of keystone species. Human ontains, such as habitat modification, selective harvest pressure, hunting, relocation of organisms, domestication and genetic modification, has had and continues to have a selective effect on the morphology of plants and animals. This can be seen, for example, in the form of a decrease in body size, changes in wing shape, altered proportions of extremities. Increase in seed size, reduction in horn size and altered growth rate. Archaeological research shows that such effects probably date back to 50,000 years BC. The intermingling of humans (as a hyperkeystone species) with non-human life has a deep history (Palumbi 2001; Wilkinson 2005; Sullivan et al. 2017: esp. Fig. 2, Fig. 4). This niche-construction approach can be combined with other approaches to material culture and the threefold "heritage" to create a model that makes the natural and cultural factors of the genesis of the Anthropocene as well as the effects of the Anthropocene understandable to some extent.

7.1 Culture Crosses Materiality—Multi-material Entanglements

The Earth has grown a nervous system: and it's us.
Daniel C. Dennett (2004)

In the world, many things are interconnected. Ecology since the time of Ernst Haeckel has made this fact particularly clear to us. The usual representation of an ecological system, prototypically a coral reef or a rainforest, shows a wide variety of organisms in diverse material and energetic interrelationships with each other and interacting with the inanimate environment. As later the theories of self-organization (for an overview, see Scheidler 2021) and cultural anthropology have shown, complex phenomena cannot be traced back to a single cause, but require explanations about their historical genesis. In biology, and especially genetics, we speak of phenomena that are caused by multiple factors (multicausality, polygeny) and of phenomena that, as a factor themselves, influence several other phenomena (pleiotropy).

7.1.1 Materiality and Practice

Materiality, sociality, and practice form a nexus even in non-human primates. However, since this is about humans, a further differentiation is called for. First of all, a distinction must be made between material goods and organismic cultural components. This is a distinction *within* the category referred to in texts on material culture as "things as substance (in nature)," which is set apart from the concept-based "things as material culture" (Hahn 2014: 10). If material culture refers to the objects that are included in the life-world of humans, thus understood as "… the sum of all objects that are used or are meaningful in a society" (Hahn 2014: 18), then domesticated animals as well as anthropogenically shaped landscape elements belong to this category. However, living beings have significantly different characteristics and existential requirements than material objects or the abiotic nature. For living beings, adaptation, reproduction, generationality, and domesticability must be taken into account. Living beings have *inalienable* characteristics that differ from the inalienable characteristics of other material things. Materiality is therefore more than just things (see Fig. 7.1).

7.1.2 Anthroecology—Multiple Materiality and Limited Agency

To take care of the Earth, humans must ontains that we are both a part of the animal kingdom and its dominant power.
Hugh Desmond (2024)

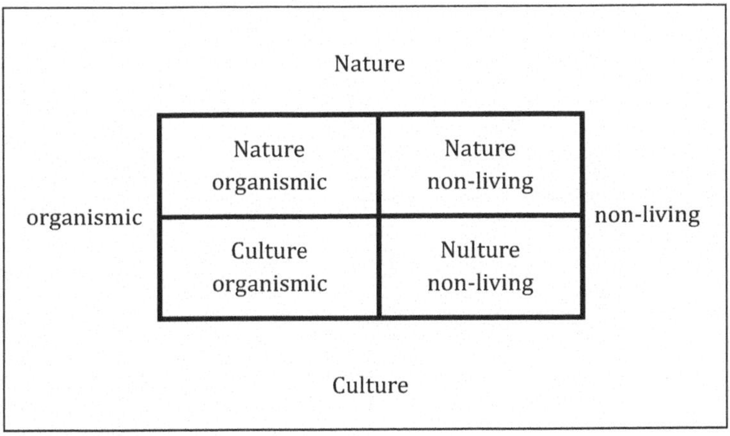

Fig. 7.1 Culture crosses material substance classes (author's own work)

In accordance with the open concept of culture explained above with regard to substance, cultivated plants, for example, are to be seen as components of culture. While the tree leaves in front of a building in Fig. 7.2 superficially appear to illustrate the contrast between nature and culture, the image can actually be seen as an example of different forms of *cultural* materiality. As part of living nature, however, they differ significantly in their characteristics, such as their generationality, from other cultural components, such as buildings, which also have their own obstinacy, for example, through their weight and statics.

This analytical distinction *within* material substance is therefore important because it influences the agency of the actors, for instance, through the different malleability of the materials. Inanimate materials enable other, often less restricted, actions than animate cultural assets do. The spectrum of *enablement* is more restricted (*constraints*, Table 7.1). Within animate beings, there are further differences that are central—and not contingent—to human use, such as in livestock. Pigs, for example, have similar requirements in terms of the origin and source of their food as humans do: carbohydrate-rich and protein-containing feed. Cattle, on the other hand, need grassland that they can graze. They are therefore much more nomadic than pigs, which tend to be more sedentary. Beef "consumes… space before it itself can be eaten" (Reichholf 2011: 83).

Due to their differently constituted materiality, cultural aspects also differ in terms of their usability as a cultural resource. A simple example: Knowledge is the only resource—unlike land, raw materials, and labor—that can be intra-socially shared (*shared*), intra-socially inherited (transmission), and passed on to other

Fig. 7.2 Materially different manifestations of culture at Bonn University, Germany. (Photo: Author)

Table 7.1 Implications of different manifestations and embodiments of culture for *agency*, transmission, and transfer

	Form of manifestation and embodiment of culture	Implications for agency	Implications for transmission and transfer	Epistemic implications for archaeology
K1	Culture manifested in a person's brain: Non-linguistic thoughts	Maximum freedom, as there are no material *constraints*	Not possible	Only introspectively accessible, forgetting
K2	Culture embodied in speech acts: *Parole*	Vocabulary differs greatly depending on education, only perceptible for co-present social partners	Limited to speech community, misunderstandings, random changes	Often short-lived and therefore not handed down
K3	Culture embodied in ontains and action: e.g., body language, routines, rituals, performance	Restricted by basic needs, viability, observable for social partners	Partly understandable even for those who speak a different language, misinterpretations	Often short-lived and not handed down, intentionality (action) hardly comprehensible
K4	Culture manifested in inanimate material things: e.g., objects, monuments, music, art (artifacts, in a strict sense)	Limited by physical and chemical conditions	Depending on the material, more or less permanent, *potential transgenerational*	Central data basis, butdegree of selection of the tradition unclear
K5	Culture manifested in material traces: e.g., work traces on artifacts	Limited by physical and chemical conditions	Transmission only if consciously noticed	Often easily accessible, but precise practical relevance unclear
K6	Culture manifested in extracorporeal storage media: Writing, IT	Capacity potentially unlimited, access by multiple users, access limited by knowledge or instruments	Varying durability depending on the medium, decay	Only in historically recent or even only in contemporary phases
K7	Culture embodied in organisms: e.g., pets, genetically modified organisms, in the ego itself	Limited by basic needs, viability	Limited by generationality	Differential transmission (Ramentation, fossilization)
K8	Culture embodied in the human organism, e.g., organ reaction to nutrition	Limited by basic needs and viability	Limited by generationality	Extremely selectively transmitted
K9	Culture embodied in *own* organism: e.g., individual stress	Limited by basic needs and viability	Limited by generationality, hardly passed on	Partly introspectively experienceable, partly observable

(continued)

Table 7.1 (continued)

	Form of manifestation and embodiment of culture	Implications for agency	Implications for transmission and transfer	Epistemic implications for archaeology
K10	Culture manifested in artificial materials in or on one's own organism: e.g., tattooing, scarification, pacemaker	Limited by basic needs by viability	Limited by generationality, extended by technical innovations	Partly only in historically contemporary phases
K11	Culture manifested in assemblages, e.g., on beaches or in landfills	Depending on the materiality assemblage, differently limited	Differently limited according to lifespan, cycles	Assignment to the place of origin of the individual parts possibly difficult
K12	Culture manifested and embodied in landscapes: e.g., anthropogenically altered biomes	Given the size, strongly limited, varies depending on ecosystem dynamics	Limited by generationality and geophysical circumstances	Tends to be permanent, up to anthromes

social groups (transfer) or spread spatially (diffusion) without being lost by the original person or community.

The renewability of resources is also closely linked to their material constitution. Ultimately, the possibility and probability of passing something on to other communities, the transfer and spatial diffusion of culture, strongly depends on materiality. However, the transfer of culture in particular is a crucial driver of cultural evolution. The actual invention of tools is rare and tied to many preconditions, of which the wheel-and-axle assembly is an example (Hodder 2018: 12–13, 61–63, Fig. 1.5). Cultural software can potentially spread more easily and quickly than hardware. In the case of technologies, cultural ideas are often adopted (*stimulus diffusion*) without the recipients being aware of the exact material basis, as is the case with porcelain, for example. Finally, the demanding character or the affordance of commodities, tools, or buildings is contingent on culture-specific knowledge. These multiple characters of material phenomena make it seem sensible to analytically distinguish forms of manifestation and/or *embodiment* of the respective culture of a community.

The mechanisms of cultural transmission are considerably more diverse in primates than in other living beings. In addition to genetic inheritance, there is non-genetic heritage from parents, relatives, non-relatives, and peers. However, in humans, in addition to the central transmission of cultural content through linguistic communication, there is also the transmission through extracorporeal culture in the form of objects and, in some cultures, writing and other extracorporeal information carriers. Some other primates also use tools (*tool using*) and even manufacture them (*tool making*), but these objects themselves hardly play any role in the transmission between generations according to current knowledge due to the lack of instruction ability.

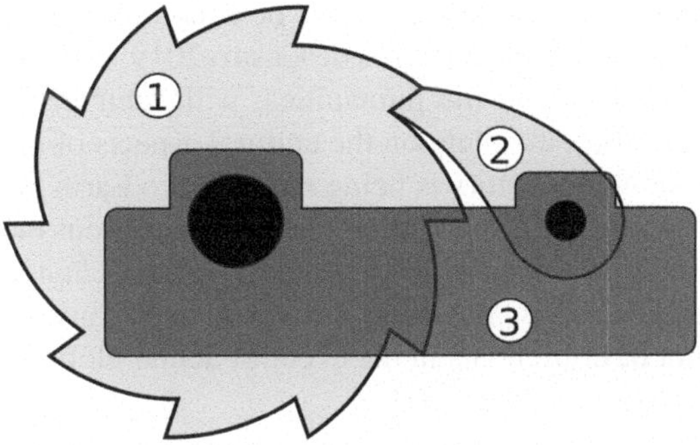

Fig. 7.3 Ratchet, consisting of a ratchet wheel (1), pawl (2), and bearing (3) https://commons.wikimedia.org/wiki/Category:Ratchets#/media/File:Ratchet_Drawing.svg

The respective materiality also strongly influences a probably unique characteristic of humans: the ability to accumulate culture (*cumulative culture*). This ability is central to explaining social evolution in terms of increasing complexity and/or adaptation (Antweiler 1990; Hodder 2018, 43; Zeder 2018; Haidle 2023). The possibility of preserving innovations and solutions as cultural achievements and building upon them (*ratchet effect,*; cf. Tomasello 2001, 2019) varies depending on the form of materiality of extracorporeal culture. To understand the transmission capability of the *chaînes opératoires* (Haidle 2012, 108–110) or Hodder's *entanglements* of biotic, social, and material things (Hodder 2024, 2018, 42, 110–112) and their relation to goals, we would actually need a precise classification of the possibilities of knowledge accumulation for each type of object. This applies to the entire continuum of ratcheting effects—between the poles of the actual mechanical ratchet and, for example, scientific progress through the ascent "on the shoulders of giants." The potential intracultural diversity of material forms and their respective forms of transmission needs to be differentiated. Even within individual cultural communities, *different* objects are handed down with *different* modes of transmission. For this purpose, one could use the multiple traditions of material culture developed by Jordan (2015) (Fig. 7.3).

7.2 Niche Construction—Cultural History Meets Natural History

We will not get rid of ourselves, but neither will we get rid of the world, no matter how often we renounce it.
Ulrich von Loyen (2019: 159)

A sociologization or culturalization of the Anthropocene also has consequences for the study of the long-term courses of human history. Linked to this is the question of whether the distinction between natural history (*res naturae*) and human history (*res humanae*) is meaningful per se, or whether a renewed, and broader, universal history is needed. Chakrabarty's "negative universal history", in which humans are only one of many species, can be interpreted as follows: "The prophecy of the 'end of the world' must therefore be announced performatively so that it is not realized" (Danowski and Viveiros de Castro 2019: 108). Chakrabarty argues that postcolonial historians and the humanities, with their insights into the Anthropocene, can no longer focus solely on the cultural aspects of human life. He goes so far as to argue that world history is being merged into Earth history. First, humans must be seen as a planetary force among others, but without them being central actors with power and thus responsibility. Second, humans cannot live free from the effects of planetary realities, not even the wealthy, because even they, as the privileged, would not have lifeboats in the event of actual global disruptions (Chakrabarty 2009: 221, see Jobson 2020: 260).

7.2.1 "Forces" of History—"Drivers" of Change

In the Anthropocene, the amalgamation of various causes is even more pronounced than in other topics of human history, for which there is hardly any vocabulary in German. The "forces of history" include the most diverse ways in which the past causally shapes the conditions and dynamics in the present. Among such causes are actions based on human intentions, that is, reasons and purposes (*forces*). In relation to the Anthropocene, this is, for example, the effect of a construction project, such as the intended asphalting of a large area. On the other hand, we also find non-volitional—for example, structurally shaped—causal relationships, that is, causes (*forcings,* Baucom 2020). In the case of climate change, these include the role of CO_2 as a greenhouse gas for the warming of the atmosphere, the causes of the monsoon, or events such as the eruptions of the Tambora volcano in 1815 and Krakatau in 1983 in Indonesia for the global change in the atmosphere.

In the Holocene, humans initially had an insignificant influence on the geosphere; they were neither *forces* nor *forcings*, but rather spectators of development. The course of the climate in the Holocene, for example, was determined by three astronomic characteristics of the Earth's orbit around the sun: the ecliptic, the tilt of the Earth's axis, and the fluctuations around this axis. The Holocene warming, which was favorable for humans, resulted from the coincidental concurrence of astronomical circumstances, which increased solar radiation about 12,000 years ago and then led to a situation that has been relatively stable until recently. In the Anthropocene, we have a new phenomenon that results from a complex interaction of systemic drivers and human forces at different scale levels (Thomas et al. 2020: 14).

From a human science perspective, the challenge is to think about both forms of causation together. Representatives of Earth sciences find this relatively easy because they are primarily interested in describing system dynamics. They therefore easily subsume both volcanic eruptions and the position of the Earth's axis as well as the consequences of human construction activity or agriculture under the label "drivers" (*drivers*). This is evident, for example, in publications by the American climate authority NOAA or the German WBGU. Although a distinction is made there between "natural" and "human" drivers, the further question on the causes of human drivers, the why (powerful forces developed), the who (benefited from it), and the how (one could get out of the situation), remains open under the term "driver."

The anthropos itself is a cumulative product of the entire evolutionary history of the planet and the comparatively young political, economic, and societal changes. All involved "drivers" have unintended effects. The aforementioned effect of blue-green algae in the formation and stabilization of the Earth's atmosphere, the "great oxygenation event" about 2.1–2.4 billion years ago, was not intentional because these bacteria cannot have intentions. Similarly, the climate effects of land acquisition and domestication by early civilizations were not intended. Finally, many of the effects of today's industrial capitalism and consumer culture on the geosphere are unintended. The tricky thing about the Anthropocene is that, unfortunately, we often have to deal with the *unintended effects of intentional actions* (McNeill 2000).

This is the insight behind Chakrabarty's program to bring Earth history and human history together. Unintended effects are a classic phenomenon in the social sciences, albeit one that is under-researched. However, it makes a big difference whether the actors involved are ones who (a) *can* have intentions and (b) *can* intentionally research these unintended effects, and (c) deliberately *want* to prevent these effects. If it is not just about description and analysis, but about intended change or desired societal transformation, the tension between intended and unintended "drivers" becomes insoluble (Thomas et al. 2020: 14). At the level of social experience and especially individual experience, however, the distinction is of central importance because people are guided by what they can change in the world; in short, by culture.

Debates on the Anthropocene, the largest mega or macro topic in the human sciences in terms of time and space, show how fruitful it can be to bring together natural history and cultural history from a materialistic perspective instead of dividing them into scientific "cultures." Colonialism is therefore not only a significant part of the history of most countries, but *also of natural history* by fundamentally shaping the texture of entire regions (Reichholf 2011; Leggewie and Hanusch 2020: N4).

7.2.2 Niche construction—Our Triple Heritage and Local Knowledge

How fruitful the model of multiple inheritance is for shedding light on macro questions of human evolution and history becomes apparent when natural history and cultural history are considered together. It shows that in the longitudinal section, the importance of genetic heritage decreases and that of social niche construction increases, which is to be expected. However, what is astonishing is that the extent of material heritage has recently been decreasing and that even the transformation of ecosystems *per capita* is decreasing (Fig. 7.4: *niche-construction intensity*). These are, of course, only very rough statements, which are partly based on hard geological data, but need to be specified in detail.

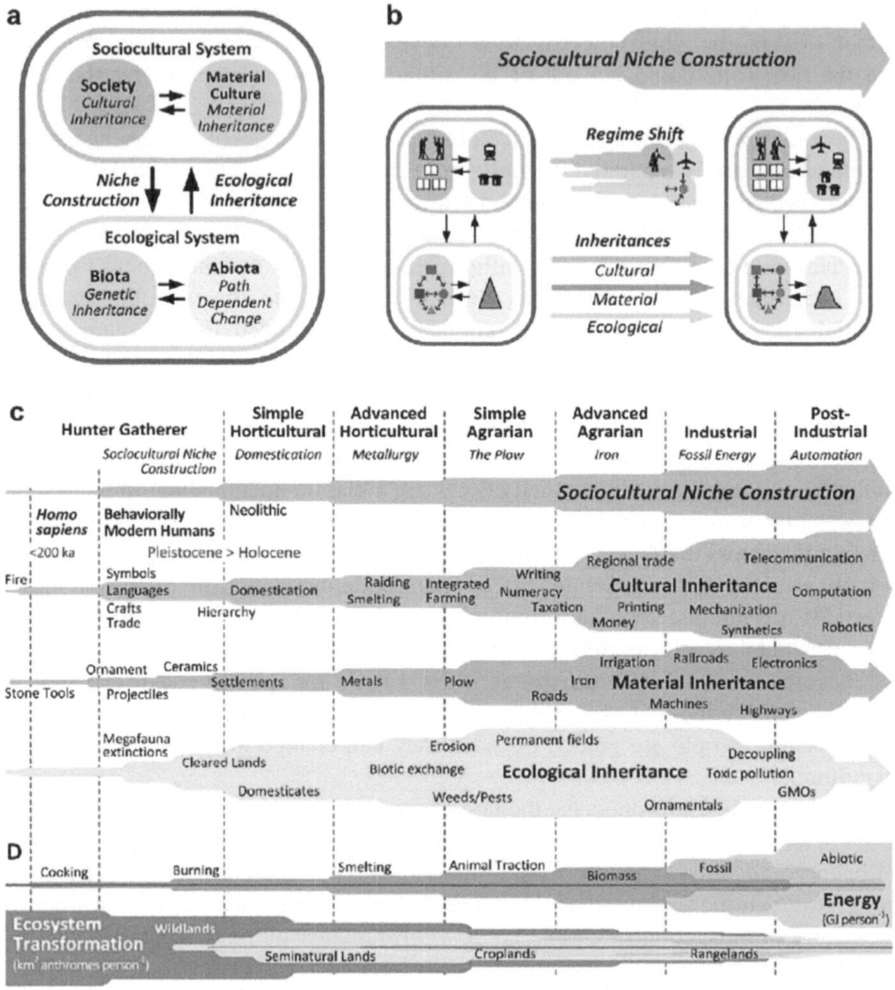

Fig. 7.4 Niche construction, material inheritance, and other modes of trans-generational transmission related to subsistence regimes (**a**) system relations, (**b**) three inheritances and regime shifts, (**c**) change through basic types of human subsistence, (**d**) energy use and ecosystem transformation per person. (Ellis et al. 2018: 122)

Empirical and theoretical findings on organic evolution can be useful for understanding the Anthropocene. This is obvious for empirical findings, because the whole idea of the Anthropocene is based on the evolutionary change of the Earth and life and on the science of fossils: paleontology. But theoretical ideas from evolutionary biology can also be useful. Examples from evolutionary ecology and paleontology are treadmills in ecosystems, denoted as the *Red Queen* phenomenon in reference to Lewis Carroll (Van Valen 1973). If your competitors improve or even just change, or if the environment changes, you have to improve or adapt just to keep up. Van Valen derived this from an evaluation of the continuity and extinction rates of fossil species.

A long phase of adaptation therefore does not yet mean a higher probability of continued existence. As an evolutionary principle, this applies equally to species, ecosystems, and academic hierarchies. In the Anthropocene, such treadmills are syndromic; they are simultaneously a prerequisite, an integral component, and the result of the runaway processes that make up the overheated world of the Anthropocene (Eriksen 2016: 23). The archaeologist Ian Hodder calls for a model of human history that recognizes the diversity of evolutionary domains, is aware of the tensions between them, and builds a theory of the realities and conjunctions of interdependencies (Hodder 2020). For him, human evolution is a "… co-creation of things that are biotic and material and mental and chemical and institutional" (Hodder 2018: 130). For this he considers the *Extended Evolutionary Synthesis* important, but not sufficient, especially because it would not do justice to the heterogeneity of relationships mediated by things. Although Hodder briefly mentions some newer approaches, in my opinion, he considerably underestimates the findings of today's interdisciplinary research on cultural evolution.

A materialistic and genuinely evolutionary, therefore non-typological, approach opens our eyes to the variability of artifacts and their role in communities of practice, which are overlooked by an idealistic-typological approach (Riede et al. 2019). In today's evolutionary biology, the Darwinian approach *clearly* involves much more than Charles Darwin's classic model… and it is anything but social Darwinist. It includes (1) Darwin's model (variation + inheritance + selection), (2) expanded since the 1970s in the so-called "modern synthesis" (Darwin's model + mutation + population genetics + speciation), and (3) expanded again since the 1990s as the so-called "extended synthesis," namely all of this expanded by multi-level selection plus epigenetics or *EvoDevo* (Pigliucci and Müller 2010; Jablonka and Lamb 2020; Lange 2020: 229–258; Belardinelli 2022: 19–22). These extensions of the Darwinian approach raise the question of whether Darwin's model represents a particular variant of a general evolutionary principle (the so-called general evolution, Toepfer 2013; Danchin 2013).

In my opinion, a particularly fruitful approach is the dual inheritance theory by Peter Richerson and Robert Boyd (Richerson and Boyd 2005). The authors assume two separate systems of intergenerational transmission of information: genetic and non-genetic "inheritance." Unlike similar approaches under the label "coevolution," they emphasize that both modes of inheritance differ significantly but are often causally intertwined (Pagel 2012: 6, Boivin et al. 2016). Such approaches provide a mechanistic but multifactorial theory of the mechanisms of historical change. In my opinion, they are particularly suitable for analytically distinguishing profound mechanisms of long-term societal change and for providing a methodically guided interpretation of empirical cases of Anthropocene change.

In cases of multiple "inheritance," material objects and their respective properties can play a significant role. Therefore, this model could contribute to the detailed elaboration of an agential realism in the sense of Karen Barad (Hoppe and Lemke 2021: 59–79). What is still rare is the systematic inclusion of material culture and the environment, the abiotic, the biotic, and especially the anthropogenically shaped environment, in a model of multilinear transmission (for an exception: Jordan 2015; Herrmann 2019). One step would be to question the material units linked in the operation chains or *tanglegrams* (for example, in Hodder 2018, 86–90, 109) as to which forms of directed learning and selective transmission (Table 7.2) they each enforce (*constraints*), suggest (*affordance*), or enable (*enablement*).

Furthermore, the model of the dual inheritance system could be expanded by referring to more recent research from cognitive evolutionary research in order to clarify the specifically human form of social learning. Cecelia Heyes argues that cultural learning is a special form of social learning using very specific *cognitive gadgets* (Heyes 2018: 80, 219–222). These are primarily selective social learning, imitation, *mindreading*, instruction, and language. Thus, cultural learning—contrary to the understanding of Tomasello as well as cultural evolutionists such as Henrich—is not just a sophisticated form of social learning, that is, learning with the help of social partners.

Simply put, according to Heyes, the special cognitive ability of humans consists less in the inherited intelligent brains of individuals than in the smartness shared in social interactions. The cognitive tools emerge primarily during childhood in a formally culture-saturated environment. They are a result of cultural adaptation and are functionally crucial for the transgenerational transmission of culture. What remains underexposed in Heyes, however, is the realization that some of the cultural learning mechanisms are based on biotically given emotional inclinations, thus ensuring copying accuracy. This has been suggested in cross-cultural anthropological studies on emotional norm socialization (Quinn 2005, for examples, see Antweiler 2019a, b; Antweiler et al. 2019). It is relevant for the archaeology of the Anthropocene, even though Heyes does not address the fact that many of these cognitive gadgets work through the mediation of *objective* culture. An example is children imitating each other and following their leader, all linked by a colorful ribbon. In human cultures, intelligence is, outside of the head, extracorporeal in two ways: in *specific* social interactions and in knowledge-containing artifacts, including entire landscapes.

Table 7.2 Factors of social evolution according to the dual inheritance approach

1 Chance
1.1 Cultural mutation: individual, for example, through false memory
1.2 Cultural drift: statistical anomalies in small populations2 Decision-making forces
2.1 Guided variation: changes during learning
2.2 Biased transmission
2.2.1 Direct bias, for example, through algorithm, cost–benefit consideration, or learning inclination
2.2.2 Frequency-dependent bias, according to the commonness of acultural pattern or rarity
2.2.3 Indirect bias: imitation of successful individuals orindividuals similar to oneself
3 Natural selection of culturally transmitted variants3.1 Selection at the individual level3.2 Group selection

The dual inheritance model has been extremely productive for research and has led to many empirical studies, theoretical works, and simulations of cultural change (Boyd 2018: 9–63). However, there is a certain gap in the precise analysis of the human-altered environment. More recent approaches in research on the mechanisms of cultural evolution in the group around Peter Richerson and Robert Boyd, however, are increasingly attempting to take greater account of the role of human intentional action as “agentive processes” in their models of ecosystemic change and, in view of the Anthropocene, particularly with regard to sustainability (Richerson et al. 2023: 26–7). There is a long tradition of these agentive processes in anthropology. Regarding this, there is a long tradition in cultural anthropology, as the anthropological concept of culture can indeed be seen as the sum of human environmental changes. But early cultural ecology also pointed out that the environment plays a strong causal role for material culture and other cultural areas. This was emphasized in cultural ecology and with regard to long-term change in neo-evolutionism (multilinear evolution). Most cultural ecology works were rather empirically oriented and focused on adaptation processes. The coevolution of societies and the abiotic and biotic environment has only recently been addressed in concrete models (Orr et al. 2015).

7.2.3 Niche Construction Knowledge as a Resource in the Anthropocene

Another argument for the causal importance of the environment for long-term change came from evolutionary biology. In 2000, Richard Lewontin, a critic of narrow Darwinian concepts, published a book with the DNA structure-alluding title “The Triple Helix,” which was translated into German with a ten-year delay (Lewontin 2000). The argument is that genetic heritage must be supplemented by an understanding of the intra-organismic environment in which the genes operate. However, even more important for our context is Lewontin’s emphasis on the transmission of the environmental situation as an independent instance of inheritance.

If we now combine Boyd and Richerson’s model of dual inheritance with Lewontin’s emphasis on the environment, three modes of inheritance can be distinguished: genetic inheritance, cultural inheritance, and environmental inheritance. This gives us a triple inheritance model, the human niche construction by John Odling-Smee (Odling-Smee 1988; Odling-Smee et al. 2003; Lange 2020, 203–227). The special feature of the model is that anthropogenic niche construction in the form of active environmental influence in one generation, t, leads to environments that co-determine the conditions of natural (sic!) selection in the next generation, t + 1 (Fig. 7.5). The classic example from the animal world are beavers, who *significantly and permanently* change their ecosystem by building dams, thereby shaping the selection of their subsequent generations and many other creatures. Their evolution thus takes place in the environment shaped by their predecessors (Fig. 7.6).

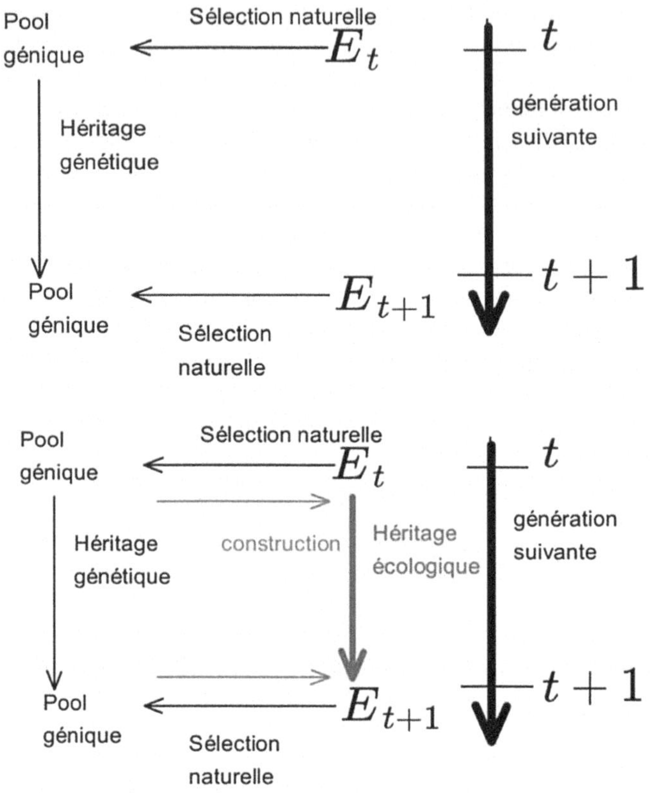

Fig. 7.5 Human niche construction in the triple inheritance model https://commons.wikimedia.org/wiki/File:Schema_construction_niche.pdf?uselang=fr

When human action significantly shapes the environmental conditions for the next generation, the respective environments embody the structures of earlier human social structures. The triple inheritance model and niche construction could be useful for applied measures against Anthropocene environmental effects if we link it with research on local knowledge (Boivin and Crowther 2021). We cannot simply return to the evolutionary normality of egalitarian small groups, but we can learn a lot from the ways in which earlier societies adapted to environmental turbulence (Ehrlich and Ehrlich 2022: 778, on climate adaptation Fagan and Durrani 2021). Regarding species loss, one of the core problems of the Anthropocene, indigenous knowledge can be helpful because the current rapid landscape change is altering not natural environments, but econiches, which have been culturally shaped for a very long time. This has been demonstrated by comparative regional studies in recent paleoecology, archaeology, and landscape history:

> With rare exceptions, current biodiversity losses are caused not by human conversion or degradation of untouched ecosystems, but rather by the appropriation, colonization, and intensification of use in lands *inhabited and used by prior societies*. Global land use history

Fig. 7.6 Triple cultural inheritance in Cologne: material culture, cultivated plants, and semi-anthropogenic atmosphere as environmental heritage. (Photo: Author)

> confirms that empowering the environmental stewardship of Indigenous peoples and local communities will be critical to conserving biodiversity across the planet (Ellis et al. 2021: 1, emphasis mine).

In my opinion, the model of niche construction is also theoretically approachable in a particularly diverse way. It allows a bridge to social theories, because, contrary to widespread perception, the core of the niche concept is *not* a spatial niche, but a functional role in the ecosystem. This makes an econiche comparable to a social role or a profession. With the emphasis on the multiple construction of the niche, a connection to (non-extreme) approaches of social constructivism is made possible. Another advantage is that the model allows for the study of systems at the macro, meso, and micro levels. The environmental aspect did not play a central role in the model by Boyd & Richerson, yet this approach, enriched by niche models, has also been further developed by Joseph Henrich's working group.

Methodologically, this includes field research, laboratory investigations as well as mathematical modeling (Chudeck et al. 2016; Creanza et al. 2017). Long-term field research is of central importance with regard to Anthropocene effects (Lansing and Fox 2011; Orr et al. 2015, 157). In addition, there are also experiments that show correlations between population size and social connectivity on the one hand, and the complexity of the material repertoire on the other (Muthukrishna et al. 2014). The precise distinction between the mechanisms of cultural transmission also allows for a more precise determination of the influences of culture on genetic evolution (*culture-driven genetic evolution*, Henrich 2016,: 59–60). The continuous use of artifacts, such as tools and weapons, results in anatomical changes in the hands, leads to new cortical connections in the brain, and changes functional coordination abilities. At the behavioral level, dexterity increases, such as throwing skills, while physical strength decreases. The intertwining of neurocultural research with the analysis of objects shows the importance of material culture for *culture-driven* genetic evolution.

7.3 Humanity as a Scale Concept—Postcolonialism Meets Geology

When it comes to long-term survival in the face of Anthropocene damage, as Dipesh Chakrabarty argues, we need a concept of humanity that—contrary to Malm, Holmberg, Swyngedouw, and the advocates of the Capitalocene concept—is explicitly *non*-differentiating (Chakrabarty 2012, 2017). I, too, advocate a strong and substantive charge for the concept of humanity. Within the Anthropocene discussion, this is represented by theorists as different as Chakrabarty and the literary scholar Timothy Morton ("Humankind" 2017). Despite all the emphasized differences—for example, those vehemently made by Malm & Hornborg, especially regarding the power and effectiveness between different categories of communities—there is a unity of humanity. In the face of global dynamics, humanity comprises a community of effects.

In view of the boundaries of the geosphere, Immanuel Kant spoke of the right to visitation to which all people are entitled, "(...) to offer themselves to society by virtue of the right of 'common possession of the surface of the Earth, on which, as a spherical surface, they cannot disperse into infinity'" (Kant 1796: 358). The novelty in the Anthropocene is that guest and host are now systemically connected instead of being opposed to each other (Leggewie and Hanusch 2020: N4). This unity depends on the "organism-like" nature of humans and their dependence on other living beings. Chakrabarty summarizes this as "boundary conditions" necessary for living beings. This is similar to the old anthropological concept of basic needs by Malinowski (Malinowski 1944). These boundary conditions affect all people on this planet. In this respect, the sum of all people forms a community of interest despite all political fragmentation, unequal living conditions, and differing concerns (Antweiler 2019a, b; *pace*, Latour 2014; Bauriedl 2015).

The reality of the Anthropocene calls for solidarity on the ethical and political level that goes beyond advocating for the interests of the poor, exploited, underprivileged, or marginalized. Chakrabarty therefore calls for solidarity among humans as such (Chakrabarty 2012, 2017). This does not exclude differentiation on a second level. Chakrabarty distinguishes three levels of personhood in humanity: (a) humans as bearers of rights, especially human rights, (b) humans categorically differentiated according to the postcolonial criteria of class, gender, ability, and *race*, and finally—which is central here—(c) humans as a collective entity, which only becomes relevant through the worldwide effects of human action.

Macro scalability is not just based on model specifications, as Tsing says: "The scale is global because the models are global" (in Haraway et al. 2016: 313).

These three aspects of being human are interconnected, but cannot be reduced to one another (Werber 2014: 71). While the third level represents the new fundamental challenge of the Anthropocene through the subject of humanity as a collective, the lessons of humanism from post-colonialist approaches and nature-oriented Marxism are relevant for the other two levels. However, they are overshadowed by the third, Anthropocene level. There is a differentiating conception, but one that emphasizes the totality of humanity. Because the anthropos in this conception is understood as a scaling term, it is particularly helpful, as Werber elaborates. On the one hand, it brings to light emergent qualities and problems that go beyond the mere sum of individuals or the accumulation of their behavioral effects. On the other hand, the approach goes beyond the characteristics or constitutional conditions of human beings that philosophical anthropology investigates, because it contains something new in terms of human history.

Bajohr has worked out why the differences in scale, in his case, "scale jumps," are so important (Bajohr 2019:71): First, humans can no longer be thought out of history in deep time. Second, humanity has only a very limited ability to counter Anthropocene effects as a collective through measures. Third, individuals, as parts of the whole, can hardly imagine this whole. Chakrabarty emphasizes that we as individuals find it difficult to perceive ourselves as members of a species. This suggests a continuity to the descriptive categories of the natural sciences and, more importantly, a use of the species concept as a desideratum for future orientation narratives:

> 'Species' is perhaps the placeholder name for a nascent, new universal history of humans, which flashes up at the moment of that danger that climate change represents (Chakrabarty 2015?, in the translation by Bajohr 2019: 71).

If the commonality between humans, in the face of the Anthropocene, takes precedence over differences (Chakrabarty 2015: 143–147), it makes sense to combine ideas of the Anthropocene and a planetocentric perspective with philosophical concepts. Consider, for example, transhistorical "anthropological constants," such as the general human capacity for reflection and decentration, as the core of a philosophically turned sociology (Chernilo 2017, 2020: 69; contributions in Bajohr 2020). Furthermore, the triply differentiated approach to the conception of humanity can tie in with findings on biotic universals, that is, features that all human individuals share and that rest on a primatological basis, such as the ability to cooperate (van Schaik 2016: 315–373; Henrich 2016). It should be added that Chakrabarty's conception can also be linked to findings on universal characteristics that distinguish all human collectives (pan-cultural universals, Antweiler 2018). All these possible connections concern the status of the Anthropocene as a concept. They imply ethical questions and political debates on how to counter Anthropocene effects. Both debates should be conducted, but in contrast to parts of Chakrabarty's work, they should be kept separate (Chernilo 2020: 57).

7.4 Eco-cosmopolitanism—Localized World Citizens?

Collective, binding decisions are how people can give the world a shape that we intend.
Jedediah Purdy (2015: 17)

Only if we all recognize each other as "earthlings," both on a personal level and on a global scale, can a new overarching solidarity emerge.
Marc Augé (2017: Flap)

Today, most people encounter more strangers on a normal day than most prehistoric people did in their entire lives. The Anthropocene makes it clear that despite all diversity, we should soon reach a worldwide consensus on mitigation measures if humanity does not want to plunge itself into crisis. What "we" now need is an intense interest in all approaches to bring people together to create a more widely shared mutual identification and solidarity, and to mobilize the appropriate political and institutional resources (Purdy 2015: 271).

We are all inhabitants of this world, but by no means all citizens of this world. In this chapter, I will outline the building blocks of cosmopolitanism as a value-based attitude of global citizenship in a culturally diverse and interconnected world (Appiah 2008; Antweiler 2012a: 61–81, 2012b, 2019a, b). Political forms of an organization of world citizenship (WBGU 2016; Schulz 2015) or political cosmopolitanism (also "cosmopolitics," Arizpe-Schlosser 2019) and speculative, epistemic, and ethical-aesthetic designs, such as Isabelle Stengers' "cosmopolitics," are only briefly mentioned.

7.4.1 Planetary Politics and a Depoliticized "New World Citizenship"?

The changes in the magnitude and speed of life constitute the decisive signature of our time. Through the discovery of this truly global problem, the Anthropocene, according to Marc Augé, brings three central priorities into focus: the individual, culture, and species (*générique*). As individuals, we continue to live in the concrete layers of time and space, but in view of the interrelationship of our lives with the geosphere, we should, or perhaps must, ask ourselves about the possibilities of what a global society could look like. From the geohistorical moment when *every* context becomes global, we can "(…) ask whether the epoch in which we live does not ultimately amount to the end of the history of humanity as a world society (…)" (Augé 2017: 9, see 29, 88, Augé 2017: 26).

To contain the Anthropocene, global cooperation and worldwide coordination are needed. The cosmopolitan ideal of an Earth-spanning socialization (*planétarisation*) would be that of a "(…) world society whose legally and factually equal citizens share the space in the interest of common benefit" (Augé 2017: 14). Since Immanuel Kant's classic theses on world citizenship in his "Ideas for a Universal History with a Cosmopolitan Purpose," global relationships have changed on several scale levels, as the anthropologist Michael Fischer notes (Fischer 2009: 253–254).

On the macro level, nation-states are playing an increasingly smaller role in solving problems, including national problems, than in the past. On the micro level, so-called new social movements have transformed into highly mediatized and volatile networks. On the middle level, the mainstream of dominant political ideas is being challenged by art, music, literature, and film. These include ideas that appear positive in the sense of a cosmopolitan orientation, but also counterforces such as religious fundamentalism or extreme nationalism. Fischer programmatically derives three temporal and also spatial relationships from his diagnosis, which must interact "as differential mutual critiques" in order to create mutual sensitization. Humanistically oriented works of literature, film, music, and philosophy, "social anthropologies" of living with diversity and alterity in different cultural worlds of experience and technoscientific knowledge and infrastructures that enable or restrict flexibility, creativity, and freedom (Fischer 2009: 270).

The realities of globalization and the Anthropocene, however, initially make the ideal of an Earth-wide socialization appear as utopian. In a three-class fragmented world, (a) poorly paid workers, marginalized states and cultures, and the excluded, (b) more or less affluent consumers, and (c) powerful states and companies face each other (Augé 2017: 14). On the other hand, the world has demonstrably become more equal in many respects, at least until the 1990s (Rosling 2018; Pinker 2018). In addition, unlike in Kant's time, there is a multitude of transnational institutions, such as the United Nations, the WTO, and the Organization of Islamic Countries, as well as regional organizations such as the EU or the ASEAN (Arizpe Schlosser 2019: 283). Augé backs the horse of global knowledge as it is the only resource that can be divided infinitely without being depleted.

The idea of world citizenship is sometimes also explicitly conceived in relation to the Anthropocene. However, one problem of world citizenship, as it is invoked in the WBGU, is when the Earth-wide dimensioning is unceremoniously transferred to the social world. Sociotopes function differently from biotopes in terms of both social mechanics and political goal-setting. Politically, the human being does not only appear there as a world citizen, but all of humanity goes on stage as a political actor, as the "world citizenry" (WBGU 2011a, b: 8 ff.). Unlike in positions of classical cosmopolitanism (see Sect. 7.4), humanity is now supposed to take collective responsibility for planetary risks according to a contractual model, for example, in order to cope with climate change (Kersten 2014: 381 ff.). Kersten (2014: 381 ff.) and Neckel (2020: 162–163) criticize such a leap in scale in the context of "scientification" and see it as the result of a scholastic error in the sense of Bourdieu, that is, as an attitude that scientists are not even aware of.

Bauriedl sharply characterizes the program of the WBGU as a "Eurocentric cosmopolitan movement" (Bauriedl 2015). In this chapter, I ask the question of what a cosmopolitan, rather than Eurocentric, movement for dealing with the Anthropocene might look like. What could a non-Eurocentric and environmentally conscious concept look like, which takes into account not only the global effects, but also the regional and local impacts and the corresponding responses? What anthropological foundations could it build on? How can networked cultures coexist on a limited planet without all having to become the same? (Antweiler 2019a, b).

7.4.2 *Planet Instead of Globe—For a Realistic Cosmopolitanism*

In light of the Anthropocene, we can ask what an eco-cosmopolitanism could look like (Heise 2015b), and for this we need knowledge about cosmopolitan traditions of thought. Recently, universalist ideas have been reconsidered as part of a revitalized discussion about world citizenship. Cosmopolitan ideas can be found throughout history in such diverse thinkers and traditions as the Stoics, Kant, Rabindranath Tagore, Sri Aurobindo, Martha Nussbaum, Thomas Pogge, Amartya Sen, and Ananta Kumar Giri. They are united by the central statement that all people belong to one world. Cosmopolitism (or cosmopolitanism) is at its core an orientation, the willingness and the openness to engage with the other, with diverging cultural experiences and, first and foremost, a search for contrast rather than uniformity (Kleingeld and Brown 2009).

Throughout the history of the quite different concepts of cosmopolitanism, two central arguments can be found: (a) Not only does human life as such have intrinsic value, but also each individual human life. This implies that we must take an interest in the practice of life and the convictions that give meaning to the life of the individual. (b) As human beings, we have duties towards other human beings beyond the members of our blood relatives, ethnicity, or nation (Bartelson 2009: 209). Cosmopolitanism as an attitude demands that people free themselves from the narrow internal view of their culture or nation and see themselves as citizens of the world. The commitment is to the world. Cultural differences are voluntarily put aside. This implies that ethnic and national loyalties must not be played off against obligations towards foreign people. In this respect, cosmopolitanism fits well with cultural anthropology, a science that has made the encounter with strangers at eye level its profession (Kahn 2003: 409–412).

The anthropologist Pnina Werbner conceives cosmopolitanism as a counter-concept to globalization, as an ethical horizon and a future-oriented "aspirational outlook." Instead of the free flow of capital and the spread of primarily Western concepts, a renewed cosmopolitanism means emphasizing empathy, tolerance, and respect for other cultures and values (Werbner 2008: 2). If a modern cosmopolitanism wants to include people of other cultures, it must not be elitist, as cosmopolitan positions often were, in that they only opened up the wider world to mostly Western elites. The uprooted and cosmopolitan attitude can quickly become elitist and androcentric. With cultural uprooting and a mobile way of life, the responsibility for local concerns is then easily left behind, just as the ontains or goal is forgotten with the orientation towards consumption (Rapport and Stade 2007: 230 ff.). In this respect, elitist cosmopolitanism can be corrected by the anthropological close perspective and responsibility for local life-worlds.

> A critical cosmopolitanism combines an ethos of independence at the macro level with a keen awareness of the irreducible peculiarities of places, characters, historical paths of development, and fateful events (Rabinow 1996: 64).

Post-colonialist proponents of a new cosmopolitanism criticize the fact that the colonial context in which cosmopolitan ideas largely developed, as well as the different experiences of different cultures, were left out of Western concepts of cosmopolitanism (Giri 2006). However, it should be noted that cosmopolitan culture, such as in Beck's cosmopolitanism, is seen as a transnational, pluralistic, and open culture in which diversity does not merge into unity. Hannerz has also recently emphasized this more strongly and speaks of a "double cosmopolitanism" (Hannerz 2010: 89, 2016: 170–186).

7.4.3 Humanity as a Community of Interest?

A central insight of the discussions about the Anthropocene is that the window of opportunity to halt or reverse key indicators of the environmental crisis is rapidly shrinking. We need a perspective that takes a longer-term view than the environmental sciences do. This means that the social sciences and history-oriented humanities, which played hardly any role in the discussion about the Anthropocene in the first ten years, are needed. Many advocates feel challenged accordingly. A guiding question could be formulated as follows: To what extent does the "more" and "greater" of all the effects of human activity also generate qualitatively special cultural phenomena and social problems? Humanity is a community of effects and, despite all its fragmentation, also a community of interests. Just as we ask about the enabling conditions of Anthropocene effects, we can look to the future for the enabling conditions that could prevent worse developments.

Most cosmopolitan positions go beyond a general attitude of openness towards the foreign, as central as this is. Today's cosmopolitanism includes responsibility and the recognition of global involvement (*global belonging*). A sophisticated humanistic understanding of cosmopolitanism states that it is possible to identify with the whole of humanity, even if it is often fragmented. It is the thesis that humanity can be conceived as a moral community. After the experiences with global wars and environmental crises, the focus today is on the fact that all people are "citizens" of one (endangered) world. Accordingly, one can look for the smallest common denominator, or a global ethos.

In recent years, there has been an increase in projects aimed at providing a visual presentation of the ways of life of humanity worldwide that do not pit diversity against unity (pioneering work by Ommer 2000). More than 20 years ago, a UNESCO-sponsored *GeoSphere Project* provided insight into data and questionnaire responses from 30 "randomly selected" families from around the world.

Such publication projects are very much welcome, but they also have pitfalls, as can be seen, for example, in the Adamistic projects of the "world family" (*Family of Man*) (Antweiler 2012a: 65–68).

Many of today's proponents of cosmopolitanism believe that an ethic of pure plural coexistence is outdated and can become dangerous in the face of global interdependencies and the worldwide effects of human action. Beck describes not only industrial societies, but the whole world as a "world risk society." This creates a "cosmopolitan view," which is primarily skeptical, self-critical, and without any illusions (Beck 2004: 7–25, 151 ff.). As different as these approaches are, they have one thing in common: humanity appears as a global community of interest.

Although a modest variant of cosmopolitanism also seeks universal norms and values, it emphasizes with Appiah that, at least at the level of everyday encounters, less is enough to advance humanity (Antweiler 2012b). The cosmopolitan curiosity about other peoples does not need to search for characteristics common to all human beings or cultures in every encounter as there are more immediate points of connection, namely basic human issues and basic problems, that all cultures have to face. Such a path to world citizenship is more realistic and more immune to the need to find the universal at all costs such that one risks falling into the trap of wishful thinking or nostrifying egalitarianism. On the other hand, it is more concrete than the approach seeking abstract principles of justice and fairness that are independent of notions of the good.

7.5 Universality Plus Difference—Appiah's Partial Cosmopolitanism

Over the past few years, Kwame Anthony Appiah has developed a concept of cosmopolitanism that takes seriously the reservations raised against such projects, but explicitly opposes the absolute relativism often associated with them. Appiah is an American philosopher who specializes in moral philosophy and practical philosophy. Raised in a family of traditional rulers in Ghana and strongly influenced by British education, he has moved towards a specific cosmopolitanism inspired by African philosophy and intercultural conceptions of philosophy. As a child of several worlds, Appiah has presented a concept of world citizenship that is not exclusively Western, and from which an inclusive humanism can learn.

A cosmopolitanism of this kind argues against the relativistic assumption of inevitable dissent in value questions or the assumption of voicelessness between cultures. Like Beck, Appiah does not conceive of world citizenship as the opposite of provincialism. He proposes the creation of a global identity, but it is a concept that must be distinguished from globality and globalism. As in Amartya Sen's "global identity," it should be possible to acquire this cosmopolitan identity without having to give up loyalty to other entities: *cultures count* (Sen 2009). Appiah speaks of "cosmopolitan patriotism" and "rooted cosmopolitanism." Despite their differences, Appiah, like Sen and Nussbaum, seriously considers the fact that humanity today lives under the condition of planetary interdependence. These concepts are "post-global" insofar as they focus less on centrality and homogenization or heterogeneity. Therefore, I see them as attempts to consider the planetary as transcending differences rather than clinging to the image of the globe with its accompanying emphasis on identity and difference. Paul Gilroy speaks of a "planetary humanism" (Gilroy 2004: 76–78; similarly, Spivak 2008 using the example of Asia).

Appiah's approach contains many ideas that are useful for the conception of an inclusive humanism. Some things initially sound exaggerated, naive, or overly idealistic. In fact, Appiah represents a very cautious concept, and he is aware of the many pitfalls in the field of intercultural encounters. For him it is about the part of interests and viewpoints that *everyone* should adopt. What is crucial for me is that Appiah's concept of world citizenship tempers the respect for cultural differences with the respect for humans. Exclusively fencing off cultural communities from others is not an option in the globally interconnected world, nor has it ever been in history.

The universal concern for others and respect for legitimate differences can collide, which is one of the reasons why cosmopolitanism is a challenge rather than offering simple solutions. The variant of cosmopolitanism presented here does not seek to belittle or even dismiss cultural differences. Approaches that speak of the whole of humanity always run the risk of becoming totalitarian. Appiah's approach differs significantly from the icy impartiality of hardened cosmopolitans and the intolerance of ultimately anti-cosmopolitan universalists of the neo-fundamentalist breed. The belief in a universal truth is central; it is a fallibilistic approach that consists in not yet believing that one already knows what this truth is. But it also differs from the partisan attitude of culturalists and ethnicizers. This concept of world citizenship is to be understood as "partial cosmopolitanism" (Appiah 2006: 13 ff., 174 f.).

Appiah assumes the normality of intercultural interaction. The growing interconnectedness is increasingly turning humanity into a kind of "global tribe" in which people and cultures come into contact with each other more than ever, can learn something about each other, and influence each other. The possible remote effects of one's own actions entail obligations towards people and cultures outside one's own life-world. Appiah's cosmopolitanism is much more moderate and less programmatic than Ulrich Beck's "cosmopolitan empathy." Appiah assumes that identity formation necessarily accompanies differentiation. As I will show, this makes him theoretically better grounded and, above all, empirically better informed than many other approaches. The challenge is to equip our thoughts and feelings, which have been shaped over thousands of years of living in small local groups during human evolution and almost to this day, with institutions and ideas that would allow us to live together on an interconnected globe (Appiah 2006: 10 f.). One goal in intercultural interaction is not to treat cultural otherness (alterity) with reverence and sublimity, nor to be confused by it. With Appiah, I consider it central to promote cross-cultural conversations and also to develop concrete forms of coexistence in which sociability between people of disparate lifestyles becomes possible ("customs for coexistence").

Appiah is therefore, in my opinion, equally cautious and realistic when he says: "Cosmopolitans assume that in the vocabulary of the *evaluative* language of all cultures, there are *sufficient overlaps* to allow for the *beginning* of a conversation" (Appiah 2006: 82, emphasis mine). Using the book of Leviticus as an example, Appiah clarifies the mixture of specific and universally recognized values that we so often find. A rejection of the specific prohibitions (sex between men) and taboos (contact with menstruating women) found there need not prevent anyone from supporting the values and virtues of a universal nature also found there, such as kindness and empathy, and branding cruelty as a vice.

7.5.1 Realistic Ethics and an Unburdened Humanism

Any meaningful ethics must consider the conditions of its own realizability. A possible realization depends primarily on whether such ethics are based on reliable knowledge about people, their actions, and the functioning of human collectives. A cosmopolitan ethics that does not want to stop at conjectures or fall prey to wishful thinking needs an empirically substantiated anthropology. It is unrealistic to hope for factually shared values worldwide. Since there are already different attitudes within every human community or society, we must allow for disagreements in the conception of each cosmopolitanism. A universal set of rules will also be difficult to build. Even if we agreed on general rules, we would have to argue about the application or implementation of these rules. Even if we shared values and agreed on their application, we may assess different values dissimilarly (Appiah 2006: 32, 90). We can live with differences of opinion, such as those concerning values. In order to understand each other, we do not need to be in complete agreement with each other (Appiah 2006: 30 ff., 105). Cosmopolitan engagement in the form of a moral conversation reckons with differences of opinion. The human universal is preserved in the differences. Cosmopolitanism, as Appiah succinctly describes it, can therefore be thought of as "universality plus difference."

What consequences do Appiah's considerations have for a non-Eurocentric humanism? Firstly, they relieve us of exaggerated demands for intercultural understanding and inflated hopes for shared values. In intercultural interaction, one does not have to immediately understand or even like each other. In conversation, it is enough to engage with each other; what is important about conversation is the chance to present oneself to another in real situations and to get used to each other. There are few or no substantial and precisely formulated common values, but we do not need them either. Differences of opinion do not need to be settled or resolved. It is enough if we come to shared decisions for action. It is not necessary to establish fixed principles, nor is it necessary to agree on a set of common universal values.

7.5.2 *Negotiated Universals and "Ethical Globalism"*

The strongest conflicts do not occur between members of cultures with completely contrasting ways of life. They arise among people who culturally have a lot to share but argue about the importance of certain values or how they are to be evaluated (Appiah 2006: 109). A lasting *modus ontain* between members of different cultures will be easier to create if we refrain from such clear-cut definitions and value justifications. Historical and current examples show this, such as the Ottoman Empire or present-day Indonesia.

Values or practices can be universalized in different ways (Kocka 2002: 124). The first way is plain imposition: universalization is achieved through power, pressure, and manipulation. Secondly, values and practices spread simply through imitation, which is a core process of cultural globalization. The third form is an extension of this adaptation, a universalization through "negotiation." In this process, an idea or practice is not only modified during its spatial expansion and cultural spread, but also substantially changed. When ideas are incorporated into a considered culture, they are selected, emphasized, diminished, and reinterpreted, and they are generally adapted in an often contested process. When such a process encompasses many or all cultures, negotiated universals can be developed (Kocka 2002: 124; Riedel 2002: 277; Jullien 2008: 182). Programmatically speaking, the concept of "negotiated universals" stands for the attitude that globally shared values should be pursued, but that these should not be imposed:

> 'Negotiated universals' stands for a view that still takes universals to be important, but does not want to impose the universals of one culture on other cultures. Rather, a deliberative process among cultures is envisaged to result in a set of values each culture can subscribe to for his own reasons (Riedel 2002: 277).

Negotiated universals do not form universalized Western values with uniform justifications, but with negotiated common orientations that communities or cultures *can subscribe to for different motives and with various justifications*. An inclusive response to the Anthropocene must not be limited to Western or Atlantic experiences and intentions. It must take the diversity of cultures seriously. An inclusive response to global problems that is not Eurocentrically limited must bring together the universalist and the cosmopolitan project (Vertovec and Cohen 2002: 6–14).

7.6 *Asianizing the Anthropocene*—An Example of Recentering

> *Presenting a neutralized view of humans without critical and political insights makes the scientism of the Anthropocene narrative a Eurocentric colonial discourse.*
> Shingo Yoneyama (2021: 261)

Further research could focus more specifically on non-western cultures and the origins and meanings of the Anthropocene for places such as Asia, Africa, Australia, central and Latin America, and Antarctica.

Carolyn Merchant (2020a, b: 25)

But if we are going to come to terms with the deep histories of global transformations, we need to (…) listen to the other half of the globe, the tribalists out there and right here, talking back.

Jeremy Adelman (2017: 5)

Asia is relevant to the Anthropocene for several reasons. Asian countries, especially China, already play a key role in the world economy and will probably also be of central importance in politics in the near future. Today, around two-thirds of humanity live in Asia. In an area that is small on a global scale, which includes East Asia, South Asia, and Southeast Asia, there are currently as many inhabitants as there are in the rest of the world. India alone has about as many people as all African countries combined and about as many people as both Americas. Over long stretches of history, southern and eastern Asia was the population center of the world. Furthermore, the earliest intensive and also long-term landscape changes were recorded in Asia, which is why cases from Asia in particular indicate an early Anthropocene. On the other hand, eco-spiritual understandings of nature and culture in Asia have a particularly long tradition, just as in Latin America. All of this should lend particular weight to experiences and voices from Asia for adequate ways of dealing with the Anthropocene in terms of vulnerability and resilient strategies (Will 2021: 267). Waste-related public and semi-piblic measures are one case in point (Fig. 7.7).

Fig. 7.7 Waste separation in a Hindu temple in Mumbai, India. (Photo: Author)

7.6.1 Correcting Eurocentrisms in the Anthropocene Debate

Both the concept and most of the known data on the Anthropocene have so far come from the Atlantic region. In some respects, however, there are clear differences between Europe and Asia, for example, which must be taken into account. Corlett shows this using the biodiversity in Southeast Asia (Corlett 2013: 1). First, the biodiversity is significantly higher than in Europe. Second, due to the small habitats necessitated by biodiversity, and in view of the strong contrast between primary and anthropogenic forests, the degree of endangerment of species in tropical forests is especially large. Third, the particularly rapid transformation of the environment poses a high risk of extinction for small populations. Fourth, the speed of this environmental change is not decreasing, unlike in Europe. Finally, the environment in Southeast Asia is significantly less well-researched than in Europe, which complicates conservation measures. All of this means that we have to deal with an increasingly patchwork-like tropical environment.

The idea of the Anthropocene originated from the Euro-American region, and the periodization proposal is heavily based on European and American data. This corresponds to the Europe-dominated chronologies in archaeology and history. However, the focus on the Anthropocene has contributed to changes in perspective (for example, Ishikawa and Soda 2020). Such problems were initially addressed by economic historians, colonial historians, and comparative sociologists, and they were given new impetus through the emergence of environmental history in the wake of global pollution from the 1970s onwards (Dukes 2011; Marks 2020: 7). The historical sciences are faced with the challenge that different dimensions of spatiality are connected with the temporalities converging in the Anthropocene, and that this has resulted in local confrontations with the phenomenon of the Anthropocene. This calls for a "transformation of scalar means and ideas" (Will 2021: 261): the planetary must be analytically linked with the global, national, and local through its spatialization (for example, Otter et al. 2018 for Great Britain). In the local appropriation of the Anthropocene, in *placing*, historical and anthropological tasks come together.

The findings on the Anthropocene in Asia (for example, Simangan 2019) have significantly contributed to a general trend in universal history and global history, in which Europe was no longer unquestioningly seen as the center (Fig. 7.8). The exploration and presentation of history would have to be decentralized. Leading historians have promoted this new orientation, foremost among them Kenneth Pomeranz (2000), Jack Goldstone (2002), John Hobson (2004), Andre Gunder Frank (1998), and Robert B. Marks (2020). Most of the works of the so-called "California School" investigated complex societies in southern and eastern Asia. They were mainly concerned with the "rise of the West," the worldwide imposition of Western societies from the late eighteenth century onwards, even though societies in parts of Asia had largely been much more developed by then.

Several of the important contributors had personal ties to Asia, such as Dipesh Chakrabarty and Prasaran Parthasarathi. There was an increase in findings

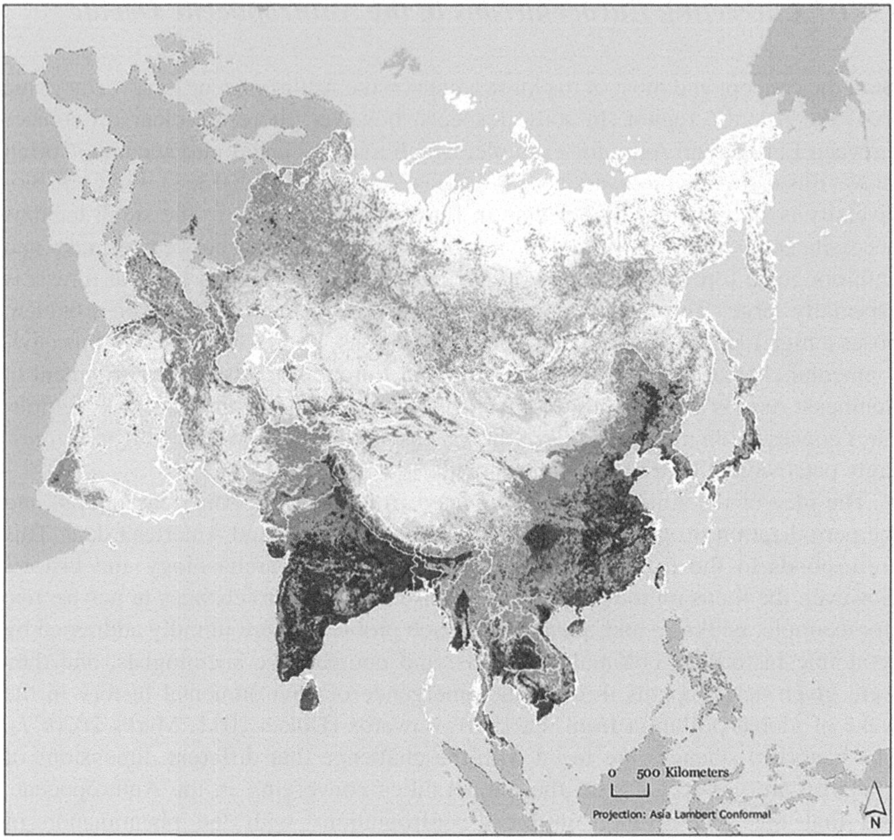

Fig. 7.8 Concentration of large Anthropogenic Biomes (Anthromes) in Southern and Eastern Asia https://commons.wikimedia.org/wiki/Category:Anthromes Alternatively. https://commons.wikimedia.org/wiki/Category:Anthromes#/media/File:Anthrome_map_v1_600.png

demonstrating that many Asian societies possessed just those characteristics that had previously been considered particularly European, such as markets, institutions, rationality, and science. This reopened the classic question of Europe's special path, of the "gap" or the "great divergence" (Pomeranz 2000) between Asia and Europe. How could it happen that by the beginning of the twentieth century, almost the whole world was ruled or dominated by Europe? Recent findings suggest that the rise of Europe is less due to fundamental cultural orientations or solely to European colonial rule, but rather to specific environmental constellations and contingent decisions. A well-known hypothesis for the rise of England over China, for example, is that China in the nineteenth century had neither colonies nor significant amounts of easily extractable coal and thus had to use labor and capital for land use, while England could use resources from the New World and had enough coal to operate steam engines continuously (Pomeranz 2000). Other authors, on the other hand, more strongly emphasize the role of state policy or institutions (Vries 2001; Parthasarathi 2011).

For example, it is evident that Asia was the central engine that drove global trade approximately between 1000 and 1800. Asia, especially China and India, formed the economic machine that was instrumental in setting in motion the exchange of ideas, new crops, and goods. Brought (Lo and Yeung 2019; Anderson 2019). From around the year 1000, the growth of China's population and economy stimulated large parts of Eurasia. Between 1400 and 1800, political power, economic innovation, and social stability were concentrated in southern and eastern Asia. Around 1400, only about 1% of humanity lived in large cities. However, only five of the twenty-five largest cities were in Europe at that time. The world's wealth was concentrated there, and particularly in China and India, there was a huge demand for silver. Western European colonial powers thus used Latin American silver to gain access to the globally coveted Asian goods, especially textiles and porcelain (Goldstone 2002; Marks 2020: 12, 24–25). In Asia, this promoted exchange and commercialization. It was not until the nineteenth century that Western European entrepreneurs managed to replace imports from Asia with their own industrial production. Faced with a lack of labor and energy, they progressively introduced machines. From this de- or re-centered perspective, the industrialization of Europe thus appears to be a catch-up development (Frank 1998).

At first glance, there appears to be a contradiction here in relation to the Anthropocene. On the one hand, the main causes of the Anthropocene are attributed to the emergence of modern capitalism in Europe; on the other hand, non-Eurocentric explanations are called for. Historian Mark Driscoll, for example, argues that the Anthropocene originated in predatory behavior during European imperialism, with the historically first assemblage of raw material extraction, trade, war, and ecological destruction. He uses China as an example to illustrate the interplay of various factors and actors, yet sharply labels the overall phenomenon as "climate caucasianism" (Driscoll 2020).

The various assessments should be seen against the background of the criticism of Eurocentrism (or Occicentrism). This critique emerged in the historical sciences in the 1980s, regarding Eurocentrism as an ideology, a master narrative, or as a theoretical explanatory model that is universal in claim but limited in fact, or as a myth, the "colonizer's model of the world" (Blaut 1993). In relation to the Anthropocene, it should be considered that through Asian examples and historical comparison, one can arrive at other explanatory factors than those in previous interpretations, which are almost exclusively based on data from and experiences in Europe:

> How can there be a non-Eurocentric explanation of a world that has European features? In short, we can find that by broadening the storyline to include parts of the world that have thus far been excluded or overlooked – we can begin and end the story elsewhere (Marks 2020: 11).

Since the Anthropocene was largely set in motion by economic growth and industrialization, a look at Eurasia is therefore appropriate in many respects. The view of writing world history more from an Asian studies perspective has now also found its way into the global histories of individual periods (for example, Osterhammel 2020), and into comprehensive accounts of world history, such as the series "A History of the World/Geschichte der Welt" edited by Akira Irye and Jürgen Osterhammel, as well as into overview works and American college textbooks on world history or global history (for example, Fernández-Armesto 2019; Lockard 2020; Marks 2020; Adelman et al. 2020). A focus on Asia has also recently been incorporated into global histories in popular science (for example, Frankopan 2015; Testot 2021).

One might think that the Anthropocene is on everyone's lips... in fact, however, it is only on almost everyone's lips, because in some regions of the world the word hardly appears in public debates or in the sciences. This can be observed, for example, in parts of East Asia and Southeast Asia, even though the issue of global change and the problems arising from it are certainly being met with strong interest there (Horn and Bergthaller 2020: 172–175; Spangenberg 2021; Antweiler 2020; Bergthaller 2020; Jobin et al. 2021). How is the idea of the Anthropocene discussed outside the Euro-Atlantic area? Although some of the authors I cite in this book come from the Global South, there are comparatively few of them. This raises the question of the extent to which the topic of the Anthropocene is discussed academically in the Global South and whether the topic is of interest to the general public. The post-colonialist theorist Achille Mbembe noted with regard to Africa:

> (...) [t]his kind of rethinking, to be sure, has been under way for some time now. The problem is that we seem to have entirely avoided it in Africa in spite of the existence of a rich archive in this regard (Mbembe 2015).

A new overview on the topic also establishes this for Latin America:

> The Anthropocene has not, as yet, become an analytic concept that is widespread in published work by scholars from the Global South (...), with the prominent exceptions of Eduardo Viveiros de Castro and Deborah Danowski in Brazil (...). A recent review of engagement with the Anthropocene by anthropologists and political ecologists in Latin America, for example, finds that they are as yet more concerned with socioenvironmental conflicts around natural resources, territory, environmental justice, and non-Western ontologies (...) (Mathews 2020: 68).

The work by Eduardo Viveiros de Castro and Deborah Danowski mentioned in the quote, which is important for debates about new ontologies, deals with ideas about the end of the world among indigenous peoples and specific forms of Amerindian perspectivism. The authors primarily associate the Anthropocene with crisis and dystopian narratives (Danowski and Viveiros de Castro 2019: 11, 22, 24, 80, 101), but only address the Anthropocene *en passant* in their somewhat cryptically formulated book.

But why is the idea of the Anthropocene, as well as the empirical findings on the Anthropocene, discussed so little both publicly and scientifically in some regions of Asia, for example, in Southeast Asia, in contrast to almost all Western countries, and if they are, then it is done in other facets (Ghosh 2016: Chap. 3; Dove 2019; Bergthaller 2020)? Given the fact that the region is of utmost importance for the Anthropocene (climate change, soils, resource extraction), the low resonance is surprising. It is also surprising in view of the fact that strong anthropogenic environmental changes precisely in Southeast Asia have been reported early on by Europeans, such as by the Sarasin brothers on Central Sulawesi in Indonesia (Schär 2016). This could be due to the fact that "nature" as an explicit everyday topic and concept that was ethnographically documented only slowly emerged in the late 1970s due to fundamental environmental changes, for example, in Taiwan and China (Weller 2006: 1–10). Or is the low interest due to the fact that the Anthropocene is perceived as a "Western concept," as seems to be the case with politicians and other decision-makers? Further assumptions can be made. The use of other terms in Asia, such as "global change," can be seen as a critique of the Anthropocene concept (see Malm and Hornberg 2014; Baskin 2015). Mark Hudson asks whether the lack of interest in Asia—as well as the Global North's lack of engagement with the Anthropocene in Asia—could be due to the fact that the Anthropocene is simply seen as another process of Westernization of Asia (Hudson 2014: 944).

The lack of consideration given to Asia in the search for causes in historical scholarship is related, among other things, to the fact that local populations have long been perceived as living in harmony with nature. Since freedom from the constraints of nature was seen as the defining feature of modern historicity, some societies in Asia appeared as communities with no history (Thomas 2017a, b), just like the "primitive peoples" in traditional cultural anthropology. However, the reason for the lack of interest in Asia itself may also lie in historical facts that are diametrically opposed to the nature thesis, namely that societies in monsoonal Asia have historically had particularly long and intensive experiences with an agriculture that has significantly reshaped the environment anthropogenically. This is particularly true for wet rice farming, and especially for terraced farming (Fuller et al. 2011; Corlett 2014; Morrison 2018).

These hypotheses would need to be developed, differentiated, and tested based on published documents and grey literature. A theoretical background is provided by the approach of provincializing knowledge ("provincializing Europe," Chakrabarty 2008, 2012). Historians in particular have devoted themselves intensively to these questions (contributions in Austin 2017; Hudson 2014), but there are also promising approaches, for example, in *Asian studies* as area studies, history, cultural anthropology, and the *environmental humanities* (Ammarell 2014; Philip 2014; Thornber 2014). Such research also exists for Latin America and Africa (Howe 2014, 2015; Issberner and Lena 2017; Hecht 2018).

The strongest Asian response to the Anthropocene to date has been in Japan. In Japan, there is a lively discussion on alternative narratives of global change. Shingo Yoneyama, for example, describes films and artworks that emerged in the wake of the minimata poisoning and the Fukushima disaster as "new animism" (Yoneyama 2019, 2021; Hudson 2019). Mark Hudson discusses forms of Japanese eco-nationalism that appear in a new form in the wake of the Anthropocene debates. This strongly concerns imaginations and concepts of nature. He asks, for example, to what extent Fujisan, the sacred mountain of Japan, which can rarely be seen clearly but is culturally represented thousands of times, can no longer be discussed as pure, separate nature in view of the Anthropocene, but is a fine example of what Tim Morton calls difference multiplication (Morton 2009): "This virus-like existence means that Fuji is a wonderfully appropriate metaphor for Nature in the Anthropocene and a way to think about shared not exclusive histories" (Hudson 2021: 37).

Historically grown constrictions of our knowledge based on predominantly Western (*WEIRD*) evidence can be expanded by non-European or non-Atlantic experiences and critically illuminated by new perspectives. A diachronic regionalization of the Anthropocene or a "local appropriation of the planetary" in non-Western space can stimulate contextualized case and regional studies (Will 2021: 259–276). For example, the environmental historian Peter Boomgaard, who focuses on regional studies, was able to show that deforestation in Southeast Asia was primarily the result of population growth and less due to industrialization (Boomgard 2017: 193). He thus places the onset of the Anthropocene in Southeast Asia in the middle of the twentieth century, at least a good 100 years later than Crutzen & Stoermer's dating and also in line with the more recent geoscientific consensus.

7.7 Anthropocene Reflexivity—For and Against an "Anthropocene Turn"

> *In any case, humans have become a planetary force, but have not yet developed a planetary reason.*
>
> Jürgen Renn (2021: 3)

> *Globally, societies are facing major upheaval and change, and the social sciences are fundamental to the analysis of these issues, as well as the development of strategies for addressing them.*
>
> Lene Pedersen and Lisa Cliggett 2021

For the humanities and cultural studies, it is paramount to determine what contribution their central concepts of society and culture bring to the broader debate and, conversely, what changes to the concepts themselves are appropriate. In relation to culture, Arizpe Schlosser's observation that the term Anthropocene became popular in science and society at about the same time as the expanded concept of culture is not only interesting in terms of the history of science. As early as the 1940s, the word "culture" became increasingly important in the international public sphere and increasingly replaced words such as "race," "humanity," and "civilization," which can be demonstrated using Google's *n-Grams*. Underpinning this was the collapse of the German concept of culture due to its association with "race" in National Socialism, the emergence of the anthropological concept of culture, and the increasing politicization of culture in the context of globalization, as well as a phase of particularly strong population growth (Arizpe Schlosser 2019: 7, 27–48).

7.7.1 *Can We Go a Size Smaller?—Enter* Humanities.

The worldwide triumph of the word "culture" and, contentwise, an expanded concept of culture from about 1950 coincides with the emergence of an understanding within the Anthropocene discourse that there is a "great acceleration." "Culture" was increasingly seen as suitable for encompassing many issues and problem areas, such that there was a simultaneous rise of the precursors of the Anthropocene concept (Arizpe Schlosser 2019: 273, 277, 281–282). The world population increased by 46% between 1950 and 1970, and the environmental effects of capitalist and socialist economies were evident especially in the Global South due to unequal ecological exchange. For this reason, these developments and the development of the geosphere should be written in conjunction with each other (Bonneuil and Fressoz 2016: 252).

From the perspective of social theory, it is still largely unclear what it means to find ourselves in the Anthropocene, provided we actually are in it. One argument in favor of including at least the living world in sociological analysis is that the habitability of the planet depends everywhere on other organisms inhabiting the same habitat. It is only through the involvement of other organisms that the conditions of the habitat of a considered organism are created, and thus also the framework conditions that ensure habitability; in other words, conditions that make a habitat habitable for humans. On the macro level, this applies to the biotic genesis of the Earth's atmosphere, the central argument of the *Gaia* idea (Latour 2017a, b), which also implies a fundamentally ecosystemic approach for the social and cultural sciences. Nevertheless,

> (…) the question arises as to whether the often admonished paradigm shift does not also have reflexive costs if holistic concepts of sociology, which intend to include the biophysical Earth system as well, have to offset the gain of a remeasurement of sociological conceptual apparatuses with the loss of analytical discernment (Neckel 2020: 157).

Ecosystem thinking and the more recent resilience thinking are important, but too limited. The majority of social scientists today assume that system boundaries are not (only) in the phenomena themselves, but are heuristic tools in the sciences and are fundamentally to be seen as the result of social construction (Pálsson et al. 2013: 7). This raises the question of the extent to which the social or the cultural needs to be geologized and, as a complement, how culture and society can be translated into planetary thinking (Adloff and Neckel 2020: 11; Clark and Szerzynsky 2021; Hanusch et al. 2021). In view of the problems entangled in the Anthropocene, Bhaskar and co-authors have called for a meta-theory of the "meta-crisis" in order to maintain the "flourishing" of the planet (Bhaskar et al. 2015; Hedlund and Esbjörn-Hargens 2023). Werber, on the other hand, comes to an extremely pessimistic conclusion about the potential of the social sciences, in this case sociology, in these debates. This is in light of the lack of consideration for the data and perspectives from the humanities and social sciences by Anthropocene proponents à la Leinfelder and Latour's statement that humanity has no alternative to our Earth—a rejection of *Gaia* and the Anthropocene concept:

> In view of the moral fervor of the Anthropocene proponents, who incidentally take the data of a differentiated scientific system for granted, for a sociology that can see in Gaia a dream of de-differentiation and in a 'global ethos' no more than a contingent, non-justiciable norm, but not the salvation of the Earth, there remains only the side of evil, the adversary (Werber 2014: 246).

The Anthropocene challenges the social sciences to rethink central concepts. The question arises as to how far the social sciences, such as cultural anthropology and sociology, should or even must open up not only to the natural sciences in general, but also to Earth system analysis in particular. Adloff & Neckel illustrate this using a central social science concept: structure (Adloff and Neckel 2020: 10). In sociology, one usually understands structure as enduring sets of rules and resources of human practices. Various practice theories and theories of materiality have made it clear that social structure is also related to material infrastructure, for example, in that some objects prescribe certain actions (Niewöhner 2013; Schatzki 2016; DeLanda 2016). Historians have often shown that certain architectural arrangements reinforce social status or gender norms.

In the face of the Anthropocene, however, it is clear that it is not only the societal and nature that interact, but that social structures are also related to structures of the Earth system. Hence, the question arises as to whether and to what extent biophysical conditions of existence should be included in the structural description of a society (Adloff and Neckel 2020: 10). Accordingly, we should no longer speak of social formations or socio-economic formations, as in Marxism, but of "geosocial formations" (Clark and Yusoff 2017).

Crises require measures, and these must be negotiated and legitimized in societies. In the case of environmental crises and epidemics, natural scientists obviously play a major role. It is about making considerations and decisions about values and value hierarchies. When it comes to the proportionality of measures, political scientists are particularly in demand because they deal with public decisions. But economists are also important because they can model fundamental dilemmas and universal problems, such as those concerning the commons and free riders. In addition, social scientists per se are also in demand, because sociologists, cultural anthropologists, and social psychologists are familiar with decision-making and assessment procedures (Renn 2021: 26).

7.7.2 Social and Cultural Sciences—Out of Marginalization!

Before the concept of the Anthropocene emerged, there was little awareness that humans play an important role in the *planetary* development. This is clearly shown by the dominant concepts of science funding on the topic of human–environment relationships. The concepts dominant in the natural sciences until the 1990s considered human activities as marginal. Up to this point, the social sciences were considered marginal by the funding bodies. Even social and cultural science funders hardly

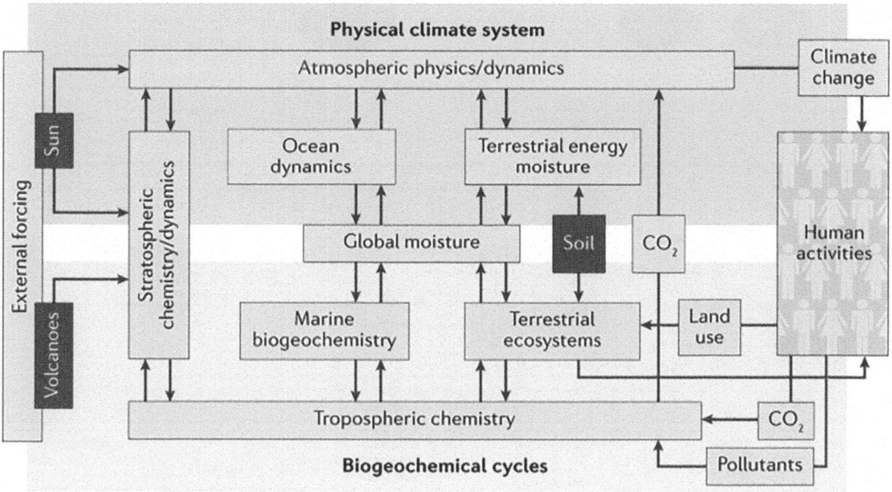

Fig. 7.9 Earth system interrelationships in early version of the NASA Bretherton Diagram (NASA, adapted from National Research Council 1986: 19) (Steffen et al. 2020, Fig. 2, here from S. J. Williams, R. Taylor (eds.), Sustainability and the New Economics, Springer https://link.springer.com/content/pdf/10.1007/978-3-030-78795-0_2.pdf)

financed any research on global change. The famous "Bretherton Diagram" (Fig. 7.9), which depicts the Earth system, summarizes them in a *single* unspecified box (National Research Councel 1988: 29–30; Pálsson 2020: 55). In later versions of the diagram, the scientific entities are more clearly broken down, while the human activities remain undifferentiated in the diagram and are later expanded in a separate diagram (Cornell et al. 2012: 9, 12 Fig. 1.3, Fig. 1.4).

The diagram of the *International Human Dimensions Programme on Global Environmental Change* was clearly different some 20 years later (IHDP 2010). Here, the social sciences were placed on an equal footing with the natural sciences, albeit in a blatantly dualistic way. In addition, "environment" only appears in the natural sciences, and the interrelationships within the "social sciences" circle remain rather vague. In this sterile polarization, it is not considered that the cultural and social sciences could also contribute to the conceptual clarification of natural science research, such as the environmental concept, or even form a responsibility-oriented meta-perspective on human-environment research (Pálsson et al. 2013: 6, see Fig. 2 there). It took about 15 years before human influence was given full weight in the latest version of the Bretherton diagram with an *equally large* box under the heading "Anthroposphere" (Fig. 7.10). This box needs to be analytically expanded, and this requires the full engagement of all human sciences (see Donges et al. 2017).

> The grand challenge for ESS is to achieve a deep integration of biophysical processes and human dynamics to build a truly unified understanding of the Earth System (Steffen et al. 2020: 54).

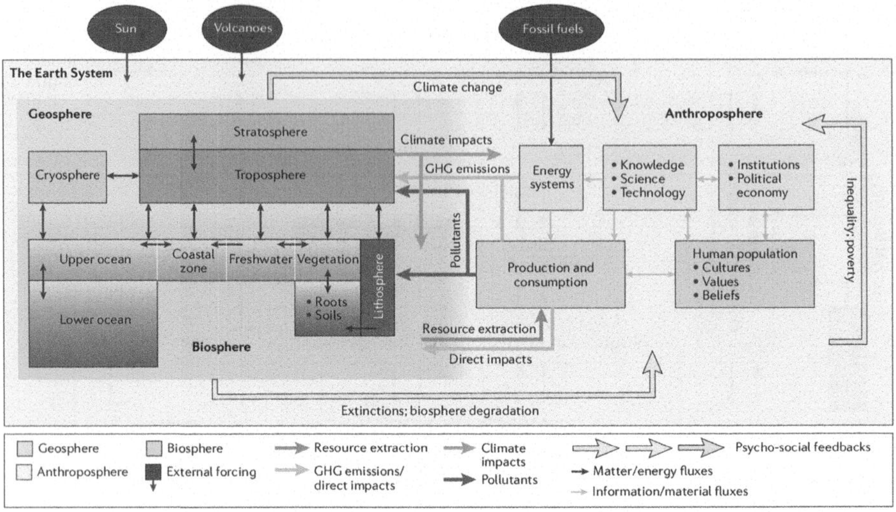

Fig. 7.10 Updated concept of the Earth system with the anthroposphere (Steffen et al. 2020, Fig. 3, here from S. J. Williams, R. Taylor (eds.), Sustainability and the New Economics, Springer https://link.springer.com/content/pdf/10.1007/978-3-030-78795-0_2.pdf)

The environmental humanities and newer, even more integrative approaches, such as the *integrative humanities* (Sörlin and Wynn 2016) and also specific forms like *energy humanities* (Szeman and Boyer 2017), have been contributing to this since around 2012. The social sciences and humanities can make multiple contributions. A central contribution, in my opinion, is the handling of the topic of directed change:

> To remedy the lack of understanding of the Anthropocene, it is essential to enhance and intensify the social sciences and humanities' work on how *directionality* could be articulated, democratically anchored, and implemented in the search for new technologies, medical knowledge, economic paradigms, and forms of social organization (Pálsson et al. 2013: 10, 11).

This sounds similar to questions in critical development research. However, in contrast to the concept of development, which refers to desired and planned improvement, the Anthropocene must focus much more on long-term and unintended, but actually effected directions of change. The humanities as well as the cultural and social sciences are needed for this, and they are also increasingly being sought out. However, we should not be too optimistic. In the interdisciplinary *Anthropocene Working Group*, for example, only 3 out of 37 scientists specialize in long-term societal change.

7.8 For a Science of the Whole Human—*Turns* and *Returns*

Some cultural or social scientists have proclaimed an "Anthropocene turn" or "Anthropocenic turn" (for example, Bonneuil 2015; Arias-Maldonado 2016; Dürbeck and Hüpkes 2020). Such a *turn* would be a programmatic turnaround due to the Anthropocene both in scientific approaches and in politics. The word would be formed according to the *linguistic turn* and the *cultural turn* as well as numerous other *turns*, such as the currently dominant ontological turn in cultural anthropology (Rheinberger 2021). In addition to the Anthropocene turn, other turns relevant to the Anthropocene theme have been suggested early on, such as a *material turn* (Coole and Frost 2010), an *environmental turn* (Sörlin 2014; Castree 2015: 301), a *geological turn* (Yusoff 2013: 779; Bonneuil 2015: 19–31), and more recently, a *planetary turn* (Clark and Szerszynski 2021: 27–32, 77–99).

The call for an environmental turn has been reflected in the institutionalization of the *environmental humanities* as an interdisciplinary approach in the Anglo-American academy. Proposals à la "Beyond the Anthropocene" also go in this direction of planetary thinking, namely to consider the outer world beyond the atmosphere, but also that in the Earth's interior, for example, in the form of its representations in human thought: "We argue for keeping 'Anthropocene' connected to its spatial absences and physical others, including those that are non-anthropos in the extreme" (Olson and Messeri 2015: 28, see Messeri 2016).

An Anthropocene turn would signal the enormous importance and urgency of thinking in planetary dimensions and would therefore be welcome. However, the word "turn" already raises my concern about such a program. Turns are used primarily in the humanities and cultural sciences as a way of organizing theoretical debates (Bachmann-Medick 2016). While sociology tends to discuss permanent paradigms such as structure, function, and evolution, the turns in the humanities and cultural sciences, especially in the *cultural studies*, usually only last until the next *turn* comes along. In the fast-paced attention cycles of the humanities and cultural sciences, the *linguistic turn* is replaced by the *cultural turn*, and the latter by the *material turn* or *practice turn*, and this by the *relational* and the *animal turn*.

A constant turning around is indeed implied by the metaphor of the turn, because it does not necessarily have to be a U-turn, but at least a departure from the established direction. If each new turn were only seen as an additional perspective, there would be no need for the term *turn*. If an orientation has only existed since the last turn, the new turn would represent a departure from the direction in which the last turn was made. In the Anthropocene literature, the term "post-Anthropocene" is occasionally used (Ferrando 2016; Fremaux 2017, 2019), which refers particularly to non-anthropocentric perspectives and sustainable solutions intended to reduce Anthropocene effects.

One problem with newly emerging turns is that they sometimes cause painstakingly obtained findings or insights to be forgotten all too quickly (Laidlaw and Heywood 2013). This applies even to insights that were only gained through earlier turns. A general example is the insight into the power of language that was gained in the *linguistic turn*... only to be forgotten in other turns. In currently popular variants of the ontological turn in cultural studies and cultural anthropology, the insights from the *material turn* into material politico-economic and politico-ecological contexts of worldviews have been again forgotten (Kleinod et al. 2022, using the example of animism in Southeast Asia). An example of a direction that, in my opinion, would be highly relevant for an appropriate global-cultural response to the Anthropocene, but seems to have been almost forgotten again, is cosmopolitanism. Just a few years ago, a "cosmopolitan turn" was being discussed in cultural anthropology (Rapport and Stade 2007), but now—at least in cultural anthropology—unfortunately, little has been heard of this cosmopolitan-oriented theoretical current, briefly outlined in Sect. 7.4.

It should also be noted that many of the entanglements between organisms and the abiotic environment that are being discussed with new concepts of the *more-than-human* have long been part of the discourse in biological ecology and also in anthropological cultural ecology, but are now not (or no longer) perceived. One variant that is less subject to this reservation is the pluralizing of an existing turn (Eitel and Meurer 2021), or modifying it rather than proclaiming a new one. Tsing, for example, suggests expanding the *ontological turn* to a *multispecies ontological turn* (Tsing 2019b). Another proposal relevant to the Anthropocene is a *decolonial turn 2.0* in cultural anthropology as an ecologically informed update of the decolonial turn in the 1990s (Todd 2018; Bendik-Keymer 2020; Jobson 2020; Simpson 2020).

In my opinion, the Anthropocene is too important a theoretical issue to leave at a turn, given that the problem it expresses is too relevant, and the word Anthropocene is too rich in content. In view of the Anthropocene, we cannot simply proclaim an "Anthropocene mainstreaming" in research and institutions because this invites routinization. What we need instead is what could be called "ecological reflexivity" (Pickering 2017; Dryzek and Pickering 2019: 34–35, 56, 79). Anthropocene reflexivity would more intensely involve a broad anthropology (Wulf 2020a, b). What we need—simply put, but difficult to do—is (a) to equally incorporate findings on the geosphere as a dynamic Earth system *and* our knowledge of the functioning of complex cultures and societies, (b) to discuss core cultural values such as sustainability and ecological justice frontally and permanently, and (c) to develop a societal ability to actively seek, perceive, and respond to early warning signals from the geosphere.

If paleoanthropology, archaeology (prehistory and early history), and cultural anthropology are taken together, anthropology can enrich our understanding of humanity in the Earth system in three ways (Thomas et al. 2020: 113, similarly Stiner et al. 2011). First, the different scales in these directions show how we are all embedded in geological and ecological systems. Secondly, all three directions have always emphasized the physical framework of human existence. Thirdly, the interest in specific life-worlds ("cultures") is balanced by the interest in universal human commonalities ("culture"). Paleoanthropologists study the evolution of all species of the genus Homo from the time it emerged about 2.8 million years ago. They bring in a deep-time perspective that links to geology. Archaeologists examine the sites of human settlement and document the early historical spread of human societies in a wide variety of environments. Cultural anthropologists research the specific systems of today's communities in their global diversity in relation to the Anthropocene:

> Anthropology plays a special role in this context – neither that of the world explainer, nor that of the moralizer, nor even the one that tirelessly points out diversity and contextuality, and certainly not the one that stages the "foreign," the "other," even the "exotic" and projects sustainability in a romanticizing, nostalgic way onto Asian or indigenous ways of thinking and their economic systems or identifies fundamental ontological differences (…) This is perhaps not directly and necessarily about very big things, but it is also and above all about everyday practice, concrete action, thoughts and feelings as well as structural and institutional embedding (Schlehe 2021: 13).

Cultural Anthropology brings in the *directly* collected experiences of present-day people. In all three anthropologies, material things play a central role, which are collected and evaluated using specific methods. These things are natural objects or the biophysical environment on the one hand, and cultural things (artifacts) on the other. But even a broad anthropology is not enough to understand the Anthropocene: it also requires the geosciences.

Chapter 8
Conclusion—We Are the Anthropocene

> *… what it means and does,*
> *when the present describes itself with*
> *the help of a geo-historical concept.*
> Andreas Folkers (2024: 3)

> *The industrial growth-based*
> *society is at an end.*
> Read and Alexander (2019: 77)

From a long-term geological perspective, humans have recently become a significant force of nature. This is the core statement of the Anthropocene debate. We are now confronted with the fact that tomorrow is today. What Martin Luther King said in 1967 in relation to the Vietnam War applies to humanity today. The central questions of this book were these: What is the Anthropocene? What is the most appropriate name for it? When did it historically begin? And what does it mean for the future? These questions could be summed up in a paradoxically sounding, but precise way (Folkers 2020: 592): *What is the Anthropocene, and what will it have been*?

I also explored the question of what the Anthropocene means for the sciences involved. The term Anthropocene confronts us with alarming facts; it shows both the seriousness of the consequences of our actions and the limits of our possible influence on long-term processes. The Anthropocene throws light on the importance of knowledge for survival and at the same time the limits of our knowledge. What can and what might the humanities, cultural studies, and social sciences add to this topic, which has its origins in the natural sciences? What can cultural anthropology contribute to the concert of many sciences in its reflection of large-scale and deep-time concepts from geology as a natural-historical science and from the Earth sciences? Conversely, how can the geosciences be culturalized in return and thus take humans seriously as a factor in the geosphere?

C. Antweiler, *Anthropology in the Anthropocene*, Anthropocene – Humanities and Social Sciences, https://doi.org/10.1007/978-3-031-74591-1_8

8.1 The Anthropocene Is Different from Other Crises

In the geological-chronostratigraphic approach, a sudden chang ("great acceleration") within a long-term, continuously *changing* Earth system is recorded as a *synchronous* timestamp by means of clear markers or geological *signals*. Contrary to common misconceptions, this *does not* mean that the geohistorically new Anthropocene Earth dynamics (new state) would be analytically decoupled from historical precursors and continuities. The new geospheric dynamics show continuities and are based on path dependencies that go back to the nineteenth century, and in some cases much further back. The diachronic onset of human influences can be integrated and analyzed in a precise timeframe that allows for consistent comparisons worldwide.

The Anthropocene is more than just another of the many crises of the twienth and twenty-first centuries. This book is a child of the changed life circumstances brought about by Covid-19, but I have deliberately made few references to the coronavirus crisis. I wanted to resist the knee-jerk temptation to make a simple connection between the Anthropocene and the pandemic. But Covid-19, the most global problem there has ever been in history, shows both how precarious growth can become and the importance of humanly shared knowledge for solutions. The truly global pandemic was fueled by Anthropocene dynamics, such as worldwide economy-driven mobility, it abruptly stopped growth in places, but at the same time, Covid-19 has resulted in international cooperation and the global spread of knowledge about causes, risks, and countermeasures against global threats.

The realization of human-induced climate change and the experiences during the Covid-19 pandemic raise the question of how to respond quickly and globally to worldwide crises. The Anthropocene has brought people together even more intensely than the climate crisis already has, it has opened up new political agendas, and expanded the scope of what is conceivable and perhaps also what is feasible (Chakrabarty 2021; Clark and Szerszynski 2021: 1, 3).

Despite their similarities, the Covid-19 crisis and the climate crisis, and especially the crisis of the Anthropocene, all differ significantly from each other. The Anthropocene reveals the long-term interconnectedness of humans with the animate and inanimate worlds, which was swiftly elucidated by the Covid-19 pandemic. Apart from these, however, there are fundamental differences, as already shown in the case of climate change (Renn 2021: 26). Firstly, unlike Covid-19, the Anthropocene is not recognizable as an immediate threat to the individual. Secondly, in contrast to Covid-19, people feel much less directly threatened and are therefore less willing to accept measures that restrict their freedom. Thirdly, if we look beyond everyday situations, measures taken against pandemics usually lead to success relatively quickly; in the case of a lockdown, after a few weeks. By contrast, if we were to reduce the entire human-produced CO_2 emissions to zero today as a measure against Anthropocene trends, the current concentration would continue to act as a greenhouse gas for about 75 years. While feedback on successful measures against pandemics can be given quickly, there will always be disputes about the "culprits" and about what brings how much and whether measures are sufficient in the fight against the Anthropocene.

The Covid-19 crisis represents a brief exceptional situation, almost like an acute illness, where one expects a quick onset and therefore (mostly) accepts the therapy. In contrast to such a kairological crisis (from Greek: *kairos*), the climate crisis is a chronological crisis (from Greek *kairos*), a long-term crisis (Nassehi 2020: 908). The Anthropocene, according to this distinction, is a very long-term, "chronological," and truly chronic crisis. We humans deal with it like a chronic illness, namely by coming to terms with it. The quick reaction that is actually necessary is constantly postponed. Solutions for kairological crises do not depend exclusively on political will, as critics of capitalism sometimes assume, or on technological possibilities, as proponents of the good Anthropocene assume. To be able to deal productively with such a complex problem, the complexity of society itself must be taken into account (Nassehi 2020: 909). If we see political conditions as the main cause of the Anthropocene and the Anthropocene represents a global threat, this will typically call for the abolition of such conditions. However, if the consequences of previous power structures leave lasting traces in the history of the Earth, they will still be of geological significance even when they have already ended in social history (Folkers 2020: 599).

I will now summarize the main statements of this book, first by citing the findings, then deriving further conclusions, and finally, by taking up positions in some theses.

8.2 Findings

The Anthropocene is both a fact-based hypothesis as well as a consequential fundamental metaphor for crisis perception. The geoscientific real concept firstly consists of a deep-time epochal notion in geology, and on the other hand, a more present-oriented Earth system concept in the Earth System Sciences. The diversity and polyphony of the use of the term testify to its enormous potential for discourses, analyses, and as a heuristic tool for reworking fundamental anthropological questions. The Anthropocene is challenging in many ways: it blurs established disciplinary and, above all, categorical boundaries, but the debate also leads to new boundaries being set.

The concept of the Anthropocene first emerged as a thesis in the Earth System Sciences and only later became a subject of geology. The Anthropocene can only be understood as a matter and a problem if geological concepts of deep time become clear. Within the geosciences, the approach of geology with its concrete field orientation and at the same time deep-historical view must be distinguished from the globally oriented and—from a geological point of view!—present-oriented Earth System Sciences.

Even if anthropogenic climate change represents the paradigmatic case of the Anthropocene, the Anthropocene should *not* be equated with climate change.

The Anthropocene is about several interrelated consequences of human activity in the geosphere. The central scientific question in the Anthropocene is how humanity is to be understood as a causal force in the Earth system and how people reflect on their own role in it. While the geosciences are mainly dedicated to exploring the Anthropocene as a cumulative effect, the main contribution of cultural anthropology—as well as history—can be cause-oriented: what social as well as cultural structures and processes typically produce Anthropocene effects?

The greatest potential for provocation for humanities scholars and cultural scientists lies in the word "Anthropocene." Nevertheless, the word is significantly more present in the humanities and cultural sciences than in the life sciences and the geosciences, even if conceptually frayed. In the geosciences and life sciences, the Anthropocene is conceptually not standardized, but it is understood much more narrowly than in the cultural sciences. The debate about the beginning of the Anthropocene opens up many more questions, more than just the dating. It opens up questions about the causal role of humans in the geosphere.

A central methodological problem of the discussion about the Anthropocene as an epoch—and a cause of notorious misunderstandings—is the discrepancy between synchronous boundaries between time units of geology and diachronic epochal boundaries in archaeology, historical anthropology, and the historical sciences. The important chronological decision here is when to let the Anthropocene begin: (a) when the human *capacity* for massive environmental change emerges *somewhere* (according to the historical and archaeological argument); or much later, namely (b), from the time when the *cumulative effects* of human activity are detectable *worldwide* (according to the majority view in geology). In any case, Quaternary geology should fully incorporate permanent effects of human activity into its subject matter, e.g. depict techno-fossils in geological maps by default. Quaternary geology should fully incorporate the permanent effects of human activity into its subject matter, for example, feature techno-fossils on geological maps by default.

Anthropological contributions to date have tended to selectively incorporate aspects of the Anthropocene concept and interpret them according to their own specific interests, thereby greatly transforming the concept. The applied anthropological question concerns what clues are provided by our knowledge about people and cultures to be able to develop alternatives to the current trend towards an uninhabitable geosphere.

The rejection of the Anthropocene as a formal geological epoch in 2024 raises the question of a meaningful term in an intensified form. Due to the very short duration of the Anthropocene as seen from a geological perspective, we cannot know whether human traces will actually consolidate into a globally synchronous *and geologically durable* marker. On the other hand, the effects of humans on the living world and especially the reduction in biological diversity can already be clearly documented, and the extinction of species is irreversible. We should therefore discuss the term anthropozoic again. As it refers to paleontological time concepts, "Anthropozoic," like the Anthropocene, implies a deep-time perspective. However, with the ending "zoic" (instead of "cene"), it is less focused on precise global temporal correlation, and at the same time, it is not only anthropocentric, but also biocentric. From an existential, psychological, and ethical point of view, Anthropozoic fits well with humans: we humans are organisms, we are animals oursell... and on these grounds, animals and plants are closer to us than stones.

8.3 For a Grounded Anthropology

We probably need to redefine rationality and take on the efforts of a new enlightenment.
Peter Finke (2022: 22)

The Anthropocene denotes an anthropogenic break in the dynamics of the geosphere, which simultaneously indicates a worldwide sociocultural crisis. The Anthropocene as a problem is too important, and the word "Anthropocene" too rich in content to be used as a shorthand for talking about globalization, global environmental problems, sustainability issues, or human-induced climate change, or to be misused for purely rhetorical maneuvers. The Anthropocene as a phenomenon is something different from globalization or globality. As a concept, the Anthropocene also represents a different idea from the theories on globalization and global history, because the Anthropocene implies much larger spatial scales and longer time scales.

The synthetic and broad concept of the Anthropocene in the humanities, cultural sciences, and social sciences is an important complement to Anthropocene concepts in Earth System Science and geology. However, it is not to be seen as a replacement. The different forms of scientific evidence in the Anthropocene discussion are important, and the debate about alternative terms is relevant as a social and cultural fact. Analytically speaking, however, it is not very fruitful because it is rarely precisely defined, and it is dominated by language games.

If the holistic concept of culture in cultural anthropology is consistently spelled out, it can make a central contribution to the understanding of the Anthropocene. The distinction between nature and culture becomes more difficult in the face of the Anthropocene; however, it is still useful as an analytical tool. From a moderately cultural materialist perspective, a refined and Anthropocene-informed nature-culture distinction is more productive than an unclear monism, such as that which dominates in the talk of the "more-than-human."

The humanities, social sciences and cultural studies criticisms of natural scientific Anthropocene concepts are politically relevant and also scientifically important. The Capitalocene and Technocene in particular are alternative concepts that should be taken seriously. The criticisms from non-scientists also raise fruitful questions for the geosciences. On the other hand, the notorious criticism from the social sciences that the concept of the Anthropocene per se negates socio-economic differences is simply a straw man. The topic is existentially too important and anthropologically too rich to misuse the word as a fashionable term or to build terminological sandcastles that soon collapse.

A view informed by the Anthropocene rather than the Holocene does not rely on permanent solutions, but on adaptation through continuous global learning from mistakes. In socio-political terms, the Anthropocene makes it abundantly clear that in times of post-factual uncertainty, a strictly knowledge-based understanding of "felt truths" is needed. In the historical sciences, there is much to suggest that the Anthropocene should be set as an epoch, albeit possibly under a different name. For me, the term "Anthropocene" seems suitable for establishing an awareness of the urgency of countermeasures in our schools and in the public sphere. As a period, the Anthropocene should be introduced into the chronologies of the natural history-informed cultural, social, and historical sciences.

With its focus on current and local—albeit intensively entangled—forms of life, cultural anthropology is challenged in the face of the Anthropocene as a temporal and spatial mega-macro phenomenon. The central contribution of cultural anthropology lies in clarifying the geospheric effects caused by the *nexus* of deep-time dynamics and recent human agency in local life-worlds by means of a realistic approach. This includes the local and partly culture-specific perceptions and narratives of the Anthropocene, not least because they are causally important for adaptive solutions. One desideratum of cultural anthropology consists in approaches to understanding stratigraphic contexts and to supplementing them from an anthropological perspective through their localization and contextualization. Here, cultural anthropology could cooperate more closely with historians and archaeologists. The applied ethnological question is what clues our knowledge of people and cultures can give us to develop alternatives to the current trend towards an uninhabitable geosphere.

8.4 Conclusion in Five Theses

> *... the current definition of "Anthropocene" is a bet on the future and, as such, its meaning and eventual formalization depend on the future development of human affairs.*
> Valentí Rull (2018a: 56)

1. *Anthropocene is a fruitful concept.*

"Anthropocene" is more suitable than any other term for drawing attention to the worldwide, long-term, and largely unintended effects of accumulated human activity on the geosphere. The concept summarises the paradox of far-reaching human intervention in the geosphere coupled with an increasing loss of control. It warns of the uncontrollable human modification of the Earth's surface and creates pressure for action in politics and society. The most important benefit of thinking about the predicaments of the Anthropocene is the long-term perspective on the Earth's surface as a system. This deep-time view enables a recognition of the long geobiological genesis of the foundations of human life. It frees us from short-term thinking and also opens up a view into the more distant planetary future, both aspects that are under-exposed when looking only at globalization.

2. *The Anthropocene contains a specific critique of society.*

Insights into the Anthropocene crisis automatically lead to criticism of the growth-orientation of almost all current lifestyles and economic forms. They imply questions about the problematic effects of consumption-oriented living and extractive activities inside and outside capitalist structures. The idea of the Anthropocene does not abolish conventional criticism of society and culture, but gives it a new framework that is not simply global and ecological, but also temporally deep. In order not to lose this specific deep-time perspective, the word "Anthropocene" should be used with caution, unlike what is currently the case in a lot of recent works that is critical of capitalist society. Even if there are connections, the Anthropocene cannot simply be equated with anthropogenic climate change, all environmental crises, all world problems, the crisis of capitalism, inequality and post-colonial or patriarchal structures.

3. *The holistic cultural concept of anthropology fits perfectly with the Anthropocene.*

A cultural concept that sees culture as the sum of what humans have made, regardless of the materiality of what has been made, fits the Anthropocene as "human time." An understanding of the Anthropocene also requires cultural concepts that emphasize systemic relationships. Cultural Anthropology is the cultural science that centers a concept of culture that is holistic and focuses on psychological, social, and material artifacts. Methodologically, cultural anthropology highlights the experiential dimensions of life. It is precisely this anthropological concept of culture and this methodological approach to local life-worlds that can contribute to "socializing" the insights of geology and the Earth System Sciences.

4. *In view of the Anthropocene analysis is just as important as synthesis.*

Any approach to the Anthropocene from the humanities, social sciences and cultural studies should seriously consider the geoscientific findings on the Earth system and, above all, the geological concepts of deep time. The analytic distinctions between nature, environment, culture and society should be constantly problematised, but without abandoning them. Talk of the "more-than-human" that is currently popular in cultural studies aims to avoid dualisms and emphasize, in a synthetic view, the entanglements of human activity with nature and thus our responsibility for nature. These are important aspects, but there is a risk of normative overload. Contributions from the humanities often show a defensive attitude towards the natural sciences and flowery to obscurantist formulations. Taken together, this hinders a factual and empirical rather than a primarily normative analysis of human-ecological contexts. An actual interdisciplinary understanding of a phenomenon as complex as the Anthropocene requires clear definitions, an analytical theoretical approach and an open attitude towards the natural sciences despite a critical stance.

5. *Geoanthropology could establish a grounded science of the human.*

The Anthropocene refers to a future problem of cultural evolution exacerbated by our current way of life. For the human sciences, I see the potential in the Anthropocene concept to establish a truly comprehensive anthropology, a human science of the whole human being, based on an urgent human problem. A scientific understanding of the Anthropocene requires knowledge about the evolutionary genesis of human cultural capacities, insights into the mechanisms of cultural evolution and facts about the long-term evolution of complex social systems. This includes the development of inequality, as well as the current neoliberal dynamics and future development of the world system. For a comprehensive understanding of the Anthropocene, we essentially need a broad anthropology, i.e. cultural anthropology supplemented above all by historical science, evolutionary research, complexity research and earth sciences—geo-anthropology!

Navigating the Anthropocene Jungle—A Media Guide

Given the overflowing literature and extensive bibliography of this book, I would now like to provide some guidance for navigating the literature jungle. These are titles that I highly recommend and, with few exceptions, those that are easily accessible, almost all in paperback. With the exception of one essay, these are books and websites. In each case I only provide the author, the main title, and the year of publication; full details can be found in the bibliography.

If you only want to read one book as an introduction to the topic, I recommend *Anthropocene* by Erle Ellis (2018). The book is an informative synthesis, written in an understandable way and penned by a geoscientist who is particularly open to cultural science questions and findings. If you want to read two books and are particularly interested in cultural aspects, you "must" read the book *The Anthropocene* for an introduction by two cultural scientists (Horn and Bergthaller 2020). This is a very clear but also demanding introduction and one of the few overviews with an explicit cultural science orientation. The authors managed to capture clearly and without jargon even the difficult trains of thought in cultural theory argumentation. For readers who want to find out more about the precarious state of the world, or more precisely the geosphere, which is heavily influenced by human activity, I recommend a compact book by the earth system scientist Mark Maslin: *How to Save Our Planet* (2021).

For all those who are looking for a philosophically informed overview of the brief history of ideas of the Anthropocene and the long history of its predecessors across a wide range of disciplines, but also in art and literature, Carolyn Merchant's short volume *The Anthropocene and the Humanities* (2020a) is highly suitable. Merchant goes particularly into topics relevant to cultural studies. With his work *In the Face of the Anthropocene* (2019), Ian Angus provides a skeptical view of the concept from a committed socialist perspective. Unlike many other critics, however, he does not throw the baby out with the bathwater and also discusses alternative terms critically. With his materialist approach, he criticizes the currently dominant but limited cultural-scientific approaches. Taken together, the

C. Antweiler, *Anthropology in the Anthropocene*, Anthropocene – Humanities and Social Sciences, https://doi.org/10.1007/978-3-031-74591-1

publications mentioned so far would be my recommendation for the core texts of an undergraduate course.

If you would like to engage scientifically with the topic and initially only want to read one or two articles, then I recommend starting with a new overview (Zalasiewicz et al. 2021a, b). This extremely rich and dense contribution addresses the different conceptions of the Anthropocene between Earth science, geology, and many other sciences, clears up many misunderstandings, and also provides the central empirical findings in graphics. It was written in collaboration with 27 scientists of various provenance and is freely available online. For a very up-to-date overview in the humanities field, including the more philosophical aspects, I additionally recommend the detailed keyword list in the *Encyclopedia of Science* by Perrin Selcer (2021), which is freely available online.

Are there any comprehensive and sophisticated overviews in book form? By far the best English-language overview currently available is the explicitly interdisciplinary book *The Anthropocene*, written by historian Julia Thomas, biologist Mark Williams, and paleobiologist and AWG chairman Jan Zalasiewicz (2020a, b). The highly informative text is based on the latest research and actually combines findings and perspectives from the natural sciences with those from the humanities and cultural sciences. Simon Lewis & Mark Maslin offer a broad and highly informative overview in *The Human Planet* (2018), which duly considers the natural science aspects but is also open to social science aspects. The overview book that most clearly addresses the geological core of the Anthropocene thesis is *The Birth of the Anthropocene* (2016), written by Jeremy Davies. This can be explored in greater depth with the stylistically convincing book *Time Consciousness* by geologist Marcia Bjornerud (Bjornerud 2018). Taken together, the texts mentioned so far comprise my suggestions for the core literature of a master's program.

If you want to delve deeply into the geoscientific and cultural-scientific argumentation of the various disciplines, the magnificent German book *Evidenz für das Anthropozän* by environmental historian Fabienne Will (2021) is the right choice. The author combines document analysis, interviews with important actors in the debate, and her own participant observations and analyzes all of this in detail. For those who want to deepen their scientific foundation and learn about the current state of geoscientific knowledge in detail, *The Anthropocene as a Geological Time Unit* (Zalasiewicz et al. 2020a, b) provides broad and in-depth information. Turner (2023) provides a well-founded overview, especially of the scientific questions and findings, based on 101 questions and answers with very good graphics and references to original literature.

The discussions outside the natural sciences are best followed in essays and anthologies. The first German-language anthologies were *Das Anthropozän* by Renn & Scherer ("The Anthropocene," 2015) and *Die Welt im Anthropozän. Erkundungen im Spannungsfeld zwischen Ökologie und Humanität* by Haber et al. ("The World in the Anthropocene. Explorations in the Tension between Ecology and Humanity," 2016), which contain contributions on art and philosophy in addition to geoscientific ones. The contributions in the volume edited by Hannes Bajohr (Bajohr 2019) are significantly more philosophical, while the essays in Adloff and

Neckel (2020) and Laux and Henkel (2018) are more sociological. The professional framework is broader in the volume edited by Stascha Rohmer and Georg Toepfer *Anthropocene—Climate Change—Biodiversity* (Rohmer and Töpfer 2021). Among the English-language anthologies, I recommend *The Anthropocene and the Global Environmental Crisis* (Hamilton et al. 2015). Hanusch et al. introduce planetary thinking in a simple way and with many illustrations in *Planetar Denken* ("Thinking Planetarily," 2021), while Clark & Szerszynski's *Planetary Social Thought* (2021) is significantly more challenging fare.

Specialist journals have existed since the 2010s, such as *Anthropocene* (Elsevier, since 2013), *The Anthropocene Review* (since 2014), and *Anthropocenes—Human, Inhuman, Posthuman* (since 2020) and *Anthropocene Science* (since 2022). Since 2012, *Environmental Humanities* has been a broadly culturally oriented journal on environmental change, and two of the most recent, interdisciplinary, and fully *Open Access* journals are *Earth's Future* (since 2019) and *Elementa: Science of the Anthropocene* (since 2021) as well as the *Nature Briefing Anthropocene* newsletter of the journal *Nature* (since 2023; go.nature.com/3parhth).

French research (such as Beau and Larrère 2022) is little recognized in the English-speaking discussion. Exceptions are the translated works of Bruno Latour and the eminently important book *The Shock of the Anthropocene* by historians Christophe Bonneuil and Jean-Baptiste Fressoz (2016). A good introduction is provided by the informative as well as opinionated volumes by the environmental scientist and publicist Nathanaël Wallenhorst (2019, 2020) and the compact overview *L'anthropocène* by the geologist Michel Magni in the series *Que-sais-je?* (Magni 2021). In the tradition of French thematic cartography, French colleagues have presented the *Atlas de l'anthropocène*, the first map series on the Anthropocene (Gemenne et al. 2021). Those who wish to delve deeper into the French discussion, which often strays far from the original concept, are well advised with the *Dictionnaire critique de l'anthropocène*, a comprehensive but affordable encyclopedia (Alexandre et al. 2020).

For those looking for a light but stimulating anthropological introduction to global problems, Marc Augé's *L'Avenir des Terriens* (*The Future of Earth Dwellers*, 2017) offers a good and concise start. The book places the Anthropocene in the context of humanity's current problems and encourages readers to develop their own position. Those who want to delve deeper into the cultural anthropology of the Anthropocene should read Thomas Hylland Eriksen's *Overheating* (2016), which brings together climate change and the Anthropocene and also introduces ethnographic case studies. If you are looking for a current anthropological case study, I recommend Amelia Moore's *Destination Anthropocene* (2020). With respect to the relevance of the anthropological concept of culture and other cultural concepts for the Anthropocene and also for international measures against the Anthropocene, those interested will find that Lourdes Arizpe Schlosser's *Culture, International Transactions and the Anthropocene* (2019) is the best choice. Anna Tsing's *Arts of Living on a Damaged Planet* (2017) also offers many visual suggestions for anthropological extensions of the concept in artistic directions. In my opinion, the best and, in many ways, most profound anthropological book on the Anthropocene is Richard Irvine's *An Anthropology of Deep Time* (2020).

In terms of popularization in German-speaking countries, the publishing house Oekom has made a name for itself with several popular science titles on the Anthropocene (e.g., Ellis 2018; Oekom 2021). There is also a book series called "Library of Sustainability: Rediscoveries for the Anthropocene," in which classic texts appear in German translation (e.g., Crutzen 2019a, b). Although the topic is present in the German-speaking academic publishing landscape, it rather tends to be in niches. The publisher Matthes & Seitz in Berlin in particular has made a significant contribution to the topic. In addition to the authoritative anthology mentioned above (Renn and Scherer 2015), several central theoretical texts have been published in German translation (Wark 2017; Morton 2017; Tsing 2019a, b, c; Moore 2019; Augé 2017; Danowski and Viveiros de Castro 2019; Bennett 2020; Bjornerud 2018, as well as recently Tsing's classic from 2005 (in 2024).

For those who do not want to or cannot spend much money and still emphatically want to read scientific books on the topic, I recommend buying the two equally up-to-date, compact, and well-illustrated overviews in the "Very Short Introductions" series from Oxford University Press, *Anthropocene* by Erve Ellis (2018) and *Earth System Science* by Tim Lenton (2016), which will only cost about 20 Euros in total. I also recommend reading the above-mentioned essay by Zalasiewicz and 25 co-authors.

For people able to read German I warmly recommend Reinhold Leinfelder's "Anthropozäniker" blog, where he explains the topic in an understandable way with the use of texts, graphics, and also self-drawn sketches (Leinfelder 2018a, b, 2019). This is well complemented by the Anthropocene Cookbook (Leinfelder et al. 2016), as well as the graphic novel anthologies by Alexandra Hamann and co-authors (Hamann et al. 2013, 2014). The often unfamiliar terms used in the debates about the Anthropocene are explained on the website of the Federal Agency for Civic Education ("Der Anthropozän-Wortschatz," Springer 2019). The Smithsonian Institution in Washington offers the informative English-language website *The Age of Humans. Living in the Anthropocene*.

A good read for laymen is Jan Zalasiewicz' book *The World After Us*. As a leading palaeontologist for the Anthropocene he provides a mixture of geoscientifically informed predictions and speculation about the future (Zalasiewicz 2009). Zalasiewicz also wrote a very descriptive and extremely well illustrated introduction to geology (Zalasiewicz 2022). For children and young people, I recommend the full-length film *The Epoch of Man* (Baichwal et al. 2018). A look at Google Earth would also certainly be enlightening for them, where it is not only possible to see almost every place on the planet near and far, but recently also to follow the change in the Earth's surface at any location by means of time travel over 37 years (Google Earth Engine 2021). If smaller children ask you about the Anthropocene, tell them the joke about the planets meeting and wait for their questions.

For me, three artists are particularly interesting and exciting in their interpretation of Anthropocene themes. One is the photographer Edward Burtynsky with his images of the Earth's transformation, which are sometimes astonishing and sometimes frightening, but also provide an aesthetic look at the Anthropocene (*The Anthropocene Project*; Burtynsky et al. 2018, 2020). Jason deCaires Taylor addresses environmental disasters with his underwater installations at coral reefs in various parts of the world—"underwater museums" in North Africa, Caribbean, and the Indian Ocean (deCaires Taylor et al. 2014, see Fig. A.1). These sculptures, modeled after local people, do not only point to coral bleaching, but also impressively highlight the Anthropocene as a problem of the human–nature relationship. One of these works exolicitly entitled "Anthropocene" depicts a VW-beetle slowly overgrown by corals under water (https://underwatersculpture.com/).

The projects of the Icelandic-Danish artist Ólafur Eliasson are also artistically stimulating, such as the giant sun in the *Tate Modern* ("The Weather Project," 2003, with Unilever), or the project "Your mobile expectations" in collaboration with BMW (Eliasson 2007), or his 2021 transformation of the Fondation Bayeler near Basel into a nature–culture nexus entitled *Life* (https://www.fondation beyeler.ch/olafur-eliasson-life), which stayed open day and night. In my opinion, the works of Alexandra Arènes (Ait-Touati et al. 2022; Arènes 2022) offer an aesthetically captivating synthesis of artistic-speculative approaches with detailed mapping. In terms of literature, I find the works of Amitav Ghosh, a novelist with an anthropological background, particularly inspiring. I can recommend his novel *The Hungry Tide*, which is set in the permanently crisis-ridden habitat of the Sundarbans in the Ganges-Brahmaputra delta in South Asia (Ghosh 2005), and also his captivating non-fiction book *The Great Derangement* (Ghosh 2016), as well as his recent work entitled *The Nutmeg's Curse* (2023), where he illustrates colonial, long-term environmental history by way of the history of the Moluccan Islands in Indonesia.

Fig. A.1 Jason De Caires Taylor: "Vicissitudes", Grenada. (Photo: Jennifer Roording) https://commons.wikimedia.org/wiki/File:Jason_deCaires_Taylor_Viccisitudes_Grenada.jpg

Glossary[1]

Abduction Inferring from observed effects or processes to presumed causes, especially in historical questions as ➔ Retrodiction, see ➔ Reverse Engineering

Actor Network Theory (ANT) An approach that emphasizes the interconnection of human actors with non-human elements of action (actants)

Actualism (Principle of Actualism, Principle of Uniformity, Uniformitarianism) A principle of geology to assume the same physical forces for the past as today, "nature makes no leaps" (steady state), against assumptions of unknown forces and against ➔ Catastrophe theories

Actualistic Paleontology Research within actual geology on the present-day embedding of dead organisms in sediments, their decay, the formation of traces, the redistribution and preservation (biostratinomy), and thus the present formation of possible geological records of the future ➔ Actualism, ➔ Technofossils

Agency The ability or possibility of individuals or corporate actors to act in their own interests, see the important distinction between human agency (agency 1) and the effects of human activity (agency 2) in the face of the Anthropocene

Agency 1 (German: "Handlungsmacht") Freedom of disposition, possibilities of action in the face of restrictive conditions (constraints)

Agency 2 (German: "Wirkmacht") Human agency through non-intentional behavior, unintended or uncontrollable side effects of accumulated intentional actions

[1] **Preliminary Remark:** In this section I explain the meaning of terms as they are used or understood in this book. In the sprawling literature on the Anthropocene, there are different definitions for many terms, even for supposedly simple words. For example, "geosphere" is understood as the entirety of the spherical surface, comprising air, rocks, soils, water, ice, and life, while in some works it is restricted to the Earth's surface. Alternative terms to "Anthropocene" ending with "-cene" are not included. Arrows (→) refer to related terms within this glossary. This should contribute to a conceptual network on the Anthropocene that emphasises the factual and conceptual connections.

C. Antweiler, *Anthropology in the Anthropocene*, Anthropocene – Humanities and Social Sciences, https://doi.org/10.1007/978-3-031-74591-1

Amerindian cosmology (Amerindian Thinking, Amerindian Perspectivism) Indigenous cosmological idea that humans existed before the (rest of the) world and thus form the primordial ground of the world or the original humanoid state of all beings

Animism Worldview according to which living and sentient beings exist outside of the organismic, sociality also exists between humans and animals, and persons also exist beyond humans, and thus there are also moral relationships between humans and animals, such as in ➔ Amerindian cosmology

Anthrome (Anthropogenic Biome, Anthroecosystem) also anthropozone or "domesticated land" Large areas of nature that have been reshaped by human land use (➔ anthropogenous) and significantly transformed by technology; a distinction is made between around 18–25 anthromes ➔ Anthropogenic form

Anthropic principle Quasi-teleological principle that the conditions of existence of the observing human being must be compatible with the observation of the universe. The factual development has made the evolution of humans possible (weak variant) or the universe is cosmologically oriented towards humans (strong, often theological and controversial variant)

Anthropocene 1, Earth scientific (Anthropocene, Anthropocène) Earth historical caesura in the status of the Earth system (new state) around the middle of the twentieth century, but other dates in discussion (➔ Geochronology), mainly due to the effects of human activity as the dominant geofactor ➔ Earth System Science

Anthropocene 2, Geological (Stratigraphic Anthropocene) Sum of the events during the very short and currently ongoing geochronological "epoch," since the mid-twentieth century, but other dates in discussion (➔ Geochronology), and all deposits during this time interval (synchronous), regardless of whether of anthropogenic, partly natural, or entirely natural genesis, as a geological "series" (➔ Chronostratigraphy); dominant view in the ➔ Anthropocene Working Group, in contrast to the minority view of ➔ Anthropocene 3

Anthropocene 3, Extended, primarily Earth scientific, archaeological, and historical (diachronous Anthropocene, diachronic Anthropocene, anthropogenic) The sum of all empirically verifiable influences of *Homo sapiens* (➔ anthropogenic) on the geosphere of all times (diachronic) and also only regionally, view of the minority in the ➔ Anthropocene Working Group (AWG), ➔ Archaeosphere

Anthropocene 4, Humanities, cultural and social sciences (consequential meta-level, the responsible Anthropocene) Synthesis term for consequences of the break caused by human-made environmental change and the corresponding reflections, especially on human responsibility and status, extremely inconsistently used for the meta-level to the analytical levels ➔ Anthropocene 1, 2, and 3

Anthropocene Science Socially engaged research on the conditions of human life in the context of the human-dominated Earth system, in contrast to methodologically similar ➔ Earth System Science explicitly value-based and focussed on scenarios and proposals for responsible ➔ Stewardship for the well-being of humans.

Anthropocene Working Group (AWG, officially the Working Group on the "Anthropocene") A multidisciplinary committee (officially in place from 2009 to 2024) of the Subcommission on Quaternary Stratigraphy of the ➔ International Commission on Stratigraphy (ICS), with the task of examining dating proposals for the Anthropocene, developing a plausible periodization, and conducting its own ➔ Stratigraphic investigations

Anthropocentrism View of humans as the center of the world, which makes them the standard of the natural order; as a moral Anthropocene (A.), the position that only or mainly humans are moral objects, against ➔ Biocentrism; A. is usually associated with the search for the difference between humans and (other) animals (anthropological difference, ➔ Speciesism), instead of thinking in a human-centered way ➔ Biocentrism ➔ Holism

Anthropogenic Form Relief form in the landscape that was created by humans (for example, field terraces, mounds) or created as a side effect of human activity, for example, through erosion (field hills, gullies), can be formed as so-called quasi-natural forms ➔ Anthrome

Anthropogenic metabolism Material and energetic flows of human societies or humanity

Anthropogenic Caused, produced, or influenced by humans, in this case, especially human-made changes in the natural environment, mostly related to landscapes, formerly often seen as the opposite of an untouched environment in equilibrium, in contrast to ➔ Anthrome

Anthropogeography (Human Geography, Cultural Geography, Geography of Humankind) Subfield of geography that deals with the spatial impact of humans or with ➔ Cultural landscapes

Anthropology Comprehensive science of the "whole human," includes biotic and cultural objects

Anthropology of ontology (Anthropology of Ontologies, Multiple Ontologies Approach) Collective term for primarily anthropological approaches to the human–nature relationship: symmetric anthropology (Latour), monistic anthropology (e.g., Descola), anthropology beyond the human (e.g., Kohn), and artifact-oriented anthropology (material studies) ➔ Ontological turn

Anthroposphere(HumanHabitat,alsoAnthrosphere,also➔Technosphere) The parts of the ➔ geosphere and ➔ biosphere that are permanently influenced by humans, see ➔ Anthrosols ➔ Anthrome

Anthrosols Soils permanently shaped by economic activity with very strong or complete changes to their original properties, e.g., through soil cultivation, influx of material (e.g., soil removal, sods, or household waste), or irrigation ➔ Technosols

Anthroturbation (Human Bioturbation) Anthropogenic stirring up of the soil, through mining, drilling, pipes, tunnel constructions, landfills, and nuclear tests, similar to the digging by other organisms in the soil, bioturbation

Anthrozoology Research field on human-influenced animals ➔ Domestication

Archaeosphere Sum of the significantly anthropogenically modified land regardless of time, similar to Anthropocene 3, in contrast, see ➔ Anthropocene 2

Archive (Memory) In contrast to concrete archives, the totality of (natural and societal) evidence of the climate and past human activity that can be found in the environment, e.g., in the form of artifact remains, traces, soils, or organisms, for the documentation of human-made changes in the broadest sense, also the collections of natural history museums

Area Studies (Regional Sciences, Transarea Studies) Interdisciplinary research fields which deal with the history and present of cultural regions (areas, cultural realms) that are larger than a country or a language area, but smaller than a continent

Assemblage A nexus of materially heterogeneous things, e.g., paper, plastic, plant remains, animal carcasses, and human excrements on garbage dumps or in hybrid, anthropogenically reshaped landscapes

Big History An approach to historical science that not only supplements the history of humanity with prehistory, but also extends it to the history of the Earth and cosmic history, back to the Big Bang, against ➔ Presentism (see ➔ Deep History)

Biocentrism (Teleological) An ethical position that all living beings are moral objects due to their vitality and purposefulness, more restricted than ➔ Biocentrism and ➔ Anthropocentrism

Biodiversity (Biological Diversity) The variety of genes (genomic variation), species (species richness, biodiversity in the strict sense), systematic units (taxonomic diversity), and ecosystems

Biological Old Regime (Biological Ancient Regime) A fundamentally limited economic system based on renewable resources that was globally dominant until the nineteenth century: agriculture based on solar energy, trees as fuel source and recycling, resulting in environmental change primarily through agriculture

Biophilia (Love of Nature) An unconscious inclination of humans to seek proximity or a close connection to plants, animals, and natural stimuli, usually assumed to be a genetically determined inclination, more specifically a preference for biotic diversity ➔ Biodiversity

Biosphere The totality of life, more precisely the organically animated parts of the outer spherical shell of the Earth, the ➔ Geosphere ➔ Critical Zone

Capacity for Culture Combined abilities for language, anticipation, tool-making, assumption of others' intentions (theory of mind), and emotional empathy ➔ Evolution of cultural capacity

Carrying Capacity Maximum number of organisms that a given ecosystem can sustainably feed

Catastrophe Theories (Catastrophism, Neo-catastrophism, Punctualism) Several theories in geology and evolutionary research that view catastrophes as a fundamental factor in evolution, in contrast to ➔ Gradualism

Chronostratigraphy Area of geological stratigraphy in which time is materially applied to geological sequences (mostly rock), here the "Anthropocene series," as opposed to the abstract "Anthropocene epoch" ➔ Geochronology

Climate Long-term state of the atmosphere near the ground, in contrast to ➔ Weather

Co-evolution Mutual adaptation of several strongly interacting species over long periods of time in phylogenetic history

Connectivity Degree of interconnectedness of elements in a network, e.g., between humans in a social network or between humans and other primates in an ecosystem ➔ Entanglement

Columbian Exchange The exchange of plants, animals, and diseases between Europe and the Americas in the wake of the exploratory expeditions

Conviviality normative orientation towards a coexistence that aims to achieve a balance based on solidarity and also includes non-human beings**; ➔ deep ecology**

Cosmopolitanism Idea or program of a cross-cultural global citizenship

Critical Zone (Earth's Critical Zone, Zone Critique) The thin, vulnerable, and highly complex boundary layer shaped by life, in which human behavior has a particularly strong impact and which extends from the treetop to the unweathered rock ➔ Anthropogenic

Cultural Anthropology (Social Anthropology, Sociocultural Anthropology) Research into contemporary ways of life (➔ Culture) worldwide using an experiential and comparative methodology, today partly similar to ➔ European Ethnology

Cultural Core Central socio-economic area of a culture that shapes other areas

Cultural Landscape Ecosystems that are primarily shaped by deforestation, drainage, irrigation, terracing, and buildings both qualitatively and quantitatively and are interconnected on a small scale as mosaic-like structures

Cultural Model Schematic representation of the world that gives meaning and guides action and is largely shared in a human collective (➔ Culture 2)

Cultural Primatology The study of cultural traditions among non-human primates, to be distinguished from ➔ Ethnoprimatology

Culture 1 (Culture with capital C) The shaping of human existence based on non-genetically (tradigenetically) transmitted information, the host of effects of human ➔ Innovations

Culture 2 (culture, a culture) The way of life of a human collective, e.g., an ➔ Ethnic Group

Culture 3 (human culture) Totality of human cultures in time and space

Decolonisation social movement and practice that analyses historical and ongoing neo- and post-colonial power relations ('coloniality') and seeks to counteract them; not to be confused with decolonization as the achievement of independence.

Deep Ecology A normative, political philosophy that aims at the best for nature and humans and strives for a comprehensive global ecosystem in which humans are integrated; ➔ Convivialism

Deep History A historical approach wherein the entire human past is considered to be historical, contrary to the dualistic concept of prehistory/history; the basis is not writing, but the evolution of anatomically modern humans (e.g., that of DNA); contrary to ➔ Presentism, see ➔ Longue Durée ➔ Big History ➔ Deep Time

Deep Time (Geological Time) Strictly speaking, time that cannot be qualitatively perceived or comprehended by humans, in contrast to ➔ Geological Time; more broadly, a geological and evolutionary long-term perspective on the past and future, which in the depth of time, clearly goes well beyond ➔ Longue Durée, contrary to ➔ Presentism,; see ➔ Deep History ➔ Big History

Degrowth (Post-Growth) Goal and societal vision of a different economy, based on a critique of the societal dominance or even hegemony of the goal economic growth

Deprovincializing Programmatic withdrawal of the characteristic overemphasis on European regions and their historical influence on non-European regions, contrary to ➔ Eurocentrism

Developmentalism critical term for ideas of directed long-term social and economic change based on an unquestioned assumption of progress

Disparity Spatial distribution differences, spatial ➔ inequality

Diversity (Heterogeneity) Horizontal diversity of cultures (cultural diversity), as opposed to vertical ➔ inequality; or diversity of life ➔ Biodiversity

Domesticated Land Landscapes dominated by humans which emerge from natural biomes such as primary forests, savannahs, and grasslands ➔ Anthrome

Domestication 1 Taming and breeding of domestic animals and cultivated plants, controlled reproduction of plants or animals through their specially provided or supervised nutrition and through their protection from damage

Domestication 2 Social imprinting of nature as part of modernization

Domino Effect Cascade-like sequence wherein the tipping of one subsystem causes the tipping of other subsystems, see ➔ Tipping Point

Dystopia Ideas of a negative future (see also ➔ Utopia)

Earth History History of the development of the planet and the corresponding field of geology and paleontology

Earth System Closed, self-regulating system of the geosphere through the interaction of non-human physical, chemical, and biological as well as human forces, including the interactions and repercussions within the ➔ Spheres on planet Earth

Earth System Science/s (ESS) Field of research on the ➔ Earth system which uses a multidisciplinary approach and large amounts of data to analyse the many factors of current global change and develop scenarios, in contrast to ➔ Geology, which mainly deals with earlier time periods

Ecocide Metaphorical term, modeled after suicide, for large-scale damage to ecosystems, also flippantly referred to as "ecological hara-kiri"

Ecocriticism Literary and cultural studies approach to investigating the relationships between humans and the environment and their transformations in texts and other media

Ecological Footprint The biologically productive area on Earth that is necessary to sustain a person's lifestyle and standard of living under current production conditions; a key indicator of ➔ Sustainability

Ecological Imperialism (Environmental Imperialism) Actions or measures imposed by external powers with negative effects on the living environment, e.g., through the introduction of diseases and animal and plant species, recently also used for the imperial appropriation of environmental goods or externally imposed environmental regulation

Ecological Novelty Fundamental newness in ecological systems

Ecomodernism (Ecomodernism, Eco-Pragmatism, Neo-Environmentalism) School of thought that advocates technological solutions for Anthropocene effects and is oriented towards positive development (➔ Good Anthropocene) ➔ Geoengineering ➔ Terraforming

Econiche (Ecological Niche) Functional position of a species in a biotope based on the interaction between species and biotope, i.e., based on reciprocal relationships rather than spatial conditions

Ecosystem Services (Natural Capital) Quantifiable contribution of a section of the environment to a larger ecosystem or to humans

Entanglement Strong relational forms of connectedness of something or relational embedding in shared practices, worlds and contexts of meaning; in the Anthropocene context, the interconnectedness between humans, other living beings and inanimate things, ➔ Holism; in other contexts, entanglements, e.g. between colonised and colonising societies

Environment Existentially relevant part of the ➔ Nature surrounding a living being and a framework for adaptation

Environmental History Area of historical science that deals with the effects of nature on humans (➔ Environment), the human-made environment, and environmental understandings and nature-related practices

Environmentality Totality of the effects of governance in relation to the environment, local environmental regulation (e.g., of forests), and environmentally conscious attitudes

Epoch Geological-stratigraphic time unit that marks major changes as an age and minor changes as a period

Ergodics (Ergodicity, location-for-time substitution) In geomorphology, the principle that observations of the average of many things at one point in space are equivalent to the observation of individual things over time, in short: spatial distribution is a surrogate of temporal sequence, time averages can be replaced by spatial averages

Essentialism Essential determination of an entity, e.g., of " man," " woman," or a culture or religion, e.g., "Islam"

Ethnocentrism (ethnocentricity, sociocentrism) Narrow view of the world from the perspective of a "we" group, generally associated with the high valuation of one's own community and the devaluation of other communities, to be distinguished from ➔ Eurocentrism

Ethnography 1 Descriptive branch of cultural anthropology, in contrast to theory-oriented or comparative ➔ Cultural Anthropology

Ethnography 2 An anthropological monograph, a book-length work about an ethnic group or a (mostly) localized topic

Ethnography 3 An approach of qualitative sociology, especially in the English-speaking world

Ethnoprimatology Study of the permanent coexistence of humans with non-human primates in what are now "normal" anthropogenically shaped environments, to be distinguished from ➔ Cultural Primatology

Eurocentrism Narrow view of European conditions, dynamics, norms, and values as "normal," "natural," or "desirable," to be distinguished from ➔ Ethnocentrism, see ➔ Nostrocentrism

European Cultural Anthropology (Empirical Cultural Science, Cultural Anthropology, Folklore, Folklore Studies) Contemporary research into human ways of life, especially in one's own society (➔ Culture), today partly similar to ➔ Cultural Anthropology in terms of subject matter, theory, and method

Event Ecology (Abductive Causal Eventism) A principle derived from pragmatist philosophy to methodically infer possible causes from an event or effect; temporally backwards and spatially inwards, or gradually outwards from micro to meso and macro contexts (progressive contextualization), similar to ➔ Geological Event; ➔ Reverse Engineering

Evidence The "how" of scientific knowledge in the form of socially negotiated knowledge that is considered to be "plausible" and thus legitimately validated; with specific evidence practices linked to it, depending on the context, e.g., stabilization

Exaptation (Exaption) Functional change of phenotypic traits from non-adaptive or neutral traits to later adaptive function due to change in the overall system

Exceptionalism critical term about the assumption that Homo Sapiens is ontologically very special; a special version of ➔ Speciesism

Extractivism An economic system based on the exploitation, use, and export of large quantities of natural resources, usually within a colonial, neo-colonial or post-colonial framework

Flat ontology Concepts of reality that, unlike hierarchical concepts, do not assume the privileged status of any entity ➔ Ontological Turn ➔ Animism

Formalism Approach of economic theory that emphasizes the universal principles of economics, especially individual interests and limited resources, in contrast to ➔ Substantivism

Frontier In history the spatially or systemically advancing boundary between "civilization" and "nature," e.g., an expanding boundary of resource extraction

Gaia The co-evolution of the atmosphere and biosphere in a system that forms a ➔ Homeostasis

Geoanthropology A recently proposed term for the study of all of the processes, mechanisms, and natural and cultural-historical trajectories that lead human societies into the Anthropocene

Geochronology An area of geological stratigraphy in which geological time is divided abstractly and hierarchically, similar to chronology in historical science, here the "Anthropocene epoch" in contrast to the materially oriented "Anthropocene series" ➔ Chronostratigraphy

Geoengineering (Climate Engineering, Geo/Climate Engineering) Earth management through deliberate intervention in the earth system, usually in the form of deliberate and large-scale climate change , e.g., by capturing carbon from industrial emissions and storing it in underground repositories (CCS, CDR) and solar radiation management (SRM)

Geoethics (1) Ethics corresponding to or suggested by the scientific findings on the Anthropocene

Geoethics (2) Ethics among geoscientists or in the geosciences

Geofactor Circumstances that form the functional structure of a landscape, including relief, climate, rock, soil, water balance, vegetation, and time; in addition to natural geofactors, factors created by humans are also increasingly being counted among them

Geohistory The history of human influence on the geosphere, in contrast to ➔ Earth History

Geological Event Drastic changes from the perspective of deep time, such as faunal changes, mass extinction of species, and sudden disturbances of the geosphere, e.g., due to asteroid impacts

Geological Time The measurable periods of time that extend significantly beyond the ➔ Longue Durée, the incomprehensibility of which is referred to as ➔ Deep Time

Geological Time Scale (GTS) Official time structure of the Earth's history, which is seamless and hierarchically structured ➔ International Chronostratigraphic Chart (ICC) ➔ Geological Time ➔ Deep Time

Geology Science of rocks and Earth history covering long geological periods of time ➔ Deep Time, as opposed to ➔ Earth System Science/-s

Geomorphology Science of the surface shape of the Earth

Geosphere Upper layer of the Earth, spherical shell, consisting of several ➔ Spheres

Global History History of worldwide connections, contacts, exchanges, transfers

Global North Wealthier countries or parts of countries that dominate the global economy ➔ Global South

Global South Poorer countries and regions that are marginalized in the context of globalization ➔ Global North, Southern Theory

Global Sustainable Development Goals (SDGs) The 17 goals that aim to achieve lasting prosperity, equality, and resource conservation for the world

Globalization (Mondialization) Worldwide spatial and structural spread of phenomena and networking as well as the awareness of these processes or the unity of the world, humanity, the planet ➔ Cosmopolitanism

Golden Spike (Golden Nail, geological signature) Informal term for Global Boundary Stratotype Sections and Point (GSSP), a plate (bronze in color) that is placed as a marker in a geological type section and thus indicates a worldwide simultaneous (isochronous) lower limit a geological time unit ➔ Geochronology ➔ Chronostratigraphy

Good Anthropocene (great Anthropocene) An optimistic school of thought that hopefully seeks solutions for a civilization that is compatible with nature and mainly considers technical measures for this purpose, prototypically a conscious influence on the atmosphere ➔ Technosphere ➔ Geoengineering

Gradualism Assumption in paleobiology and evolutionary biology that evolution usually proceeds evenly ➔ Actualism, as opposed to punctualism, saltationism or neocatastrophism ➔ Catastrophe Theories

Great Acceleration The phase of high industrialization since about 1950, with exponentially increasing socio-economic trends in socio-economic developments and changes in the planetary Earth system

Great Transformation 1 According to Polanyi, the conflictual transition from agrarian society to industrial society in the nineteenth century, to be distinguished from the normative project ➔ Great Transformation 2

Great Transformation 2 (also the "Third Great Transformation") normative programme or project of a radical social upheaval by means of a restructuring of technology, economy, and society in the context of the multiple crises of the twentieth century towards a society embedded in natural cycles, more specifically with solidary relationships with the non-human living world as a socio-ecological transformation ➔ WBGU ➔ Great Transformation 1

Habitability Permanent habitability of the Earth or certain biomes or habitats for all living beings, qualitatively different from an anthropocentric ➔ Sustainability

Heterogeneity Diverseness between species, genes, and ecosystems (➔ Biodiversity) and cultures and within cultural collectives; horizontal difference in order to be differentiated from ➔ Inequality

Holism 1 (in Cultural Anthropology) View of the components of ➔ Culture as a comprehensive system

Holism 2 (in Environmental Ethics) Position that everything in ➔ Nature is a moral object; in contrast to ➔ Anthropocentrism and ➔ Biocentrism

Holobiont (metaorganism) large organism, e.g. that of humans, understood as a whole organism, as a collective or symbiosis of different organisms, e.g. in the immune system or in the form of bacterial microbiota in the human gut ➔ Assemblage

Holocene Geological warm period of the last approximately 11,700 years, relatively stable in temperature, popularly known as the "post-glacial period" ➔ Pleistocene

Homeostasis (Homoiostasis) Steady-state equilibrium, dynamic equilibrium in temporal change, in the organism, transferred to larger systems ➔ Gaia

Hubris (human mastery) human overconfidence in the Anthropocene, an unreflected and infinite confidence in technical solutions to global environmental problems, underpinned by a quasi-religious exaltation of science

Human exceptionalism view of humans as unique, as an ontological exception, often in connection with the separation of culture (1, 2, 3) from nature, which is to be conquered prometheically, a view contradicting a ➔ biocentric or evolutionary worldview

Human-Animal Studies Research into the human–animal relationship, in which animals are considered as living beings with their own agency ➔ Agency ➔ Multispecies Ethnography

Hyperobject An object that cannot be experienced directly due to its wide spatial distribution, but is designed in such a way that it can be measured and future developments can be modeled;

Index Fossil (Marker) Fossils that characterize a specific geological layer ➔ Stratigraphy ➔ Golden Spike

Inequality Differences between the top and bottom, between the more powerful and wealthier and the less powerful and poorer people or collectives, in contrast to horizontal differences ➔ Diversity ➔ Heterogeneity

Infrastructure Large-scale and ecosystemically consequential landscape changes, e.g., dams and offshore oil drilling stations, in contrast to the conventional and Marxist understanding of infrastructure

Innovation (Novelty) A socially accepted invention

Interdisciplinarity Cooperation of different scientific disciplines, occasionally synonymous with multidisciplinarity, to be distinguished from ➔ Transdisciplinarity

Intergovernmental Panel on Climate Change (IPCC, informally known as World Climate Council) Institution for summarizing the state of research on climate change as a basis for science-based but political decisions

Intergovernmental Science-Policy Platform on Biodiversity and Ecosystem Services (IPBES, informally World Biodiversity Council) United Nations organization for providing the international community with scientific knowledge ➔ IPCC

International Chronostratigraphic Chart (ICC) (International Chronostratigraphic Scale) Hierarchical structure of Earth's history in ➔ Chronostratigraphy, which forms the basis of the official geological chronology ➔ Geological Time Scale

International Commission on Stratigraphy (ICS) Sub-commission of the ➔ IUGS, a advisory body of the Geological Society of London responsible for the classification of the geological eras ➔ AWG

International Geosphere-Biosphere Programme (IGPB) International research program on Global Change from 1986 to 2015

International Union of Geological Sciences (IUGS) International association based in London, responsible for the geological time scale ➔ ICC, whose stratigraphic sub-commission is the ➔ ICS

Invasive Species (Non-native Species, Neobiota) Referring to the current habitat of alien organisms that have existed in other regions of the world over a long period of time

Just-so Story (Conjecture) Critical term for speculative causal justifications and speculative, non-testable, origin or sequence narratives especially in the evolutionary sciences

Knowledge Economy Totality of the knowledge-related institutions and processes of one or more societies

Longue Durée (La longue durée) History of long-term historical processes instead of a chronology of ongoing events; greatly extended in terms of time in the concepts of ➔ Deep Time ➔ Deep History and ➔ Big History

Majority World The totality of the world's population centers in Asia, Africa, and Latin America, roughly equivalent to the former term "Tricont" for these three continents, as an alternative to terms such as "Third World" or "Developing World" or "Global South," in contrast, see ➔ WEIRD

Mass Extinction Sudden extinction of many species in evolution, a central criterion for differentiating between geological epochs

More-than-human (other-than-human, beyond-the-human) A cross-disciplinary approach that emphasizes the interconnectedness of humans as organisms, their culture, and other living beings or even inanimate things ➔ Multispecies Ethnography ➔ Human-Animal Studies

Multispecies Ethnography (Interspecies Ethnography) Anthropological approaches that consider the exchange and interaction of humans and other living beings (and in some cases, also material phenomena) as fundamental for cultural theory and therefore for ethnographies ➔ Human-Animal Studies

Narrative (a "story") Collective, meaningful, and politically or socially mobilizing narrative with protagonists, chain of events, plot, causes, effects, morals as well as a specific spatial and temporal structure

Natural Experiment (Quasi-experiment) Method of retrospectively evaluating events and processes that have occurred in reality and which, as a field of variation, resemble a controlled experiment

Nature All parts of the world not made by humans; one part of nature is the human ➔ Environment

Negative Anthropology The conscious rejection of the possibility of characterizing "the" human being, in contrast to ➔ Essentialism

Neoliberalism A variant of market-oriented economic principles through the free market, free trade, and private property, for which the state only provides the appropriate framework

Neolithic Revolution (Agricultural Revolution, Transition) Transition from a mobile way of life to permanent settlements and somewhat later from a hunter-gatherer economy to agriculture and domestication and stockpiling; starting around 11,000 BC in present-day Iraq, 9500 BC in China, 5500 BC in Mexico, and 4500 BC in eastern North America

Niche Construction (Human Niche Construction) Process in which environmental changes by one generation t become partly permanent and thus form a basis for the supposedly "natural" living conditions of the next generation t+1 ➔ Econiche

Nostrocentrism inadmissible generalization of one's own experiences and worldviews to other cultures or times, exaggerated universalization or absolutization that presents one's own as "unmarked," similar to ➔ Eurocentrism

Ontological Turn Utilization of knowledge from non-Western worldviews, e.g., the so-called ➔ Perspectivism of Amerindian societies, especially the human-culture-environment nexus, sometimes propagated as an alternative to "Western" nature-culture concepts ➔ Animism ➔ More-than-human ➔ Pluriverse

Paleontology The science of past life, the study of fossils of animal (paleozoology) and of plant origin (paleobotany)

Palimpsest Multiple overwriting of earlier, partially still-existing writings by other writings in a document ➔ Path Dependency

Panarchy (Scale Problem) Hierarchically structured complex systems in which functional interactions and mutual influences take effect between the hierarchical levels

Path Dependency (Historical Hang-Over, Legacy) Past events, processes, or decisions that have a channeling or solidifying effect through feedback, thus limiting flexible adaptation

Perspectivism A worldview that is particularly prevalent in Amazonian societies, according to which there are different realities instead of just different views of one reality

Planetary Boundaries (Planetary Limits, Planetary Thresholds) Earth system limits of loads and material flows that should not be exceeded in order to continuously secure human life ➔ Planetary Guardrails

Planetary Guardrails (Planetary Boundaries) Metaphorical expression for political goals to limit change in order to enable long-term survival ➔ Planetary Boundaries ➔ WBGU ➔ Great Acceleration

Planetary Health Addressing the generally negative impact of humans' expansive lifestyle on human health, especially through environmental pollution, species loss, and anthropogenic climate change; metaphorically also for the problems of a "sick" Earth

Pleiotropy One cause has many effects ➔ Polygenesis

Pleistocene The first period of the Quaternary, beginning 2.7 million years BP, replaced around 11,700 years BP by the ➔ Holocene

Plural Ecologies Coexistence of different ways of dealing with the environment in a collective, e.g., extractive usage orientation alongside preserving attitude, also for pluralistic scientific conceptions thereof

Pluriverse (Pluriversum, Pluriversality, Multiverse, Multinaturalism, Earthly multitudes) View of humanity and nature as inherently diverse and thus in no way reducible to any kind of unity; in contrast to a universalistic "one-world" perspective and generalizing statements

Politics of Scale Changes in societal power relations in the process of globalization through changes in the relationship between the spaces of regulation

Polygenesis One effect arises from multiple causes, syndrome ➔ Syndromes of Global Change ➔ Pleiotropy

Posthumanism (in variants, also ecocentrism, new animism, politics of nature, ➔ more-than-human) Scientific and ideological direction that emphasizes the limits of human action and cognition and wants to overcome the essence of being human, in contrast to ➔ Anthropocentrism, and generally contrary to any ➔ Speciesism ➔ Transhumanism

Presentism (Short Termism) Excessive concentration on time periods close to the present, a form of chronocentrism

Provincializing A historicization of Europe as an unquestioned prototype in social theory and in cultural comparisons, which is intended to enable self-reflection and criticism of thoughtless universal categories, not to be confused with ➔ Cultural Relativism

Proxy (Marker) Indirect indication for determining a global period transition, layer (primary marker), or, in this case, the onset of Anthropocene effects through technofossils (secondary marker)

Quasi-object objects that represent hybrids of nature and society, often in a materially distributed, difficult to grasp form, e.g. ozone in the atmosphere; ➔ hyperobject

Quaternary Geological period beginning about 2.59 million years ago, as the last phase of the Cenozoic (Cainozoic Era)

Ratchet Effect Cumulative cultural evolution in humans through precise cultural transmission and further development based on it, according to the principle of the American car jack

Red Queen Dynamics (Treadmill Syndrome) The constant need for change in competitive relationships, usually improved performance and adaptation, just to keep up in case competitors change or improve at the same time or if the environment changes

Relationship with Nature (German: "Naturverhältnisse") Socio-politically shaped interactions between societies and nature, which are constitutive for these societies and their history, above all as a relationship of domination

Resilience The ability of a system to absorb disturbances and to withstand them while maintaining its basic structure or returning to equilibrium

Retrodiction (Postdiction, Retrognosis, Post-Prediction, Subsequent Explanation) Statements about (unknown) facts or processes in the past (antecedents) which are derived from later states, events, or sequences, thus a past-oriented post-factual speculation that is similar to a diagnosis, a variant of ➔ Abduction ➔ Reverse Engineering

Reverse Engineering (Back Engineering, Backwards Engineering) A step-by-step reconstruction of the historical genesis of an evolved system back to earliest causes ➔ Retrodiction

Scalability Change in system properties due to size changes in space, time, and scope ➔ Scaling

Scaling (Upscaling, Downscaling) An expansion or enlargement (or reduction) of a measure in space or systemic scope without having to qualitatively change or improve the input ➔ Scalability

Science and Technology Studies (STS) Social and cultural studies on the real-world development of scientific knowledge

Sink Spaces that absorb more of a substance than they emit, in ecosystems, the place where substances remain, e.g., tropical rainforests as well as temperate and boreal forests as carbon sinks

Sixth Species Extinction (Sixth Mass Extinction) Term used to describe today's human-induced extinction of species, in reference to the five major extinction events in the history of life on Earth

Shifting Baseline Syndrome (generational environmental amnesia) a persistent downgrading of perceived 'normal' environmental conditions with every sequential generation due to a lack of experience and/or memory and/or knowledge of its past condition and thus a gradual change in the soceitally accepted norms:, a societal forgetiimg of fundamental changes in the environment

Social Metabolism Entanglement of social and value-based action with natural resources ➔ Relationship with Nature

Solastalgia Feeling of loss in the face of the major change or destruction of one's own habitat, a kind of homesickness while being at home, similar to nostalgia

Southern Theory An attitude that is both knowledge-oriented as well as a political project, seeing contributions from the ➔ Global South as an important source of knowledge for cultural and social science theory production

Speciesism Discrimination based on species affiliation, usually with an emphasis on the uniqueness or remarkability of humans compared to other species (human ➔ exceptionalism) ➔ Anthropocentrism

Spheres (in the geosciences) Interacting spherical surfaces of the air (atmosphere, parts of it, e.g., the stratosphere), rocks (lithosphere), soils (pedosphere), water (hydrosphere), ice (cryosphere), and life (biosphere), collectively forming the ➔ geosphere ➔ Earth system ➔ Anthroposphere

Stewardship Ecological responsibility to actively control future environmental dynamics, derived from the perception of catastrophic changes in the living world

Stratigraphy Part of geology that deals with layer sequences in rocks, their correlation, and their relative dating using ➔ Index Fossils, as well as methods of absolute age determination

Substantivism Approach of economic theory that emphasizes the social and thus also culture-specific basis of all economic activity, both factually and normatively, as opposed to ➔ Formalism

Succession (Ecological Succession) A directed directional sequence of colonization and extinction in plant communities determined by climate and soils, leading to a stable so-called climax community

Superorganism An older interpretation of biotic communities as an organism of closely connected species with a common evolutionary history

Sustainability Consumption and use of natural resources that take into account the interests of future generations of humans, in contrast to ➔ Habitability and ➔ Viability; ➔ Sustainable Livelihood

Sustainable Livelihood A concept that focuses on local living conditions, basic needs, economic security, and solutions such as locally sustainable farming methods, bioregionalism, and indigenous environmental concepts, contrary to sustainable development

Syndromes of Global Change (Global Syndromes) Characteristic negative developments or degradation patterns of ecosystems that can be identified in many regions of the world and are undesirable; they arise from typical constellations of natural and/or civilizational effects: one example is the "overexploitation syndrome" or the "green revolution syndrome," introduced by the German ➔ WBGU

Technosols Soils that are primarily made of materials like concrete, glass, bricks, rubble, plastic, household waste, and industrial waste ➔ Anthrosols

Technosphere Totality of material culture in the geosphere

Telecoupling Linking of human and natural phenomena over very large distances through flows of matter, people, other organisms, or information

Terraforming Expansion of human settlement by creating life-friendly structures on other planets ➔ Ecomodernism

Theory of World Society Theory stating that due to global networking, there is now only one communicative large collective in the world that no longer has a societal outside ➔ World System Theory ➔ World Culture

Tipping Point Sudden turnaround or crisis-like break in the temporal development of a complex system after a long latency period by exceeding limit and threshold values ➔ Planetary Boundaries

Tradition Persistent difference in behavior or action between populations of a species across generations

Transdisciplinarity Cooperation between different disciplines (➔ interdisciplinarity) plus inclusion of societal knowledge and non-academic actors in the design of research and teaching

Transformation Guiding concept in social, environmental, and spatial science for comprehensive and targeted changes to society and the living environment, largely for planned system change ➔ Great Transformation

Transhumanism Scientific and ideological direction that aims to modify, develop, and improve humans through technological transformation ➔ Posthumanism

Ultrasociality social life in populous to extremely large societies whose members almost all do not know each other personally

United Nations Environmental Programme (UNEP) Environmental sub-organization of the United Nations responsible for providing scientific knowledge to the international community ➔ IPBES

Universals 1 (Human Universals, Bio-Universals) Commonalities of all healthy human individuals, in contrast to ➔ Universals 2

Universals 2 (Cultural Universals, Human Universals, Pancultural Patterns) Commonalities of all human collectives, causally partly based on ➔ Universals 1, but analytically distinguishable

Viability Long-term livability due to an economy and environment that enable human life, see ➔ Sustainability and ➔ Habitability

Vulnerability Societal condition characterized by susceptibility, insecurity, and defenselessness

WBGU, Wissenschaftlicher Beirat der Bundesregierung Globale Umweltveränderungen German Advisory Council on Global Change Permanent advisory body of the German federal government on the environment and sustainability

WEIRD (Acronym for Western, Educated, Industrial, Rich, Democratic) A critical expression –ironically playing with the English word "weird" – for supposedly general knowledge in social sciences and psychology, which is in fact almost exclusively based on Atlanto-centric data, thus omitting large parts of the ➔ Majority World

Wicked problem Problem for which there is no certain recipe for a solution, often being a symptom of other problems

World Culture A system of cultural ideas and standardizations that permeates the cultures of the world, as well as global activities, such as sporting events, in contrast, see ➔ Theory of World Society and ➔ World System Theory

World System Theory (Theory of the World System) Theory which states that for centuries, there has only been one single economic system in the world, in contrast to ➔ Theory of World Society

Worlding (Colonized Space) Transformation of colonized landscapes through Western cartography, surveying, descriptions, and forms of travel, especially in the tropics of Asia

Bibliography[1]

AAA (American Anthropological Association) (2014). *Statement on Humanity and Climate Change*. (http://practicinganthropology.org/docs/01-29-15_AAA_CCS.pdf).
Abram, David (1996). *The Spell of the Sensuous. Perception and Language in a More-Than-Human World*. New York: Pantheon.
Acciaioli, Greg & Akal Sabharwal (2017). Frontierization and Defrontierization: Reconceptualizing Frontier Frames in Indonesia and India. In: Jaime Tejada & Bradley Tatar (eds.). *Transnational Frontiers of Asia and Latin America since 1800*. London: Routledge: 431–346.
Ackerman, Diane (2014). *The Human Age. The World Shaped by Us*. New York, N.Y.: Norton.
Adams, Richard Newbold (1978). Man, Energy, and Anthropology. *American Anthropologist* 80: 297–309.
Adelman, Jeremy (2017). What Is Global History Now? (https://aeon.co/essays/is-global-history-still-possible-or-has-it-had-its-moment).
Adelman, Jeremy, Elizabeth Pollard, Clifford Rosenberg & Robert Tignor ([3]2020). *Worlds Together, Worlds Apart. A History of the World from the Beginnings of Humankind to the Present*. New York: W. W. Norton.
Adloff, Frank (2020). „It's the End of the World as We Know It": Sozialtheorie, symbiotische Praktiken und Imaginationen im Anthropozän. In: Frank Adloff & Sighard Neckel (eds.). *Gesellschaftstheorie im Anthropozän*. Frankfurt/New York: Campus: 95–121.
Adloff, Frank & Tanja Busse (eds.) (2021). *Welche Rechte braucht die Natur? Wege aus dem Artensterben*. Frankfurt am Main: Campus (Zukünfte der Nachhaltigkeit, 3).
Adloff, Frank & Sighard Neckel (eds.) (2020). *Gesellschaftstheorie im Anthropozän*. Frankfurt & New York: Campus (Zukünfte der Nachhaltigkeit, 1).
Adloff, Frank, Benno Fladvad, Martina Hasenfratz & Sighard Neckel (eds.) (2020). *Imaginationen von Nachhaltigkeit. Katastrophe. Krise. Normalisierung*. Frankfurt & New York: Campus (Zukünfte der Nachhaltigkeit, 2).
Ahmann, Chloe (2019). Waste to Energy: Garbage Prospects and Subjunctive Politics in Late-Industrial Baltimore. *American Ethnologist* 46 (3). 328–342.
Ait-Touati, Frederique, Alexandra Arènes & Axelle Gregoire (2022): *Terra Forma. A Book of Speculative Maps*. Cambridge, Mass. & London: The MIT Press.
Albrecht, Glenn A. (2019). *Earth Emotions. New Words for a New World*. New Haven, Conn.: Cornell University Press.

[1] All electronic sources were checked on 1.8.2024.

C. Antweiler, *Anthropology in the Anthropocene*, Anthropocene – Humanities and Social Sciences, https://doi.org/10.1007/978-3-031-74591-1

Alexandre, Frédéric, Fabrice Argounès, Rémi Bénos, David Blanchon et al. (comitee de pilotage (2020). *Dictionnaire critique de l'anthropocène*. Paris: CNRS Éditions.

Almond, Rosamunde E. A., Monique Grooten & Tanya Petersen (eds.) (2020). *Living Planet Report 2020. Bending the Curve of Biodiversity Loss*. Gland: Worldwide Fund for Nature (WWF).

Altvater, Elmar (2013). Wachstum, Globalisierung, Anthropozän. Steigerungsformen einer zerstörerischen Wirtschaftsweise. *Emanzipation* 3(1). 71–88.

Altvater, Elmar (2017). Der Kapitalismus schreibt Erdgeschichte. *Luxemburg. Gesellschaftsanalyse und linke Praxis* 2/3 (Marx´te nochmal?!). 108–117 (https://www.zeitschrift-luxemburg.de/lux/wp-content/uploads/2018/01/Luxemburg_2017_2-3_Marx200.pdf).

Altvater, Elmar (2018). Beschleunigung und Expansion im Erdzeitalter des Kapitals. In: Rüdiger Dannemann, Henry W. Pickford & Hans-Ernst Schiller (eds.). *Der aufrechte Gang im windschiefen Kapitalismus. Modelle kritischen Denkens*. Berlin: Springer: 227–241.

Ameli, Katharina (2021). *Multispezies-Ethnographie. Zur Methodik einer ganzheitlichen Erforschung von Mensch, Tier, Natur und Kultur*. Bielefeld: Transcript (Kultur und soziale Praxis).

American Anthropologist (2019). Commentaries on "The Case for Letting Anthropology Burn". https://www.americananthropologist.org/online-content/blog-post-title-three-e323d.

Ammarell, Gene (2014). Whither Southeast Asia in the Anthropocene? Comments on the Papers from the 2014 Roundtable JAS at AAS: Asian Studies and Human Engagement with the Environment. *The Journal of Asian Studies* 73(4). 1005–1007.

Anders, Günther (2018). *Die Weltfremdheit des Menschen. Schriften zur Anthropologie*. ed by Christian Dries. München: C. H. Beck (orig. 1982).

Anderson, David E., Andrew S. Goudie, Adrian G. Parker (^{2}2013). *Global Environments through the Quaternary. Exploring Evironmental Change*. Oxford etc.: Oxford University Press.

Anderson, Eugene N. (2019). *The East Asian World-System. Climate and Dynastic Change*. Cham: Springer Nature (World-Systems Evolution and Global Futures).

Angerer, Marie-Luise Angerer & Naomie Gramlich (eds.) (2020). *Feministisches Spekulieren. Genealogien, Narrationen, Zeitlichkeiten*. Berlin: Kulturverlag Kadmos.

Angus, Ian (2016). *Facing the Anthropocene. Fossil Capitalism and the Crisis of the Earth System*". New York: Monthly Review Press.

Anonymous (2011). *The human epoch. Official recognition for the Anthropocene would focus minds on the challenges to come*. Nature 473, 19. May, 254.

Anonymus (2024). Are we in the Anthropocene yet? *Nature* 627: 466.

Anselm, Sabine & Christian Hoiß (eds.) (2017). *Crossmediales Erzählen vom Anthropozän. Literarische Spuren in einem neuen Zeitalter*. München: Oekom.

Anthony, David W. (2007). *The Horse, the Wheel, and Language. How Bronze-Age Riders from the Eurasian Steppes Shaped the Modern World*. Princeton, N.J.: Princeton University Press.

Anthropocene Working Group AWG (2019). Results of Binding Vote by AWG Released 21st May 2019. http://quaternary.stratigraphy.org/working-groups/anthropocene/

Anthropocene Working Group AWG (2023). Executive Summary. The Anthropocene Epoch and Crawfordian Age: proposals by the Anthropocene Working Group. Submitted to the ICS Subcommission on Quaternary Stratigraphy on October 31st, 2023. https://eartharxiv.org/repository/view/6853/

Antweiler, Christoph (1988). *Kulturevolution als transgenerationaler Wandel. Probleme des neueren Evolutionismus und Lösungsansätze dargestellt unter besonderer Berücksichtigung der angloamerikanischen Diskussion um sogenannte kulturelle Selektion*. Berlin: Dietrich Reimer Verlag (Kölner Ethnologische Studien, 13).

Antweiler, Christoph (1990). Das eine und die vielen Gesichter kultureller Evolution. Eine Orientierung zum begrifflichen Handwerkszeug des Neoevolutionismus. *Anthropos* 85(4–6). 383–405.

Antweiler, Christoph (1991). On Natural Experiments in Social Evolution: The Case of Oceania. In: Christoph Antweiler & Richard N. Adams (eds.). *Social Reproduction, Cultural Selection, and the Evolution of Social Evolution*, Leiden: E.J. Brill: 158–171.

Antweiler, Christoph (2007). Wissenschaft quer durch die Kulturen. Wissenschaft und lokales Wissen als Formen universaler Rationalität. In: Hamid Reza Yousefi, Klaus Fischer, Rudolf Lüthe & Peter Gerdsen (eds.). *Wege zur Wissenschaft. Eine interkulturelle Perspektive. Grundlagen, Differenzen, Interdisziplinäre Dimensionen*. Nordhausen: Verlag Traugott Bautz: 67–94.

Antweiler, Christoph (2008). Evolutionstheorien in den Sozial- und Kulturwissenschaften. Zusammenhangs- und Analogiemodelle. In: Christoph Antweiler, Nicole Thies & Christoph Lammers (eds.). *Die unerschöpfte Theorie. Evolution und Kreationismus in den Wissenschaften*. Aschaffenburg: Alibri Verlag: 115–141.

Antweiler, Christoph (2009). *Heimat Mensch. Was uns alle verbindet*. Hamburg: Murmann Verlag.

Antweiler, Christoph (2012a). *Inclusive Humanism. Anthropological Basics for a Realistic Cosmopolitanism*. Göttingen: V+R Unipress & Taipeh: National Taiwan University Press (Reflections on (In) Humanity, 4).

Antweiler, Christoph (2012b). Neuer Kosmopolitismus. Appiah weiter denken im Anthropozän. *FIPH-Journal* (Forschungsinstitut für Philosophie Hannover). 26–27.

Antweiler, Christoph (2017). Zur Universalität von Emotionen. Befunde und Kritik kulturvergleichender Ansätze. In: Tobias Kienlin & Leonie C. Koch (eds.). *Emotionen – Perspektiven auf Innen und Außen*. Bonn: Verlag Dr. Rudolf Habelt: 125–147 (Universitätsforschungen zur Prähistorischen Archäologie, 305, Kölner Beiträge zu Archäologie und Kulturwissenschaften, Cologne Contributions to Archaeology and Cultural Studies 2).

Antweiler, Christoph (2018). Urbanization and Urban Environments. In: Hilary Callan (general ed.). *The International Encyclopedia of Anthropology*. London etc: Wiley-Blackwell: 1–10. https://doi.org/10.1002/9781118924396.wbiea1585.

Antweiler, Christoph (2019a). Ethnologie braucht die Praxis. Der Beitrag der angewandten Ethnologie für die akademische Ethnologie. In: Sabine Klocke-Daffa (eds.). *Angewandte Ethnologie – Perspektiven einer anwendungsorientierten Wissenschaft*. Wiesbaden: Springer-Verlag: 99–116.

Antweiler, Christoph (2019b). Kosmopolitismus – Weltbürgertum im Anthropozän jenseits von Wunschdenken. In: Gerd Jüttemann (eds.). *Menschliche Höherentwicklung. Das Zusammenspiel von Soziogenese und Psychogenese*. Lengerich: Pabst Science Publishers (Psychogenese der Menschheit, 7). 109–120.

Antweiler, Christoph (2020). Southeast Asia as a Litmus Test for Grounded Area Studies. *International Quarterly of Asian Studies* 51 (3–4). 67–86 (https://crossasia-journals.ub.uni-heidelberg.de/index.php/iqas/article/view/13426).

Antweiler, Christoph (2021a). Soziale Kohäsion. Mechanismen sozialen Zusammenhalts und Beispiele aus Asien. In: Ulrike Niklas, Heinz Werner Wessler, Peter Wyzlic & Stefan Zimmer (eds.). *„Das alles hier". Festschrift für Konrad Klaus zum 65. Geburtstag*. Heidelberg: X-Asia Books: 1–21.

Antweiler, Christoph (2021b). Transdisziplinarität als ko-produktives Scheitern? Autoethnographische Reflexionen zu einem Projekt in Thailand. *Zeitschrift für Ethnologie* 145(1). 131–151.

Antweiler, Christoph (2022a). *Anthropologie im Anthropozän. Theoriebausteine für das 21. Jahrhundert*. Darmstadt: WBG (WBG Academic).

Antweiler, Christoph (2022b). Ethnologie im Anthropozän. Eine postulierte Megamakroepoche und ein lokal orientiertes wie gegenwartsbezogenes Fach. In: Roland Hardenberg, Josephus Platenkamp & Thomas Widlok (eds.). *Ethnologie als angewandte Wissenschaft. Das Zusammenspiel von Theorie und Praxis* (Studien zur Kulturkunde, 136) Berlin: Dietrich Reimer Verlag: 361–383.

Antweiler, Christoph (2024). Nischenkonstruktion im Anthropozän. Elemente einer geoanthropologischen Synthese. In: Manfred Hammerl, Sascha Schwarz & Kai P. Willführ (eds.). *Evolutionäre Sozialwissenschaften. Ein Rundgang*. Berlin: Springer VS: 15–37.

Antweiler, Christoph & Fabienne Will (2024). *Wir sind das Anthropozän. Schrecken und Schönheit globalen Wandels* (Foto- Exhibition). München: Deutsches Museum (22.4. -16.6. 2024).

Antweiler, Christoph, Michi Knecht, Ehler Voss & Martin Zillinger (eds.) (2019). *What's in a Name? Die Kontroverse um die Umbenennung der Deutschen Gesellschaft für Völkerkunde*. Bonn, Bremen, Köln, Siegen: boasblogs (Boasblogs Papers, 1). https://boasblogs.org/wp-content/uploads/2019/12/boasblogs-papers-1-2.pdf

Appel, Hannah (2012). Offshore Work: Oil, Modularity, and the How of Capitalism in Equatorial Guinea. *American Ethnologist* 39(4). 692–709.

Appadurai, Arjun (2013). *The Future as Cultural Fact. Essays on the Global Condition*. London: Verso.

Appiah, Kwame Anthony (2006). *Cosmopolitanism. Ethics in a World of Strangers*. New York & London: Norton.

Appiah, Kwame Anthony (2008). *Experiments in Ethics. Mary Flexner Lectures*. Harvard University Press.

ArchaeoglobeProject (2019). Earth's Early Transformation through Land Use. *Science* 365 (6456), 897–902

Archer, David (2009). *The Long Thaw. How Humans Are Changing the Next 100,000 Years of Earth's Climate*. Princeton etc.: Princeton University Press (Princeton Science Library).

Arendt, Hannah (1963). *Eichmann in Jerusalem*. New York: The Viking Press.

Arendt, Hannah (1998). *The Human Condition*, Chicago: The University of Chicago Press. (zuerst 1958).

Arènes, Alexandra (2022). *Design at the time of the Anthropocene: Reporting from the Critical Zone*. Ph.D. Thesis, Manchester, School of Environment, Education and Development (SEED), Department of Architecture.

Arènes, Alexandra, Bruno Latour & Jérôme Gaillardet (2018). Giving Depth to the Surface – an Exercise in the Gaia-graphy of Critical Zones. *Anthropocene Review* 5: 120–135

Arias-Maldonado, Manuel (2016). The Anthropocenic Turn: Theorizing Sustainability in a Postnatural Age. *Sustainability* 8,´(10). 1–17, https://doi.org/10.3390/su8010010.

Arias-Maldonado, Manuel & Zev Matthew Trachtenberg (eds.) (2019). *Rethinking the Environment for the Anthropocene. Political Theory and Socionatural Relations in the New Geological Epoch*. Milton Park & New York: Routledge.

Arizpe-Schlosser, Lourdes (2019). *Culture, International Transactions and the Anthropocene*. Berlin etc.: Springer (The Anthropocene: Politik –Economics –Society – Science, 17).

Arizpe, Lourdes (ed.) (1996). *The Cultural Dimensions of Global Change*. Paris: United Nations Publising (Culture and Development Series).

Asafu-Adjaye, John, Linus Blomquist, Stewart Brand, Barry Brook, Ruth DeFries, Erle Ellis, Christopher Foreman, David Keith, Martin Lewis, Mark Lynas, Ted Nordhaus, Roger Pielke, Rachel Pritzker, Joyashree Roy, Mark Sagoff, Michael Shellenberger, Robert Stone & Peter Taege (2015). *An Ecomodernist Manifesto*. www.ecomodernism.org/manifesto-english/

Asayama, Shinichiro, Masahiro Sugiyama, Atsushi Ishii & Takanobu Kosugi (2019). Beyond Solutionist Science for the Anthropocene: To Navigate the Contentious Atmosphere of Solar Geoengineering. *The Anthropocene Review* 6 (1–2). 19–37.

Atran, Scott (1998). Folk Biology and the Anthropology of Science. Cognitive Universals and Cultural Particulars. *Behavioral and Brain Sciences* 21:547–609.

Atran, Scott & Douglas L. Medin (2008). *The Native Mind and the Cultural Construction of Nature*. Cambridge, Mass. & London: The MIT Press (A Bradford Book) (Life and Mind: Philosophical Issues in Biology and Psychology).

Augé, Marc ([4]2014). *Orte und Nicht-Orte. Vorüberlegungen zu einer Ethnologie der Einsamkeit*. München: C.H. Beck.

Augé, Marc (2017). *L'Avenir des Terriens. Fin de la préhistoire de l'humanité comme société planétaire*. Paris: Albin Michel.

Augé, Marc (2019). *Die Zukunft der Erdbewohner. Ein Manifest*. Berlin: Matthes & Seitz (orig. »L'Avenir des Terriens. Fin de la préhistoire de l'humanité comme société planétaire", Paris: Albin Michel, siehe 2017).

Austin, Gareth (ed.) (2017). *Economic Development and Environmental History in the Anthropocene: Perspectives on Asia and Africa*. London etc.: Bloomsbury Academic.
Autin, Whitney J. & John M. Holbrook (2012). Is the Anthropocene an Issue of Stratigraphy or Pop Culture? *GSA Today* 22(7).60–61.
Bachmann-Medick, Doris (2016). *Cultural Turns. New Orientations in the Study of Culture*. Berlin: de Gruyter (orig. "Cultural Turns. Neuorientierungen in den Kulturwissenschaften", Reinbek bei Hamburg: Rowohlt Taschenbuch Verlag, Rowohlts Enzyklopädie).
Bänzinger, Peter Paul (2019). Weltgeschichte oder Anthropozän? Die ökologische Frage zwischen Dualismen und Verschwörungstheorie. In. *Geschichte der Gegenwart*. (https://geschichtedergegenwart.ch/weltgeschichte-oder-anthropozaen-die-oekologische-frage-zwischen-dualismen-und-verschwoerungstheorie/9.
Baer, Hans A. (2017). Anthropocene or Capitalocene? Two Political Ecological Perspectives. *Human Ecology* 45: 433–435.
Baer, Hans A. & Merrill Singer (eds.) ([2]2018). *The Anthropology of Climate Change. An Integrated Critical Perspective*. London etc.: Routledge.
Bätzing, Werner (2020). *Landleben. Geschichte und Zukunft einer gefährdeten Lebensform*. München: C.H. Beck.
Bätzing, Werner (2023). *Homo Destructor. Eine Mensch-Umwelt-Geschichte*. München. C. H. Beck.
Baichwal, Jennifer, Nicholas de Pencier & Edward Burtynsky (Regie) (2018): Die Epoche des Menschen. Film. 132 Min. (orig. „Anthropocene. The Human Epoch", 87 Min., Mongrel Media; (https://theanthropocene.org/film/).
Bajohr, Hannes (2019). Keine Quallen. Anthropozän und negative Anthropologie. *Merkur* 849 (5). 63–74 (https://www.merkur-zeitschrift.de/2019/05/02/keine-quallen-anthropozaen-und-negative-anthropologie/#enref-12523-38).
Bajohr, Hannes (eds.) 2020: *Der Anthropos im Anthropozän. Die Wiederkehr des Menschen im Moment seiner vermeintlich endgültigen Verabschiedung*. Berlin: Walter de Gruyter.
Bajohr, Hannes & Sebastian Edinger (eds.) (2021). *Negative Anthropologie. Ideengeschichte und Systematik einer unausgeschöpften Denkfigur*. Berlin: Walter de Gruyter (Philosophische Anthropologie, 12).
Baldwin, Andrew & Bruce Erickson (2021). Introduction: Whiteness, coloniality, and the Anthropocene. *Environment and Planning D: Society and Space* 38(1).3–11. (https://www.semanticscholar.org/paper/Introduction%3A-Whiteness%2C-coloniality%2C-and-the-Baldwin-Erickson/9d3ee6ed7e7020c4c1daa6a6854e8c6b39fe9cd8
Balée, William (2006). The Research Program of Historical Ecology. *Annual Review of Anthropology* 35:75–98
Balée, William (2013). *Cultural Forests of the Amazon. A Historical Ecology of People and Their Landscapes*. Tuscaloosa, Al.: University of Alabama Press.
Balmes, Hans Jürgen (2021). *Der Rhein. Biographie eines Flusses*. Frankfurt am Main: S. Fischer.
Balter, Michael (2013). Archaeologists Say the 'Anthropocene' Is Here – But It Began Long Ago. *Science* 340:261–262.
Bammé, Arno (2016). *Geosoziologie. Gesellschaft neu denken*. Weimar bei Marburg: Metropolis-Verlag für Ökonomie, Gesellschaft und Politik.
Bancone, C. E. P., Turner, S. D., Ivar do Sul, J. A., & Rose, N. L. (2020). The Palaeoecology of Microplastic Accumulation. *Frontiers in Environmental Science* 8: 574008.
Barad, Karen (2010). Quantum Entanglements and Hauntological Relations of Inheritance: Dis/continuities, SpaceTime Enfoldings, and Justice-to-Come. *Derrida Today* 3(2). 240–268.
Barca, Stefania (2020). *Forces of Reproduction. Notes for a Counter-Hegemonic Anthropocene*. Cambridge etc.: Cambridge University Press (Cambridge Elements. Elements in Environmental Humanities).
Barker J. P. & J. M. Barker (1988). Geoanthropology. In: *General Geology. Encyclopedia of Earth Science*. Boston: Springer (https://doi.org/10.1007/0-387-30844-X_34).

Barnes, Jessica & Michael R. Dove (eds.) (2015). *Climate Cultures. Anthropological Perspectives on Climate Change*. New Haven, Conn.: Yale University Press.

Barnosky, Anthony D. (2008). Megafauna Tradeoff as a Driver of Quarternary and Future Extinctions. *Proceedings of the National Academy of Sciences of the United States of America* 105: 11543–48.

Barnosky, Anthony D., Elizabeth A. Hadly, Jordi Bascompte, Eric L. Berlow, James H. Brown, Mikael Fortelius, Wayne M. Getz, John Harte, Alan Hastings, Pablo A. Marquet, Neo D. Martinez, Arne Mooers, Peter Roopnarine, Geerat Vermeij, John W. Williams, Rosemary Gillespie, Justin Kitzes, Charles Marshall, Nicholas Matzke, David P. Mindell, Eloy Revilla & Adam B. Smith (2012). Approaching a State Shift in Earth´s Biosphere. *Nature* 486: 52–58.

Bar-On, Y. M., R. Phillips & R. Milo (2018). The Biomass Distribution On Earth. *Proceedings of the National Academy of Sciences* 115(25). 6506.

Barrios, Roberto E. (2017). What Does Catastrophe Reveal for Whom? The Anthropology of Crises and Disasters at the Onset of the Anthropocene. *Annual Review of Anthropology* 46: 151–166.

Bartelson, Jens (2009). *Visions of World Community*. Cambridge: Cambridge University Press.

Barton, C. Michael, Isaac I.T. Ullah, Sean M. Bergin, Helena Mitasovac & Hessam Sarjoughian (2012). Looking for the Future in the Past: Long-Term Change in Socioecological Systems. *Ecological Modeling* 241:42–53.

Bartz, Dietmar & Carolin Sperk (2015). Die Archive des Anthropozän. In: Christine Chemnitz & Jes Weigelt (Ltg.). *Bodenatlas. Daten und Fakten über Acker, Land und Erde*. Berlin: Heinrich-Böll-Stiftung: 10–11.

Baskin, Jeremy (2015). Paradigm Dressed as Epoch: The Ideology of the Anthropocene. *Environmental Values* 24(1). 9–29.

Baskin, Jeremy (2019). *Geoengineering, the Anthropocene and the End of Nature*. Cham: Palgrave Macmillan.

Bassermann, Markus (2017). *Marxistisches Framework des Anthropozän. Ansätze zur Erfassung und Erklärung: Mit uns die Sintflut?* München: Grin Publishing.

Bateson, Gregory (1985). *Ökologie des Geistes. Anthropologische, psychologische, biologische und epistemologische Perspektiven*. Berlin: Suhrkamp (orig: "Steps to an Ecology of Mind: Collected Essays in Anthropology, Psychiatry, Evolution, and Epistemology", Chicago: The University of Chicago Press, 1972).

Baucom, Ian (2020). *History 4° Celsius: Search for a Method in the Age of the Anthropocene*. Durham & London: Duke University Press (Theory in Forms).

Bauer, Andrew M. & Mona Bhan (2018). *Climate Without Nature. A Critical Anthropology of the Anthropocene*. Cambridge: Cambridge University Press.

Bauer, Andrew M. & Erle C. Ellis (2018). The Anthropocene Divide. Obscuring Understanding of Social-Environmental Change. *Current Anthropology* 59(2). 209–227.

Baur, Bruno (2021). *Naturschutzbiologie*. Bern: Haupt Verlag.

Bauriedl, Sybille (eds.) (2015). *Wörterbuch Klimadebatte*. Bielefeld: Transcript.

Bayer, Anja & Daniela Seel (eds.) (2016). *„all dies hier, Majestät, ist deins": Lyrik im Anthropozän. Anthologie*. Berlin: kookbooks.

Beau, Rémi & Catherine Larrère (dir.) (2022). *Penser l'Anthropocène*. Paris: Les Presses Science Po.

Beck, Ulrich 2004: *Der kosmopolitische Blick oder: Krieg ist Frieden*. Frankfurt am Main: Suhrkamp Verlag (Edition Zweite Moderne).

Beck, Ulrich, Scott Lash & Anthony Giddens (1994). *Reflexive Modernization*. Cambridge: Polity Press.

Becker, Egon, Diana Hummel & Thomas Jahn (2011). Societal Relations to Nature as a Common Frame of Reference for Integrated Environmental Research. In: Matthias Groß (eds.). *Handbuch Umweltsoziologie*. Wiesbaden: VS Verlag für Sozialwissenschaften, 75–96.

Becker, Egon, Thomas Jahn & Diana Hummel (2006). Naturverhältnisse. In: Egon Becker & Thomas Jahn (eds.). *Soziale Ökologie. Grundzüge einer Wissenschaft von den gesellschaftlichen Naturverhältnissen*. Frankfurt am Main: Campus Verlag: 174–197.

Begon, Michael & Colin R. Townsend ([5]2021). *Ecology. From Individuals to Ecosystems.* Hoboken, NY. & Chichester: John Wiley & Sons.
Behringer, Wolfgang ([6]2022). *Kulturgeschichte des Klimas. Von der Eiszeit bis zur globalen Erwärmung*. München: C.H. Beck.
Bélanger, Pierre (2007). Underground Landscape: The Urbanism and Infrastructure of Toronto's Downtown Pedestrian Network. *Tunnelling and Underground Space Technology* 22: 272–292.
Bélanger, Pierre (2021). Airport Landscape. Altitudes of Urbanization. Harvard University Graduate School of Design (https://www.gsd.harvard.edu/exhibition/airport-landscape-altitudes-of-urbanization/).
Belardinelli, Sofia (2022). The Human-Nature Relationship in the Anthropocene: A Science-based Philosophical Perspective. *Azimuth*, October: 19–33.
Bendik-Keymer, Jeremy David (2020). *Involving Anthroponomy in the Anthropocene: On Decoloniality*. London etc.: Routledge (Routledge Research in the Anthropocene).
Benediktsson, Karl & Katrin Anna Lund (eds.) (2010). *Conversations with Landscape*. Farnham & Burlington: Ashgate (Anthropological Studies of Creativity and Perception).
Benjamin, Craig, Esther Quaedakers & David Baker (eds.) (2020). *The Routledge Handbook of Big History.* Oxon: Taylor & Francis (Routledge Companions).
Benjamin, Walter (1982). *Das Pasasagen-Werk.* Frankfurt am Main: Suhrkamp Verlag.
Bennardo, Giovanni (ed.). 2019: *Cultural Models of Nature. Primary Food Producers and Climate Change.* London & New York: Routledge (Routledge Studies in Anthropology, 52).
Benner, Susanne, Gregor Lax Paul J. Crutzen, Ulrich Pöschl, Jos Lelieveld & Hans Günter Brauch (eds.) (2021). *Paul J. Crutzen and the Anthropocene: A New Epoch in Earth'sHistory*. Cham: Springer Nature (TheAnthropocene: Politik – Economics – Society – Science).
Bennett, Carys E., Richard Thomas, Mark Williams, Jan Zalasiewicz, Matt Edgeworth, Holly Miller, Ben Coles, Alison Foster, Emily J. Burton & Upenyu Marume (2018). The Broiler Chicken as a Signal of a Human Reconfigured Biosphere. *Royal Society Open Science* 5: 180325: 1–11 https://royalsocietypublishing.org/doi/pdf/10.1098/rsos.180325.
Bennett, Jane (2020). *Vibrant Matter. A Political Ecology of Things*. Durham, NC.: Duke University Press, 2010, John Hope Franklin Center Books.
Benson, Melinda Harm & Robin Kundis Craig (2017). *The End of Sustainability. Resilience and the* Future of *Environmental Governance in the Anthropocene*. University Press of Kansas (Environment and Society).
Bergthaller, Hannes (2020). Thoughts on Asia and the Anthropocene. In: Gabriele Dürbeck & Phillip Hüpkes (eds.). *The Anthropocenic Turn. The Interplay Between Disciplinary and Interdisciplinary Responses to a New Age*. New York: Routledge (Routledge Interdisciplinary Perspectives on Literature, 117). 77–90.
Bergthaller, Hannes & Peter Mortensen (eds.) (2018). *Framing the Environmental Humanities*. Leiden: E. J. Brill.
Berkes, Fikret (2003). Rethinking Community-based Conservation. *Conservation Biology* 18: 621–30.
Berkes, Fikret, J. F. Colding & Carl Folke (eds.) (2003). *Navigating Nature's Dynamics. Building Resilience for Complexity and Change*. Cambridge: Cambridge University Press.
Berlin, Brent (1992). *Ethnobiological Classification. Principles of Categorization of Plants and Animals in Traditional Societies*. Princeton, NJ: Princeton University Press.
Berry, Thomas (1998). *The Dream of the Earth*. San Francisco: Sierra Club Books.
Bertelmann, Brigitte & Klaus Heidel (eds.) (2018). *Leben im Anthropozän. Christliche Perspektiven für eine Kultur der Nachhaltigkeit*. München: Oekom Verlag.
Bessire, Lucas & David Bond (2014). Ontological Anthropology and the Deferral of Critique. *American Ethnologist* 41(3). 444–456.
Bhaskar, Ron, S. Esbjörn-Hargens, N. Hedlund and Hartwig, M. (eds.) (2015). *Metatheory for the 21st-Century: Critical Realism and Integral Theory in Dialogue*. London: Routledge,
Bhattacharyya, Debjani (2018). *Empire and Ecology in the Bengal Delta. The Making of Calcutta*. Cambridge etc.: Cambridge University Press (Studies in Environment and History).

Bhattacharyya, Debjani & Maria J. Santos (2023). „Wir verändern unseren Planeten nachhaltig" (Interview mit Thomas Gull & Stefan Stöcklin). Zürich: ETH (https://www.news.uzh.ch/de/articles/news/2023/anthropozaen.html).

Biermann, Frank (2014). *Earth System Governance. World Politics in the Anthropocene*. Cambridge, Mass. & London: The MIT Press

Biermann, Frank, K. Abbott, S. Andresen, K. Bäckstrand, S. Bernstein, M. M. Betsill, H. Bulkeley, B. Cashore, J. Clapp, C. Folke, A. Gupta, J. Gupta, P. M. Haas, A. Jordan, N. Kanie, T. Kluvánková-Oravská, L. Lebel, D. Liverman, J. Meadowcroft, R. B. Mitchell, P. Newell, S. Oberthür, L. Olsson, Philipp Pattberg, R. Sánchez-Rodríguez, H. Schroeder, A. Underdal, S. Camargo Vieira, C. Vogel, O. R. Young, A. Brock, R. Zondervan (2012). Navigating the Anthropocene: Improving Earth System Governance. *Science* 335: 1306–1307.

Biermann, Frank & Eva Lövbrand (eds.) (2019). *Anthropocene Encouters. New Directions in Green Political Thinking*. Cambridge etc.: Cambridge University Press (Earth System Governance).

Bird, Rebecca Bliege (2015). Disturbance, Complexity, Scale: New Approaches to the Study of Human–Environment Interactions. *Annual Review of Anthropology* 44: 241–257

Bjornerud, Marcia (2018). *Timefulness. How Thinking Like a Geologist Can Help Save the World*. Princeton & Oxford: Princeton University Press.

Blackbourn, David (2006). *The Conquest of Nature. Water, Landscape and the Making of Modern Germany*. London: Jonathan Cape.

Blaikie, Piers M. (1985). *The Political Economy of Soil Erosion in Developing Countries*. New York: Longman

Blake, Jonathan S. & Nils Gilman (2024). *Children of a Modest Star. Planetary Thinking for an Age of Crises*. Stanford, Cal.: Stanford University Press.

Blaser, Mario & Marisol De La Cadena (2018). Pluriverse. A Proposal for a World of Many Worlds. In Marisol De La Cadena, & Mario Blaser (eds.). *A World of Many Worlds*. Durham, NC: Duke University Press.: 1–21.

Blaut, James Morris (1993). *The Colonizer´s Model of the World. Geographic Diffusionism and Eurocentric History*. New York & London: The Guilford Press.

Bloch, Maurice (2012). *Anthropology and the Cognitive Challenge*. Cambridge: Cambridge Univesity Press.

Block, Katharina (2018). Was würde Helmuth Plessner wohl zu einer Anthropozänikerin sagen? Ein kleiner Essay zu einer großen Verantwortung. In: Hennning Laux & Anna Henkel (eds.). *Die Erde, der Mensch und das Soziale. Zur Transformation gesellschaftlicher Naturverhältnisse im Anthropozän*. Bielefeld: Transcript: 51–64.

Block, Katharina (2021). Sozialtheorie im Anthropozän. In: SONA - Netzwerk Soziologie der Nachhaltigkeit (eds.). *Soziologie der Nachhaltigkeit*. Bielefeld: Transcript (Soziologie der Nachhaltigkeit, 1). 203–229.

Blok, Anders & Caspar Bruun Jensen (2019). The Anthropocene event in social theory: On ways of problematizing nonhuman materiality differently. *The Sociological Review* 67(6). 1195–1211.

Blute, Marion (2010). *Darwinian Sociocultural Evolution. Solutions to Dilemmas in Cultural and Social Theory*. Cambridge etc.: Cambridge University Press.

Bobbette, Adam (2023). *The Pulse of the Earth. Political Geology in Java*. Durham & London: Duke University Press.

Bobbette, Adam & Amy Donovan (eds.) (2018). *Political Geology. Active Stratigraphies and the Making of Life*. Cham: Palgrave Macmillan.

Bodley, John H. ([6]2012). *Anthropology and Contemporary Human Problems*. Lanham, Mad. etc.: Altamira Press.

Bodley, John H. ([6]2016). *Victims of Progress*. Lanham etc.: Rowman & Littlefield.

Bodley, John H. ([7]2020). *Cultural Anthropology. Tribes, States, and the Global System*. Lanham, Mad. Etc.: Rowman & Littlefield.

Böhme, Hartmut (2020). Über den Menschen, der kein Tier sein will, und den Menschen auf Verwandtensuche. *Paragrana. Internationale Zeitschrift für Historische Anthropologie* 29(1). 97–113.

Boellstorff, Tom (2016). For Whom the Ontology Turns. Theorizing the Digital Real. *Current Anthropology* 57(4). 387–398.

Bogusz, Tanja (2022). Fieldwork in the Anthropocene: On the Possibilities of Analogical Thinking between Nature and Society. *Science & Technology Studies* 36(2). 3–25. https://doi.org/10.23987/sts.111538.

Boivin, Nicole (2008). *Material Cultures, Material Minds. The Impact of Things on Human Thought, Society, and Evolution.* Cambridge etc: Cambridge University Press.

Boivin, Nicole & Alison Crowther (2021). Mobilizing the Past to Shape a Better Anthropocene. Nature Ecology & Evolution 5: 273–284.

Boivin, Nicole, Melinda A. Zeder, Dorian Q. Fuller, Alison Crowther, Greger Larson, Jon M. Erlandson, Tim Denham & Michael D. Petraglia (2016). Ecological Consequences of Human Niche Construction: Examining Long-Term Anthropogenic Shaping of Global Species Distributions. *Proceedings of the National Academy of Sciences* 113 (236): 6388–6396.

Bollig, Michael (2014). Resilience – Analytical Tool, Bridging Concept or Development Goal? Anthropological Perspectives on the Use of a Border Object. *Zeitschrift Für Ethnologie* 139(2). 253–279.

Bollig, Michael & Franz Krause (2024): *Environmental Anthropology. Current Issues and Fields of Engagement.* Bern: Haupt Verlag.

Bonneuil, Christophe (2015). The Geological Turn. Narratives of the Anthropocene. In: Clive Hamilton, Chrstophe Bonneuil & François Gemenne (eds.). *The Anthropocene and the Global Environmental Crisis: Rethinking Modernity.* London: Routledge: 15–31.

Bonneuil, Christophe & Jean-Baptiste Fressoz (2016). *The Shock of the Anthropocene. The Earth, History, and Us.* London & New York: Verso. (orig. fr. "L'Événement Anthropocène. La Terre, l'histoire et nous". Paris: Points [1]2013, [2]2016) (Points histoire).

Boomgard, Peter (2017). The Forests of Southeast Asia. Forest Transition Theory, and the Anthropocene, 1500–2000. In: Austin, Gareth (ed.) (2017). *Economic Development and Environmental History in the Anthropocene. Perspectives on Asia and Africa.* London etc.: Bloomsbury Academic: 179–197.

Bornemann, Basil, Katharina Glaab & Lena Partzsch (2019). Politikwissenschaftliches Interpretieren des Anthropozäns – Zur Bedeutung interpretativer Politikforschung im Umgang mit Objektivität, Normativität und Performativität von Meta-Konzepten. *Zeitschrift für Politikwissenschaft* 29: 325–344.

Bostic, Heidi & Meghan Howey (2017). To address the Anthropocene, engage the liberal arts. *Anthropocene* 18:105–110.

Boulding, Kenneth E. (2006a). The Economics of Coming Spaceship Earth. In: Henry Jarrett (ed.). *Environmental Quality in a Growing Economy. Essays from the Sixth RFF Forum on Environmental Quality.* Baltimore, Mad.: Resources for the Future/Johns Hopkins University Press: 3–14.

Boulding, Kenneth E. (2006b). Spaceship Earth Revisited. In: Herman E. Daly, Kenneth N. Townsend, (eds.). *Valuing the Earth. Economics, Ecology, Ethics.* Cambridge, Mass. & London: The MIT Press: 311–313.

Bourdieu, Pierre (2000). *Die zwei Gesichter der Arbeit. Interdependenzen von Zeit- und Wirtschaftsstrukturen am Beispiel einer Ethnologie der algerischen Übergangsgesellschaft.* Konstanz: UVK Verlagsgesellschaft (Édition Discours, 25) (orig. „Algérie 60. Structures économiques et structures temporelles", Paris: Les Éditions de Minuit, 1977, Reihe Les Sens Commun, tw. zuerst als "Travail et Travallieurs en Algérie", Paris, 1963).

Boyd, Robert (2018). *A Different Kind of Animal. How Culture Transformed Our Species.* Princeton, N.J.: Princeton University Press (The University Center for Human Values Series).

Bradley, Raymond S. ([3]2015). *Paleoclimatology. Reconstructing Climates of the Quaternary.* Oxford etc. Academic Press.

Bradtmöller, Marcel, Sonja Grimm & Julien Riel-Salvatore (2017). Resilience Theory in Archaeological Practice: An Annotated Review. *Quaternary International* 466: 3–16.

Braje, Todd J. & Jon M. Erlandson (2013). Human Acceleration of Animal and Plant Extinctions: A Late Pleistocene, Holocene, and Anthropocene Continuum. Anthropocene. https://doi.org/10.1016/j.ancene.2013.08.003

Braje, Todd J. (2015). Earth Systems, Human Agency, and the Anthropocene: Planet Earth in the Human Age. *Journal of Archaeological Research* 23: 369–396.

Braje, Todd J. (2018). Comment on Bauer & Ellis 2018. *Current Anthropology* 59(2). 215–216.

Braje, Todd, Thomas P. Leppard, Scott M. Fitzpatrick & Jon M. Erlandsson (2017). Archaeology, historical ecology and anthropogenic island ecosystems. *Environmental Conservation.* i:https://doi.org/10.1017/S0376892917000261.

Brand, Karl-Werner (2018). Zum historischen Wandel gesellschaftlicher Naturverhältnisse in der kapitalistischen Moderne. In: Henning Laux & Anna Henkel (eds.). *Die Erde, der Mensch und das Soziale. Zur Transformation gesellschaftlicher Naturverhältnisse im Anthropozän.* Transcript: Bielefeld: 91–122.

Brand, Ulrich & Markus Wissen (2017). *Imperiale Lebensweise. Zur Ausbeutung von Mensch und Natur in Zeiten des globalen Kapitalismus.* München: Oekom Verlag.

Brand, Stewart (1968). *Whole Earth Catalog. Access to Tools.* o.O.: Portola Institute

Brand, Stewart (2010). *Whole Earth Discipline. Why Dense Cities, Nuclear Power, Transgenic Crops, Restored Wildlands, Radical Science, and Geoengineering are Necessary.* New York: Penguin Books.

Brauch, Hans Günter (2022). The Anthropocene Concept in the Natural and Social Sciences, the Humanities and Law – A Bibliometric Analysis and a Qualitative Interpretation (2000–2020) In: Benner, Susanne, Gregor Lax Paul J.Crutzen, Ulrich Pöschl, Jos Lelieveld & Hans Günter Brauch (eds.) (2021): *Paul J. Crutzen and the Anthropocene: A New Epoch in Earth's History.* Cham: Springer Nature (The Anthropocene: Politik – Economics – Society – Science): 289–439.

Braudel, Fernand (1981). *Civilisation and Capitalism. 15th to 18th Centuries. Vol. 1: The Structures of Everyday Life.* New York, N.Y.: Harper & Row.

Bredekamp, Horst (2011). Blue Marble. Der blaue Planet. In: Christoph Markschies, Ingeborg Reichle, Jochen Brüning & Peter Deuflhard (eds.) u. Mitarbeit von Steffen Siegel & Achim Spelten: *Atlas der Weltbilder.* Berlin: Akademie-Verlag (Interdisziplinäre Arbeitsgruppen, Forschungsberichte, 25). 366–375.

Bremer, Scott, Werner Krauß & Diana Wildschut (eds.) (2021). How Narratives of Change Influence Local Climate Risk Governance. *Climate Risk Management.* (https://www.sciencedirect.com/journal/climate-risk-management/special-issue/10MD6WT077N).

Bretherton, F. P. (1985). Earth System Science and Remote Sensing. *Proceedings of the IEEE,* 71(6). 1118–1127.

Bridge, Gavin (2009). Material Worlds: Natural Resources, Resource Geography and the Material Economy. *Geography Compass* 3(3). 1217–1244.

Brightman, Marc & Jerome Lewis (eds.) (2017). *The Anthropology of Sustainability. Beyond Development and Progress.* New York, N.Y.: Palgrave Macmillan (Palgrave Studies in Anthropology of Sustainability).

Bröckling, Ulrich (2017). *Resilienz. Über einen Schlüsselbegriff des 21. Jahrhunderts.* https://soziopolis.de/daten/kalenderblaetter/beobachten/kultur/artikel/resilienz/

Broecker, Wallace S. (1987). Unpleasant Surprises in the Greenhouse. *Nature* 328 (9 July) 123–6.

Brondizio, Eduardo S., Karen O'Brien, Xuemei Bai, Frank Biermann, Will Steffen, Frans Berkhout, Christophe Cudennec, Maria Carmen Lemos, Alexander Wolfe, Jose Palma-Oliveira, Chen-Tung Arthur Chen (2016). Re-conceptualizing the Anthropocene: A Call for Collaboration. *Global Environmental Change* 39: 318–327.

Brooke, John L. (2014). *Climate Change and the Course of Global History. A Rough Journey.* Cambridge etc.: Cambridge University Press (Studies in Environment and History).

Brosius, Christiane & Ulrike Gerhard (2020). *Natur in den Städten. Das apokalyptische Narrativ.* Nachdenken: Nord-Süd: 89–97. https://heiup.uni-heidelberg.de/journals/index.php/rupertocarola/article/view/23573/17303

Brown, Antony G., Stephen Tooth, Joanna E. Bullard, David S. G. Thomas, Richard C. Chiverrell, Andrew J. Plater, Julian Murton, Varyl R. Thorndycraft, Paolo Tarolli, James Rose, John Wainwright, Peter Downs & Rolf Aalto (2016). The Geomorphology of the Anthropocene: *Emergence, Status and Implications. Earth Surface Processes and Landforms*. 42: 71–90.

Brown, Peter G. & Peter Timmerman (eds.) (2015). *Ecological Economics for the Anthropocene*. New York: Columbia University Press.

Brown, Nina, Thomas McIlwraith & Laura Tubelle de González (eds.) (^{2}2020). *Perspectives. An Open Invitation to Cultural Anthropology*. Arlington, Virg.: American Anthropological Association.

Bruck, Andreas (1985). *Funktionalität beim Menschen. Ein konstruktiv-systematischer Überblick*. Frankfurt am Main etc.: Peter Lang (Europäische Hochschulschriften, Reihe XXII, Soziologie, 112).

Brunnengräber, Achim (^{2}2019). *Ewigkeitslasten. Die „Endlagerung" radioaktiver Abfälle als soziales, politisches und wissenschaftliches Projekt – eine Einführung*. Bonn: Bundeszentrale für politische Bildung.

Brunnengräber, Achim (2021). Das Kapitalozän ist das eigentliche Problem. *Politische Ökologie* 39(167). 51–57.

Bruns, Antje (2019). Das Anthropozän und die große Transformation – Perspektiven für eine kritische raumwissenschaftliche Governance- und Transformationsforschung. In: Milad Abassiharofteh, Jessica Baier, Angelina Göb, Insa Thimm, Andreas Eberth, Falco Knaps, Vilja Larjosto & Fabiana Zebner (eds.). *Räumliche Transformation. Prozesse, Konzepte, Forschungsdesigns*. Hannover: Akademie für Raumfirschung und Landesplanung. (Forschungsberichte der ARL, 10). 53–64.

Bruns, Antje (2020). Provincializing Degrowth. Alternativen zu Entwicklung und der Globale Süden. In: Bastian Lange, Martina Hülz , Benedikt Schmid & Christian Schulz (eds.) (2020). *Postwachstumsgeographien. Raumbezüge diverser und alternativer Ökonomien*. Bielefeld: Transcript (Sozial- und Kulturgeographie, 38). 242–256.

Bryant, Rebbecca & Daniel M. Knight (2019). *The Anthropology of the Future*. Cambridge etc.: Cambridge University Press (New Departures in Anthropology).

Bubandt, Nils Ole (2018). Anthropocene Uncanny: Nonsecular Approaches to Environmental Change. In: Niels Ole Bubandt (ed.). *Non-secular Anthropocene: Spirits, Specters and Other Nonhumans in a Time of Environmental Change*. Aarhus: Aarhus University Research on the Anthropocene (AURA). (More-than-Human Working Papers, 3). 2–15.

Bubandt, Nils Ole (2019). Of Wildmen and White Men: Cryptozoology and Inappropriate/d Monsters at the Cusp of the Anthropocene. *Journal of the Royal Anthropological Institute* 25(2).223–240.

Buchmann, Sabeth & Mercedes Bunz (2013). *Whole Earth. Kalifornien und das Verschwinden des Außen*. Berlin: Sternberg Press.

Bubenzer, Olaf, Hans Gebhardt & Frank Keppler (2021). Das Zeitalter des Anthropozäns. Der Mensch als geologische Kraft. *Ruperto Carola* 15:25–33.

Buck-Morss, Susan (2002). *Dreamworld and Catastrophe. The Passing of Mass Utopia in East and West*. Cambridge, MA: MIT Press.

Büscher, Bram & Robert Fletcher (2020). *The Conservation Revolution. Radical Ideas for Saving Nature Beyond the Anthropocene*. London & New York: Verso.

Buhr, Lorina (2022).Gesellschaft, Natur und Erde: Elemente für eine politische Theorie im Anthropozän. Eine Literatursichtung. *Zeitschrift für politische Theorie* 13(1–2). 311–326.

Burchfield, Joe D. (1998). The Age of the Earth and the Invention of Geological Time. In: Derek J. Blundell & Andrew C. Scott (eds.). *Lyell. The Past is the Key to the Present*. London: 137–143.

Burgess, Seth (2019). Deciphering Mass Extinction Triggers. Science 363(6429). 815–816.

Burtynsky, Edward, Jennifer Baichwal & Nicholas de Pencier (2018). *Anthropocene*. Toronto: Art Gallery of Ontario & Goose Lane Editions.

Burtynsky, Edward, Jennifer Baichwal & Nicholas de Pencier (2020). Anthropocene. Wien: Steidl.

Busche, Detlef, Jürgen Kempf & Ingrid Stengel ([2]2021). Landschaftsformen der Erde. Bildatlas der Geomorphologie. Darmstadt: WBG.

Buttimer, Anne (2015). Diverse Perspectives on Society and Environment. In: Frauke Kraas, Dietrich Soyez, Carsten Butsch, Franziska Krachten, & Holger Kretschmer (eds.). *IGC Cologne 2012. Down to Earth. Documenting the 32rd International Geographical Congress in Cologne 26–30 August 2012*. Cologne: 41–52 (Kölner Geographische Arbeiten, 95).

Butzer, Karl W. & Georgina H. Endfield (2012). Critical Perspectives on Historical Collapse. *Proceedings of the National Academy of Sciences* 109: 3628–3631.

Canaparo, Claudio (2021). *El mundo de atrás. Efecto antropoceno y especulación en los ámbitos periféricos*. Frankfurt am Main: Peter Lang Verlag.

Candea, Matei (2010). "I Fell in Love With Carlos the Meerkat: Engagement and Detachment in Human–Animal Relations." *American Ethnologist* 37(2). 241–258.

Caniglia, Beth Schaefer, Andrew Jorgenson, Stephanie A. Malin, Lori Peek, David N. Pellow & Xiaorui Huang (eds.) 2021: *Handbook of Environmental Sociology.* Cham: Springer Nature (Handbooks of Sociology and Social Research).

Caputi, Jane (2020). *Call Your "mutha". A Deliberately Dirty-Minded Manifesto for the Earth Mother in the Anthropocene*. Oxford etc.: Oxford University Press (Heretical Thought).

Carey, John (2016). Core Concept: Are We in the Anthropocene? Proceedings of the National Academy of Sciences 113(5). 3908–3909.

Carrasco, L. Roman, Joleen Chan, Francesca L. McGrath & Le T. P. Nghiem (2017). Biodiversity Conservation in a Telecoupled World. Ecology and Society 22(3). Art. 24. (https://www.ecologyandsociety.org/vol22/iss3/art24/)

Carrier, James G. & Deborah Gewertz (eds.) 2013: *Handbook of Sociocultural Anthropology*. London etc.: Bloomsbury.

Carrithers, Michael, L. J. Bracken & S. Emery (2011) Can a species be a person? A trope and its entanglements in the anthropocene era. *Current Anthropology* 52(5). 661–685.

Carson, Rachel (1962). *Silent Spring*. New York: Houghton Mifflin.

Casimir, Michael (2003). *Flocks and Food. Biocultural Aspects of Foodways*. Berlin: Reimer Verlag.

Casimir, Michael (2021). *Mobility and Territoriality. Social and Spatial Boundaries Among Foragers, Fishers, Pastoralists and Peripatetics*. Oxford & New York: Berghahn (Explorations in Anthropology).

Castree, Noel (2014a). The Anthropocene and Geography I: The Back Story. *Geography Compass* 8(7). 436–449.

Castree, Noel (2014b). The Anthropocene and Geography II: The Back Story. *Geography Compass* 8(7). 450–463.

Castree, Noel (2014c). The Anthropocene and Geography III: The Back Story. *Geography Compass* 8(7). 464–476.

Castree, Noel (2015). Anthropocene and the Environmental Humanities: Extending the Conversation. *Environmental Humanities* 5:233–260.

Castree, Noel (2017). Anthropocene: Its Stratigraphic Basis. *Nature* 541: 289.

Ceballos, Gerardo, Paul E. Ehrlich & Peter H. Raven (2020). Vertebrates on the Brink as Indicators of Biological Annihilation and the Sixth Mass Extinction. *Proceedings of the National Academy of Sciences* 117:13596–13602.

Cera, Agostino (2020). O Antropoceno ou o "fim" do Imprativo responsabilidade. The Anthropocene or the "End" of the Imperative Responsibiolity. *Pensando – Revista de Filosofia* 11(24).31–43.

Cera, Agostino (2022). The Stratigraphic Fallacy or the Anthropocene as an Epistemic Question. In: *Acta Philosophica Fennica* 97: 129–152.

Cera, Agostino (2023). *A Philosophical Journey into the Anthropocene. Discovering Terra Incognita*. Lanham etc.: Lexington Books.

Certini, Giacomo & Riccardo Scalenghe (2011). Anthropogenic Soils Are the Golden Spikes for the Anthropocene. *The Holocene* (https://journals.sagepub.com/doi/10.1177/0959683611408454.

Chakrabarty, Dipesh (2000). *Provincializing Europe. Postcolonial Thought and Historical Difference*. Princeton: Princeton University Press.

Chakrabarty, Dipesh ([2]2008). *Provincializing Europe. Postcolonial Thought and Historical Difference*. Princeton: Princeton University Press ([1]2000).
Chakrabarty, Dipesh (2009). The Climate of History: Four Theses. *Critical Inquiry* 35 Winter: 197–222.
Chakrabarty, Dipesh (2012). Postcolonial Studies and the Challenge of Climate Change. *New Literary History* 43(1). 1–18.
Chakrabarty, Dipesh (2014) Climate and Capital. On Conjoined Histories. *Critical Inquiry* 35(2). 197–222.
Chakrabarty, Dipesh (2015). The Human Condition in the Anthropocene. The Tanner Lectures in Human Values. New Haven: Yale University Press.
Chakrabarty, Dipesh (2017). The Politics of Climate Change Is More Than the Politics of Capitalism. *Theory, Culture & Society* 34(2–3). 25–37.
Chakrabarty, Dipesh (2018). Anthropocene Time. *History and Theory* 57 (1). 5–32.
Chakrabarty, Dipesh (2019). The Planet: An Emergent Humanist Category. *Critical Inquiry* 46 (1). 1–31.
Chakrabarty, Dipesh (2020a). Die Zukunft der Geisteswissenschaften im Zeitalter des Menschen: Eine Notiz. In: Hannes Bajohr (eds.). *Der Anthropos im Anthropozän. Die Wiederkehr des Menschen im Moment seiner vermeintlich endgültigen Verabschiedung*. Berlin: Walter de Gruyter: 233–237.
Chakrabarty, Dipesh (2020b). Der Planet als neue humanistische Kategorie. In: Frank Adloff & Sighard Neckel (eds.). *Gesellschaftstheorie im Anthropozän*. Frankfurt a.M./New York: Campus: 23–53.
Chakrabarty, Dipesh (2021). The Climate of History in a Planetary Earth. Chicago & London: The University of Chicago Press.
Chandler, David (2017). Securing the Anthropocene? International Policy Experiments in Digital Hacktivism: A Case Study of Jakarta. *Security Dialogue* 48(2). 113–130.
Chandler, David (2018). *Ontopolitics in the Anthropocene. An Introduction to Mapping, Sensing and Hacking*. London etc: Routledge (Critical Issues in Global Politics, 9).
Chandler, David (2020). Resilience in the Anthropocene. Governance and Politics at the End of the World. New York: Routledge (Routledge Research in the Anthropocene).
Chandler, David, Franziska Müller & Delef Rothe (eds.) (2021). *International Relations in the Anthropocene. New Agendas, New Agencies and New Approaches*. London etc.: Palgrave Macmillan.
Chandler, David & Jonathan Pugh (2020). Islands of Relationality and Resilience: The Shifting Stakes of the Anthropocene. *Area: Journal of the Royal Geographical Society* 52 (1). 65–72. https://doi.org/10.1111/area.12459
Chandler, David & Julian Reid (2019). *Becoming Indigenous. Governing Imaginaries in the Anthropocene*. Rowman & Littlefield.
Chao, Sophie, Wendy Wolford, Andrew Ofstehage, Shalmali Guttal, Euclides Gonçalves & Fernanda Ayala (2023). The Plantationocene as Analytical Concept: a Forum for Dialogue and Reflection. *The Journal of Peasant Studies* 1–23 (https://doi.org/10.1080/03066150.2023.2228212).
Chernilo, Daniel (2017). *Debating Humanity. Towards a Philosophical Sociology*. Cambridge: Cambridge University Press.
Chernilo, Daniel (2020). Die Frage nach dem Menschen in der Anthropozändebatte. In Bajohr (eds.). 55–76.
Chew, Sing C. (2000). *World Ecological Degradation. Accumulation, Urbanization, and Deforestation, 3000 B.C. – A.D. 2000*. Walnut Creek, Cal. Altamira Press.
Choy, Timothy K., Lieba Faier, Michael J. Hathaway, Miyako Inoue, Shiho Satsuka & Anna Tsing. (2009). A New Form of Collaboration in Cultural Anthropology: Matsutake Worlds. *American Ethnologist* 36 (2). 380–403.
Christian, David ([2]2011). *Maps of Time. An Introduction to Big History*. Berkeley: University of California Press. (California World History Library, 2).

Christian, David (2018). *Origin Story. A History of Everything*. New York: Little, Brown & Company.

Christian, David (2019). The Anthropocene Epoch. The Background to Two Transformative Centuries. In: Felipe Fernández-Armesto(ed.).*The Oxford Illustrated History of the World*. Oxford: Oxford University Press: 339–373.

Christian, David (2020). What is Big History? In: Craig Benjamin, Esther Quaedakers and David Baker (eds.). *The Routledge Companion to Big History*. Milton Park & New York: Routledge (Routledge Companions). 16–34.

Christian, David, Cynthia Stokes Brown & Craig Benjamin (2014). *Big History. Between Nothing and Everything*. New York McGraw-Hill Education

Chua, Liana & Hannah Fair (2019). Anthropocene. In: Felix Stein, Sian Lazar, Matei Candea, Hildegard Diemberger, Joel Robbins, Andrew Sanchez & Rupert Stasch (eds.). *The Open Encyclopedia of Anthropology* (https://www.anthro encyclopedia.com/entry/anthropocene).

Chudeck, M. Muthukrishna & Joseph Henrich (2016). Cultural Evolution. In: David Buss (ed.). *The Handbook of Evolutionary Psychology*. New York, NY: Wiley & Sons: 749–769.

Chwałczyk, Franciszek (2020). Around the Anthropocene in Eighty Names–Considering the Urbanocene Proposition. *Sustainability* 12, 4458, doi:https://doi.org/10.3390/su12114458

Clark, Nigel & Yasmin Gunaratnam (2017). Earthing the Anthropos? From 'Socializing the Anthropocene' to Geologizing the Social. *European Journal of Social Theory* 20(1). 146–163.

Clark, Nigel & Bronislaw Szerszynski (2021). *Planetary Social Thought. The Anthropocene Challenge to the Social Sciences*. London etc.: Polity Press.

Clark P. U, J. D. Shakun, S. A. Marcott et al. (2016). Consequences of Twenty-first-century Policy for Multimillennial Climate and Sea-level Change. *Nature Climate Change* 6: 360–369.

Clark, Timothy (2015). *Ecocriticism at the Edge. The Anthropocene as a Threshold Concept*. London & New York: Bloomsbury Academic.

Clark, Timothy (2019). *The Value of Ecocriticism*. Cambridge etc.: Cambridge University Press.

Clark, Timothy & Kathryn Yusoff (2017). Geosocial Formations in the Anthropocene. *Theory, Culture and Society* 34(2–3). 3–23.

Clarke, Bruce (2020). *Gaian Systems. Lynn Margulis, Neocybernetics, and the End of the Anthropocene*. Minneapolis: University of Minnesota Press.

Cohen, Jeffrey Jerome (2015). *Stone. An Ecology of the Inhuman*. Minneapolis & London: University of Minnesota Press.

Colebrook, Claire (2017). We Have Always Been Post-Anthropocene. The Anthropocene:The Anthropocene Counterfactual. In: Richard Grusin (ed.). *Anthropocene Feminism*. Minneapolis & London: University of Minnesota Press: 1–20 (Center for 21st Century Studies).

Coleman, Simon & P. Hellerman (eds.) (2011). *Multi-sited Ethnography. Problems and Possibilities in the Translocation of Research Methods*. London & New York: Routledge.

Collard, Rosemary-Claire, Jessica Dempsey & Juanita Sundberg (2015). A Manifesto for Abundant Futures. *Annals of the Association of American Geographers* 105(2). 322–330.

Comos, Gina & Caroline Rosenthal (eds.) (2019). *Anglophone Literature and Culture in the Anthropocene*. Newcastle upon Tyne: Cambridge Scholars Publishing

Conceição, Pedro (lead author) et al. (2020). *The next Frontier. Human Development and the Anthropocene. Human Development Report* 2020. New York: United Nations Development Programme.

Conklin, Harold (1955). Hanunóo color categories. *Southwestern Journal of Anthropology* 11(4).339–44.

Connolly, William E. (2011). A *World of Becoming*. Durham, N.C.: Duke University Press.

Conrad, Sebastian (2016). *What is Global History?* Princeton, N.J.: Princeton University Press.

Coole, Diana & Samantha Frost (eds.) (2010). *New Materialisms. Ontology, Agency, and Politics*. Durham & London: Duke University Press.

Costanza, Robert, Lisa Graumlich, Will Steffen, Carole Crumley, John Dearing, Kathy Hibbard, Rik Leemans, Charles Redman & David Schimel (2007). Sustainability or Collapse: What Can We Learn from Integrating the History of Humans and the Rest of Nature? *Ambio A Journal of the Human Environment* 36(7). 522–527.

Constanza, Robert , Lisa J. Graumlich & Will Steffen (eds.) (2007). *Sustainability or Collapse? An Integrated History and Future of People on Earth*. Cambridge, Mass. & London: The MIT Press (Dahlem Workshop Reports).
Conversi, Daniele (2020). The Ultimate Challenge: Nationalism and Climate Change. *Nationalities Papers* 48(4). 625–636. (https://doi.org/10.1017/nps.2020.18).
Cook, Brian, Lauren Rickards & Ian D. Rutherfurd (2015). Geographies of the Anthropocene. *Geographical Research* 53(3). 231–243.
Cooper, Anthony H. , Teresa J. Brown, Simon J. Price, Jonathan R. Ford & Colin N. Waters (2018). Humans Are the Most Significant Global Geomorphological Driving Force of the 21st Century. *Anthropocene Review* 5(3). 222–229.
Corlett, Richard T. (2013). Becoming Europe: Southeast Asia in the Anthropocene. *Elementa: Science of the Anthropocene* 1: 000016 (https://doi.org/10.12952/journal.elementa.000016).
Corlett, Richard T. ([2]2014). *The Ecology of Tropical East Asia*. Oxford: Oxford University Press.
Corlett, Richard T. (2015a). Becoming Europe. Southeast Asia in the Anthropocene. *Elementa. Science* of the Anthropocene 1: Art. 16 (https://doi.org/10.12952/journal.elementa.000016.)
Corlett, Richard T. (2015b). The Anthropocene Concept in Ecology and Conservation. *Trends in Ecology & Evolution* 30(1). 36–41.
Cornell, Sarah Elizabeth, Catherine J. Downy, Evan D. G. Fraser & Emily Boyd (2012). Earth System Science and Society: A Focus on the Anthroposphere. In: Sarah E. Cornell, I. Colin Prentice, Joanna I. House & Catherine J. Downy (eds.). *Understanding the Earth System. Global Change Science for Application*. Cambridge: Cambridge University Press: 1–38.
Cosgrove, Denis (1994). Contested Global Visions: One-World, Whole-Earth, and the Apollo Space Photographs. *Annals of the Association of American Geographers* 84(2). 270–294.
Cosgrove, Denis (2001). *Apollo´s Eye. Anthropocene Cartographic Genealogy of Earth in the Western Imagination*. Baltimore: Johns Hopkins University Press.
Craps, Stef (2023). Guilty Grieving in an Age of Ecocide. *Parallax* 29(3). 323–342.
Craps, Stef (2024 Lost Words and Lost Worlds: Combatting Environmental Generational Amnesia. *Memory Studies Review* (2024) 1–20 (https://doi.org/10.1163/29498902-20240001).
Craps, Stef, Rick Crownshaw, Jennifer Wenzel, Rosanne Kennedy, Claire Colebrook, Vin Nardizzi (2018). Memory studies and the Anthropocene: A roundtable. *Memory Studies* 11(4) 498–515.
Crate, Susan A. (2009). Gone the Bull of Winter? Contemplating Climate Change's Cultural Implications in Northeastern Siberia, Russia. In: Susan A. Crate & Mark Nuttall (eds.). *Anthropology and Climate Change: From Encounters to Actions*. Walnut Creek, Cal: Left Coast Press: 139–152.
Crate, Susan A. (2011). Climate and Culture: Anthropology in the Era of Contemporary Climate Change. *Annual Review of Anthropology* 40: 175–194.
Crate, Susan A. & Mark Nuttall (eds.) ([2]2016). *Anthropology and Climate Change. From Actions to Transformations*. New York & London: Routledge.
Creanza, Nicole, O. Kolodny, Markus Feldman (2017). Cultural Evolutionary Theory: How Culture Evolves and Why it Matters. *Proceedings of the National Academy of Sciences* 114: 7782–7789.
Crist, Eileen (2013). On the Poverty of Our Nomenclature. *Environmental Humanities* 3(1). 129–147.
Crist, Eileen (2020). *Schöpfung ohne Krone. Warum wir uns zurückziehen müssen, um die Artenvielfalt zu bewahren*. München: Oekom (orig "Abundant Earth: Toward an Ecological Civilization"., Chicago: The University of Chicago Press, 2019).
Critical Zones. Observatories for Earthly Politics. Karlsruhe: Center for Art and Media (ZKM) https://zkm.de/en/exhibition/2020/05/critical-zones.
Crivellari, Fabio, Bernhard Kleeberg & Tilman Walter 2005: Urmensch und Wissenschaftskultur: Einleitung. In: Bernhard Kleeberg, Tilman Walter & Fabio Crivellari (eds.). *Urmensch und Wissenschaften. Eine Bestandsaufnahme*. Darmstadt: Wissenschaftliche Buchgesellschaft:7–24
Cronin, Thomas M. (2009). *Paleoclimates. Understanding Climate Change Past and Present*. New York: Columbia University Press-

Cronon, William (1991). *Nature's Metropolis. Chicago and the Great West.* New York & London: W. W. Norton.
Cronon, William (1996). The Trouble With Wilderness: Or, Getting Back to the Wrong Nature. *Environmental History* 1(1).7–28.
Cronon, William (ed.) (1995). *Uncommon Ground: Rethinking the Human Place in Nature.* New York: W. W. Norton & Co.
Crosby, Alfred W. ([2]2003). *The Columbian Exchange. Biological and Cultural Consequences of 1492.* Westport, Conn. & London: Praeger ([1]1972) (Contributions in American Studies, 2).
Crosby, Alfred W. (2015). *Ecological Imperialism: The Biological Expansion of Europe, 900–1900. New Edition.* Cambridge etc.: Cambridge University Press (Studies in Environment and History) (Canto Classics Edition).
Crumley, Carole L. (ed.) (1994). *Historical Ecology. Cultural Knowledge and Changing Landscapes.* Santa Fe, N.M.: School of American Research Press.
Crumley, Carole L. (2007). Historical Ecology: Integrated Thinking at Multiple Temporal and Spatial Scales. In: Alf Hornborg & Carole Crumley (eds.). *The World System and the Earth System. Global Socio-Environmental Change and Sustainability Since the Neolithic.* Walnut Creek, Cal.: Left Coast Press: 15–28.
Crumley, Carole L. (2019). New Paths into the Anthropocene. In: Christian Isendahl & Daryl Stump (eds.). *Oxford Handbook of Historical Ecology and Applied Archaeology.* Oxford: Oxford University Press: 6–20.
Crumley, Carole L., Jan C. A. Kolen, Maurice de Kleijn & Niels van Manen (2017). Studying Long-term Changes in Cultural Landscapes: Outlines of a Research Framework and Protocol : Research Framework and Protocol. *Landscape Research*, https://doi.org/10.1080/01426397.2017.1386292
Crutzen, Paul J. (2000). Anthropocene. The Geology of Humanity. *Global Change Magazine* 78. Stockholm
Crutzen, Paul J. (2006a). Albedo Enhancement by Stratospheric Sulfur Injections: A Contribution to Resolve a Policy Dilemma? *Climate Change*, 77 (3–4). 211–220.
Crutzen Paul J. (2006b). Albedo enhancement by stratospheric sulfur injections: A contribution to resolve a policy dilemma? An Editorial Essay. *Climatic Change* 77(3–4). 211–219.
Crutzen, Paul J. (2019a). Eine kritische Analyse der Gaia-Hypothese als Modell für die Wechselwirkung zwischen Klima und Biosphäre. In: Ders. *Das Anthropozän. Schlüsseltexte des Nobelpreisträgers für das neue Erdzeitalter.* München: Oekom Verlag: 175–204; zuerst in *Gaia* 11(2). 96–103, 2002.
Crutzen, Paul J. (Michael Müller, eds.) (2019b). *Das Anthropozän. Schlüsseltexte des Nobelpreisträgers für das neue Erdzeitalter.* München: Oekom Verlag (Bibliothek der Nachhaltigkeit: Wiederentdeckungen für das Anthropozän).
Crutzen, Paul J. (2002). Geology of Mankind. *Nature* 415:23.
Crutzen, Paul J. & Hans Günter Brauch (eds.) (2014). *Paul J. Crutzen. A Pioneer on Atmospheric Chemistry and Climate Change in the Anthropocene.* Cham: Springer (Springer Briefs on Pioneers in Science and Practice, 50).
Crutzen, Paul J., Mike Davies, Michael D. Mastrandrea, Stephen H. Schneider & Peter Sloterdijk (2011). *Das Raumschiff Erde hat keinen Notausgang.* Berlin: Suhrkamp Verlag (edition unseld).
Crutzen, Paul J. & Christian Schwägerl (2011). Living in the Anthropocene. Toward a New Global Ethos. *Yale Environment* 360, 24 January (http://e360.yale.edu/feature/living_in_the_anthropocene_toward_a_new_global_ethos/2363).
Crutzen, Paul J. & Will Steffen (2003). How Long Have We Been in the Anthropocene Era? An Editorial Comment. *Climate Change* 61:251–57.
Crutzen, Paul J. & Eugene F. Stoermer (2000). The „Anthropocene". IGBP Global Change Newsletter (41):17–18.
Cunha, Daniel (2015a). The Anthropocene as Fetishism. *Mediations* 28: 65–77.
Cunha, Daniel (2015b). The Geology of the Ruling Class? *The Anthropocene Review* 2(3). 262–266.

Cunsolo, Ashlee & Karen Landman (2017). Introduction: To Mourn beyond the Human. In: Dies. (eds.). *Mourning Nature. Hope at the Heart of Ecological Loss and Grief.* Montreal & Kingston: McGill-Queen's University Press: 3–26.

Dalby, Simon (2007). Ecological Intervention and Anthropocene Ethics. *Ethics & International Affairs* 21 (3) (http://216.70.70.40/?p=1131).

Dalby, Simon (2017). Environment: From Determinism to the Anthropocene. In: John Agnew, Virginie Mamadouh, Anna J. Secor & Joanne Sharp (eds.). *The Wiley Blackwell Companion to Political Geography* (Hoboken, N.J. & Chichester, West Sussex: Wiley Blackwell (Wiley Blackwell Companions to Geography). 451–461.

Dalby, Simon (2020). Anthropocene Geopolitics. Globalization, Security, Sustainability. Ottawa: University of Ottawa Press (Politics and Public Policy).

D'Angelo, Lorenzo & Robert Jan Pijpers (2022). The Anthropology of Resource Extraction. New York: Routledge (Routledge Studies of the Extractive Industries and Sustainable Development).

DellaSala, Dominick A. & Michael I. Goldstein (chief. eds.) (2018a). *Encyclopedia of the Anthropocene*. Dordrecht etc.: Elsevier

DellaSala, Dominick A., Michael I. Goldstein, S. A. Elias, B. Jennings, T. E. Lacer, Jr., P. Mineau & S. Pyare (2018b). The Anthropocene: How the Great Acceleration Is Transforming the Planet at Unprecedented Levels. In: Dominick,DellaSala, A. & Michael I. Goldstein (chief. eds.) (2018). *Encyclopedia of the Anthropocene*. Dordrecht etc.: Elsevier: 1–7.

Danchin, Étienne (2013). Avatars of Information. Towards an Inclusive Evolutionary Synthesis. *Trends in Ecology & Evolution* 28:351–358.

Danowski, Déborah & Eduardo Batalha Viveiros de Castro (2019). *In welcher Welt leben? Ein Versuch über die Angst vor dem Ende*. Berlin: Matthes & Seitz (orig. brasil. Port.: "Há mundo por vir? Ensaio Sobre os medos e os fins", 2014/2015, auch engl. "The Ends of the World", Cambridge: Polity Press, 2017).

Danowski, Déborah, Eduardo Viveiros de Castro & Bruno Latour (2014). Position Paper: The Thousand Names of Gaia. From the Anthropocene to the Age of the Earth. https://thethousandnamesofgaia.files.wordpress.com/2014/07/positionpaper-ingl-para-site.pdf (Accessed 23.11.2020).

Danowski, Déborah & Eduardo Batalha Viveiros de Castro (2017). The Ends of the World. Cambridge: Polity Press.

Dartnell, Lewis (2018). *Origins. How the Earth Made Us*. London: The Bodley Head.

Dasmann, R. F. (1989). Toward a biosphere consciousness. In D. Worster (ed.): *The ends of the earth: Perspectives on modern environmental history*. Cambridge: Cambridge University Press: 277–288.

Daston, Lorraine (2021). *Unruly Nature*. New Haven, Cambridge, Mass. & London: The MIT Press. (Untimely Meditations, 17, orig. "Gegen die Natur", Berlin: Matthes & Seitz, Fröhliche Wissenschaft, 141).

Davies, Jeremy (2016). *The Birth of the Anthropocene*. Oakland, Cal.: The University of California Press.

Davis, Heather & Zoe Todd (2017). On the Importance of a Date, or Decolonizing the Anthropocene. *ACME: An International E-Journal for Critical Geographies* 16(4). 761–80.

Davis, Heather & Etienne Turpin (eds.) (2015). *Art in the Anthropocene. Encounters Among Aesthetics, Politics, Environments and Epistemologies*. London: Open Humanities Press (Critical Climate Change).

Davis, Janae, Alex Moulton, Levi Van Sant & Brian Williams (2019). Anthropocene, Capitalocene,… Plantationocene?: A Manifesto for Ecological Justice in an Age of Global Crises. *Geography Compass* (13). 1–15.

Davis, Robert V. (2011). Inventing the Present: Historical Roots of the Anthropocene. *Earth Sciences History* 30(1). 63–84.

Dawkins, Richard (2016). *The Extended Phenotype. The Long Reach of the Gene*. Oxford: Oxford University Press (Oxford Landmark Science).

Deane-Drummond, Celia, Sigurd Bergmann & Markus Vogt (eds.) (2017). *Religion in the Anthropocene*. Eugene, Or.: Cascade Books.

Dearing, J., Wanga, R., Zhang, K., James, G., Dyke, J.G., Haberl, H., Hossain, M.S., Langdon, P.G., Lenton, T.M., Raworth, K., Brown, S., Carstensen, J., Cole, M.J., Cornell, S.E., Dawson, T.P., Doncasterm, P., Eigenbrodm, F., Florke, M., Jeffers, E., Mackay, A.W., Nykvist, B., Poppy, G.M., (2014). Safe and Just Operating Spaces for Regional Social-ecological Systems. *Global Environmental Change* 28, 227–238.

deCaires Taylor, Jason, Carlo McCormick & Helen Scales (2014). *The Underwater Museum. The submerged Sculptures of Jason deCaires Taylor*. San Francisco: Chronicle Books.

Dech, Stefan, Rüdiger Glaser & Robert Meisner (2008). *Globaler Wandel. Die Erde aus dem All.* München: Frederking & Thaler.

De La Cadena, Marisol (2010): Indegenous Cosmopolitics in the Andes. Conceptual reflectiones beyond "Politics". *Cultural Anthropology* 25(2): 334–370.

De La Cadena, Marisol (2015). *Earth Beings. Ecologies of Practice Across Andean Worlds*. Durham, N.C.: Duke University Press (Lewis Henry Morgan Lectures).

De La Cadena, Marisol & Mario Blaser (eds.) (2018). *A World of Many Worlds*. Durham, NC: Duke University Press.

Delanda, Manuel (1992). Nonorganic Life. In: Jonathan Crary & Sanford Kwinter (eds.). *Incorporations*. New York: Zone: 129–167 (Zone, 6).

DeLanda, Manuel (2016). *Assemblage Theory*. Edinburg: Edinburg University Press (Speculative Realism).

Delanty, Gerard & Aurea Mota (2017). Governing the Anthropocene: Agency, Governance, Knowledge. *European Journal of Social Theory* 20: S. 9–38.

Deleuze, Gilles & Félix Guattari (1992). *Tausend Plateaus*. Berlin: Merve Verlag (orig. „Capitalisme et schizophrenie 2: Mille plateaux", Paris: Minuit, 1980).

DellaSala, Dominick A. & Michael I. Goldstein (eds.). (2018). *Encyclopedia of the Anthropocene*. Dordrecht etc.: Elsevier Publications.

DeLoughrey, Elizabeth (ed.) (2019). Allegories of the Anthropocene. Durham, N.C & London: Duke University Press.

DeMello, Margo ([2]2021). *Animals and Society. An Introduction to Human-Animal Studies*. New York: Columbia University Press.

Demos, T. J. (2015). *Anthropocene, Capitalocene, Gynocene. The Many Names of Resistance*. https://www.fotomuseum.ch/de/2015/06/12/anthropocene-capitalocene-gynocene-the-many-names-of-resistance/

Demos, T. J. (2017). *Against the Anthropocene. Visual Culture and Environment Today*. Berlin: Sternberg.

Denevan, W.M. (1992). The Pristine Myth. The Landscape of the Americas in 1492. *Annals of the Association of American Geographers* 82: 369–358.

Denizen, Seth (2013). Three Holes: In the Geologic Present. In: Etienne Turpin (ed.). Architecture in the Anthropocene. Encounters Among Design, Deep Time, Science and Philosophy. Ann Arbor: Open Humanities Press: 25–42.

Dennett, Daniel C. (2004). *Freedom Evolves*. New York: Viking.

De Pina-Cabral, João (2017). *World. An Anthropological Examination*. Chicago, Ill.: HAU Books (The Malinowski Monographs Series, 1).

Desmond, Hugh (2024). Dominion. To take care of the Earth, humans must recognise that we are both a part of the animal kingdom and its dominant power. *Aeon*. (https://aeon.co/essays/human-dominance-is-a-fact-not-a-debate).

Descola, Philippe (2013). *The Ecology of Others*. Chicago: Prickly-Paradigm Press.

Descola, Philippe (2015). *Jenseits von Natur und Kultur*. Frankfurt am Main: Suhrkamp (orig. „Par-delà nature et culture", Paris: Gallimard, 2005).

Descola, Philippe & Gísli Pálsson (1996a). Introduction. In: Dies. (eds.). *Nature and Society. Anthropological Perspectives*. London & New York: Routledge: 1–22.

Descola, Philippe & Gísli Pálsson (eds.) (1996b). *Nature and Society. Anthropological Perspectives.* London & New York: Routledge (European Association of Social Anthropologists Series).

DeVries, Jop (2017). Bruno Latour, a Veteran of the 'Science Wars,' Has a New Mission. https://www.sciencemag.org/news/2017/10/bruno-latour-veteran-science-wars-has-new-mission.

De Vries, Bert & Johan Goudsblom (eds.) (2012). *Mappae Mundi. Humans and Their Habitats in a Long-Term Socio-Ecological Perspective. Myths, Maps, and Models.* Amsterdam: Amsterdam University Press.

DGSKA (Deutsche Gesellschaft für Sozial- und Kulturanthropologie). 2021 GAA Conference 2021, 27 to 30 September 2021 at the University of Bremen „Worlds. Zones. Atmospheres. Seismographies of the Anthropocene".

Dewey, Scott Hamilton (2016). Review zu Purdy 2015. *The American Historical Review* __ :1329–1330.

Diamond, Jared (2005). *Collapse. How Societies Choose to Fail or Succeed.* New York: Viking

Dibley, Ben (2012). 'The Shape of Things to Come": Seven Theses on the Anthropocene and Attachment. *Australian Humanities Review* 52: 139–153.

Di Chiro, Giovanna (2017). Welcome to the White (M) Anthropocene? A feminist-environmentalist critique. In: Sherilyn MacGregor (ed.). *Routledge Handbook of Gender and Environment.* New York: Routledge: 427–504 (Routledge Environment and Sustainability Handbooks).

Dickel, Sascha (2021). Der „Technological Fix". Zur Kritik einer kritischen Semantik. In: SONA - Netzwerk Soziologie der Nachhaltigkeit (eds.). *Soziologie der Nachhaltigkeit.* Bielefeld: Transcript (Soziologie der Nachhaltigkeit, 1). 271–284.

Diederichsen, Diedrich & Anselm Franke (eds.) (2013). *The Whole Earth. Kalifornien und das Verschwinden des Außen.* Berlin: Sternberg Press.

Dietzsch, Ina (2017). Klimawandel. Kulturanthropologische Perspektiven darauf, wie ein abstrakter Begriff erfahrbar gemacht wird. *Schweizerisches Archiv für Volkskunde/Archives Suisses desTraditions Populaires* 113(1). 21–39.

Dikau, Richard, Katharina Eibisch, Jana Eichel, Karoline Meßenzehl & Manuela Schlummer-Held (2019). *Geomorphologie.* Berlin: Springer Spektrum.

Domańska, Ewa (2020). History, Anthropogenic Soil, and Unbecoming Human. In: Saurabh Dube, Sanjay Seth & Ajay Skaria (eds.). *Dipesh Chakrabarty and the Global South. Subaltern Studies, Postcolonial Perspectives, and the Anthropocene.* London & New York: Routledge (Postcolonial Politics). 201–214.

Donges, Jonathan F., Ricarda Winkelmann, Wolfgang Lucht, Sarah E Cornell, James G Dyke, Johan Rockström, Jobst Heitzig & Hans Joachim Schellnhuber (2017). Closing the Loop: Reconnecting Human Dynamics to Earth System Science. *The Anthropocene Review* 4(2). 151–157.

Doughty, C.E., A. Wolf & C.B. Field (2010). Biophysical Feedbacks Between the Pleistocene Megafauna Extinction and Climate: The First Human-Induced Global Warming? *Geophysical Research Letters* 37: L15703.

Dove, Michael R. (2019). Plants, Politics, and the Imagination Over the Past 500 Years in the Indo-Malay Region. *Current Anthropology* 60(suppl. 20). S309–S320.

Dove, Michael R. (ed.) (2014). *The Anthropology of Climate Change. A Historical Reader.* Malden, Mass.: Wiley-Blackwell.

Dries, Christian (2018). Von der Weltfremdheit zur Antiquiertheit des Menschen. Günther Anders' negative Anthropologie. In: Günther Anders: *Die Weltfremdheit des Menschen. Schriften zur Anthropologie.* eds. v. Christian Dries. München: C. H. Beck: 437–533.

Driscoll, Catherine (2015). Neither Adaptive Thinking nor Reverse Engineering: Methods in the Evolutionary Social Sciences. *Biology and Philosophy* 30(1). 59–75.

Driscoll, Mark W. (2020). *The Whites Are Enemies of Heaven. Climate Caucasianism and Asian Ecological Protection.* Durham, N.C.: Duke University Press.

Dryzek, John S. & Andrew Pickering (2019). *The Politics of the Anthropocene.* Oxford etc.: Oxford University Press.

Dube, Saurabh, Sanjay Seth & Ajay Skaria (eds.) (2020). *Dipesh Chakrabarty and the Global South. Subaltern Studies, Postcolonial Perspectives, and the Anthropocene*. London & New York: Routledge (Postcolonial Politics).

Dubeau, Mathieu (2018). Reclaiming Species-Being: Toward an Interspecies Historical Materialism. *Rethinking Marxism* 30(2).186–207.

Dubos, René (1968). *So Human an Animal*. New York: Scribner.

Duile, Timo, Kristina Großmann, Michaela Haug and Guido Sprenger (eds.) (2023). *Plural Ecologies in Southeast Asia: Hierarchies, Conflicts, and Coexistence. London etc.: Routledge* (Routledge Contemporary Southeast Asia Series).

Dürbeck, Gabriele (2015). Das Anthropozän in geistes- und kulturwissenschaftlicher Perspektive. In: Dies. & Urte Stobbe (eds.). *Ecocriticism. Eine Einführung*, Köln: Böhlau: 107–119.

Dürbeck, Gabriele (2018a). Das Anthropozän erzählen: Fünf Narrative. *Aus Politik und Zeitgeschichte* 21–23/11–17.

Dürbeck, Gabriele (2018b). Narrative des Anthropozän –Systematisierung eines interdisziplinären Diskurses. *Kulturwissenschaftliche Zeitschrift* 2(1). 1–20.

Dürbeck, Gabriele (2021). Ansichtssache. Narrative und ihr Mobilisierungspotential. *Politische Ökologie* 39(167). 31–38.

Dürbeck, Gabriele & Philip Hüpkes (eds.) (2020). *The Anthropocenic Turn. Interplay Between Disciplinary and Interdisciplinary Responses to a New Age*. London & New York: Routledge (Routledge Interdisciplinary Perspectives on Literature, 117).

Dürbeck, Gabriele & Philip Hüpkes (eds.) (2022). *Narratives of Scale in the Anthropocene. Imagining Human Responsibility in an Age of Scalar Complexity*. London & New York: Routledge (Routledge Interdisciplinary Perspectives on Literature, 139).

Dürbeck, Gabriele & Jonas Nesselhauf (eds.) (2019). *Repräsentationsweisen des Anthropozän in Literatur und Medien. Representations of the Anthropocene in Literature and Media*. Berlin etc.: Peter Lang (Studies in Literature, Culture, and the Environment, 5).

Dürbeck, Gabriele & Urte Stobbe (eds.) (2015). *Ecocriticism. Eine Einführung*, Köln: Böhlau (Böhlau Studienbücher).

Duflo, Esther & Rohini Pande (2007). Dams. *Quarterly Journal of Economics* 122(2). 601–646.

Dukes, Paul (2011*). Minutes to Midnight. History and the Anthropocene Era from 1763*. London & New York: Anthem Press.

Dunbar, Robin I. M. (1993). Coevolution of Neocortical Size, Group Size and Language in Humans. *Behavioral and Brain Sciences* 16(4). 681–735.

Eagleton, Terry (2000). *The Idea of Culture*. Oxford: Blackwell.

Ebron, Paulla & Anna Lowenhaupt Tsing (2017). Feminism and the Anthropocene: Assessing the Field Through Recent Books. *Feminist Studies* 43: 658–83.

Edgeworth, Matt (2014). *The Relationship Between Archaeological Stratigraphy and Artificial Ground and Its Significance in the Anthropocene*. London: Geological Society. Special Publications 395(1). 91–108.

Edgeworth, Matt (2018a). Review of 'Climate Without Nature: A Critical Anthropology of the Anthropocene' by Andrew Bauer & Mona Bhan. *American Antiquity* 83(4). 758–759.

Edgeworth, Matt (2018b). More Than Just a Record: Active Ecological Effects of Archaeological Strata. In: Marcos Andre Torres de Souza & Diogo Menezes Costa (eds.). Historical Archaeology and Environment. Cham, Switzerland: Springer: 19–40.

Edgeworth, Matt (2018c). *River Deltas* in the Context of the Anthropocene Debate. In Franz Krause (ed.). *Delta Methods. Reflections on researching hydrosocial lifeworlds*. Cologne University of Cologne (Cologne Working Papers in Cultural and Social Anthropology No. 7 Special issue). 22–27.

Edgeworth, Matt, Dan deB. Richter, Colin Waters, Peter Haff, Cath Neal & Simon James Price (2015). Diachronous Beginnings of the Anthropocene: The Lower Bounding Surface of Anthropogenic Deposits. *The Anthropocene Review* 2:33–58.

Edgeworth, Matt, Erle C. Ellis, Philip Gibbard, Cath Neal & Michael Ellis (2019). The Chronostratigraphic Method is Unsuitable for Determining the Start of the Anthropocene. *Progress in Physical Geography: Earth and Environment* 43(3). 334–344.

Edwards, Paul N. (2015). Wissenschaftsinfrastrukturen für das Anthroppzän. In Jürgen Renn & Bernd Scherer (eds.). *Das Anthropozän. Zum Stand der Dinge*. Berlin: Matthes & Seitz. 242–255.

Eggert, Manfred K. H. (2012). Cultural Anthropology, Archaeology and Sociocultural Evolution: Exploring Submerged Territory. In: Tobias L. Kienlin & Andreas Zimmermann (eds.). *Beyond Elites. Alternatives to Hierarchical Systems in Modelling Social Formations*. Bonn: Habelt, 91–104 (Universitätsforschungen zur Prähistorischen Archäologie, 215).

Egner, Heike & Horst Peter Groß (eds.) (2019). *Das Anthropozän. Interdisziplinäre Perspektiven auf eine Krisendiagnostik*. München & Wien: Profil Verag (Klagenfurter Interdisziplinäres Kolleg).

Egner, Heike & Moremi Zeil (2019). Das Anthropozän – ein begriffliches Erdbeben (nicht nur für die Geographie). In: Heike Egner & Horst Peter Groß (eds.) (2019). *Das Anthropozän. Interdisziplinäre Perspektiven auf eine Krisendiagnostik*. München & Wien: Profil Verag (Klagenfurter Interdisziplinäres Kolleg). 15–32.

Ehlers, Eckart (2008). *Das Anthropozän. Die Erde im Zeitalter des Menschen*. Darmstadt: WBG Wissenschaftliche Buchgesellschaft.

Ehlers, Eckart (2015). Down to Earth – Geography in the Anthropocene. In: Frauke Kraas, Dietrich Soyez, Carsten Butsch, Franziska Krachten, & Holger Kretschmer (eds.). *IGC Cologne 2012. Down to Earth. Documenting the 32nd International Geographical Congress in Cologne 26–30 August 2012*. Cologne: 26–32 (Kölner Geographische Arbeiten, 95).

Ehlers, Eckart & Thomas Krafft (eds.) (2010). *Earth System Science in the Anthropocene. Emerging Issues and Problems*. Berlin & Heidelberg: Springer.

Ehrlich, Paul R. & Anne H. Ehrlich (2008). *The Dominant Animal. Human Evolution and the Environment*. Washington, D.C. etc: Island Pres/ Shearwater Books.

Ehrlich, Paul R. & Anne H. Ehrlich (2022). Returning to „Normal"? Evolutionary Roots of the Human Prospect. *BioScience* 72(8). 778–788.

Eisenstadt, Shmuel Noah (2002). *Multiple Modernities*. New Brunswick, NJ: Transaction Publishers.

Eisl, Markus, Judith Grubinger-Preiner, Gerald Mansberger, Paul Schreilechner & Sonja Staudenherz (2011). *Human Footprint. Satellitenbilder dokumentieren menschliches Handeln*. Salzburg: Ecovision.

Eitel, Kathrin & Michaela Meurer (2021). Introduction. Exploring Multifarious Worlds and the Political Within the Ontological Turn(s). In: Michaela Meurer & Kathrin Eitel (eds.). Ecological Ontologies .Approaching Human-Environmental Engagements. Berliner Blätter 84: 3–19.

Elhacham, Emily, Liad Ben-Uri, Jonathan Grozovski, Yinon M. Bar-On & Ron Milo (2020). Global Human-Made Mass Exceeds All Living Biomass. Nature 588: 442–444.

Ellen, Roy (1982). *Environment, Subsistence and System. The Ecology of Small-Scale Social Formations*. New York: Cambridge University Press.

Ellen, Roy (1999). Models of Subsistence and Ethnobiological Knowledge: Between Extraction and Cultivation in Southeast Asia. In: Scott Atran (ed.). *Folk Biology*. Boston: The MIT Press: 91–117.

Ellen, Roy (2021). Introduction: Nature Beyond the 'Ontological Turn'. In: Ders.: *Nature Wars. Essays around a Contested Concept*. New York, Oxford: Berghahn.: 1–27 (Environmental Anthropology and Ethnobiology, 27).

Eliasson, Ólafur (2007). *Your Mobile Expectations. BMW H2R Project. Exhibition Booklet*. Berlin: Studio Olafur Eliasson.

Ellis, Erle C. (2015). Ecology in an Anthropogenic Biosphere. *Ecological Monographs* 85: 287–331.

Ellis, Erle C. (2018). *Anthropocene. A Very Short Introduction*. Oxford: Oxford University Press (Very Short Introductions).

Ellis, Erle C. (2023). The Anthropocene Condition: Evolving Through Socio-Ecological Transformations. *Philosophical Transactions of the Royal Society B* 379(1893) (https://doi.org/10.1098/rstb.2022.0255).

Ellis, Erle C. (2024). The Anthropocene is not an Epoch – but the Age of Humans is most definitely underway. *The Conservation* (https://anthroecology.org/wp-content/uploads/2024/03/ellis_2024b.pdf).

Ellis, Erle C. & Peter K. Haff (2009). Earth Science in the Anthropocene: New Epoch, New Paradigm, New Possibilities. *Eos. Transactions American Geophysical Union* 90(49). 473.

Ellis, Erle C., Arthur H.W. Beusen & Kees Klein Goldewijk (2020). Anthropogenic Biomes: 10,000 BCE to 2015 CE. Land 9(129). 1–19 (https://doi.org/10.3390/land9050129).

Ellis, Erle C., Jed O. Kaplan, Dorian Q. Fuller, Steve Vavrus, Kees Klein Goldewijk & Peter H. Verburg. (2013). Used Planet: A Global History. *Proceedings of the National Academy of Sciences* 110, no. 20: 7978–7985.

Erle C. Ellis, Kees Klein Goldewijk, Stefan Siebert, Deborah Lightman and Navin Ramankutty (2010). Anthropogenic transformation of the biomes, 1700 to 2000. *Global Ecology and Biogeography*, 19: 589–606

Erle C. Ellis, Nicholas R. Magliocca Chris J. Stevens and Dorian Q. Fuller (2018). Evolving the Anthropocene: linking multi-level selection with long-term social–ecological change. *Sustainability Science* 13:119–128.

Ellis, Erle C., Nicolas Gauthier, Kees Klein Goldewijk, Rebecca Bliege Bird, Nicole Boivin, Sandra Díazi , Dorian Q. Fuller, Jacquelyn L. Gilll, Jed O. Kaplan, Naomi Kingston, Harvey Locke, Crystal N. H. McMichael, Darren Ranco, Torben C. Rick, M. Rebecca Shaw, Lucas Stephens, Jens-Christian Svenning & James E. M. Watson (2021). People Have Shaped Most of Terrestrial Nature for at Least 12,000 years. *Proceedings of the National Academy of Sciences* 118(17). 1–8. (https://www.pnas.org/content/pnas/118/17/e2023483118.full.pdf)

Ellis, Erle C., Mark Maslin, Nicole Boivin & Andrew Bauer (2016a). Involve Social Scientists in Defining the Anthropocene. *Nature* 540:192–193.

Ellis, Erle C., Nicholas R. Magliocca, Chris J. Stevens & Dorian Q. Fuller (2016b). Evolving the Anthropocene: Linking Multi-Level Selection With Long-Term Social–Ecological Change. Sustainability Science 13:119–128.

Ellis, Erle C. & Navin Ramankutty (2008). Putting People in the Map: Anthropogenic Biomes of the World. Frontiers in Ecology and the Environment 6: 439–447.

Ellsworth, Elizabeth & Jamie Kruse (eds.) (2013). *Making the Geologic Now. Responses to Material Conditions of Contemporary Life*. New York: Punktum.

Emmett, Robert & Thomas Lekan (eds.) (2016). *Whose Anthropocene? Revisiting Dipesh Chakrabarty's "Four Theses"*. Munich: Rachel Carson Center for Environment and Society (RCC Perspectives, Transformations in Environment and Society, 2).

Engle Merry, Sally (2011). Measuring the World: Indicators, Human Rights, and Global Governance. *Current Anthropology* 52(S3). S83–S95.

Eriksen, Thomas Hylland ([2]2014). *Globalization. The Key Concepts*. London etc. Bloomsbury (The Key Concepts Series).

Eriksen, Thomas Hylland ([5]2023). *Small Places, Large Issues: An Introduction to Social and Cultural Anthropology*. London, East Haven, Conn.: Pluto Press (Anthropology, Culture and Society) ([1]1996)

Eriksen, Thomas Hylland (2016). *Overheating: An Anthropology of Accelerated Change*. London & New York: Pluto Press.

Eriksen, Thomas Hylland (2018). *Cooling down the Overheated Anthropocene. Lessons from Anthropology and Cultural History*. 2018 Gutorm Gjessing Lecture, Oslo: Department of Social Anthropology, University of Oslo (Museum of Cultural History, University of Oslo Working Paper 1).

Eriksen, Thomas Hylland & Elizabeth Schober (2016). Economies of Growth or Ecologies of Survival? *Ethnos* 83 (3). 415–422.

Eriksen, Thomas Hylland & Astrid B. Stensrud (2018). *Climate Capitalism and Communities. An Anthropology of Environmental Overheating*. London & New York: Pluto Press.

Erlandson, Jon M. & Todd J. Braje (2014). Archeology and the Anthropocene. *Anthropocene*. https://www.sciencedirect.com/science/article/abs/pii/S2213305414000265

Ernstson, Henrik & Erik Swyngedouw (eds.) (2019). *Urban political ecology in the Anthropo-obscene*. London: Routledge.
Escobar, Arturo (2016). Thinking-Feeling With the Earth. Territorial Struggles and the Ontological Dimension of the Epistemologies of the South. *AIBR, Revista de Antropología Iberoamericana* 11(1). 11–32.
Escobar, Arturo (2018). *Designs for the Pluriverse. Radical Interdependence, Autonomy, and the Making of Worlds*. Durham, NC.: Duke University Press (New Ecologies for the Twenty-first Century).
Etelain, Jeanne (2023). Critical Zone. In: Nathanaël Wallenhorst & Christoph Wulf (eds.). *Handbook of the Anthropocene. Humans between Heritage and Future*. Berlin etc.: Springer: 1611–1615.
Evans-Pritchard, Edward Evan (1940). *The Nuer. A Description of the Modes of Livelihood and Political Institutions of a Nilotic People*. Oxford: Clarendon Press.
Fabian, Johannes (2014). *Time and the Other. How Anthropology Makes its Object*. New York: Columbia University Press (reprint, orig. 1983).
Fagan, Brian (2020). Big History and Archaeology. Archaeology is Big History. In: Craig Benjamin, Esther Quaedackers and David Baker (eds.). *The Routledge Companion to Big History. Milton Park & New York: Routledge* (Routledge Companions). 156–169.
Fagan, Brian & Nadia Durrani (2021). *Climate Chaos. Lessons on Survival from our Ancestors*. New York: Public Affairs, Hachette Group, Perseus Books.
Fairhead, James & Melissa Leach (1996). *Misreading the African Landscape. Society and Ecology in a Forest-Savanna Mosaic*. Cambridge, UK: Cambridge Univ. Press
Falb, Daniel (2015). *Anthropozän. Dichtung in der Gegenwartsgeologie* Berlin: Verlagshaus Berlin (Edition Poeticon).
Fardon, Richard, Olivia Harris, Trevor H. J. Marchand, Mark Nuttall, Cris Shore, Veronica Strang & Richard A. Wilson (eds.) (2012). *The Sage Handbook of Social Anthropology: 2 Volumes*. Los Angeles: Sage Publications.
Farrell, Justin (2015). Corporate Funding and Ideological Polarisation About Climate Change. *Proceedings of the National Academy of Sciences* 113(1). 201509433.
Fassin, Didier (ed.) (2012). *A Companion to Moral Anthropology*. Malden, Mass. etc.: Wiley-Blackwell (Blackwell Companions to Anthropology, 20).
Federau, Alexander (2017). *Pour une philosophie de l'anthropocène*. Paris: Presses Universitaires de France (Écologie en questions).
Felgentreff, Carsten & Thomas Glade (eds.) (2008). *Naturrisiken und Sozialkatastrophen*. Heidelberg: Spektrum Akademischer Verlag.
Fenske, Michaela & Arnika Peselmann (eds.) (2020). *Wasser, Luft und Erde. Gemeinsames Werden in NaturenKulturen*. Würzburg: Königshausen und Neumann (Beiträge zur Europäischen Ethnologie/Volkskunde).
Ferdinand, Malcolm (2022). *Decolonial Ecology. Thinking Ecology from the Carribean World*. London: John Wiley & Sons (frz. Orig. "Une écologie décoloniale. Penser l´écologie depuis le monde caribéen", Paris: Seuil.
Fernández-Armesto, Felipe (ed.) (2019). *The Oxford Illustrated History of the World*. Oxford: Oxford University Press.
Ferrando, Francesca (2016). The Party of the Anthropocene: Post-humanism, Environmentalism and the Post-anthropocentric Paradigm Shift. *Relations. Post-Anthropo-centrism* 4(2). 159–173.
Ficek, Rosa E. (2019). Cattle, Capital, Colonization: Tracking Creatures of the Anthropocene in and out of Human Projects. *Current Anthropology* 60(suppl. 20).S260–S271.
Figueroa, Adolfo (2017). *Economics of the Anthropocene Age*. New York: Palgrave Macmillan.
Filipi, Ângela Marques (2011). Through the Looking-Glass: a Critical Review of Sociology and Medicine towards the Diagnosis of ADHD. *Configurações. Revista de sociologia* 8 (Cultura, Tecnologia e Identidade). 83–86.
Finke, Peter L. W. (2022). *Mut zum Gaiazän. Das Anthropozän hat versagt*. München: Oekom Verlag.

Finney, Stanley C. (2018). Comment on Bauer & Ellis 2018. *Current Anthropology* 59(2). 216–217.

Finney, Stanley C. & Lucy E. Edwards (2016). The "Anthropocene" Epoch: Scientific Decision or Political Statement? *Geological Society of America Today* 26(2). 4–10.

Finney, Stanley C. & Philipp L. Gibbard (2023). The Humanities are invited to the Anthropocene Event but not to the Anthropocene Series/Epoch: a response to Chvostek (2023). *Journal of Quaternary Science* 38(4). 461–462.

Finney, Stanley C. (2014). The Anthropocene as a ratified unit in the ICS International Chronostratigraphic Chart: fundamental issues that must be addressed by the Task Group. In: Colin N. Waters, Jan A. Zalasiewicz, M. Williams Erle & A.M. Snelling (eds.). *A Stratigraphical Basis for the Anthropocene*. London: Geological Society (Special Publications, 395). 23–28.

Fischer, Michael (2009). Anthropological Futures. Durham, NC.: Duke University Press.

Fischer-Kowalski, Marina & Helmut Haberl (1993). Metabolism and Colonization: Modes of Production and the Physical Exchange between Societies and Nature. *Innovation: The European Journal of Social Science Research* 6: 415–442.

Fischer-Kowalski, Marina, Fridolin Krausmann & Irene Pallua (2014). A sociometabolic reading of the Anthropocene: Modes of subsistence, population size and human impact on Earth. *The Anthropocene Review* 1(1) 8–33.

Fischer-Kowalski, Marina & Helmut Haberl (eds.) (2017). *Socioecological Transitions and Global Change. Trajectories of Social Metabolism and Land Use*: Cheltenham: Edward Elgar.

Fladvad, Benno (2021). Von Technoutopien und Endzeitszenarien: Zur Bedeutung der Dialektik der Aufklärung für eine kritisch-emanzipatorische Humangeographie im Anthropozän. *Geographische Zeitschrift* 109(2–3).144–163.

Flitner, Michael (2014). Global Change. In: Julia Lossau, Tim Freytag & Roland Lippuner (eds.). *Schlüsselbegriffe der Kultur- und Sozialgeographie*. Stuttgart: Verlag Eugen Ulmer: 81–93.

Foley, Robert (1987). *Another Unique Species. Patterns in Human Evolutionary Ecology*. Essex, England: Longman.

Foley, Stephen F., Detlef Gronenborn, Meinrat O. Andreae, Joachim W. Kadereit, Jan Esper, Denis Scholz, Ulrich Pöschl, Dorrit E. Jacob, Bernd R. Schöne, Rainer Schreg, Andreas Vött, David Jordan, Jos Lelieveld, Christine G. Weller, Kurt W. Alt, Sabine Gaudzinski-Windheuser, Kai-Christian Bruhn, Holger Tost, Frank Sirocko & Paul J. Crutzen (2014). The Palaeoanthropocene. The Beginnings of Anthropogenic Environmental Change. *Anthropocene* 3: 83–88.

Folke, Carl (2006) Resilience: The Emergence of a Perspective for Social–Ecological Systems Analyses. *Global Environmental Change* 16: 253–67.

Folke, Carl (2016). Resilience. *Ecology and Society* 21(4). 44. https://doi.org/10.5751/ES-09088-210444.

Folke, Carl , Stephen Polasky, Johan Rockström, Victor Galaz, Frances Westley, Michèle Lamont, Marten Scheffer, Henrik Österblom, Stephen R. Carpenter, F. Stuart Chapin III, Karen C. Seto, Elke U. Weber, Beatrice I. Crona, Gretchen C. Daily, Partha Dasgupta, Owen Gaffney, Line J. Gordon, Holger Hoff, Simon A. Levin, Jane Lubchenco, Will Steffen & Brian H. Walker (2021). Our Future in the Anthropocene Biosphere. *Ambio. A Journal of the Human Environment* 50: 834–896.

Folkers, Andreas (2017). Politik des Lebens jenseits seiner selbst. Für eine ökologische Lebenssoziologie mit Deleuze und Guattari. *Soziale Welt* 68: 365–384.

Folkers, Andreas (2020). Was ist das Anthropozän und was wird es gewesen sein? Ein kritischer Überblick über neue Literatur zum kontemporären Erdzeitalter. *N.T.M. Zeitschrift für Geschichte der Wissenschaften, Technik und Medizin* 28(4): 589–604.

Folkers, Andreas (2024). Anthropozän. In: Marco Sonnberger, Alena Bleicher & Matthias Groß (eds.). *Handbuch Umweltsoziologie*. 2. Auflage. Berlin: Springer VS: 657–669.

Folkers, Andreas & Nadine Marquardt (2017). Die Kosmopolitik des Ereignisses. Gaia, das Anthropozän und die Welt ohne uns. In: Corinna Bath, Hanna Meißner, Stephan Trinkaus & Susanne Völker (eds.). *Verantwortung und Un/Verfügbarkeit. Impulse und Zugänge eines (neo) materialistischen Feminismus*. Reihe der DGS-Sektion Frauen und Geschlechterforschung, Münster.

Folkers, Andreas & Nadine Marquardt (2018). Die Verschränkung von Umwelt und Wohnwelt. Grüne smart homes aus der Perspektive der pluralen Sphärologie. *Geographica Helvetica* 73: 79–93.

Fortun, Kim (2001). *Advocacy After Bhopal. Environmentalism, Disaster, New World Orders*. Chicago: University of Chicago Press.

Fortun, Kim, James Adams, Tim Schütz, Scott Gabriel Knowles (2021). Knowledge infrastructure and research agendas for quotidian Anthropocenes: Critical localism with planetary scope. *The Anthropo*cene Review 8(2). 169–182 (Special Issue: The Mississippi Papers).

Foster, Jay (2018). Let's not talk about the Anthropocene. *Analecta Hermeneutica* 10:1–22.

Foster, John Bellamy (2020). *The Return of Nature. Socialism and Ecology*. New York, N.Y.: Monthly Review Press.

Foster, John Bellamy (2022). *Capitalism in the Anthropocene. Ecological Ruin or Ecological Revolution*. New York, N.Y.: Monthly Review Press.

Foster, John Bellamy, Brett Clark & Richard York (2010). *The Ecological Rift. Capitalism's War on the Earth*. New York: Monthly Review Press.

Foster, John Bellamy & Brett Clark (2020). *The Robbery of Nature. Capitalism and the Ecological Rift*. New York, N.Y.: Monthly Review Press.

Foster, John Bellamy & Ian Angus (2016). Marxism in the Anthropocene: Dialectical Rifts on the Left. *International Critical Thought* 6(3). 393–421.

Foucault, Michel (1974). *Die Ordnung der Dinge. Eine Archäologie der Humanwissenschaften*. Frankfurt am Main: Suhrkamp. (orig. „Les mots et les choses", Paris: Éditions Gallimard, 1966).

Fourault, Véronique (2020). Antropocène. En Géographie francaise. In: Frédéric Alexandre, Fabrice Argounès, Rémi Bénos & David Blanchon (2020). *Dictionnaire critique de l'anthropocène-* Paris: CNRS Éditions.

Frake, Charles Oliver (1962). The Ethnographic Study of Cognitive Systems. In: Thomas Gladwin, & W. C. Sturtevant (eds.). *Anthropology and Human Behavior*. Washington, D.C.: Soc. Wash. 72–93

Frank, Andre Gunder (1998). *Re-Orient. Global Econonmy in the Asian Age*. Berkeley: The University of California Press.

Frank, Andre Gunder und Barry K. Gills (eds.) (^{2}1996). *The World System. Five Hundred Years or Five Thousand?* London & New York: Routledge.

Frankopan, Peter (^{11}2015). *The Silk Roads. A New History of the World*. London: Bloomsbury.

Frankopan, Peter (2023). *The Earth Transformed. An Untold History*. London: Bloomsbury.

Fremaux, Anne (2017). Towards a Critical Theory of the Anthropocene and a Life-affirming Politics. A Post-Anthropocentric, Post-Growth, Post-(neo)Liberal Green Republican Analysis.n Ph.D-Thesis, Belfast: Queen's University, School of History, Anthropology, Philosophy and Politics.

Fremaux, Anne (2019). For a Post-Anthropocentric Socio-Nature Relationship in the Anthropocene after the Anthropocene: In: Dies. (ed.). *After the Anthropocene. Green Republicanism in a Post-Capitalist World*. Palgrave Macmillan (Environmental Politics and Theory). 119–163.

Frisch, Max (1979). *Der Mensch erscheint im Holozän. Eine Erzählung*. Frankfurt am Main: Suhrkamp.

Fuentes, Agustín (2006). Human-Nonhuman Primate Interconnections and Their Relevance to Anthropology. *Ecological and Environmental Anthropology* (University of Georgia). 1.

Fuentes, Agustín & Marcus Baynes-Rock (2017). Anthropogenic Landscapes, Human Action and the Process of Co-Construction With Other Species: Making Anthromes in the Anthropocene. *Land* 2017, 6, 15, https://doi.org/10.3390/land6010015

Fuhr, Harald (2021). The Rise of the Global South and the Rise in Carbon Emissions. *Third World Quarterly* 42(11). 2724–2746.

Fuller, Richard Buckminster (1963). *Operating Manual for Spaceship Earth*. New York: E. P. Dutton & Co.

Fuller, Richard (Buckminster 1998). *Bedienungsanleitung für das Raumschiff Erde und andere Schriften*. Amsterdam/Dresden: Verlag der Kunst (zuerst Reinbek bei Hamburg, Rowohlt, 1973).

Fuller, Dorian Q., Jacob van Etten, Katie Manning, Christina Kobo Castillo, Eleanor Kingwell-Banham, Alison Weisskopf, Ling Qin, Yo-Ichiro Sato & Robert J. Hijmans (2011). The Contribution of Rice Agriculture and Livestock Pastoralism to Prehistoric Methane Levels: An Archaeological Assessment. *The Holocene* 21: 743–759 (https://doi.org/10.1177/0959683611398052).

Future Earth (2013). *Future Earth Initial Science Report*. Paris: International Council of Science.

Galaz, Victor (2014). *Global Environmental Governance, Technology and Politics. The Anthropocene Gap*. Cheltenham & Northhampton, Mass.: Edward Elgar.

Gamble, Clive (2013). *Settling the Earth. The Archaeology of Deep Human History*. Cambridge etc.: Cambridge University Press.

Gamble, Clive (2021). *Making Deep History. Zeal, Perseverance, and the Time Revolution of 1859*. Oxford: Oxford University Press.

Gan, Elaine, Anna Tsing & Daniel Sullivan (2018). Using Natural History in the Study of Industrial Ruins. *Journal of Ethnobiology* 38(1). 39–54.

Gan, Elaine, Anna Lowenhaupt Tsing, Heather Swanson & Nils Bubandt (2017). Haunted Landscapes of the Anthropocene. In: Anna Lowenhaupt Tsing, Heather Swanson, Elaine Gan & Nils Bubandt (eds.) (2017). *Arts of Living on a Damaged Planet. Ghosts of the Anthropocene. Monsters of the Anthropocene*. Minneapolis: University of Minnesota Press: G1–G14.

Garcés, Marina (2019). *Neue radikale Aufklärung*. Wien: Verlag Turia + Kant, Wien, 2019 (Orig. span. „Nova il lustració radicale", Barcelona, 2017).

García-Moreno, Olga, Diego Álvarez-Laó , Miguel Arbizu , Eduardo Dopico , Eva García-Vázquez , Joaquín García Sansegundo, Montserrat Jiménez-Sánchez, Laura Miralles, Ícaro Obeso, Ángel Rodríguez-Rey, Marco de la Rasilla Vives, Luis Vicente Sánchez Fernández, Luis Rodríguez Terente, Luigi Toffolatti, Pablo Turrero (2020). The Little Big History of the Nalón River, Asturias, Spain. In: Craig Benjamin, Esther Quaedackers and David Baker (eds.). *The Routledge Companion to Big History*. Milton Park & New York: Routledge (Routledge Companions). 300–319.

Gardner, Katy & David Lewis (2015). *Anthropology and Development. Challenges for the Twenty-First Century*. London, Chicago, Ill: Pluto Press (Anthropology, Culture and Society).

Geertz, Clifford (1963). *Agricultural Involution. The Process of Ecological Change in Indonesia*. Berkeley: University of California Press.

Geertz, Clifford (1977). Religion as a Cultural System. In: Ders.: *The Interpretation of Cultures: Selected Essays*. New York: Basic Books: 87–125.

Geertz, Clifford James (1992). Kulturbegriff und Menschenbild. In: Habermas, Rebekka & Nils Minkmar (eds.). *Das Schwein des Häuptlings. Sechs Aufsätze zur Historischen Anthropologie*. Berlin: Verlag Klaus Wagenbach: 56–82 (Wagenbachs Taschenbuch, 212).

Gebhardt, Hans (2016). Das „Anthropozän" – zur Konjunktur eines Begriffs. In: Michael Wink & Joachim Funke (eds.). *Heidelberger Jahrbücher* 1: 28–42. (https://doi.org/10.17885/heiup.hdjbo.23557).

Geiger, Danilo (2009). *Turner in the Tropics: The Frontier Concept Revisited*. Luzern: Universität, Ph.-D. Dissertation.

Geiger, Danilo (ed.) (2008). Frontier Encounters: Indigenous Communities and Settlers in Asia and Latin America. Geneva: IWGIA (Iwgia Document, 120).

Gellner, Ernest (1983). *Nations and Nationalism*. Oxford: Blackwell.

Gell, Alfred (1992). *The Anthropology of Time. Cultural Costructions of Temporal Maps and Images*. Oxford: Berg.

Gemenne, François & Aleksandar Rankovic, Thomas Ansart, Benoît Martin, Patrice Mitrano, Antoine Rio & Atelier de Cartographie de Sciences Po ([2]2021). *Atlas de l'anthropocène. Deuxième édition actualisée et augmentée*. Paris: Presses de Siences Po.

Gerasimov, Innokenti P. (1979). Anthropogene and its major problem. *Boreas* 8: 23–30.

Gertenbach, Lars (2019). Postkonstruktivismus in der Kultursoziologie. In: Stephan Moebius et al. (eds.). *Handbuch Kultursoziologie, Band 2: Theorien – Methoden – Felder*. Wiesbaden: Springer VS: 53–76.

Gesing, Friderike, Katrin Amelang, Michael Flitner & Michi Knecht (2019). NaturenKulturen-Forschung. Eine Einleitung. In: Dies (eds.). *NaturenKulturen*. Bielefeld: transcript: 7–50.
Ghosh, Amitav (2005). *The Hungry Tide*. London: Harper Collins.
Ghosh, Amitav (2016). *The Great Derangement. Climate Change and the Unthinkable*. Chicago: Chicago University Press.
Ghosh, Amitav (2023). *The Nutmeg's Curse. Parables for a Planet in Crisis*. Chicago: The University of Chicago Press.
Gibbard, Philipp L. & Mike J. C. Walker (2014). The Term "Anthropocene" in the Context of Formal Geological Classification. In: Colin Waters, Jan Zalasiewicz, M. Williams, M. Ellis & A. Snelling (eds.). A stratigraphic basis for the Anthropocene. London: Geological Society, Special Publications: 29–37.
Gibbard, Philipp L., Andrew M. Bauer, Matthew Edgeworth, William F. Ruddiman, Jacquelyn L. Gill, Dorothy J. Merritts, Stanley C. Finney, Lucy E. Edwards, Michael J. C. Walker, Mark Maslin & Erle C. Ellis (2021). A practical solution: the Anthropocene is a Geological Event, not a formal Epoch. *Episodes* 45(4). 349–357.
Gibson, Katherine, Deborah Bird Rose, Ruth Fincher (eds.) (2015). *Manifesto for Living in the Anthropocene*. Brooklyn, NY: Punctum Books.
Gibson, Hannah & Sita Venkateswar (2015). Anthropological Engagement with the Anthropocene. A Critical Review. *Environment and Society: Advances in Research* 6: 5–27 (https://doi.org/10.3167/ares.2015.060102).
Gilroy, Paul (2004). *After Empire. Melancholia or Convivial Culture?* Abingdon: Routledge, auch New York: Columbia University Press.
Giri, Ananta Kumar (2006). *Cosmopolitism*. Vortrag, Universität Trier, Zentrum für Ostasien- und Pazifikstudien (ZOPS), 27.6.2006, Mitschrift CA.
Gilbert, Scott F. (2017). Holobiont by Birth. Multilineage Individuals as the Concretion of Cooperative Processes. In: Anna Lowenhaupt Tsing et al. (eds.). *Arts of Living on a Damaged Planet. Ghosts and Monsters of the Anthropocene*. Minneapolis: University of Minnesota Press: 73–89.
Glabau, Danya (2017). Feminists Write the Anthropocene: Three Tales of Possibility in Late Capitalism. *Journal of Cultural Economy* 10(6). 541–548.
Glaeser, Bernhard (1996). Humanökologie: Der sozialwissenschaftliche Ansatz. *Naturwissenschaften*. 83(4). 145–152.
Glaser, Bruno & Willliam I. Woods (2004). *Amazonian Dark Earths. Explorations in Space and Time*. Berlin & Heidelberg: Springer.
Glaser, Marion, Gesche Krause, Beate M. W. Ratter & Marion Welp (eds.) (2012). *Human-Nature Interactions in the Anthropocene*. New York: Routledge (Routledge Studies in Environment, Culture, and Society, 1).
Glaser, Rüdiger (2017). *Physische Geographie kompakt*. Berlin: Springer (zuerst Heidelberg: Spektrum Akademischer Verlag, 2010).
Glaser, Rüdiger, u. Mitarb. von Elke Schliermann-Kraus (2014). *Global Change. Das neue Gesicht der Erde*. Darmstadt: Primus, WBG Wissenschaftliche Buchgesellschaft.
Glaubrecht, Matthias (2019). *Das Ende der Evolution. Der Mensch und die Vernichtung der Arten*. München: C. Bertelsmann.
Glikson, Andrew Y. (2013). Fire and Human Evolution: The Deep-Time Blueprints of the Anthropocene. *Anthropocene* 3: 89–92.
Glikson, Andrew Y. & Colin Groves (2016). *Climate, Fire and Human Evolution. The Deep Time Dimensions of the Anthropocene*. Cham: Springer (Modern Approaches in Solid Earth Sciences, 10).
Goeke, Pascal (2022). Anthropozän. Von einer Beobachtungskategorie zur kategorialen Weltbeobachtungsformel. *Geographische Zeitschrift* 110(2). 88–07.
Görg, Christoph (2003a). *Regulation der Naturverhältnisse*. Münster: Westfälisches Dampfboot.
Görg, Christoph (2003b). Gesellschaftstheorie und Naturverhältnisse. Von den Grenzen der Regulationstheorie. In: Ulrich Brand & Werner Raza (eds.). *Fit für den Postfordismus?*

Theoretisch-Politische Perspektiven des Regulationsansatzes. Münster: Westfälisches Dampfboot 175–194.

Görg, Christoph (2015). Anthropozän. In: Sybille Bauriedl (eds.). *Wörterbuch Klimadebatte*. Bielefeld: Transcript: 29–36 (Edition Kulturwissenschaft).

Görg, Christoph (2016). Zwischen Tagesgeschäft und Erdgeschichte: Die unterschiedlichen Zeitskalen in der Debatte um das Anthropozän. *Gaia* 25(1). 9–13.

Görg, Christoph (2019). Naturverhältnisse. In: Claudia von Braunmühl, Heide Gerstenberger, Ralf Ptak & Christa Wichterich (Hg.). *ABC der globalen (Un)Ordnung. Von „Anthropozän" bis „Zivilgesellschaft"*. Hamburg: Verlag VSA: 172–173.

Görg, Christoph, Christina Plank, Dominik Wiedenhofer, Andreas Mayer, Melanie Pichler, Anke Schaffartzik & Fridolin Krausmann (2020). Scrutinizing the Great Acceleration: The Anthropocene and its Analytic Challenges for Social-ecological Transformations. *The Anthropocene Review* 7(1) 42–61.

Goldstone, Jack A. (2002). Efflorescence and Economic Growth in World History: Rethinking the "Rise of the West" and the Industrial Revolution. *Journal of World History* 13(2, Fall). 323–398.

González-Ruibal, Alfredo (2018). Beyond the Anthropocene: Defining the Age of Destruction. *Norwegian Archaeological Review* 51: 10–21.

Goody, Jack (2010). *The Eurasian Miracle*. Cambridge: Polity Press.

Google Earth Engine (2021). engine.google.com/timelapse.

Gordon, Helen (2021). *Notes from Deep Time. A Journey Through Our Past and Future Worlds*. London: Profile Books.

Goren-Inbar, Naama, Nira Alperson, Mordechai E. Kislev, Orit Simchoni, Yoel Melamed, Adi Ben-Nun & Ella Werker (2004). Evidence of Hominin Control of Fire at Gesher Benot Ya`aqov, Israel. *Science* 304(5671). 725–727.

Goudie, Andrew S. ([8]2019). *Human Impact on the Natural Environment. Past, Present, and Future*. London & New York: Blackwell.

Goudie, Andrew S. & Heather A. Viles (2016). *Geomorphology in the Anthropocene*. Cambridge: Cambridge University Press.

Gould, Stephen Jay (1987). *Time's Arrow. Time's Cycle. Myth and Metaphor in the Discovery of Geological Time*. Cambridge, Mass.: Harvard University.

Graedel, Thomas E & Paul J. Crutzen (1996). *Atmosphere, Climate, and Change*. New York: W. H. Freeman, 1996 (Scientific American Library, 1996).

Graeber, David & David Wengrow (2021). *The Dawn of Everything. A New History of Humanity*. New York: Farrar, Straus & Giroux.

Graham, Stephen (2018). *Vertical. The City from Satellites to Bunkers*. London & New York: Verso.

Gramlich, Naomie (2020). Feministisches Spekulieren. Einigen Pfaden folgen. In: Marie-Luise Angerer & Naomie Gramlich (eds.). *Feministisches Spekulieren. Genealogien, Narrationen, Zeitlichkeiten*. Berlin: Kulturverlag Kadmos: 9–31.

Grana-Behrens, Daniel (2021). "Big Data" and Alexander von Humboldt´s Approach to Science. *German Life and Letters* __: 1–22.

Grana-Behrens, Daniel (in prep.). *Anthropologie der Natur. Beziehung von Dingen als Weltverständnis– Zur Kritik an der ontologischen Wende*. unpubl. Ms.

Grant, Benjamin (2016). *Overview. A New Perspective of Earth*. New York: Preface Publishing.

Grant, Benjamin & Timothy Dougherty (2020). *Timelapse. How We Change the Earth*. San Francisco: Ten Speed Press.

Green, John (2021). *Tbbhe Anthropocene Reviewed*. London: Dutton.

Greenspoon, David (2016). *Earth in Human Hands. Shaping Our Planet´s Future*. New York: N.Y. & Boston: Grand Central Publishing.

Greenwood, Davydd J. (1984). *The Taming of Evolution. The Persistence of Nonevolutionary Views in the Study of Humans*. Ithaca, N.J. and London.

Greenwood, Davydd J. & William A. Stini (1977). *Nature, Culture, and Human History. A Bio-Cultural Introduction to Anthropology*. New York: Harper & Row, Publishers.

Gren, Martin & Edward Huijbens (eds.) (2015). *Tourism and the Anthropocene*. London etc.: Routledge.

Grinevald, Jacques (2012). *Le concept d'Anthropocène et son contexte historique et scientifique*. Séminaire du 11 mai 2012. Paris: Institut Momentum (L´Anthropocène et ses issues).

Gronenborn, Detlef (2024) Das Anthropozän (Kommentar). *Archäologie in Deutschland* (2). 50–51.

Grusin, Richard A. (2017). *Anthropocene Feminism*. Minneapolis & London: University of Minnesota Press (Center for 21st Century Studies).

Guiliani, Gaia (ed.) 2021: *Monsters, Catastrophes and the Anthropocene. A Postcolonial Critique*. Milton Park, Abingdon & New York: Routledge (Routledge Environmental Humanities).

Guyot-Téphanie, Josselin (2020). Antropocène. Histoire de Concept. In: Alexandre, Frédéric, Fabrice Argounès, Rémi Bénos & David Blanchon (dir.) *Dictionnaire critique de l'anthropocène*. Paris: CNRS Éditions.

Haardt, Oliver F. R. (2022). *Industrielle Revolution 2.0. Eine historische Navigationshilfe*. Darmstadt : WBG (Wissenschaftliche Buchgesellschaft).

Haber, Heinz (1965). *Unser blauer Planet*. München: Deutsche Verlags-Anstalt.

Haber, Wolfgang, Martin Held & Markus Vogt (eds.) (2016). *Die Welt im Anthropozän. Erkundungen im Spannungsfeld zwischen Ökologie und Humanität*. München: Oekom.

Haberl, Helmut, Dominik Wiedenhofer, Stefan Pauliuk, Fridolin Krausmann, Daniel B. Müller & Marina Fischer-Kowalski (2019). Contributions of Sociometabolic Research to Sustainability Science. *Nature and Sustainability* 2: 173–184.

Haeckel, Ernst (1886). *Generelle Morphologie der Organismen*. Berlin: Walter de Gruyter.

Haeckel, Ernst (2019). *Generelle Morphologie der Organismen*. Berlin: Walter de Gruyter (reprint, original 1886).

Head, Martin J. & Philip Gibbard (2015). Formal subdivision of the quaternary system/period: past, present, and future. *Quaternary International* 383:4–35.

Haenn, Nora & Richard Wilk (eds.) (2006). *The Environment in Anthropology. A Reader in Ecology, Culture, and Sustainable Living*. New York: New York University Press.

Haff, Peter K. (2013). Technology as a Geological Phenomenon: Implications for Human Well-Being. In: Waters, C. N., Zalasiewicz, J. A., Williams, M., Ellis, M. A. & Snelling, A. M. (eds). A *Stratigraphical Basis for the Anthropocene* (Geological Society, London, Special Publications, 395). https://doi.org/10.1144/SP395.4

Haff, Peter K. (2018). Breaking the Anthropocene Illusion, in: *Being Human in the Anthropocene*. blog, 2018. https://perma.cc/Z4CT-V7L9.

Haff, Peter (2019). The Technosphere and Its Relation to the Anthropocene. In: Jan Zalasiewicz et al. (eds.). *The Anthropocene as a Geological Time Unit. A Guide to the Scientific Evidence and Current Debate*. Cambridge: Cambridge University Press: 138–143.

Hahn, Hans Peter ([2]2014). *Materielle Kultur. Eine Einführung*. Berlin: Dietrich Reimer Verlag (Ethnologische Paperbacks).

Haidle, Miriam (2012). *How to Think Tools? A Comparison of Cognitive Aspects in Tool Behaviour of Animals and During Human Evolution*. Habilitationsschrift, Universität Tübingen (Cognitive Aspects in Tool Behaviour 1)(urn:nbn:de:bsz:21-opus-60146).

Haidle, Miriam Noel (2023). Cognigrams: systematically reconstructing behavioral architectures as basis for cognitive archaeology. In: T. Wynn, K.A. Overmann, and F.L. Coolidge (eds.): *Oxford Handbook of Cognitive Archaeology*. Oxford, Oxford University Press, C12S1–C12S8. https://doi.org/10.1093/oxfordhb/9780192895950.013.12.

Haidle, Miriam & Oliver Schlaudt (2019). Where Does Cumulative Culture Begin? A Plea for a SociologicallyInformed Perspective. *Biological Theory* (https://doi.org/10.1007/s13752-020-00351-w).

Haines, Andy & Howard Frumkin (2021). *Planetary Health. Safeguarding Human Health and the Environment in the Anthropocene*. Cambridge: Cambridge University Press.

Hall, M. (2005). *Earth Repair. A Transatlantic History of Environmental Restoration.* Charlottesville, Virg.: University of Virginia Press.
Halle, Clemence & Anne-Sophie Milon (2020). The Infinity of the Anthropocene: A (Hi)story With a Thousand Names. In: Bruno Latour & Peter Weibel (eds.). *Critical Zones: The Science and Politics of Landing on Earth.* Karlsruhe & Cambridge, Mass: ZKM Center for Art and Media & MIT Press: 44–49.
Haller, Tobias (2007). Is There a Culture of Sustainability? What Social and Cultural Anthropology Has to Offer 15 Years after Rio. In P. Burger & R. Kaufmann- Hayoz (eds.). *15 Jahre nach Rio – Der Nachhaltigkeitsdiskurs in den Geistes- und Sozialwissenschaften: Perspektiven – Leistungen – Defizite.* Bern: Schweizerische Akademie der Geistes- und Sozialwissenschaften SAGW: 329–356.
Haller, Tobias (2019). The Different Meanings of Land in the Age of Neoliberalism: Theoretical Reflections on Commons and Resilience Grabbing from a Social Anthropological Perspective. *Land* 8 (https://boris.unibe.ch/141094/1/land-08-00104-v2.pdf).
Halvaksz, Jamon Alex (2021). Environmental Anthropology. In: Lene Pedersen & Lisa Cligett (eds.). *The Sage Handbook of Cultural Anthropology.* London: Sage Publications: 245–260.
Hamann, Alexandra, Claudia Zea-Schmid & Reinhold Leinfelder (eds.) (2013). *Die große Transformation. Klima – Kriegen wir die Kurve?* Berlin: Verlag Jacoby & Stuart.
Hamann, Alexandra, Reinhold Leinfelder, Helmuth Trischler, Henning Wagenbreth (eds.) (2014). *Anthropozän: 30 Meilensteine auf dem Weg in ein neues Erdzeitalter – Eine Comic-Anthologie.* München: Deutsches Museum.
Hamilton, Clive (2013). *Earthmasters . The Dawn of the Age of Climate Engineering.* Sydney etc.: Yale University Press.
Hamilton,Clive(2014).EcologistsButtOut:YouAreNotEntitledtoRedefinetheAnthropocene.https://clivehamilton.com/ecologists-butt-out-you-are-not-entitled-to-redefine-the-anthropocene/.
Hamilton, Clive (2015). The Theodicy of the "Good Anthropocene." *Environmental Humanities* 7:233–238.
Hamilton, Clive (2016a). Define the Anthropocene in Terms of the Whole Earth. *Nature News* 536 (7616). 251.
Hamilton, Clive (2016b). The Anthropocene as Rupture. *The Anthropocene Review* 3(2). 93–106.
Hamilton, Clive (2017). *Defiant Earth. The Fate of Humans in the Anthropocene.* Cambridge & Malden, Mass: Polity Press.
Hamilton, Clive & Jacques Grinevald (2015). Was the Anthropocene Anticipated? *The Anthropocene Review* 2: 59–72.
Hamilton, Clive, Christophe Bonneuil & François Gemenne (eds.) (2015). *The Anthropocene and the Global Environmental Crisis. Rethinking Modernity in a New Epoch.* London & New York: Routledge (Routledge Environmental Humanities).
Hamilton, Rebecca, Jesse Wolfhagen, Noel Amano, Nicole Boivin, David Max Findley, José Iriarte, Jed O. Kaplan & Janelle Stevenson & Patrick Roberts (2021). Non-uniform tropical forest responses to the 'Columbian Exchange' in the Neotropics and Asia-Pacific. *Nature, Ecology and Evolution.* https://doi.org/10.1038/s41559-021-01474-4.
Hann, Chris (ed.) (1994). *When History Accelerates: Essays on Rapid Social Change, Complexity and Creativity.* London: Athlone Press.
Hann, Chris (2015). Backwardness Revisited: Time, Space and Civilization in Rural Eastern Europe. *Comparative Studies in Society and History* 57(4). 881–911.
Hann, Chris (2016a). The Anthropocene and Anthropology: Micro and Macro Perspectives. *European Journal of Social Theory.* https://doi.org/10.1177/1368431016649362).
Hann, Chris (2016b). A Concept of Eurasia. *Current Anthropology* 57(1). 1–27.
Hann, Chris & Keith Hart (2011). *Economic Anthropology. History, Ethnography, Critique.* Cambridge & Malden, Mass.: Polity Press
Hannah, Michael (2021). *Extinctions. Living and Dying in the Margin of Error.* Cambridge etc.: Cambridge University Press.

Hannerz, Ulf (2010). *Anthropology's World. Life in a Twenty-First-century Discipline*, London/New York: Pluto Press (Anthropology, Culture and Society).

Hannerz, Ulf (2016). *Writing Future Worlds. An Anthropologist Explores Global Scenarios.* Basingstoke: Palgrave Macmillan (Palgrave Studies in Literary Anthropology).

Hansen, Klaus P. ([4]2011). *Kultur und Kulturwissenschaft. Eine Einführung*. Tübingen und Basel: A. Francke Verlag.

Hanusch, Frederic, Claus Leggewie & Erik Meyer (2021). *Planetar denken. Ein Einstieg*. Bielefeld: Transcript (X-Texte zu Kultur und Gesellschaft).

Haraway, Donna J. (1991). *Simians, Cyborgs, and Women: The Reinvention of Nature*. New York: Routledge

Haraway, Donna J. (2003). *The Companion Species Manifesto. Dogs, People, and Significant Otherness. Vol 1.* Chicago: Prickly Paradigm Press.

Haraway, Donna J. (2015). Anthropocene, Capitalocene, Plantationocene, Chthulucene: Making Kin. *Environmental Humanities* 6: 159–165.

Haraway, Donna J. (2016). *Staying With the Trouble. Making Kin in the Chthulucene.* Durham, N.C.: Duke University Press.

Haraway, Donna J. (2020). Chthulucene. In: Krogh, Marianne (ed.). *Connectedness. An Incomplete Encyclopedia of Anthropocene*. Copenhagen: Strandberg Publishing: 100–103.

Haraway, Donna, Noboru Ishikawa, Scott F. Gilbert, Kenneth Olwig, Anna L. Tsing & Nils Ole Bubandt (2016). Anthropologists Are Talking – About the Anthropocene. *Ethnos. Journal of Anthropology* 81(3). 535–564. https://doi.org/10.1080/00141844.2015.1105838.

Harper, Kyle, Lynn K. Nyhart, Joanna Radin, Julia A. Thomas, Russell H. Tuttle & Jonathan Lyon (2016). Bio-History in the Anthropocene: Interdisciplinary Study on the Past and Present of Human Life. *Chicago Journal of History* 7: 5–19.

Hartigan Jr., John (2021). Knowing Animals: Multispecies Ethnography and the Scope of Anthropology. *American Anthropologist* 123(4). 846–860.,

Hartung, Gerald (2022). Willkommen im Anthropozän. Ein Menschenbild für das neue Erdzeitalter. *Die politische Meinung* 66(571). 14–18.

Harris, Andrew (2016). Vertical Urbanisms. Opening Up Geographes of the Three-Dimensional City. *Progress in Human Geography* 39(5). 601–620.

Hastrup, Kirsten (2016). Climate Knowledge. Assemblage, Anticipation, Action. In: Susan A. Crate & Mark Nuttall (eds.). Anthropology and Climate Change. From Actions to Transformations. New York: Routledge: 35–57.

Haumann, Sebastian (2019). Zwischen „Nachhaltigkeit" und „Anthropozän". Neue Tendenzen in der Umweltgeschichte. *Neue Politische Literatur* 64: 295–326.

Haus der Kulturen der Welt (2013). *The Anthropocene Curriculum.* (https://www.anthropocene-curriculum.org/about).

Haus der Kulturen der Welt (2014). *The Anthropocene Project. Kulturelle Grundlagenforschung mit den Mitteln der Kunst und der Wissenschaft* 2013/2014: (http://www.hkw.de/de/programm/projekte/2014/anthropozaen/anthropozaen_2013_2014.php).

Hazen, Robert M., Edward S. Grew, Marcus J. Origlieri & Robert T. Downs (2017). On the Mineralogy of the "Anthropocene Epoch". *American Mineralogist* 102(3). 595–611 (Outlooks in Earth and Planetary Materials).

Head, Martin J. (2019). Formal Subdivision of the Quaternary System/Period: Present Status and Future Directions. *Quaternary International* 500: 32–51.

Headrick, Daniel R. (2021). *Humans versus Nature. A Global Environmental History.* New York: Oxford University Press.

Hecht, Gabrielle (2018). Interscalar Vehicles for an African Anthropocene: On Waste, Temporality, and Violence. *Cultural Anthropology* 33(1). 109–141.

Hedlund, Nicholas & Sean Esbjörn-Hargens (eds.) (2023). *Big Picture Perspectives on Planetary Flourishing. Metatheory for the Anthropocene 1.* London & New York: Routledge (Routledge Studies in Critical Realism, 1).

Heichele, Thomas (eds.) (2020). *Mensch – Natur – Technik. Philosophie für das Anthropozän*. Münster: Aschendorff Verlag (Studien zur systematischen Theologie, Ethik und Philosophie, 19).

Heise, Ursula K. (2008). *Sense of Place and Sense of Planet*. Oxford: Oxford University Press.

Heise, Ursula K. (2015a). Posthumanismus. Den Menschen neu denken. In: Nina Möllers, Christian Schwägerl & Helmuth Trischler (eds.). *Willkommen im Anthropozän. Unsere Verantwortung für die Zukunft der Erde*. München: Deutsches Museum Verlag: 38–42.

Heise, Ursula K. (2015b). Ökokosmopolitismus. In Gabriele Dürbeck & Urte Stobbe (eds.). *Ecocriticism. Eine Einführung*, Köln: Böhlau: 121–32.

Henningsen, Dierk (2009). Aktualismus in den Geowissenschaften – die Gegenwart als Schlüssel zur Vergangenheit. *Naturwissenschaftliche Rundschau* 62(5). 229–232.

Henrich, Joseph (2016). *Secret of Our Success. How Culture is Driving Human Evolution, Domesticating Our Species, and Making Us Smarter*. Princeton & Oxford: Princeton University Press.

Henrich, Joseph (2020). *The WEIRDest People in the World. How the West Became Psychologically Peculiar and Particularly Prosperous*. New York: Macmillan.

Henrich, Joseph, Steven Heine & Aram Norenzayan (2010). The Weirdest People in the World. *Behavioral and Brain Sciences* 33: 61–135.

Hernes, Gudmund (2012). *Hot Topic–Cold Comfort: Climate Change and Attitude Change*. Oslo: Nordforsk Press.

Herrmann, Bernd (2014). Einige umwelthistorische Kalenderblätter und Kalendergeschichten. In: Manfred Jakubowski-Tiessen& Jana Sprenger (eds.). *Natur und Gesellschaft. Perspektiven der interdisziplinären Umweltgeschichte*. Göttingen: Universitätsverlag Göttingen: 7–58.

Herrmann, Bernd (2015). Sind Umweltkrisen Krisen der Natur oder Krisen der Kultur? Zur Einführung. In: Ders. (eds.). *Sind Umweltkrisen Krisen der Natur oder der Kultur?* Berlin: Springer: 1–10.

Herrmann, Bernd ([2]2016). *Umweltgeschichte. Eine Einführung in Grundbegriffe*. Berlin & Heidelberg: Springer Spektrum.

Herrmann, Bernd (2019). *Das menschliche Ökosystem. Ein humanökologisch-erkenntnistheoretischer* Essay. Wiesbaden: Springer Spektrum (Essentials).

Herrmann, Bernd, Bernhard Glaeser & Thomas Potthast (2021) *Humanökologie*. Springer Spektrum, Wiesbaden (Essentials).

Hetzel, Andreas, Eva Schürmann & Harald Schwaetzer (eds.) (2020). *Widerstand und ziviler Ungehorsam im Anthropozän*. Frankfurt: Fromann-Holzboog (*Allgemeine Zeitschrift für Philosophie*: Heft 45.2).

Heyes, Cecilia (2018). *Cognitive Gadgets. The Cultural Evolution of Thinking*. Cambridge, Mass & London: The belknap Press of Harvard University Press.

Heymann, Matthias & Amy Dahan Dalmedico (2019). Epistemology and Politics in Earth System Modeling: Historical Perspectives. *Journal of Advances in Modeling Earth Systems*, 11: 1139–1152.

Hickmann, Thomas, Lena Partzsch, Philipp Pattberg & Sabine Weiland (eds.) (2019). *The Anthropocene Debate and Political Science*. London etc: Routledge (Routledge Research in Global Environmental Governance).

Hinrichsen, Jan, Reinhard Johler & Sandro Ratt (eds.) (2020). *Katastrophen/Kultur: Beiträge zu einer interdisziplinären Begriffswerkstatt*. Tübingen: Tübinger Vereinigung für Volkskunde (Studien und Materialien des Ludwig-Uhland-Instituts der Universität Tübingen, 5).

Hilberg, Sylke ([2]2022). *Umweltgeologie. Eine Einführung in Grundlagen und Praxis*. Berlin: Springer Spektrum.

Hirsbrunner, Simon David (2021). *A New Science for Future. Climate Impact Modeling and the Quest for Digital Openness*. Bielefeld: Transcript Verlag (Locating Media).

Hobson, John M. (2004). *The Eastern Origins of Western Civilization*. Cambridge: Cambridge University Press.

Hockings, Kimberley J., Matthew R. McLennan, Susana Carvalho, Marc Ancrenaz, René Bobe, Richard W. Byrne, Robin I.M. Dunbar, Tetsuro Matsuzawa, William C. McGrew, Elizabeth

A. Williamson, Michael L. Wilson, Bernard Wood, Richard W. Wrangham, and Catherine M. Hill (2015). Apes in the Anthropocene: Flexibility and Survival. *Trends in Ecology & Evolution* 30(4). 215–222.
Hodder, Ian (2012). *Entangled. An Archaeology of the Relationships Between Humans and Things*. Oxford: Wiley-Blackwell.
Hodder, Ian (2018). *Where Are We Heading? The Evolution of Humans and Things*. New Haven, Conn.: Yale University Press.
Hodder, Ian (2020). The Paradox of the Longterm: Human Evolution and Entanglement. *The Journal of the Royal Anthropological Institute* 26: 389–411.
Hodder, Ian (^{2}2024). *Entangled. An Archaeology of the Relationships Between Humans and Things*. Hoboken, N.J.: Wiley-Blackwell.
Höhler, Sabine (2015). *Spaceship Earth in the Environmental Age, 1960–1990*. London & New York: Routledge History and Philosophy of Technoscience, 4).
Höhler, Sabine & Fred Luks (2006a). Kenneth Bouldings „Raumschiff Erde" – 40 Jahre danach. In: Dies (eds.). Beam Us Up, Boulding! 40 Jahre "Raumschiff Erde". O.O.: Vereinigung für Ökologische Ökonomie (Beiträge & Berichte, 7). 5–8.
Höhler, Sabine & Fred Luks (eds.) (2006b). Beam Us Up, Boulding! 40 Jahre "Raumschiff Erde". O.O.: Vereinigung für Ökologische Ökonomie (Beiträge & Berichte, 7).
Hoelle, Jeffrey & Nicholas C. Kawa (2021). Placing the Anthropos in the Anthropocene. *Annals of the American Association of Geographers* 111(3). 655–662.
Hörl, Erich (2013). Tausend Ökologien. Der Prozess der Kybernetisierung und die Allgemeine Ökologie. In: Diedrich Diederichsen & Anselm Franke (eds.). *The Whole Earth. Kalifornien und das Verschwinden des Außen*. Berlin: Sternberg Press: 121–130.
Hoffmann, Christoph (2020). Spontane Wissenschaftsphilosophien und Philosophie der Wissenschaftler. *Merkur* 74(859). 80–87.
Hoffman, Susanna M., Thomas Hylland Eriksen & Paulo Mendes (eds.) (2022). *Cooling Down. Local Responses to Global Climate Change*. Oxford & New York: Berghahn.
Hoggenmüller, Sebastian W. (2016). Die Welt im (Außen-)Blick. Überlegungen zu einer ästhetischen RelKonstruktionsanalyse am Beispiel der Weltraumfotografie „Blue Marble". *Zeitschrift für Qualitative Forschung* 17(1–2). 11–40.
Hoiß, Christian (2017). Anthropozän. Spuren einer Narration. In: Anselm, Sabine & Christian Hoiß (eds.) : *Crossmediales Erzählen vom Anthropozän. Literarische Spuren in einem neuen Zeitalter*. München: Oekom: 13–37.
Hoiß, Christian (2019). *Deutschunterricht im Anthropozän. Didaktische Konzepte einer Bildung für nachhaltige Entwicklung*. Dissertation, München: LMU, Fakultät für Sprach- und Literaturwissenschaften, (https://edoc.ub.uni-muenchen.de/24608/).
Højrup, Mathilde & Heather Anne Swanson (2018). The Making of Unstable Ground: The Anthropogenic Geologies of Søby, Denmark. *Journal of Ethnobiology* 38(1). 24–38.
Holbraad, Martin and Morten Axel Pedersen (2017). *The Ontological Turn: An Anthropological Exposition*. Cambridge: Cambridge University Press (New Departures in Anthropology).
Holling, C. S. (1973). Resilience and Stability of Ecological Systems, Annual Review of Ecology and Systematics, 4/1, 1–23, https://doi.org/10.1146/annurev.es.04.110173.000245
Holling, Crawford Stanley & L. H. Gunderson (eds.) (2010). *Foundations of Ecological Resilience*. Washington, D.C.: Island Press.
Hollis, Martin (1997). The Limits of Irrationality. In: Bryan R. Wilson (ed.). *Rationality*. London: Blackwell und Oxford & New York: Harper & Row: 214–220 (Key Concepts in the Social Sciences).
Holm, Nicholas & Sy Taffel (eds.) (2016). *Ecological Entanglements in the Anthropocene*. Lanham etc.: Lexington Books (Ecocritical Theory and Practice).
Hommels, Anique (2005). Studying Obduracy in the City: Toward a Productive Fusion between Technology Studies and Urban Studies. *Science Technology and Human Values* 30(3). 323–351.
Hooke, Roger Le B. (2000). On the History of Humans as Geomorphic Agents. *Geology* 28(9). 843–846.

Hoppe, Katharina (2021). *Die Kraft der Revision. Epistemologie, Politik und Ethik bei Donna Haraway*. Frankfurt & New York: Campus.

Hoppe, Katharina (2022). Das Anthropozän kompostieren: Speziesübergreifende Verwandtschaft und sozial-ökologische Transformation. *INSERT. Artistic Practices as Cultural Inquiries, 2: senseABILITIES – auf der Suche nach einem anderen Erzählen im Anthropozändiskurs* (https://insert.art/ausgaben/senseabilities/das-anthropozaen-kompostieren, https://doi.org/10.5281/zenodo.6772311 .

Hoppe, Katharina & Thomas Lemke (2021). *Neue Materialismen zur Einführung*. Hamburg: Junius Verlag (Junius Einführungen).

Horkheimer, Max & Theodor W. Adorno (2006). *Dialektik der Aufklärung: philosophische Fragmente*. Frankfurt am Main: Fischer Verlag.

Horn, Eva (2014). *Die Zukunft als Katastrophe*. Frankfurt am Main: S. Fischer Verlag.

Horn, Eva (2016). Klimatologie um 1800. Zur Genealogie des Anthropozäns. *Zeitschrift für Kulturwissenschaften 1*: 87–202.

Horn, Eva (2017). Jenseits der Kindeskinder. Nachhaltigkeit im Anthropozän. *Merkur* 71 (814). 5–17.

Horn, Eva, im Gespräch mit Michael Reitz (2020). *Folgen einer Epoche. Der Mensch erscheint im Anthropozän*. Deutschlandfunk. (https://www.deutschlandfunk.de/der-mensch-erscheint-im-anthropozaen-folgen-einer-neuen-100.html).

Horn, Eva & Hannes Bergthaller (2020). *The Anthropocene. Key Issues for the Humanities*, London & New York: Routledge, Earthscan.

Horn, Eva & Hannes Bergthaller ([3]2022). *Anthropozän zur Einführung*. Hamburg: Junius (Junius Einführungen).

Hornborg, Alf (2014). Technology as Fetish. Marx, Latour, and the Cultural Foundatons of Global Capitalism. *Theory, Culture & Society* 31(4), 119–140.

Hornborg, Alf (2015). The Political Ecology of the Technocene: Uncovering Ecologically Unequal Exchange in the World-System. In: Clive Hamilton, Christophe Bonneuil & François Gemenne (eds.). *The Anthropocene and the Global Environmental Crisis: Rethinking Modernity in a New Epoch*. London: 57–69.

Hornborg, Alf (2017a). Artifacts Have Consequences, Not Agency: Toward a Critical Theory of Global Environmental History. *European Journal of Social Theory* 20: 95–110.

Hornborg, Alf (2017b). Dithering While the planet Burns: Anthropologists' Approaches to the Anthropocene. *Reviews in Anthropology* 46(2–3). 61–77.

Hornborg, Alf (2020a). Anthropology in the Anthropocene, Guest Editorial. *Anthropology Today* 36(2). 1–2.

Hornborg, Alf (2020b). Anthropocene, The. In: Hilary Callan (ed.). *The International Encyclopedia of Anthropology*. London etc: Wiley-Blackwell: 1–10. https://doi.org/10.1002/9781118924396.wbiea2386

Hornidge, Anna-Katharina & Christoph Antweiler (eds.) (2012). *Environmental Uncertainty and Local Knowledge. Southeast Asia as a Laboratory of Global Change*. Bielefeld: Transcript (Global Studies).

Howard, Peter, Ian Thompson Emma Waterton & Mick Atha (eds.) ([2]2019). *The Routledge Companion to Landscape Studies*. London & New York: Routledge (Routledge Campanions).

Howe, Cymene (2014). Anthropocene Ecoauthority: The Winds of Oaxaca. *Anthropological Quarterly* 87: 381–404.

Howe, Cymene (2015). Latin America in the Anthropocene: Energy Transitions and Climate Change Mitigations. *The Journal of Latin Amnerican and Caribbean Anthropology* 2082). 231–241.

Howe, Cymene & Anand Pandian (eds.) (2020). *Anthropocene Unseen. A Lexicon*. Brooklyn, N.Y.: Punctum Press. https://culanth.org/fieldsights/series/lexicon-for-an-anthropocene-yet-unseen.

Hubatschke, Christoph (2020). Anthropozän/Kapitalozän/Chthulucene. In: *diebresche*, 24. 9. 2020. www.diebresche.org/anthropozaenkapitalozaen-chthulucene/

Hübner, Andreas (2022). „Schwellenwerte“: Das Anthropozän erfahren. In: Jörg van Norden & Lale Yildirim (eds.). *Historische Erfahrung*. Frankfurt: Wochenschau Verlag: 265–278 (Geschichtsdidaktik theoretisch, 1).

Hüpkes, Philip (2019). “Anthropocenic Earth Mediality”. On Scaling and Deep Time in the Anthropocene. In: Gina Comos & Caroline Rosenthal (eds.). *Anglophone Literature and Culture in the Anthropocene*. Newcastle upon Tyne: Cambridge Scholars Publishing: 196–213.

Hüpkes, Philipp (2020a). Gaia. In: *Grundbegriffe des Anthropozän. Ein interdisziplinäres Lexikon des DFG-Forschungsprojekts „Narrative des Anthropozän in Wissenschaft und Literatur“* (www.uni-vechta.de/anthropozaen).

Hüpkes, Philipp (2020b). Tiefenzeit. In: *Grundbegriffe des Anthropozän. Ein interdisziplinäres Lexikon des DFG-Forschungsprojekts „Narrative des Anthropozän in Wissenschaft und Literatur“* (www.uni-vechta.de/anthropozaen).

Hüther, Gerald, Harald Lesch & Abraham Maslow (2020). *Bewusstsein, Liebe & Kreativität. Human- und Kulturpotenziale im Anthropozän*. Berlin & Pommritz: Pikok-Verlag.

Hudson, Mark J. (2014). Placing Asia in the Anthropocene: Histories, Vulnerabilities, Responses. *The Journal of Asian Studies* 73(4). 941–962.

Hudson, Mark J. (2019). Review of S. Yoneyama, Animism in Contemporary Japan: Voices for the Anthropocene from Post-Fukushima Japan. *Anthropological Notebooks* 25: 189–192.

Hudson, Mark J. (2021). *Conjuring Up Prehistory. Landscape and the Archaic in Japanese Nationalism*. Oxford: Archaeopress.

Hughes, J. Donald ([2]2009). *An Environmental History of the World. Humankind's Changing Role in the Community of Life*. Milton Park & New York: Routledge.

Hupke, Klaus-Dieter ([2]2020). *Naturschutz. Eine kritische Einführung*. Berlin & Heidelberg: Springer Spektrum.

Hurston, Zora Neale (1942). *Dust Tracks on a Road. An Autobiography*. Philadelphia: J. B. Lippincott.

Hutton, M.D. James (1788). *Theory of the Earth; or an Investigation of the Laws Observable in the Composition, Dissolution, and Restoration of Land Upon the Globe*. Transactions of the Royal Society of Edinburgh.

Huxley, Julian (1966). *The Future of Man. Evolutionary Aspects*. New York.

ICS (International Commission on Stratigraphy) (2020): *Chronostratigraphic Diagram* (www.stratigraphy.org/index.php/ics-chart-timescale).

IHDP (2010). *International Human Dimensions Programme on Global Environmental Change*. Bonn: United Nations University. (https://www.researchgate.net/profile/Carole_Crumley/publication/257588246/figure/fig3/AS:661747459452930@1534784297893/Beyond-the-Bretherton-diagram.png) .

Ingold, Tim (2011a). *Being Alive. Essays on Movement, Knowledge and Description*. London: Routledge.

Ingold, Tim (2011b). *Redrawing Anthropology. Materials, Movements, Lines*. Aldershot, UK: Ashgate.

Ingold, Tim (2018). *Anthropology. Why it Matters*. Cambridge & Madford, Mass.: Polity Press (Why it Matters Series).

Ingold, Tim & Gísli Pálsson (eds.) (2013). *Biosocial Becomings. Integrating Social and Biological Anthropology*. Cambridge: Cambridge University Press.

Ingwersen, Moritz & Sina Steglich (2022). Neue Zeitichkeiten und das Anthropozän. *Sciendo*. Sonderheft 5: 1–11; Open Access.

IPBES (2019). Summary for Policymakers of the IPBES Global Assessment Report Biodiversity and Ecosystem Services. Bonn: Intergovernmental Science-Policy Platform on Biodiversity and Ecosystem Services (https://ipbes.net/sites/default/files/inline/files/ipbes_global_assessment_report_summary_for_policymakers.pdf).

IPCC (Masson-Delmotte, V., P. Zhai, A. Pirani, S.L. Connors, C. Péan, S. Berger, N. Caud, Y. Chen, L. Goldfarb, M.I. Gomis, M. Huang, K. Leitzell, E. Lonnoy, J.B.R. Matthews, T.K. Maycock, T. Waterfield, O. Yelekçi, R. Yu, and B. Zhou, eds.) (2021). *Climate Change 2021: The*

Physical Science Basis. Contribution of Working Group I to the Sixth Assessment Report of the Intergovernmental Panel on Climate Change []. Cambridge: Cambridge University Press.

Irvine, Richard D. G. (2020). *An Anthropology of Deep Time. Geological Temporality and Social Life*. Cambridge etc.: Cambridge University Press (New Departures in Anthropology).

Isager, Lotte, Line Vestergaard Knudsen, Ida Theilade (2021). A New Keyword in the Museum: Exhibiting the Anthropocene. *Museum & Society* 19(1) 88–107.

Isenberg, Andrew C. (2014). *The Oxford Handbook of Environmental History*. Oxford & New York: Oxford University Press.

Isendahl, Christian (2010). The Anthropocene Forces us to Reconsider Adaptationist Models of human-environment interactions. *Environmental Science & Technology* 16: 6007.

Isendahl, Christian & Daryl Stump (eds.) (2019). Conclusion: Anthropocentric Historical Ecology, Applied Archaeology, and the Future of a Usable Past. In: Dies. (eds.). *Oxford Handbook of Historical Ecology and Applied Archaeology*. Oxford: Oxford University Press: 581–598

Ishikawa, Noboru & Ryoji Soda (eds.) (2020). *Anthropogenic Tropical Forests. Human–Nature Interfaces on the Plantation Frontier*. Singapore: Springer Nature (Advances in Asian Human-Environmental Research).

Issberner, Liz-Rejane & Philippe Léna (eds.) (2017). *Brazil in the Anthropocene. Conflicts between Predatory Development and Environmental Policies*. Milton Park & New York: Routledge (Routledge Environmental Humanities).

IUGS, International Union of Geological Sciences (2024). The Anthropocene. https://www.iugs.org/_files/ugd/f1fc07_ebe2e2b94c35491c8efe570cd2c5a1bf.pdf?index=true&trk=public_post_comment-text

Jablonka, Eva & Marion Lamb (2020). *Inheritance Systems and the Extended Synthesis*. Cambridge: Cambridge University Press.

Jackson, Jr., John P. (2010). Definitional Argument in Evolutionary Psychology and Cultural Anthropology. *Science in Context* 23(1). 121–150.

Jahn, Thomas, Diana Hummel & Engelbert Schramm (2016). *Sustainable Science in the Anthropocene*. Frankfurt am Main: Institut für sozial-ökologische Forschung (ISOE) (ISOE-Diskussionspapiere, 49).

Jamieson, Dale (2014). *Reasons in a Dark Time. Why the Struggle against Climat Change Failed, and What it Means for our Future*. Oxford: Oxford University Press.

Jansen, Markus (2023): *Mensch oder Erde. Ökologische Aufklärungen*. Marburg: Büchner Verlag.

Jasanoff, Sheila (2004a). The Idiom of Co-Production. In: Sheila Jasanoff (ed.). States of Knowledge. The Co-Production of Science and Social Order. London & New York: Routledge: 1–11. (International Library of Sociology).

Jasanoff, Sheila (2004b). Heaven and Earth. The Politics of Environmental Images. In: Sheila Jasanoff & Marybeth Long Martello (eds.). *Earthly Politics. Local and Global in Environmental Governance*. Cambridge: MIT Press: 31–52.

Jasanoff, Sheila (2010). A New Climate for Society. *Theory Culture and Society* 27(2–3). 233–253.

Jessop, Bob, Neil Brenner & Martin Jones (2008). Theorizing Sociospatial Relations. *Environment and Planning D: Society and Space* 26(3). 389–401.

Jobin, Paul, Ming-sho Ho & Hsin-Huang Michael Hsiao (eds.) (2021). *Environmental Movements and Politics of the Asian Anthropocene*. Singapore: ISEAS – Yusof Ishak Institute.

Jobson, Ryan Cecil (2020). The Case for Letting Anthropology Burn: Sociocultural Anthropology in 2019. *American Anthropologist* 122(2). 260–271.

Johnson, Amy, Chris Hebdon, Paul Burow, Deepti Chatti & Michael Dove (2022). Anthropocene. In: *Oxford Research Encyclopedias Anthropology* (https://doi.org/10.1093/acrefore/9780190854584.013.295).

Jon, Inhi (2021). *Cities in the Anthropocene. New Ecology and Urban Politics*. London: Pluto Press.

Jonas, Hans (2020). *Das Prinzip Verantwortung. Versuch einer Ethik für die technologische Zivilisation*. Berlin: Suhrkamp (zuerst 1979).

Jones, Lizzie P., Samuel T. Turvey, Dario Massimino & Sarah K. Papworth (2020): Investigating the implications of shifting baseline syndrome on conservation. *People and Nature* 2(4):1131–1144.

Jordan-Bychkov, Terry G, Mona Domosh & Lester Rowtree ([9]2003). *The Human Mosaic. A Thematic Introduction to Cultural Geography*. New York: W. H. Freeman.

Jordan, Peter (2015). *Technology as Human Social Tradition. Cultural Transmission among Hunter-Gatherers*. Berkeley, Cal.: University of California Press (Origins of Human Behavior and Culture, 7).

Jordheim, Helge & Einar Wigen (2018). Conceptual Synchronization. From Progress to Crisis. *Millennium. Journal of International Studies* 46(3). 421–439.

Jørgensen, Dolly (2014). Not by Human Hands: Five Technological Tenets for Environmental History in the Anthropocene. *Environment and History* 20(4). 20th Anniversary Issue: 479–489.

Jørgensen, Finn Arne & Dolly Jørgensen (2016). The Anthropocene as a History of Technology (Exhibit Review). *Technology and Culture* 57(1). 231–237.

Jørgensen, Peter Søgaard, Vanessa P. Weinberger & Timothy M. Waring (2023). Evolution and sustainability: gathering the strands for an Anthropocene synthesis. *Philosophical Transactions Royal Society B* 379: 20220251. (https://doi.org/10.1098/rstb.2022.0251).

Jullien, François (2008). *De l'universel, de l'uniforme, du commun et du dialogue entre les cultures*. Paris: Éditions Fayard.

Kahn, Joel S. (2003). Anthropology as Cosmopolitan Practice. *Anthropological Theory* 3(4). 403–415.

Kaika, Maria (2018). Between the Frog and the Eagle: Claiming a 'Scholarship of Presence' for the Anthropocene. *European Planning Studies* 26(9). 1714–1727.

Kallis, Giorgios, Susan D. Paulson, Giacomo D'Alisa & Federico Demaria (2020). *The Case for Degrowth*. London: Polity Press (The Case For Series).

Kaltmeier, Olaf, Eleonora Rohland, Gerardo Cham & Susana Herrera Lima (2024a). *Handbook of the Anthropocene in Latin America* (6 Vols.). Bielefeld: Transcript.

Kaltmeier, Olaf, Eleonora Rohland, Gerardo Cham, Susana Herrera Lima, Antoine Acker, León Enrique Ávila Romero, Juan Arturo Camacho Becerra, Virginia García Acosta, Anthony Goebel McDermott, Ricardo Gutiérrez, Regina Horta Duarte, Cecilia Ibarra, María Fernanda López Sandoval, Sofía Mendoza Bohne, José Augusto Pádua, Elissa Rashkin, Heidi V. Scott, Javier Taks, Helge Wendt & Adrián Gustavo Zarrilli (2024b): The Anthropocene as Multiple Crisis. Latin American Perspectives on Biodiversity. In: Olaf Kaltmeier, Antoine Acker, León Enrique Ávila Romero & Regina Horta Duarte (eds.) (2024b): *Biodiversity – Handbook of the Anthropocene in Latin America II*. Bielefeld: Transcript: 15–46.

Kampourakis, Kostas (2021). *Understanding Evolution*. Cambridge etc.: Cambridge University Press. (Understanding Life Series).

Kant, Immanuel ([2]1796). *Zum ewigen Frieden Ein philosophischer Entwurf. Neue vermehrte Auflage*. Königsberg: Friedrich Nicolovius.

Kapferer, Bruce & Marina Gold (2016). Introduction: Reconceptualizing the Discipline. In: *Moral Anhropology. A Critique*. New York & Oxford: Berghahn (Critical Interventions. A Forum for Social Analysis, 16). 1–24.

Kaplan, Jed O. (2018). The Importance of a Reference Frame (Kommentar zu Bauer & Ellis 2018). *Current Anthropology* 59(2). 217–218.

Kaplan, Jed O., Kristen M. Krumhardt & Niklaus Zimmermann (2009). The Prehistoric and Preindustrial Deforestation of Europe. *Quaternary Science Reviews* 28 (27–28).3016–3034.

Kaplan, Jed O., Kristen M. Krumhardt, Erle C. Ellis, William F. Ruddiman, Carsten Lemmen, & Kees Klein Goldewijk (2011). Holocene Carbon Emissions as a Result of Anthropogenic Land Cover Change. *The Holocene* 21(5). 775–791.

Kapp, Karl William (1963). Social Costs of Business Enterprise. New York.

Kaube, Jürgen (2017). *Die Anfänge von Allem*. Berlin: Rowohlt Berlin Verlag.

Kawa, Nicholas C. (2016). *Amazonia in the Anthropocene. People, Soils, Plants, Forests*. Austin: University of Texas Press.

Kellert, Stephen R. 1993: Introduction. In: Stephen R. Kellert & Edward Osborne Wilson (eds.). *The Biophilia Hypothesis*. Washington, D.C.: Island Press & Shearwater Books: 20–27.

Kelley, Kevin W. (eds.) (1988). *The Home Planet*. New York: Addison-Wesley & Moskau: Mir.

Kelly, Jason M. (2019). Anthropocenes – A Fractured Picture. In: Jason M. Kelly, Philip V. Scarpino, Helen Berry, James Syvitski & Michel Meybeck (eds.). *Rivers of the Anthropocene*. Berkeley: University of California Press.

Kelly, Luke T., Katherine M. Giljohann , Andrea Duane, Núria Aquilué, Sally Archibald, Enric Batllori, Andrew F. Bennett , Stephen T. Buckland, Quim Canelles, Michael F. Clarke, Marie-Josée Fortin, Virgilio Hermoso, Sergi Herrando, Robert E. Keane, Frank K. Lake, Michael A. McCarthy, Alejandra Morán-Ordóñez, Catherine L. Parr, Juli G. Pausas, Trent D. Penman, Adrián Regos, Libby Rumpff , Julianna L. Santos, Annabel L. Smith, Alexandra D. Syphard, Morgan W. Tingley & Lluís Brotons (2020). Fire and Biodiversity in the Anthropocene. *Science* 370: 1–10.

Kelly, Alice B. & Nancy Lee Peluso (2015). Frontiers of Commodification. *Society & Natural Resources* 28: 473–495.

Kersten, Jens (2013). The Enjoyment of Complexity: A New Political Anthropology for the Anthropocene? In: Helmuth Trischler (ed.). *Anthropocene: Envisioning the Future of the Age of Humans*, (RCC Perspectives 2013(3)). 39–55.

Kersten, Jens (2014). *Das Anthropozän-Konzept. Kontrakt – Komposition – Konflikt*. Baden-Baden: Nomos.

Kerner, Charlotte (2024). *We are Volcanoes. Die Öko-Visionärinnen Rachel Carson, Lynn Margulis, Donna Haraway*. Neu-Isenburg: Westend Verlag.

Keys, P. W, L. Badia & R. Warrier (2023). The Future in Anthropocene Science. *Earth's Future* 12: 1–13.

Khan, Naveeda (2019). At Play With the Giants: Between the Patchy Anthropocene and Romantic Geology. *Current Anthropology* 60(suppl. 20).S333–S341.

Kidner, David W. (2014). Why "Anthropocentrism" is not Anthropocentric. *Dialectical Anthropology* 38(4). 465–480.

King, Martin Luther Jr. (1967). Beyond Vietnam - A Time to Break Silence. Speech at a Meeting of Clergy and Laity Concerned About Vietnam at Riverside Church in New York City, https://www.lebenshaus-alb.de/magazin/005584.html.

Kirchhoff, Thomas, Nicole C. Karafyllis, Dirk Evers, Brigitte Falkenburg, Myriam Gerhard, Gerald Hartung, Jürgen Hübner, Kristian Köchy, Ulrich Krohs, Thomas Potthast, Otto Schäfer, Gregor Schiemann, Magnus Schlette, Reinhard Schulz, & Frank F. Vogelsang (eds.) ([2]2020). *Naturphilosophie. Ein Lehr – und Studienbuch*. Tübingen: Mohr Siebeck.

Kirksey, S. Eben (2015). *Emergent Ecologies*. Durham, N.C.: Duke University Press.

Kirksey, S. Eben & Stefan Helmreich (2010). The Emergence of Multispecies Ethnography. *Cultural Anthropology* 25(4). 545–576.

Kirsch, Stuart (2014). *Mining Capitalism. The Relationships Between Corporations and Their Critics*. Berkeley, Cal.: The University of California Press.

Kirsch, Stuart (2018). *Engaged Anthropology. Politics beyond the Text*. Berkeley: University of California Press.

Klass, Morton (2003). *Mind over Mind. The Anthropology and Psychology of Spirit Possession*. Lanham, Mad. etc.: Rowman & Littlefield Publishers.

Klein, George Defries (2015). The "Anthropocene": what is its geological utility? (Answer: It has none!). *International Union of Geological Sciences* 38(3). 218–218.

Klein, Naomi (2014). *This Changes Everything. Capitalism vs. Climate*. New York: Simon & Schuster.

Kleingeld, Pauline & Eric Brown (2009). Cosmopolitanism. In: Edwin N. Zalta (ed.). *The Stanford Encyclopedia of Philosophy* (http://plato.stanford.edu/archives/sum2009/3ntries/cosmpolitanism).

Kleinod, Michael, Timo Duile & Christoph Antweiler (2022). Outwitting the Spirits? Political-Ecological Implications of Southeast Asian Animism. *Berlin Journal of Critical Theory* 6(1). 127–174.
Klingan, Katrin & Christoph Rosol (eds.) (2019). *Technosphäre*. Berlin: Matthes & Seitz (Bibliothek 100 Jahre Moderne, 12).
Klingan, Katrin, Ashkan Sepahvand, Christoph Rosol & Bernd M. Scherer (eds.) (2015). *Textures of the Anthropocene, Grain, Vapour, Ray*. 4 Vols. Cambridge, Mass. & London: The MIT Press.
Knauß, Stefan (2018). Conceptualizing Human Stewardship in the Anthropocene: The Rights of Nature in Ecuador, New Zealand and India. *Journal of Agricultural and Environmental Ethics* 31: 703–722.
Knopf, Thomas (2017). *Ressourcennutzug und Umweltverhalten prähistorischer Bauern. Eine Analyse archäologischer und ethnographischer Untersuchungen*. Tübingen: Universität Tübingen (RessourcenKulturen, 3).
Knoll, Andrew H. (2021). *A Brief History of the Earth. Four Billion Years in Eight Chapters*. New York: N.Y.: Custom House.
Knox, Paul L. & Sallie A. Marston ([7]2016). *Human Geography. Places and Regions in Global Context*. Boston etc.: Pearson.
Koch, Alexander, Chris Brierley, Mark M. Maslin & Simon L. Lewis (2019). Earth System Impacts of the European Arrival and Great Dying in the Americas after 1492. *Quarternary Science Reviews* 207: 13–36.
Kocka, Jürgen (2002). Multiple Modernities and Negotiated Universals. In: Dominic Sachsenmaier, & Jens Riedel, with Shmuel Noah Eisenstadt (eds.). *Reflections on Multiple Modernities. European, Chinese and Other Interpretations*. Leiden: Brill Academic: 119–128.
Köhler, Bettina (2015). Raumschiff Erde. In: Sybille Bauriedl (eds.). *Wörterbuch Klimadebatte*. Bielefeld: Transcript: 245–252 (Edition Kulturwissenschaft).
Kohn, Eduardo (2013). *How Forests Think. Toward an Anthropology Beyond the Human*. Berkeley, Cal. Etc.: University of California Press.
Kohn, Eduardo (2015). Anthropology of Ontologies. *Annual Review of Anthropology* 44: 311–327.
Kohr, Leopold (1957). *The Breakdown of Nations*. New York: Dutton.
Kolbert, Elizabeth (2011). Enter the Anthropocene. Age of Man. *National Geographic Magazine*. 219(3). 60–85.
Kolbert, Elizabeth (2014). *The sixth extinction. An unnatural history*. London: Bloomsbury, New York: Holt.
Kolbert, Elizabeth (2021). *Under a White Sky. The Nature of the Future*. New York: Crown.
Kolbert, Elizabeth (2024): The "Epic Row" Over a New Epoch. *The New Yorker*, 20. April 2024 (https://www.newyorker.com/news/the-weekend-essay/the-epic-row-over-a-new-epoch).
Kopnina, Helen & Eleanor Shoreman-Oumet (eds.) (2017). *Routlege Handbook of Environmental Anthropology*. London & New York: Routledge & Earthscan (Routledge International Handbooks).
Kosiba, Steve (2019). New Digs: Networks, Assemblages, and the Dissolution of Binary Categories in Anthropological Archaeology. *American Anthropologist* 121(2). 147–163.
Kosseleck, Reinhart (1989). *Vergangene Zukunft. Zur Semantik geschichtlicher Zeiten*. Frankfurt am Main: Suhrkamp.
Kosseleck, Reinhart ([6]2013). *Zeitschichten. Studien zur Historik*. Frankfurt am Main: Suhrkamp.
Kothari, Ashish, Ariel Salleh, Arturo Escobar & Federico Demaria (eds.) (2019). *Pluriverse. A Post-Development Dictionary*. New Delhi: Tulika Books & Authorsupfront.
Kottak, Conrad Phillip ([19]2021). *Anthropology. The Exploration of Human Diversity*. Boston, Mass. etc.: McGraw-Hill.
Krämer, Anna Kalina (2016). *Das „Anthropozän" als Wendepunkt zu einem neuen wissenschaftlichen Bewusstsein? Eine Untersuchung aus ethnologischer Perspektive zur Bedeutung und Verwendung des Konzeptes*. Köln: (Kölner ethnologische Beiträge, 45).

Kramer, Caroline (federf. eds.). (2018). *Landschaftsforschung 50 Jahre nach Kiel – Bestandsaufnahme und Perspektiven*. Leipzig: Selbstverlag Akademie für Landeskunde e.V. (Berichte. Geographie und Landeskunde 92(3–4).

Krause, Franz (ed.) (2018). *Delta Methods. Reflections on Researching Hydrosocial Lifeworlds*. Cologne: University of Cologne (Kölner Arbeitspapiere zur Ethnologie /Cologne Working Papers in Cultural and Social Anthropology, KAE, 7, Special issue).

Krause, Karen & Andreas Reitz (2020). Die Bedeutung der Raumfahrt für das Konzept der Identität einer allumfassenden Menschheit. In: Gerd Jüttemann (eds.). *Psychologie der Geschichte*. Lengerich: Pabst Science Publishers: 228–234.

Krauß, Werner (2015a). Anthropology in the Anthropocene: Sustainable Development, Climate Change and Interdisciplinary Research. In: Heike Greschke & Julia Tischler (eds.). *Grounding Global Climate Change. Contributions from the Social and Cultural Sciences*. Dordrecht: Springer Scienece and Busines Media: 59–76.

Krauß, Werner (2015b). Klimawissenschaften. In: Sybille Bauriedl (eds.). *Wörterbuch Klimadebatte*. Bielefeld: Transcript: 201–208 (Edition Kulturwissenschaft).

Krauß, Werner ([2]2019). Postenvironmental Landscapes. In: Howard, Peter, Ian Thompson, Emma Waterton & Mick Atha (eds.). *The Routledge Companion to Landscape Studies*. London & New York: Routledge (Routledge Campanions).

Krauß, Werner (2020). Narratives of Change and the Co-Development of Climate Services for Action. *Climate Risk Management* 28: 100217.

Krauß, Werner & Scott Bremer (2020). The Role of Place-Based Narratives of Change in Climate Risk Governance. *Climate Risk Management* 28: 1–7 (https://doi.org/10.1016/j.crm.2020.100221).

Krebber, André (2019). Human-Animal Studies. Tiere als Forschungserspektive. In: Elke Diehl & Jens Tuider (eds.). *Haben Tiere Rechte? Aspekte und Dimensionen der Mensch-Tier-Beziehung*. Bonn: Bundeszentrale für politische Bildung (Schriftenreihe, 10450). 310–322.

Krech, Shepard (1999). *The Ecological Indian. Myth and History*. New York: W.W. Norton.

Kreff, Fernand, Eva-Maria Knoll & Andre Gingrich (eds.) (2011). *Lexikon der Globalisierung*. Bielefeld: Transcript Verlag.

Kroeber, Alfred Louis (1917). The Superorganic. *American Anthropologist* 19(2). 163–213.

Kroeber, Alfred Louis (2003). *Anthropology. Race, Language, Culture, Psychology, Prehistory*. New York: Crown (1948, 1956).

Krogh, Marianne (ed.) (2020). *Connectedness. An Incomplete Encyclopedia of the Anthropocene*. Copenhagen: Strandberg Publishing.

Kropp, Cordula (2024). StadtNaturen: Urbane Assemblagen und ihre Transformation. In: Marco Sonnberger, Alena Bleicher & Matthias Groß (eds.). *Handbuch Umweltsoziologie*. Berlin: Springer: 319–332.

Kropp, Cordula & Marco Sonneberger (2021). *Umweltsoziologie*. Baden-Baden: Nomos (Studienkurs Soziologie).

Krüger, Fred, Greg Bankoff, Terry Cannon, Benedikt Orlowski & E. Lisa F. Schipper (eds.) (2012). *Cultures and Disasters. Understanding Cultural Framings in Disaster Risk Reduction*. London etc.: Routledge.

Kruse, Jamie & Elizabeth Ellsworth (2011). *Geologic City. A Field Guide to the Geoarchitecture of New York*. New York: Smudge Studio.

Kueffer, Christoph (2013). Ökologische Neuartigkeit: die Ökologie des Anthropozäns. *ZiF-Mitteilungen* 1/2013: 21–30.

Kühnhardt, Ludger & Tilman Meyer (eds.) (2017). *Bonner Enzyklopädie der Globalität. 2 Bände*. Wiesbaden: Springer VS.

Kühne, Olaf (2018). Die Landschaften 1, 2 und 3 und ihr Wandel. Perspektiven für die Landschaftsforschung in der Geographie – 50 Jahre nach Kiel. In: Caroline Kramer (federf. eds.). *Landschaftsforschung 50 Jahre nach Kiel – Bestandsaufnahme und Perspektiven*. Leipzig: Selbsrtvberalg Akademie für Landeskunde e.V. (Berichte. Geographie und Landeskunde 92(3–4). 217–231.

Küster, Hansjörg (2012). *Die Entdeckung der Landschaft. Einführung in eine neue Wissenschaft.* München: C. H .Beck (Beck´sche Reihe, 6061).
Kuper, Adam (2003). *Culture. An Anthropological Account.* Cambridge, Mass.: Harvard University Press.
Kupper, Patrick (2021). *Umweltgeschichte.* Göttingen: Vandenhoeck & Ruprecht (Einführungen in die Geschichtswissenschaft, 3).
Kutschera, Ulrich (2013). The Age of Man: A Father Figure. *Science* 340: 1287.
Laboratory for Anthropogenic Landscape Ecology (2020). https://ecotope.org/anthromes/
La Danta Las Canta (2017). *El Faloceno: Redefinir el Antropoceno desde una mirada ecofeminista.* Grupo Ecofeminista de Investigación y Acción. http://ladantalascanta.blogspot.com/2017/07/articulo-el-faloceno-redefinir-el.html
Lahsen, Myanna (2005). Seductive Simulations? Uncertainty Distribution around Climate Models. *Social Studies of Science* 35: 895–922.
Laidlaw, James & Paolo Heywood (2013). One More Turn and You're There. In: *Anthropology of This Century* 7, o.S. http://aotcpress.com/articles/turn.
Laland, Kevin (2017). *Darwin's Unfinished Symphony. How Culture Made the Human.* Princeton & Oxford: Princton University Press.
Laland, Kevin and Lynn Chiu (2020). Evolution's Engineers. *Aeon* (https://aeon.co/essays/organisms-are-not-passive-recipients-of-evolutionary-forces?utm_source=Aeon+Newsletter&utm_campaign=476fb16283).
Laland, Kevin, John Odling-Smee & Marcus Feldman (2000). Niche Construction, Biological Evolution, and Cultural Change. *Behavioural and Brain Sciences* 23: 131–175.
LALE, Laboratory for Anthropogenic Landscape Ecology (2020)
Lampert, Irene & Kai Niebert (2019). Den globalen Wandel verstehen: Vorstellungen zur Stabilität und Instabilität der Erdsysteme. *Zeitschrift für Didaktik der Biologie (ZDB) –Biologie Lehren und Lernen* 23: 39–58.
Lang, Bernhard & Christof Mauch (2024). *Wie der Mensch die Erde verändert. Die Verwandlung der Welt.* München: Frederking & Thaler.
Lange, Bastian, Martina Hülz, Benedikt Schmid & Christian Schulz (eds.) (2020). *Postwachstumsgeographien. Raumbezüge diverser und alternativer Ökonomien.* Bielefeld: Transcript (Sozial- und Kulturgeographie, 38).
Lange, Axel (2020). *Evolutionstheorie im Wandel. Ist Darwin überholt?* Berlin: Springer
Lansing, J. Stephen & Murray P. Cox (2019). *Islands of Order. A Guide to Complexity Modeling for the Social Sciences.* Princeton & Oxford: Princeton University Press (Princeton Studies in Complexity).
Lansing, Stephen J. & Katherine Fox (2011). Niche Construction on Bali: The Gods of the Countryside. *Philosophical Transactions of the Royal Soceity* B 366: 927–34.
Latour, Bruno (2004). Why Has Critique Run Out of Steam? From Matters of Fact to Matters of Concern. *Critical Inquiry* 30 (2). xx–yy.
Latour, Bruno (2010). An Attempt to a 'Compositionist Manifesto'. *New Library History* 41(3). 471–490).
Latour, Bruno (2014). Agency at the Time of the Anthropocene. *New Literary History* 45(1). 1–18.
Latour, Bruno (2015). Telling Friends from Foes in the Time of the Anthropocene. In: Clive Hamilton, François Gemenne & Christophe Bonneuil (eds.). *The Anthropocene and the Global Environmental Crisis. Rethinking Modernity in a New Epoch.* London & New York: Routledge (Routledge Environmental Humanities). 145–155.
Latour, Bruno (2016). *Existenzweisen. Eine Anthropologie der Modernen.* Berlin: Suhrkamp. (orig. „Enquêtes sur les modes d'existence. Une anthropologie des Modernes", Paris: Éditions La Decouverte, 2012).
Latour, Bruno (2017a). Anthropology at the Time of the Anthropocene: A Personal View of What Is to Be Studied. In: Marc Brightman & Jerome Lewis (eds.). *The Anthropology of Sustainability. Beyond Development and Progress.* New York, N.Y.: Palgrave Macmillan (Palgrave Studies

in Anthropology of Sustainability). 35–49 (Distinguished lecture at the 113th American Anthropological Association Annual Meeting, Washington, D.C., December 2014).

Latour, Bruno (2017b). *Kampf um Gaia. Acht Vorträge über das neue Klimaregime.* Berlin: Suhrkamp (orig. frz. "Face à Gaïa. Huite conférences sur le Nouveau Regime Climatique", Paris: Éditions La Decouverte, 2015, Les empêcheurs de penser en rond).

Latour, Bruno (2018). *Das terrestrische Manifest.* Berlin: Suhrkamp. (original "Où atterrir? Comment s´orienter en politique", Paris: La Découverte, 2017).

Latour, Bruno & Peter Weibel (eds.) (2020). *Critical Zones. The Science and Politics of Landing on Earth.* Karlsruhe & Cambridge, Mass: ZKM Center for Art and Media & MIT Press: 44–49.

Launder, Brian & J. Michael T. Thompson (eds.) (2010). *Geo-Engineering Climate Change. Environmental Necessity or Pandora's Box?* Cambridge: Cambridge University Press.

Laux, Henning & Anna Henkel (eds.) (2018). *Die Erde, der Mensch und das Soziale. Zur Transformation gesellschaftlicher Naturverhältnisse im Anthropozän.* Bielefeld: Transcript (Reihe Sozialtheorie).

Laux, Henning (2020). Grundrisse einer Theorie der CO2-Verhältnisse. In: Frank Adloff, & Sighard Neckel (eds.) (2020). *Gesellschaftstheorie im Anthropozän.* Frankfurt & New York: Campus (Zukünfte der Nachhaltigkeit, 1). 123–155.

Lavenda, Robert H. & Emily A. Schultz (⁷2020). *Core Concepts in Cultural Anthropology.* Oxford etc.: Oxford University Press

Law, John (2015). What's Wrong With a One-World World? *Distinktion: Journal of Social Theory* 16(1). 126–139.

Lawrence, Susan, Peter Davies & Jodi Turnbull (2016). The Archaeology of Anthropocene Rivers: Water Management and Landscape Change in 'Gold Rush' Australia. *Antiquity* 90 (353). 1348–1362.

Lazier, Benjamin (2011). Earthrise, or, the Globalization of the World Picture. *American Historical Review* 116: 602–630.

Lazrus, H. Eather (2012). Sea Change: Island Communities and Climate Change. *Annual Review of Anthropology* 41: 285–301.

Leach, Edmund R. (1970). *Political Systems of Highland Burma. A Study of Kachin Social Structure.* London: G. Bell (orig. London: Athlone Press, 1954).

LeCain, Timothy J. (2015). Against the Anthropocene. A Neo-Materialist Perspective. *International Journal for History, Culture and Modernity* 3(1). 1–28.

LeCain, Timothy J. (2017). *The Matter of History. How Things Create the Past.* Cambridge: Cambridge University Press (Studies in Environment and History).

Leggewie, Claus & Frederic Hanusch (2020). Einstieg ins planetare Denken. *Frankfurter Allgemeine Zeitung* vom 5.8.2020, S. N4.

Le Guin, Ursula K. (1947). *The Disposessed.* New York: Scribner.

Le Guin, Ursula K. (2020). *The Carrier Bag Theory of Fiction.* New York: Ignota Books (Terra Ignota, 1).

Leibenath, Markus, Stefan Heiland, Heiderose Kilper, Sabine Tzschaschel (eds.) (2012). *Wie werden Landschaften gemacht? Sozialwissenschaftliche Perspektiven auf die Konstituierung von Kulturlandschaften.* Bielefeld: Transcript.

Leinfelder, Reinhold (2011). *Das Anthropozän – von der Umwelt zur Unswelt.* scilogs.spektrum.de/der-anthropozaeniker/umwelt-unswelt-anthropozaen/.

Leinfelder, Reinhold (2012). Paul Joseph Crutzen: The ´Anthropocene´. In: Claus Leggewie, Darius Zifonun, Anne-Katrin Lang, Marcel Siepmann & Johanna Hoppen (eds.). *Schlüsselwerke der Kulturwissenschaften.* Bielefeld: Transcript: 257–260.

Leinfelder, Reinhold (2015a). Welcome to the Anthropocene! Keynote by R. Leinfelder. Conference: Transformation Festival at Copernicus Science Center, Warszawa, 4. 9.2015. https://www.researchgate.net/publication/283345525_Welcome_to_the_Anthropocene_Keynote_by_R_Leinfelder

Leinfelder, Reinhold (2015b). 'Die Zukunft war früher auch besser' – Neue Herausforderungen für die Wissenschaft und ihre Kommunikation. In: Nina Möllers, Christian Schwägerl & Helmuth

Trischler (eds.). *Willkommen im Anthropozän. Unsere Verantwortung für die Zukunft der Erde.* München: Deutsches Museum Verlag: 97–102.
Leinfelder, Reinhold (2017a). Das Anthropozän verständlich und spannend erzählen – Ein neues Narrativ für die globalen Herausforderungen? https://scilogs.spektrum.de/der-anthropozaeniker/narrative/
Leinfelder, Reinhold (2017b). „Die Erde wie eine Stiftung behandeln" – Ressourcenschutz und Rohstoffeffizienz im Anthropozän. In: DWA-BW (ed.). *Tagungsband 2017, Im Dialog: Phosphor-Rückgewinnung: 3. Kongress Phosphor – Ein kritischer Rohstoff mit Zukunft*: Stuttgart: 11–25
Leinfelder, Reinhold (2018a). Meghalayan oder Anthropozän? In welcher erdgeschichtlichen Zeit leben wir denn nun? *Der Anthropozäniker* vom 20. Juli 2018. https://scilogs.spektrum.de/deranthropozaeniker/meghalayan-oder-anthropozaen.
Leinfelder, Reinhold (2018b). Nachhaltigkeitsbildung im Anthropozän – Herausforderungen und Anregungen. In: LernortLabor – Bundesverband der Schülerlabore e.V. (Hg). *MINT-Nachhaltigkeitsbildung in Schülerlaboren – Lernen für die Gestaltung einer zukunftsfähigen Gesellschaft*: 130–141.
Leinfelder, Reinhold (2019). Das Anthropozän. Die Erde in unserer Hand. In: Elke Schwinger (eds.): *Das Anthropozän im Diskurs der Fachdisziplinen*. Marburg: Metropolis-Verlag: 23–46.
Leinfelder, Reinhold (2024). Anthropozän – Anatomie eines Falls. *SciLogs. Der Anthropozäniker.* https://scilogs.spektrum.de/der-anthropozaeniker/anatomie-eines-falls/.
Leinfelder, Reinhold, Alexandra Hamann und Jens Kirstein (2015). Wissenschaftliche Sachcomics. Multimodale Bildsprache, partizipative Wissensgenerierung und raumzeitliche Gestaltungsmöglichkeiten. In: Horst Bredekamp & Wolfgang Schäffner (eds.). *Haare hören – Strukturen wissen – Räume agieren. Berichte aus dem Interdisziplinären Labor Bild Wissen Gestaltung*. Bielefeld: Transcript: 45–59.
Leinfelder, Reinhold, Alexandra Hamann, Jens Kirstein & Marc Schleunitz (eds.) (2016). *Die Anthropozän-Küche. Matooke, Bienenstich und eine Prise Phosphor – in zehn Speisen um die Welt*. Berlin: Springer.
Leinfelder, Reinhold & S. Schwaderer (2020). "Wir sollten die Erde wie eine Stiftung behandeln". In: Stiftung Nantesbuch (ed.). *Notizen aus dem Anthropozän*. München: Stiftung Nantesbuch: 70–77.
Lenton, Tim (2016). *Earth System Science*. Oxford: Oxford University Press (Very Short Introductions).
Lenton, Timothy M. & Bruno Latour (2016). Gaia 2.0. Science 361: 1066–1068.
Lenton, Timothy M., Johan Rockstrom, Owen Gaffney, Stefan Rahmstorf, Katherine Richardson, Will Steffen & Hans Joachim Schellnhuber (2019). Climate Tipping Points – Too Risky to Bet Against. *Nature* 575 593–596.
Lesch, Harald & Klaus Kamphausen ([6]2017). Die Menschheit *schafft sich ab. Die Erde im Griff des Anthropozän*. München: Knaur.
Lévi-Strauss, Claude (1973). *Das wilde Denken*. Frankfurt am Main: Suhrkamp (zuerst 1968, orig. „La Pensée Sauvage", Paris, Librairie Plon, 1962).
Lewis, Simon L. & Mark A. Maslin (2015a). Defining the Anthropocene. *Nature* 519: 171–80.
Lewis, Simon L. & Mark A. Maslin (2015b). Geological Evidence for the Anthropocene. *Science* 349: 246–247.
Lewis, Simon L. & Mark A. Maslin (2018). *The Human Planet. How We Created the Anthropocene.* New Haven & London: Yale University Press (also London: Pelican Books).
Lewis, Sophie (2017). Cthulhu Plays No Role For Me. *Viewpoint Magazine*, 08.05.2017, (https://www.viewpointmag.com/2017/05/08/cthulhu-plays-no-role-for-me/ (Zugriff: 1.03.2024).
Lewontin, Richard (2000). *The Triple Helix. Gene, Organism, and Environment*. New Haven, Conn.: Harvard University Press.
Li, Tania Murray (2014). *Land's End: Capitalist Relations on an Indigenous Frontier*. Durham, N.C.: Duke University Press.

Li, Tania Murray & Pujo Semedi (2021). *Plantation Life. Corporate Occupation in Indonesia's Oil Palm Zone*. Durham & London: Duke University Press.

Li, Xiaoqiang, John Dodson, Jie Zhou & Xinying Zhou (2009). Increases of Population and Expansion of Rice Agriculture in Asia, and Anthropogenic Methane Emissions Since 5000 BP. *Quaternary International* 202: 41–50.

Lidskog, Rolf & Claire Waterton (2016). Anthropocene – a cautious welcome from environmental sociology? *Environmental Sociology*, 2(4). 395–406.

Lien, Marianne Elisabeth & Gísli Pálsson (2019). Ethnography Beyond the Human: The 'Other-than-Human' in Ethnographic Work. *Ethnos* 86(3). 1–20.

Lightfoot, Kent G., Lee M. Panich, Tsim D. Schneider & Sara L. Gonzalez (2013). European Colonialism and the Anthropocene: A view from the Pacific Coast of North America. *Anthropocene* 4: 101–115.

Lindhof, Matthias (2019). *Internationale Gemeinschaft. Zur politischen Bedeutung eines wirkmächtigen Begriffs*. Wiesbaden: Nomos (Rekonstruktive Weltpolitikforschung, 2)

Lippuner, Roland, Johannes Wirths & Pascal Goeke (2015). Das Anthropozän: eine epistemische Herausforderung für die spätmoderne Sozialgeographie.) http://www.raumnachrichten.de/diskussionen/1988-roland-lippuner-johannes-wirths-und-pascal-goeke-das-anthropozaen).

Little, Paul (1995). Ritual, Power and Ethnography at the Rio Earth Summit. *Critique of Anthropology*. 15(3), 265–288.

Liu, J., Dou, Y., Batistella, M., Challies, E., Connor, T., Friis, C., et al. (2018). Spillover systems in a telecoupled Anthropocene: Typology, methods, and governance for global sustainability. *Current Opinion in Environmental Sustainability* 33: 58–69.

Lo, Kwai-Cheung & Jessica Yeung (eds.) (2019). *Chinese Shock of the Anthropocene: Image, Music and Text in the Age of Climate Change*. Singapore: Palgrave Macmillan.

Lockard, Craig A. ([4]2020). *Societies, Networks, and Transitions. A Global History*. Stamford, Ct.: Cengage Learning.

Lockyer, Joshua & James R. Veteto (2013). *Environmental Anthropology Engaging Ecotopia. Bioregionalism, Permaculture, and Ecovillages*. New York & Oxford: Berghahn (Environmental Anthropology and Ethnobiology, 17).

Lövbrand, Eva, Silke Beck, Jason Chilvers, Tim Forsyth, Johan Hedrén, Mike Hulme, Rolf Lidskog & Eleftheria Vasileiadou (2015). Who Speaks for the Future of Earth? How Critical Social Science Can Extend the Conversation on the Anthropocene. *Global Environmental Change* 32 (Suppl. C). 211–218.

Lövbrand, Eva, Malin Mobjörk & Rickard Söder (2020). The Anthropocene and the Geo-Political Imagination: Re-writing Earth as Political Space. *Earth System Governance* 4: 1–8. (https://www.sciencedirect.com/science/article/pii/S2589811620300100?via%3Dihub).

Loh, Janina (2018). *Trans- und Posthumanismus zur Einführung*. Hamburg: Junius (Junius Einführungen).

Lomolino, Mark V. (2020). *Biogeography. A Very Short Introduction*. Oxford: Oxford University Press (Very Short Introductions).

Long, Norman & Ann Long (eds.) (1992). *Battlefields of Knowledge. The Interlock of Theory and Practice in Social Research and Development*. London & New York: Routledge.

Lopez, Mario, et al. (2013). The Focus: Sustainable Humanosphere Studies. *The Newsletter* 66, Winter International Institute for Asian Studies (ICAS), pp. 23–34. http://www.iias.nl/the-newsletter/newsletter-66-winter-2013

Lorimer, Jamie (2012). Multinatural Geographies for the Anthropocene. *Progress in Human Geography* 36(5). 593–612.

Lorimer, Jamie (2015). *Wildlife in the Anthropocene. Conservation After Nature*. Minnesota University Press.

Lorimer, Jamie (2017). The Anthropo-Scene: A Guide for the Perplexed. *Social Studies of Science* 47(1). 117–42.

Lorimer, Jamie (2020). *The Probiotic Planet. Using Life to Manage Life*. Minneapolis & London: University of Minnesota Press (Posthumanities, 59).

Lovelock, James (1979). *Gaia. A New Look at Life on Earth.* Oxford etc.: Oxford University Press.
Lovelock, James (1988). *The Ages of Gaia. A Biography of Our Living Earth.* Oxford: Oxford University Press.
Lovelock, James, with Bryan Appleyard (2020). *Novacene. The Coming Age of Hyperintelligence.* London: Allen Lane.
Luciano, Eugenio (2022a). Is 'Anthropocene' a Suitable Chronostratigraphic Term? *Anthropocene Science* 1: 29–41.
Luciano, Eugenio (2022b). The shape of Anthropocene: The early contribution of the water science. *The Anthropocene Review* 10(3): 1–18 (https://doi.org/10.1177/20530196221140l).
Luke, Timothy W. (2018). Tracing race, ethnicity, and civilization in the Anthropocene. *Environment and Planning D: Society and Space* 38(1).129–146.
Lundershausen, Johannes (2018). The Anthropocene Working Group and its (Inter-)Disciplinarity. *Sustainability: Science, Practice and Policy* 14(1). 31–45.
Lussem, Felix (2020). Monströs oder gespenstisch? Fragen von Schuld, Verantwortung und Solidarität in Zeiten der Corona-Pandemie (#WitnessingCorona) (https://www.medizinethnologie.net/monstroes-oder-gespenstisch/).
Lyle, Paul (2016). *The Abyss of Time. A Study in Geological Time and Earth´s History.* Edinburgh: Dunedin Academic Press.
Maasen, Sabine, Barbara Sutter & Laura Trachte (2018). Was bio bedeutet. Soziomaterielle Konfigurationen von TechnoNatures. In: Bernhard Gill, Franziska Torma & Karin Zachmann (eds.). *Mit Biofakten leben. Sprache und Materialität von Pflanzen und Lebensmitteln*, Baden-Baden: Nomos: 177–198.
Macfarlane, Robert (2016). *Underland. A Deep Time Journey.* London: Hamish Hamilton, Penguin.
Macfarlane, Robert (2019). *Underland. A Deep Time Journey.* London: Hamish Hamilton, Penguin.
MacLeod, Norman (2013). The Geat Extinctions. London: Natural History Museum.
Maffi, Liisa (2001). *On Biocultural Diversity. Linking Language, Knowledge and the Environment.* Berkeley, Cal.: University of California Press.
Magni, Michel (2019). *Au racines de l´Anthropocène. Une crise écologique reflet d´une crise de l´homme.* Lormont: le Bord de l´Eau.
Magni, Michel (2021). *L´anthropocène.* Paris: Que sais-je?/Humensis (Que sais-je?)
Mahnkopf, Birgit (2019). Anthropozän. In: Claudia von Braunmühl, Heide Gerstenberger, Ralf Ptak & Christa Wichterich (eds.). *ABC der globalen (Un)Ordnung. Von „Anthropozän" bis „Zivilgesellschaft".* Hamburg: Verlag VSA: 12–13.
Mainzer, Klaus (2020). Vom Anthropozän zur Künstlichen Intelligenz. Herausforderungen von Mensch und Natur durch Technik im 21. Jahrhundert. In: Thomas Heichele (eds.). *Mensch – Natur – Technik. Philosophie für das Anthropozän.* Münster: Aschendorff Verlag (Studien zur systematischen Theologie, Ethik und Philosophie, 19). 155–168.
Malafouris, Lambros (2013). *How Things Shape the Mind. A Theory of Material Engagement.* Cambridge, Mass.: The MIT Press.
Malafouris, Lambros (2019). Mind and Material Engagement. *Phenomenology and the Cognitive Sciences* 18: 1–17.
Malhi, Yadvinder (2017). The Concept of the Anthropocene. *Annual Review of Environment and Resources* 42:25.1–25.28.
Malinowski, Bronislaw Kaspar (1944). *A Scientific Theory of Culture and Other Essays.* Chapel Hill, N.C.: The University of North Carolina Press.
Malm, Andreas (2016). *Fossil Capital. The Rise of Steam Power and the Roots of Global Warming.* London: Verso Books.
Malm, Andreas (2018). *The Progress of this Strom. Nature and Society in a Warming World.* London & New York: Verso Books.
Malm, Andreas (2021). *How to Blow Up a Pipeline. Learning to Fight in a World on Fire.* London: Verso.

Malm, Andreas & Alf Hornborg (2014). The Geology of Mankind? A Critique of the Anthropocene Narrative. *The Anthropocene Review* 1(1). 62–69.

Malone, Nicholas (2022). *The Dialectical Primatologist. The Past, Present and Future of Life in the Hominoid Niche*. London & New York: Routledge (New Biological Anthropology).

Malone, Nicholas & Kathryn Ovenden (2017). Natureculture. In: Agustín Fuentes (ed.). *The International Encyclopedia of Primatology*. London: John Wiley & Sons. https://doi.org/10.1002/9781119179313.wbprim0135.

Manemann, Jürgen (2014). *Kritik des Anthropozäns. Plädoyer für eine neue Humanökologie*. Bielefeld: Transcript (X-Texte zu Kultur und Gesellschaft).

Mann, Geoff & Joel Wainwright (2018). *Climate Leviathan. A Political Theory of Our Planetary Future*. London & New York: Verso.

Mann, Charles. C.. (2011). 1493. Uncovering the New World Columbus Created. New York: Alfred A. Knopf.

Mann, Charles C. (2024). The Dawn of the Homogenocene. *Orion. Nature and Culture* (https://orionmagazine.org/article/the-dawn-of-the-homogenocene/).

Mann, Michael (2021). *The New Climate War. The Fight to Take Back Our Planet*". New York, N.Y.: Hachette Book Group.

Manning, Patrick (2020a). *Methods for Human History. Studying Social, Cultural, and Biological Evolution*. Palgrave Macmillan.

Manning, Patrick (2020b). A *History of Humanity. The Evolution of the Human System*. Cambridge etc.: Cambridge University Press.

Marcus, George E. & Michael M. J. Fischer (eds.) ([2]1999). *Anthropology as Cultural Critique. An Experimental Moment in the Human Sciences*. Chicago: The University of Chicago Press ([1]1986).

Marcus, George E. & Marcelo Pisarro (2008). The End(s) of Ethnography: Social/Cultural Anthropology's Signature form of Producing Knowledge in Transition. *Cultural Anthropology* 23(1).1–14.

Margulis, Lynn (2017). *Der symbiotische Planet oder Wie die Evolution wirklich verlief*. Frankfurt am Main: Westend Verlag (orig: „Symbiotic Planet“, London: Weidenfeld & Nicolson, 1998).

Markl, Hubert (1986). *Natur als Kulturaufgabe. Über die Beziehung des Menschen zur lebendigen Natur*. München. Deutsche Verlags-Anstalt.

Markl, Hubert (1998). *Homo Sapiens. Zur fortwirkenden Naturgeschichte des Menschen*. Münster. Gerda Henkel Vorlesung.

Marks, Robert B. (2015). *China. An Environmental History*. Lanham, Mad. Etc.: Rowman & Littlefield.

Marks, Robert B. ([4]2020). *The Origins of the Modern World. A Global and Environmental Narrative from the Fifteenth to the Twenty-First Century*. Lanham etc.: Rowman & Littlefield.

Marris, Emma (2013). *Rambunctious Garden. Saving Nature in a Post-Wild World*. New York, N.Y.: Bloomsbury.

Marsh, George Perkins (2021). *Man and Nature: Or, Physical Geography as Modified by Human Action*. Dover Publications (Dover Thrift Editions, firstly New York: Charles Scribner, 1864).

Marshall, George (2014). *Don't Even Think About It. Why Our Brains Are Wired to Ignore Climate Change*. London etc.: Bloomsbury.

Marsiske, Hans-Artur (2005). *Heimat Weltall. Wohin soll die Raumfahrt führen?* Frankfurt am Main: Suhrkamp.

Marston, Sallie A., John Paul Jones III & Keith Woodward (2005). Human Geography Without Scale. *Transactions of the Institute of British Geographers* 30(4). 416–432.

Martens, R. L. & Bii Robertson (2021). *How the Soil Remembers Plantation Slavery*. https://edgeeffects.net/soil-memory-plantationocene/.

Martin, Paul S. & Herbert Edgar Wright (eds.) (1967). *Pleistocene Extinctions. The Search for a Cause*. New Haven, Conn.: Yale University Press.

Marx, Karl (1867). *Das Kapital. Kritik der politischen Ökonomie. Band. I, Der Produktionsprocess des Kapitals* . Hamburg: Verlag Otto Meißner.

Masco, James (2010). Bad Weather: On Planetary Crisis. *Social Studies of Science* 40(1). 7–40.
Masco, James (2006). *The Nuclear Borderlands. The Manhattan Project in Post-Cold War New Mexico*. Princeton: Princeton University Press.
Masco, James ([2]2020). *The Nuclear Borderlands. The Manhattan Project in Post-Cold War New Mexico*. Princeton: Princeton University Press.
Maslin, Mark A. (2018a). The Anthropocene. In: Noel Castree, Mike Hulme & James D. Proctor (eds.). *Companion to Environmental Studies*. Abingdon & New York: Routledge: 144–150.
Maslin, Mark A. (2018b). *The Cradle of Humanity. How the Changing Landscape of Africa Made Us So Smart*. Oxford: Oxford University Press.
Maslin, Mark (2021). *How to Save Our Planet. The Facts*. London etc.: Penguin Life..
Mattheis, Nikolas (2022). Making Kin, not Babies? Towards Childist Kinship in the "Anthropocene". *Childhood* 29(4). 512–528.
Mauch, Christof & Christian Pfister (eds.) (2009). *Natural Disasters, Cultural Responses. Case Studies Toward a Global Environmental History*. Lanham etc.: Lexington Books (International Environmental History. Publications of the German Historical Institute).
Mauelshagen, Franz (2012)., Anthropozän'. Plädoyer für eine Klimageschichte des 19. und 20. Jahrhunderts. Zeithistorische Forschungen 9: 131–137 (http://www.zeithistorische-forschungen.de/1-2012/id=4596).
Mauelshagen, Franz (2013) Ungewissheit in der Soziosphäre: Risiko und Versicherung im Klimawandel. In: Roderich von Detten, Fenn Faber & Martin Bemmann (eds.). *Unberechenbare Umwelt*. Wiesbaden: Springer: 253–269.
Mauelshagen, Franz (2019). Anthropozän. In: Staatslexikon online (https://www.staatslexikon-online.de/Lexikon/Anthropoz%C3%A4n).
Mauelshagen, Franz (2020). The Dirty Metaphysics of Fossil Freedom. In: Gabriele Dürbeck & Philip Hüpkes (eds.). *The Anthropocenic Turn. Interplay Between Disciplinary and Interdisciplinary Responses to a New Age*. London & New York: Routledge (Routledge Interdisciplinary Perspectives on Literature).
Mauelshagen, Franz (2023). *Geschichte des Klimas. Von der Steinzeit bis zur Gegenwart*. München: C. H. Beck (Beck'sche Reihe).
Matejovski, Dirk (2016). *Anthropozän und Apokalypse. Zum Verhältnis von Ökologie, Theoriedesign und kommunikativer Hegemonie*. Düsseldorf: Chemie-Stiftung Sozialpartner-Akademie (CSSA Discussion Paper 2016/3).
Mathews, Andrew S. (2018). Landscapes and Throughscapes in Italian Forest Worlds: Thinking Dramatically About the Anthropocene. *Cultural Anthropology* 33(3).386–414.
Mathews, Andrew S. (2020). Anthropology and the Anthropocene: Criticisms, Experiments, and Collaborations. *Annual Review of Anthropology* 49:67–82
Matsutake Worlds Research Group (2009). A New Form of Collaboration in Cultural Anthropology. *American Ethnologist* 36, no. 2: 380–403.
Mbembe, Achille (2015). Decolonizing Knowledge and the Question of the Archive. Vortrag, University of Witwatersand, 22 April 2015.
McAfee, Kathleen (2016). The Politics of Nature in the Anthropocene. In: Robert Emmett & Thomas Lekan (eds.). *Whose Anthropocene? Revisiting Dipesh Chakrabarty's "Four Theses"*. Munich: Rachel Carson Center for Environment and Society (RCC Perspectives, Transformations in Environment and Society, 2). 65–72.
McAnany P. A. & Norman Yoffee (2010). *Questioning Collapse. Human Resilience, Ecological Vulnerability, and the Aftermath of Empire*. New York: Cambridge University Press.
McCorriston, Joy & Julie Field (2020). *Anthropocene. A New Introduction to World Prehis*tory. London: Thames & Hudson (auch als "World Prehistory and the Anthropocene. An Introduction to Human History", Thames & Hudson, 2019)
McHale, John (1970). *The Ecological Context*. London: George Braziller.
McKibben, Bill (1989). *The End of Nature*. New York: Anchor.
McKibben, Bill (2010). *Eearth. Making a Life on a Tough New Planet*. New York: Times Books.

McKinney Michael L. (2005). New Pangea: Homogenizing the Future Biosphere. *Proceedings of the California Academy of Sciences* 56, Supplement I, (11). 119–129.

McKittrick, Katherine (2013). Plantation Futures. *Small Axe: Caribbean Journal of Criticism* 17(3, # 42). 1–15.

McNeill, Desmond (2017). Safety in Numbers: Why Everyone Listens to Economists. In: Thomas Hylland Eriksen & Elisabeth Schober (eds.). *Knowledge and Power in an Overheated World.* Oslo: Department of Social Anthropology University of Oslo: 126–140. (https://www.sv.uio.no/sai/english/research/projects/overheating/news/2017/knowledge-and-power-in-an-overheated-world.pdf).

McNeill, John R. (2000). *Something New Under the Sun. An Environmental History of the Twentieth-Century World.* New York & London: W.W. Norton (Global Century Series).

McNeill, John R. (2015). Global Environmental History: The First 150,000 Years. In: John R. McNeill & Erin Stewart Mauldin (eds.). *A Companion to Global Environmental History.* Blackwell Companions to World History). 3–17.

McNeill, John R. (2016). Introductory Remarks: The Anthropocene and the Eighteenth Century. *Eighteenth-Century Studies* 49(2).117–28.

McNeill, John R. (2019). The Industrial Revolution and the Anthropocene in: Jan Zalasiewicz, Colin Waters, Mark Williams & Colin P. Summerhayes (eds.) (2019). *The Anthropocene as a Geological Time Unit: A Guide to the Scientific Evidence and Current Debate.* Cambridge etc.: Cambridge University Press: 250–254.

McNeill, John R. & Erin Steward Mauldin (eds.) (2015). A *Companion to Global Environmental History.* Chichester & Malden, Mass.: Wiley-Blackwell (Wiley Blackwell Companions to World History).

McNeill, John R. & Peter Engelke (2013). Mensch und Umwelt im Zeitalter des Anthropozän. In: Akira Irye (eds.). *Geschichte der Welt. 1945 bis heute – Die globalisierte Welt.* München: C. H. Beck & Cambridge, Mass. & London: The Belknap Press of Harvard University Press (Geschichte der Welt. A History of the World). 356–534.

McNeill, John R. & Peter Engelke (2016). *Great Acceleration. An Environmental History of the Anthropocene since 1945.* Cambridge, Mass.: The Belknap Press of Harvard University Press.

McNetting, Robert (1977). *Cultural Ecology.* Reading, Mass.: Cummings Publishing Company.

McPhearson, Timon, Christopher M. Raymond, Natalie Gulsrud, Christian Albert, Neil Coles, Nora Fagerholm, Michiru Nagatsu, Anton Stahl Olafsson, Niko Soininen & Kati Vierikko (2021). Radical changes are needed for transformations to a good Anthropocene. *Urban Sustainability* 1(1). 1–13.

McPhee, John (1981). *Basin and Range.* New York: Farrar, Straus and Giroux (Annals of the Former World, 1).

Mead, Margaret (1971). *Der Konflikt der Generationen. Jugend ohne Vorbild.* Olten: Walter Verlag (orig, "Culture and Commitment. The New Relationships Between the Generations in the 1970s", New York: N.Y.: Columbia University Press).

Meadows, Dennis H., Donella L. Meadows, J. Randers & W.W. Behrens III (1972). *The Limits to Growth: A Report for the Club of Rome's Project on the Predicament of Mankind.* London: Earth Island Press.

Meadows, Donella L., Jørgen Randers & Dennis H. Meadows (2012). *Limits to Growth: The 30-Year Update.* London etc.: Chelsea Green Publishing.

Medin, Douglas L., Norbert Ross, Scott Atran, Russel C. Burnett & Sergey V. Blok (2002). Categorization and Reasoning in Relation to Culture and Expertise. *The Psychology of Learning and Motivation* 41:1–41.

Meier, Andreas (2011). Natur war gestern. *Die Zeit* Nr. 13 vom 24.3.2011: 49.

Meiske, Martin (2021). *Die Geburt des Geoengineerings. Großbauprojekte in der Frühphase des Anthropozäns.* München: Wallstein (Deutsches Museum. Abhandlungen und Berichte, N.F. 34).

Meneganzin, Andra, Telmo Pievani and Stefano Caserini (2020). Anthropogenic climate change as a monumental niche construction process: background and philosophical aspects. *Biology & Philosophy* 35(38) (https://doi.org/10.1007/s10539-020-09754-2).

Mentz, Steve (2019). *Break Up the Anthropocene*. Minneapolis: University of Minnesota Press (Forerunners: Ideas First).

Merchant, Carolyn (2020a). *Anthropocene and the Humanities. From Climate Change to a New Age of Sustainability*. New Haven & London: Yale University Press.

Merchant, Carolyn ([3]2020b). *The Death of Nature. Women, Ecology, and the Scientific Revolution*. San Francisco: Harper Collins (Fourtieth Anniversary Edition, [1]1990).

Messeri, Lisa (2016). *Placing Outer Space. An Earthly Ethnography of Other Worlds*. Durham, NC.: Duke University Press (Experimental Futures: Technologocal Lives, Scientific Arts, Anthropological Voices).

Messner, Dirk ([2]2016). Globaler Wandel – Herausforderung an unsere Lernfähigkeit. In: Jörg-Robert Schreiber Hannes Siege (Zus.-stellung und Bearb.). *Orientierungsrahmen für den Lernbereich Globale Entwicklung im Rahmen einer Bildung für nachhaltige Entwicklung*. Bonn: Engagement Global und Cornelsen Verlag: 22.

Meyer, Diethard E. (2022). *Geofaktor Mensch. Eingriffe in die Umwelt und* ihre Folgen. Berlin: Springer Spektrum

Meyer, William B. (2016). *The Progressive Environmental Prometheans: Left-Wing Heralds of a "Good Anthropocene"*. o.O. (Basingstoke). Palgrave Macmillan.

Middleton, Guy D. (2017). The Show Must Go On: Collapse, Resilience, and Transformation in 21st-Century Archaeology. *Reviews in Anthropology* 46(2–3). 78–105, https://doi.org/10.1080/00938157.2017.1343025

MilNeil, Christian (2013). Inner-City Glaciers. In: Ellsworth, Elizabeth & Jamie Kruse (eds.). *Making the Geologic Now. Responses to Material Conditions of Contemporary Life*. New York: Punktum: 79–81.

Milton, Kay (1996). *Environmentalism and Cultural Theory. Exploring the Role of Anthropology in Environmental Discourse*. London & New York: Routledge (Environment and Society).

Milton, Kay (2002). *Loving Nature. Towards an Ecology of Emotion*. London & New York: Routledge

Mintz, Sydney (1985). *The Sweetness of Power. The Place of Sugar in Modern History*. London.

Mirzoeff, Nicholas (2014). Visualising the Anthropocene. *Public Culture* 26(2). 213–232.

Mirzoeff, Nicholas (2018). It's not the anthropocene, it's the white supremacy scene; or, The geological color line. In: Richard Grusin (ed.). *After Extinction*. Minneapolis: University of Minnesota Press: 123–149.

Mitchell, Logan et al. (2013). Constraints on the Late Holocene Anhropogenic Contribution to the Atmospheric Methane Budget. *Science* 342: 946–966.

Mitchell, W. J. T. (2005). *What do Pictures want? The Lives and Loves of Images*. Chicago: Chicago University Press.

Mittelbach, Gary G. & Brian J. McGill ([2]2019). *Community Ecology*. Oxford: Oxford University Press.

Mittelstaedt, Werner (2020). *Anthropozän und Nachhaltigkeit. Denkanstöße zur Klimakrise und für ein zukunftsfähiges Handeln*. Berlin etc.: Peter Lang Verlag.

Mittelstraß, Jürgen (1992). *Leonardo-Welt. Über Wissenschaft, Forschung und Verantwortung*. Frankfurt am Main: Suhrkamp Verlag.

Moebius, Stephan, Frithjof Nungesser & Katharina Scherke (eds.) (2019). *Handbuch Kultursoziologie. Bd. 1: Band 1: Begriffe – Kontexte – Perspektiven – Autor_innen; Band 2: Theorien – Methoden – Felder*. Wiesbaden: Springer VS.

Möllers, Nina (2013) Cur(at)ing the Planet– How to Exhibit the Anthropocene and Why. In: Helmuth Trischler (ed.). Anthropocene. Envisioning the Future of the Age of Humans. Munich: Rachel Carson Center for Environment and Society (RCC Perspectives, 3). 57–65.

Möllers, Nina, Christian Schwägerl & Helmuth Trischler (eds.) (2015). *Willkommen im Anthropozän. Unsere Verantwortung für die Zukunft der Erde*. München: Deutsches Museum Verlag und Rachel Carson Center.

Mönnig, E. (2016). Anthropozän. Meinungen „Zur Anthropozän-Debatte", 2 Seiten. (http://www.stratigraphie.de/ergebnisse/Texte/Ergebnisse_1_4_pdf5.pdf).

Monastersky, Richard (2015). Anthropozän: Zeitalter des Menschen. *Spektrum der Wissenschaft*. https://www.spektrum.de/news/zeitalter-des-menschen/1341897, zuerst als "Anthropocene: The Human Age", *Nature* 519(7542). 144–147, https://www.nature.com/news/anthropocene-the-human-age-1.17085).

Monbiot, George (2014). *Destroyer of Worlds* (http://www.monbiot.com/2014/03/24/destroyer-of-worlds/)

Montanarella, Luca & Panao Panagos (2015). Policy Relevance of Critical Zone Science. *Land Use Policy* 49: 86–91.

Moore, Amelia (2015a). Anthropocene Anthropology: Reconceptualizing Contemporary Global Change. *Journal of the Royal Anthropological Institute* 22(1). 27–46.

Moore, Amelia (2015b). Islands of Difference: Design, Urbanism, and Sustainable Tourism in the Anthropocene Caribbean. *Journal of Latin American and Caribbean Anthropology* 20: 513–32.

Moore, Amelia (2019). *Destination Anthropocene. Science and Tourism in the Bahamas*. Berkeley, Cal.: University of California Press (Critical Environments: Nature, Science, and Politics, 7).

Moore, Kathleen Dean (2013). Anthropocene is the Wrong Word. In *Earth Island Journal*. https://www.earthisland.org/journal/index.php/magazine/entry/anthropocene_is_the_wrong_word/.

Moore, Jason W. (2014a). *The Capitalocene, Part I: On the Nature and Origins of Our Ecological Crisis* (unpublished paper). Fernand Braudel Center, Binghamton University.

Moore, Jason W. (2014b). *The Capitalocene, Part II: Abstract Social Nature and the Limits to Capital* (unpublished paper). Fernand Braudel Center, Binghamton University.

Moore, Jason W. (2019). *Capitalism in the Web of Life. Ecology and the Accumulation of Capital*. London: Verso.

Moore, Jason W. (2020). „Die Macht des einen Prozents gerät unter Druck". Interview mit Raul Zelik. *WOZ: Die Wochenzeitung* Nr. 23 vom 4.6.2020. https://jasonwmoore.com/wp-content/uploads/2021/04/Moore-Die-Macht-des-einen-Prozents-gerat-unter-Druck-WOZ-June-2020.pdf

Moore, Jason W. (2021a). Unthinking the Anthropocene: Man and Nature in the Capitalocene. *Global Dialogue* 11(3). 36–37.

Moore, Jason W. (2021b). Opiates of the Environmentalists? Anthropocene Illusions, Planetary Management & the Capitalocene Alternative. *Abstrakt* (http://www.abstraktdergi.net/opiates-of-the-environmentalists-anthropocene-illusions-planetary-management-the-capitalocene-alternative/).

Moore, Jason W. (ed.) (2016). *Anthropocene or Capitalocene? Nature, History, and the Crisis of Capitalism*. Oakland, Cal.: PM Press, Kairos (http://arena-attachments.s3.amazonaws.com/772469).

Moore, Sophie Sapp, Monique Alewaert, Pablo F. Gómez, & Gregg Mittman (2021). Plantation Legacies. (https://edgeeffects.net/plantation-legacies-plantationocene/).

Moran, Emilio F. ([3]2008). *Human Adaptability. An Introduction to Ecolocgical Anthropology*. Boulder, Col.: Westview Press.

Moran, Emilio F. (2010). *Environmental Social Science. Human - Environment Interactions and Sustainability*. Malden: Wiley-Blackwell.

Moran, Emilio F. ([2]2016). *People and Nature. An Introduction to Human Ecological Relations*. Malden, Mass.: Blackwell Publishing.

Morcillo, Jesús Muñoz (2022). *Anthropozän? Die ökologische Frage und der Mensch, der sie stellt*. Baden-Baden: Tectum Verlag.

Morris, Ian (2015). *Foragers, Farmers, and Fossil Fuels. How Human Values Evolve*. New York & Oxford: Princeton University Press.

Morrison, Kathleen D. (2015). Provincializing the Anthropocene. *Seminar. A Journal of Germanic Studies* 673:75–80.
Morrison, Kathleen D. (2018). Provincializing the Anthropocene: Eurocentrism in the Earth System. In Gunnel Cederlöf & Maheesh Rangarjan (eds.). *At Nature´s Edge. The Global Present and the Long-Term History*. New Delhi: Oxford University Press.
Morrison, Philipp & Phylis Morrison (1992). *Zehn Hoch. Dimensionen zwischen Quarks und Galaxien*. Heidelberg: Spekrum Akademischer Verlag.
Morton, Timothy (2009). *Ecology Without Nature. Rethinking Environmental Aesthetics*. Cambridge, Mass.: Harvard Univ. Press.
Morton, Timothy (2013). *Hyperobjects. Philosophy and Ecology after the End of the World*. Minneapolis: University of Minnesota Press (Posthumanities, 27).
Morton, Timothy (2014). How I Learned to Stop Worrying and Love the Term "Anthropocene". *Cambridge Journal of Postcolonial Literary Inquiry* 2:257–264.
Morton, Timothy (2017). *Humankind. Solidarity with Nonhuman People*. London: Verso.
Mosley, Stephen (^{2}2024). *The Environment and World History*. Abingdon: Routledge (Themes in World History).
MPI Max-Planck-Institut für Wissenschaftsgeschichte (2017). IV. IV. *Anthropocene Formations*. https://www.mpiwg-berlin.mpg.de/project/knowledge-anthropocene
Müller, Burghard (2021). Von wegen Anthropozän. *Merkur* 75(865). 5–16.
Müller, Michael (2021a). Gesellschaftskritisches Denken wiederbeleben. Anthropozän und kritische Theorie. *Politische Ökologie* 39(167). 73–80.
Müller, Michael (2021b). Das Anthropozän oder: Wie wir die Erde verkonsumieren. Das Erbe des Jahrhundertwissenschaftlers Paul J. Crutzen. *Blätter für deutsche und internationale Politik* 3/2021: 107–112.
Müller, Michael, Jörg Sommer & Pierre L. Ibisch (Red.) (2022). *Das Zeitalter der Städte: Die entscheidende Kraft im Anthropozän*. Stuttgart: S. Hirzel Verlag (Jahrbuch Ökologie 2022).
Müller-Mahn, Detlev, Jonathan Everts & Christiane Stephan (2018). Riskscapes Revisited - Exploring the Relationship between Risk, Space and Practice. *Erdkunde* 72(3). 197–213.
Mulch, Andreas & George Zizka (2021) Erdsystemforschung. Eine Nummer kleiner geht es nicht!. *Senckenberg Natur – Forschung – Museum* 151(1–3). 18–21.
Munn, Nancy D. (1992). The Cultural Anthropology of Time: A Critical Essay. *Annual Review of Anthropology* 21: 93–123.
Murphy, Michael A. & Amos Salvador (2000). International subcommission on stratigraphic classification of IUGS International Commission on Stratigraphy: international stratigraphic guide – an abridged version. *GeoArabia* 5(2). 231–266.
Muthukrishna, Michael, Ben W. Shulman, Vlad Vasilescu &Joseph Henrich (2014). Sociality Influences Cultural Complexity. *Proceedings of the Royal Society B* 281: 1.8.
Naess, Arne (eds. David Rothenberg) (2013). *Die Zukunft in unseren Händen. Eine tiefenökologische Philosophie*. Stuttgart: Edition Trickster im Peter Hammer.
Naess, Arne (2016). *Ecology of Wisdom. Writings by Arne Naess*. Ed. by Alan Drengson & Bill Devall. London: Penguin (Penguin Modern Classics).
Nassehi, Armin (2020). Komplexitätsprobleme. Klimawandel aus soziologischer Perspektive. *Forschung & Lehre* (11). 908–909.
National Research Council (1986). *Earth System Science: Overview. A Program for Global Change*. Washington, D.C.: The National Aeronautics Corporation. (https://doi.org/10.17226/19210).
National Research Council (1988). *Earth System Science: A Closer View*. Washington, D.C.: The National Aeronautics Corporation (https://doi.org/10.17226/19088).
Nazarea, Virginia D. (ed.) (1999). *Ethnoecology. Situated Knowledge/Located Lives*. Tuscon: University of Arizona Press.
Neckel, Sighard (2020). Scholastische Irrtümer? Rückfragen an das Antrropozän. In: Frank Adloff, & Sighard Neckel (Hg.) (2020). *Gesellschaftstheorie im Anthropozän*. Frankfurt & New York: Campus (Zukünfte der Nachhaltigkeit, 1): 157–168.

Neimanis, Astrida, Cecilia Åsberg and Johan Hedrén (2015). Four Problems, Four Directions for Environmental Humanities: Toward Critical Posthumanities for the Anthropocene. *Ethics and the Environment* 20(1). 67–97.

Nevle, Richard J., Dennis K. Bird, William F. Ruddiman & Richard A. Dull (2011). Neotropical Human-Landscape Interactions, Fire, and Atmospheric CO2 During European Conquest. *The Holocene* (Special Isue).1–12.

Niedenzu, Heinz-Jürgen (2012). *Soziogenese der Normativität. Zur Emergenz eines neuen Modus der Sozialorganisation*. Weilerswist: Velbrück.

Niewöhner, Jörg (2013). Natur und Kultur im Anthropozän – eine sozialanthropologische Perspektive auf gesellschaftliche Transformation. *Humboldt-Universitäts-Gesellschaft. Jahresgabe* 2012/2013: 43–49.

Nicholson, Simon & Sikina Jinnah (2016). Living on a New Earth. In: Dies. (eds.). *New Earth Politics. Essays from the Anthropocene*. Cambridge, Mass. & London: The MIT Press: 1–16.

Niederberger, Andreas & Philipp Schink (eds.) (2011). *Globalisierung. Ein interdisziplinäres Handbuch*. Stuttgart & Weimer: Verlag J. B. Metzler

Nitzke, Solvejg & Nicolas Pethes (2017). Visions of the "Blue Marble". Technology, Philosophy, Fiction. In: Dies. (eds.). *Imagining Earth. Concepts of Wholeness in Cultural Constructions of Our Home Planet*. Bielefeld: Transcript: 7–23.

Nixon, Rob (2011). *Slow Violence and the Environmentalism of the Poor*. Cambridge, Mass.: Harvard University Press.

Nixon, Rob (2014). The Anthropocene: The Promise and Pitfalls of an Epochal Idea. *Edgeeffects* (https://edgeeffects.net/anthropocene-promise-and-pitfalls/).

Nixon, Rob (2016). The Anthropocene and Environmental Justice. In: Jennifer Newell, Libby Robin & Kirsten Wehner (eds.). *Curating the Future. Museums, Communities and Climate Change*. Abingdon: Routledge & Earthscan: 23–31.

Noller, Jörg (2023). *Ethik des Anthropozäns. Überlegungen zur dritten Natur*. Basel: Schwabe Verlag (Schwabe Reflexe, 79).

Ockenfels, Axel (2020). Größtes Kooperationsproblem der Menschheitsgeschichte. Was hilft und was nicht im Kampf gegen den Klimawandel? *Forschung & Lehre*_(11). 906–907.

Odling-Smee, John (1988). Niche-Constructing Phenotypes. In: H. C. Plotkin (ed.). *The Role of Behavior in Evolution*. Cambridge: The MIT Press: 73–132.

Odling-Smee, John, Kevin Laland & Marcus Feldman (2003). *Niche Construction. The Neglected Process in Evolution*. Princeton, N.J.: Princeton University Press.

Odum, Eugene P. & Josef H. Reichholf (1980). *Ökologie. Die Brücke zwischen Natur- und Sozialwissenschaften*. München: BLV.

Oekom e.V. – Verein für ökologische Kommunikation (eds.) (2021). *Menschengemacht. Streifzüge durch das Anthropozän*". München: Oekom Verlag (= Politische Ökologie 39,167).

Oeser, Erhard (1987). *Psychozoikum. Evolution und Mechansmus der menschlichen Erkenntnisfähigkeit*. Berlin & Hamburg: Verlag Paul Parey (Reihe Biologie und Evolution interdisziplinär).

Ogden, Laura, Nik Heynen, Ulrich Oslender, Page West, Karim-Aly Kassam& Paul Robbins (2013). Global Assemblages, Resilience, and Earth Stewardship in the Anthropocene. *Frontiers in Ecology and the Environment* 11(7). 341–347.

Ogle, Vanessa (2013). Whose Time Is It? The Pluralization of Time and the Global Condition, 1870s–1940s. *American Historical Review* 118(5). 1376–1402.

Oguz, Zeynep (2020). Geological Anthropology. Theorizing the Contemporary. *Fieldsights*, September22.(https://www.academia.edu/44165760/Introduction_Geological_Anthropology).

Oliver-Smith, Anthony & Susanna A. Hoffmann (eds.) ([2]2020). *The Angry Earth. Disaster in Anthropological Perspective*. Milton Park & New York: Routledge.

Olson, Valerie & Lisa Messeri (2015). Beyond the Anthropocene: Un-Earthing an Epoch. *Environment and Society* 6: 28–47.

Ommer, Uwe (2000). *1000 Families. Das Familienalbum des Planeten Erde. The Family Album of Planet Earth. L'album de famille de la planète Terre*. Köln: Taschen Verlag.

Omura, Keiichi, Grant Jun Otsuki, Shiho Satsuka & Atsuro Morita (eds.) (2019). *The World Multiple. The Quotidian Politics of Knowing and Generating Entangled Worlds*. London & New York: Routledge (Routledge Advances in Sociology).

Ong, Aihwah & Stephen J. Collier (eds.) (2005). *Global Assemblages. Technology, Politics, and Ethics as Anthropological Problems*. Malden, Mass.: Blackwell Publishers.

Oppermann, Serpil (2023). *Ecologies of a Storied Planet in the Anthropocene*. Morgantown: West Virginia University Press (Salvaging the Anthropocene).

O'Reilly, Jessica, Cindy Isenhour, Pamela McElwee & Ben Orlove (2020). Climate Change: Expanding Anthropological Possibilities. *Annual Review of Anthropology* 49: 13–29.

Oreskes, Naomi & Erik M. Conway (2010). *Merchants of Doubt. How a Handful of Scientists Obscured the Truth on Issues from Tobacco Smoke to Global Warming*. New York: Bloomsbury Press.

Oreskes, Naomi & Erik M. Conway (2014). *The Collapse of Western Civilization. A View from the Future*. New York: Columbia University Press.

Orlove, Benjamin S. (1980). Ecological Anthropology. In: *Annual Review of Anthropology* 9:235–273.

Orr, Yancey, J. Stephen Lansing & Michael R. Dove (2015). Environmental Anthropology: Systemic Perspectives. *Annual Review of Anthropology* 44: 153–168.

Ortner, Sherry B. (2018). Dark Anthropology and Its Others, Theory Since the Eighties. *Hau: Journal of Ethnographic Theory* 6(1). 47–73

Osborn, Henry Fairfield Jr. (1948). *Our Plundered Planet*. London: Faber & Faber.

Osborne, Mike, M. Traer & L. Chang (2013). *Generation Anthropocene: Stories About Planetary Change*. online: (http://www.stanford.edu/group/anthropocene/cgi-bin/wordpress/, Zugang 5.2.2021).

Oschmann, Wolfgang (2018*). Leben der Vorzeit Grundlagen der Allgemeinen und Speziellen Paläontologie*. Bern: Haupt Verlag.

Osterhammel, Jürgen (2017). *Die Flughöhe der Adler. Historische Essays zur globalen Gegenwart*. München Verlag C.H. Beck.

Osterhammel, Jürgen ([6]2020). *Die Verwandlung der Welt. Eine Geschichte des 19. Jahrhunderts*. München: Verlag C.H. Beck (Historische Bibliothek der Gerda Henkel Stiftung).

Ostheimer, Jochen (2016). Die Renaissance der Geisteswissenschaften in der Ära des Menschen – die Rolle der angewandten Ethik im Anthropozän-Diskurs. In: Matthias Maring (eds..). *Zur Zukunft der Bereichsethiken – Herausforderungen durch die Ökonomisierung der Welt*. Schriftenreihe des Zentrums für Technik- und Wirtschaftsethik am Karlsruher Institut für Technologie. Karlsruhe: KIT Scientific Publishing.

Ostrom, Elinor (1999). Revisiting the Commons: Local Lessons, Global Challenges. *Science* 284: 278–282.

Ott, Konrad (2016). Verantwortung im Anthropozän und Konzepte von Nachhaltigkeit. In: Markus Patenge, Roman Beck & Markus Luber (eds.). *Schöpfung bewahren –Theologie und Kirche als Impulsgeber für eine nachhaltige Entwicklung*. Regensburg: Verlag Friedrich Pustet Verlag: 64–104 (Weltkirche und Mission, 7).

Ott, Konrad (2019). Praktische Diskurse im Anthropozän und die Hierarchie der Gründe. In: Michelle Borelli, Francesca Caputo & Reinhard Hesse (eds.). *Topologik Sonderheft: Karl-Otto Apel – Leben und Denken*. Cleto: Luigi Pellegrini Editore: 205–225.

Ott, Konrad (2020). Umweltethik. In: Thomas Kirchhoff (eds.). *Online Encyclopedia Philosophy of Nature / Online Lexikon Naturphilosophie*. ISSN 2629-8821. (https://doi.org/10.11588/oepn.2020.0.68742).

Otter, Chris, Alison Bashford, John L. Brooke, Fredrik Albritton Jonsson & Jason M. Kelly (2018). In Roundtable: The Anthropocene in British History. *Journal of British Studies* 57(3). 568–596.

Paál, Gábor (2016). Meinung: Das Anthropozän muss wissenschaftlich bleiben. *Spektrum der Wissenschaft* 20: 6–10. http://www.spektrum.de/news/meinung-das-anthropozaen-muss-wissenschaftlich-bleiben/1347395, Aufruf (9.11.20).

Pagel, Mark (2012). *Wired for Culture. Origins of the Human Social Mind.* New York & London: W. W. Norton & Co.

Pagli, Carolina & Freysteinn Sigmundsson (2008). Will Present Day Glacier Retreat Increase Volcanic Activity? Stress Induced by Recent Glacier Retreat and Its Effect on Magmatism at the Vatnajo¨ kull Ice Cap, Iceland. *Geophysical Research Letters* 35: 1–5.

Palmer, Christian T. ([2]2020). Culture and Sustainability: Environmental Anthropology in the Anthropocee. In: Nina Brown, Thomas McIlwraith & Laura Tubelle de González (eds.). *Perspectives. An Open Invitation to Cultural Anthropology.* Arlington, Virg.: American Anthropological Association: 357–381.

Pálsson, Gísli (2020). *The Human Age. How We Created the Anthropocene Epoch and Caused the Climate Crisis.* London: Welbeck.

Pálsson, Gísli, Bronislaw Szerszynski, Sverker Sörlin, JohnMarks, Bernard Avrile, Carole Crumley, Heide Hackmann, Poul Holm, John Ingram, Mercedes Pardo Buendía, Rifka Weehuizen (2013). Reconceptualizing the 'Anthropos' in the Anthropocene: Integrating the Social Sciences and Humanities in Global Environmental Change Research. *Environmental Science and Policy* 28: 3–13.

Palumbi, S. R. (2001). Humans as World's Greatest Evolutionary Force. *Science* 293: 1786–1790.

Panakhyo, Maria & Stacy McGrath (2000). Ecological Anthropology. In: *Anthropological Theories: A Guide Prepared by Students for Students* (University of Alabama). https://anthropology.ua.edu/theory/ecological-anthropology/

Pandian, Anand (2019). *A Possible Anthropology. Methods for Uneasy Times.* Durham: Duke University Press.

Pare, Christopher (2008). Archaeological Periods and Their Purpose. In: A. Lehoerff (dir.). *Construire le temps. Histoire et méthodes des chronologies et calendriers des derniers millénaires avant notre ère en Europe occidentale. Actes du XXXe colloque international de Halma-Ipel, UMR 8164 (CNRS, Lille 3, MCC), 7-9 décembre 2006, Lille.* Glux-en-Glenne: Bibracte: 69–84 (Bibracte, 16).

Park, Sung-Joon (2020). Thinking -With the Favourite Reads in the Anthropology of Global Health and Environmental Health. *Curare. Zeitschrift für Medizinethnologie. Journal of Medical Anthropology* 42(1–2). 96–100.

Parenti, Christian (2016). Environment-Making in the Capitalocene. Political Ecology of the State. In: Jason W. Moore (ed.). *Anthropocene or Capitalocene? Nature, History, and the Crisis of Capitalism.* Oakland, Cal.: PM Press, Kairos: 166–184.

Pattberg, Philipp & Michael Davis-Venn (2020). Dating the Anthropocene. In: Gabriele Dürbeck & Phillip Hüpkes (eds.). *The Anthropocenic Turn. The Interplay Between Disciplinary and Interdisciplinary Responses to a New Age.* New York: Routledge (Routledge Interdisciplinary Perspectives on Literature). 130–149.

Parikka, Jussi (2015a). A *Geology of Media.* Minneapolis & London: University of Minnesota Press (Electronic Mediations, 46).

Parikka, Jussi (2015b). *The Anthrobscene.* Minneapolis: University of Minnesota Press (Forerunners: Ideas First).

Parthasarathi, Prassanan (2011). *Why Europe Grew Rich and Asia Did Not. Global Economic Divergence, 1600–1850.* Cambridge etc.: Cambridge University Press.

Parzinger, Hermann (2019). "Die Mühlen der Zivilisation": Ganz früher war alles besser. Rezension zu Scott 2019. *Die Zeit* Nr. 32 vom 1. August 2019 (https://www.zeit.de/2019/32/die-muehlen-der-zivilisation-james-c-scott/komplettansicht).

Patel, Raj (Rajeev Charles) & Jason W. Moore (2017). *A History of the World in Seven Cheap Things.* Oakland, Cal.: University of California Press.

Pattberg, Philipp & Fariborz Zelli (eds.) (2016). *Environmental Politics and Governance in the Anthropocene. Institutions and Legitimacy in a Complex World.* London & New York: Routledge (Routledge Research in Global Environmental Governance).

Pedersen, Lene & Lisa Cliggett (eds.) 2021: *The Sage Handbook of Cultural Anthropology.* London etc.: Sage Publications (Sage Handbook of the Social Sciences).

Perfecto, Ivette, M. Estelí Jiménez-Soto & John Vandermeer (2019). Coffee Landscapes Shaping the Anthropocene: Forced Simplification on a Complex Agroecological Landscape. *Current Anthropology* 60(suppl. 20). S236–S250.

Pfister, Christian (eds.) (1995). *Das 1950er-Syndrom. Der Weg in die Konsumgesellschaft*. Bern: Paul Haupt (Publikation der Akademischen Kommission der Universität Bern, o.N.).

Pfister, Christian & Heinz Wanner (2021). *Klima und Gesellschaft in Europa. Die letzten tausend Jahre*. Bern: Paul Haupt Verlag.

Philip, Kavita (2014). Doing Interdisciplinary Asian Studies in the Age of the Anthropocene. *The Journal of Asian Studies* 73(4). 975–987.

Pickering, Andrew (2017). The Ontological Turn: Taking Different Worlds Seriously. *Social Analysis: The International Journal of Anthropology* 61(2).134–150.

Pielke, Jr., Roger A. (2014). *The Rightful Place of Science. Disasters and Climate Change*. Tempe, Ar.: Consortium for Science, Policy & Outcomes, Arizona State University.

Pigliucci, Massimo & Gerd B. Müller (eds.) (2010). *Evolution. The Extended Synthesis*. Cambridge, Mass. & London: The MIT Press.

Pinker, Steven (2018). Enlightenment Now. The Case for Reason, Science, Humanism, and Progress. New York: Viking.

Platenkamp, Josephus D. M. (1999). Natur als Gegenbild der Gesellschaft. Einige Betrachtungen zu einer paradoxen Idee. In: Ruth-Elisabeth Mohrmann (eds.). *Argument Natur – Was ist natürlich?* Münster: Lit Verlag (Worte – Werke – Utopien. Thesen und Texte Münsterscher Gelehrter, 7).5–16.

Polanyi, Karl (1944). *The Great Transformation. The Political and Economic Origins of our Times*. Boston, Mass.: Beacon.

Pomeranz, Kenneth (2000). The Great Divergence. China, Europe, and the Making of the Modern World Economy. Princeton & Oxford: Princeton University Press (Princeton Economic History of the Western World).

Poole, Robert (2008). *Earthrise. How Man First Saw the Earth*. New Haven: Yale University Press.

Postill, John (2002). Clock and Calendar Time: A Missing Anthropological Problem. *Time and Society* 11(2–3). 251–270.2002

Potts, Richard (2013). Hominin Evolution in Settings of Strong Environmental Variability. *Quarternary Science Review* 73: 1–13.

Pottage, Alain (2019). Holocene Jurisprudence. *Journal of Human Rights and the Environment* 10: 153–175.

Potthast, Ulrich (1999). *Die Evolution und der Naturschutz. Zum Verhältnis von Evolutionsbiologie, Ökologie und Naturethik*. Frankfurt am Main: Campus Verlag (Campus Forschung).

Povinelli, Elizabeth A. (2014). The Four Figures of the Anthropocene. *Anthem*, 25 April.

Povinelli, Elizabeth A. (2016). *Geontologies. A Requiem to Late Liberalism*. Durham & London: Duke University Press (http://societyandspace.com/2014/05/02/elizabeth-povinelli-the-four-figures-of-the-anthropocene) .

Preston, Christopher J. (2019). *The Synthetic Age. Outdesigning Evolution, Resurrecting Species, and Reorganizing Our World*. Cambridge, Mass.: The MIT Press.

Price, Simon J., Jonathan R. Ford, Anthony H. Cooper & Catherine Neal (2011). Humans As Major Geological and Geomorphological Agents in the Anthropocene: The Significance of Artificial Ground in Great Britain. *Philosophical Transactions of the Royal Society A: Mathematical, Physical and Engineering Sciences* 369(1938). 1056–1084.

Pries, Ludger (2021). *Verstehende Kooperation. Herausforderungen für Soziologie und Evolutionsforschung im Anthropozän*. Frankfurt & New York: Campus.

Probst, Simon (2020). Geologische Poetik. In: *Grundbegriffe des Anthropozän. Ein interdisziplinäres Lexikon des DFG-Forschungsprojekts „Narrative des Anthropozän in Wissenschaft und Literatur*. Vechta: Universität Vechta: 1–6. (https://voado.uni-vecha.de/bitstream/handle/21.11106/297/Probst_Simon_GeologischePoetik_2020_korr.pdf?sequence=2&isAllowed=y).

Puig de la Bellacasa, Maria (2017). *Matters of Care. Speculative Ethics in More than Human Worlds*. Minneapolis: University of Minnesota Press (Posthumanities, 41).
Purdy, Jedediah (2015). *After Nature. A Politics for the Anthropocene*. Cambridge, Mass.: Harvard University Press.
Purdy, Jedediah (2019). *This Land Is Our Land. The Struggle for a New Commonwealth*. Princeton: Princeton University Press.
Putz, Francis E. (1998). Halt the homogeocene: a frightening future filled with too few species. *The Palmetto* 18:7–10.
Pye, Oliver (2021). *Labour, Nature, and Development. The Social Relations of Palm Oil in Southeast Asia*. Unveröff.. Habilitationsschrift. Bonn: Universität Bonn, Philosophische Fakultät.
Pyne, Stephen J. ([2]2019). *Fire. A Brief History*. Seattle & London: University of Washington Press.
Pyne, Stephen J. (2020). From Pleistocene to Pyrocene: Fire Replaces Ice. *Earth´s Future*. (https://agupubs.onlinelibrary.wiley.com/doi/10.1029/2020EF001722)-
Quaedackers, Esther (2020). A Case for Little Big Histories. In: Craig Benjamin, Esther Quaedackers and David Baker (eds.). *The Routledge Companion to Big History*. Milton Park & New York: Routledge (Routledge Companions). 279–299.
Queloz, Matthieu (2021). *The Practical Origins of Ideas Genealogy as Conceptual Reverse-Engineering*. Oxford etc.: Oxford University Press.
Quinn, Naomi (2005). Universals of Child Rearing. *Anthropological Theory* 5(4). 477–516.
Rabinow, Paul (1996). *Essays in the Anthropology of Reason*. Princeton, N.J.: Princeton University Press.
Rabinow, Paul M. (2008). *Marking Time. On the Anthropology of the Contemporary*. Princeton: University Press.
Rabinow, Paul M. & George E. Marcus, with James Faubion & Tobias Rees (2008). *Designs for an Anthropology of the Contemporary*. Durham, N.C.: Duke University Press (John Hope Franklin Center Book).
Radkau, Joachim (2000). *Natur und Macht. Ine Weltgeschichte der Umwelt*. München: C. H. Beck.
Radkau, Joachim (2011). *Die Ära der Ökologie: Eine Weltgeschichte*. München: C. H. Beck.
Radcliffe-Brown, Alfred Reginald (1940). On Social Structure. *Journal of the Royal Anthropological Institute of Great Britain and Ireland* 70(1). 1–12.
Raffnsøe, Sverre (2016). *Philosophy of the Anthropocene. The Human Turn*. Basingstoke: Palgrave Macmillan (Palgrave Pivot).
Radjawali, Irendra, Oliver Pye & Michael Flitner (2017). Recognition through reconnaissance? Using drones for counter-mapping in Indonesia. *The Journal of Peasant Studies* 44(4). 817–833.
Randeria, Shalini (1999). Jenseits von Soziologie und soziokultureller Anthropologie: zur Verortung der nichtwestlichen Welt in einer zukünftigen Sozialtheorie. *Soziale Welt* 50 (4). 373–382.
Rapp, Friedrich (1994). *Die Dynamik der modernen Welt. Eine Einführung in die Technikphilosophie*. Hamburg: Junius.
Rappaport, Roy A. (1968). *Pigs for the Ancestors. Ritual in the Ecology of a New Guinea People*. New Haven, Conn: Yale University Press.
Rappaport, Roy A. (1993). Distinguished Lecture in General Anthropology: The Anthropology of Trouble. *American Anthropologist* 95: 295–303.
Rapport, Nigel J. & Ronald Stade (2007). A Cosmopolitan Turn – or Return? (Debate Section). *Social Anthropology* 15(2).223–235.
Rathjens, Carl (1979). *Die Formung der Erdoberfläche unter dem Einfluß des Menschen. Grundzüge der Anthropogenetischen Geomorphologie*. Tübingen: B.G. Teubner (Teubner Studienbücher der Geographie).
Raup, David M. (1992). Krisen der Vielfalt in erdgeschichtlicher Vergangenheit. In: Edward O. Wilson (eds.). *Ende der biologischen Vielfalt. Der Verlust an Arten, Genen und Lebensräumen und die Chancen für eine Umkehr*, Heidelberg: Spekrum Akademischer Verlag: 69–75.
Raup, David M. & John Sepkoski (1982). Mass Extinctions in the Marine Fossil Record Source: *Science* 215 (4539). 1501–1503.

Raworth, Kate (2018). *Doughnut Economics. Seven Ways to Think Like a 21st-Century Economist.* White River Junction, Vt.: Chelsea Green Publishing, New York, Random House.
Ray, Gene (2024). Surviving the Globe: An Introduction to Planetary Politics. In: *Searcch #1.* Genève: Haute Ecole d´Art et de Design (HEAD).
Read, Rupert & Samuel Alexander (2019). *This civilisation is finished. Conversatons on the end of Empire –and what lies behind.* Melbourne: Simplicity Institute.
Redman, Charles L., Steven James, Paul Fish & J. Daniel Rogers (eds.) (2004). *The Archaeology of Global Change. The Impact of Humans on their Environment.* Washington, D.C: Smithsonian Press.
Reichholf, Josef H. ([3]2011). *Der Tanz um das goldene Kalb. Der Ökokolonialismus Europas.* Berlin: Verlag Klaus Wagenbach.
Renfrew, Colin & Paul Bahn ([8]2020). *Archaeology. Theories, Methods, and Practice.* London: Thames & Hudson.
Renn, Jürgen (2020). *The Evolution of Knowledge. Rethinking Science for the Anthropocene.* Princeton University Press.
Renn, Jürgen & Bernd Scherer ([2]2017a). Einführung. In: Dies. (eds.). *Das Anthropozän. Zum Stand der Dinge.* Berlin: Matthes & Seitz: 7–23.
Renn, Jürgen & Bernd Scherer (eds.) ([2]2017b). *Das Anthropozän. Zum Stand der Dinge.* Berlin: Matthes & Seitz.
Renn, Ortwin (2021). Die Zukunft muss nicht schlechter werden (Interview). *ZiF.Mitteilungen* 17 2021: 25–28
Renn, Ortwin ([3]2023). *Gefühlte Wahrheiten. Orientierung in Zeiten postfaktischer Verunsicherung.* Opladen etc.: Verlag Barbara Budrich.
Resnick, Eleana (2021). The Limits of Resilience: Managing Waste in the Racialized Anthropocene. *American Anthropologist* 123(2). 222–236.
Reuter, Thomas A. (2010a). Anthropological Theory and the Alleviation of Anthropogenic Climate Change: Understanding the Cultural Causes of Systemic Change Resistance. *World Anthropology Network E-Journal* 5: 5–27 (http://www.ram-wan.net/documents/05_e_Journal/journal-5/2-reuter.pdf).
Reuter, Thomas A. (2010b). In Response to a Global Environmental Crisis: How Anthropologists Are Contributing Toward Sustainability and Conservation. In: Ders. (ed.). *Averting a Global Environmental Collapse: The Role of Anthropology and Local Knowledge.* Cambridge: Cambridge Scholars Publishing: 1–19.
Reuter, Thomas A. (ed.) (2015). *Averting a Global Environmental Collapse: The Role of Anthropology and Local Knowledge.* Cambridge: Cambridge Scholars Publishing.
Revkin, Andrew (1992). *Global Warming. Understanding the Forecast.* New York: American Museum of Natural History, Environmental Defense Fund & Abbeville Press.
Reszitnyk, Andrew (2020). The Descent into Disanthropy: Critical Theory and the Anthropocene. *Telos* 190: 9–27.
Rheinberger, Hans-Jörg (2021). The "Material Turn" and the "Anthropocenic Turn" from a History of Science Perspective. In: Gabriele Dürbeck & Philip Hüpkes (eds.). *The Anthropocenic Turn.* London: Routledge. Routledge (Routledge Interdisciplinary Perspectives on Literature). 27–36.
Ribeiro, Gustavo & Arturo Escobar (eds.). (2006). *World Anthropologies. Disciplinary Transitions Within Systems of Power.* Oxford etc.: Berg.
Ribot, Jesse (2018). Ontologies of Occlusion in the Anthropocene (Kommenater zu Bauer & Ellis 2018). *Current Anthropology* 59(2). 218–219.
Richards, John F. (2003). *The Unending Frontier. An Environmental History of the Early Modern World.* Berkeley, Cal. etc.: University of California Press (California World History Library, 1).
Richardson, Katherine, Will Steffen, Wolfgang Lucht, Jørgen Bendtsen, Sarah E. Cornell, Jonathan F. Donges, Markus Drüke, Ingo Fetzer, Govindasamy Bala, Werner von Bloh, Georg Feulner, Stephanie Fiedler, Dieter Gerten, Tom Gleeson, Matthias Hofmann, Willem Huiskamp, Matti Kummu, Chinchu Mohan, David Nogués-Bravo, Stefan Petri , Miina Porkka, Stefan Rahmstorf, Sibyll Schaphoff, Kirsten Thonicke, Arne Tobian, Vili Virkki, Lan Wang-Erlandsson, Lisa

Weber & Johan Rockström (2023). Earth beyond Six of Nine Planetary Bondaries. *Science Advances* 9(37). (doi.org/10.1126/sciadv.adh2458).
Richerson, Peter J. & Robert Boyd (2005). *Not by Genes Alone. How Culture Transformed Human Evolution*. Chicago, IL: The University of Chicago Press.
Richerson, Peter J., Robert T. Boyd & Charles Efferson (2023). Agentic Processes in Cultural Evolution: Relevance to Anthropocene Sustainabiity. *Philosophical Transactions of the Royal Society B*: 379: 2020252.
Richter Jr., Daniel deB. (2020). Game Changer in Soil Science: The Anthropocene in Soil Science and Pedology. *Journal of Plant Nutrition and Soil Science* 183: 5–11.
Richter Jr., Daniel deB. & Sharon A Billings (2015). 'One Physical System: Tansley's Ecosystem as Earth's Critical Zone. *New Phytologist* 206: 900–912.
Rickards, L. (2015). Metaphor and the Anthropocene: presenting humans as a geological force. *Geographical Research* 53: 280–287.
Riede, Felix (2011). Adaptation and Niche Construction in Human Prehistory: A Case Study From the Southern Scandinavian Late Glacial. *Philosophical Transactions of the Royal Society B* 366: 793–808.
Riede, Felix (2019). Niche Construction Theory and Human Biocultural Evolution. In: A. M. Prentiss (eds.). *Handbook of Evolutionary Research in Archaeology*. Cham: Springer 2019, 337–358.
Riede, Felix, Christian Hoggard & Stephen Shennan (2019). Reconciling Material Cultures in Archaeology Withgenetic Data Requires Robust Cultural Evolutionarytaxonomies. *Palgrave Communications* 5/55. (https://doi.org/10.1057/s41599-019-0260-7).
Riede, Felix, Christina Vestergaard & Kristoffer H. Fredensborg (2014). A Field Archaeological Perspective on the Anthropocene. *Antiquity* 90: 1–5.
Riedel, Jens (2002). The Multiple Modernities Perspective: Enriching Business Strategy. In: Dominic Sachsenmaier und Jens Riedel, with Shmuel Noah Eisenstadt (eds.). *Reflections on Multiple Modernities. European, Chinese and Other Interpretations*. Leiden: ___ : 271–292.
Rigg, Jonathan (2007). *An Everyday Geography of the Global South*. London & New York: Routledge.
Riley, Erin P. (2020). *The Promise of Contemporary Primatology*. New York & London: Routledge (New Biological Anthropology)
Rispoli, Giulia (2020). Genealogies of Earth System Thinking. *Nature Reviews Earth & Environment* 1(1). 4–5.
Ritzer, George ([10]2021). *The Mcdonaldization of Society. Into the Digital Age*. Los Angeles: Sage Publications.
Rival, Laura (2020). Arguing for a Systems Change in the Anthropocene. *St. Antony´s International Review* 15(2). 109–119.
Rival, Laura (2021). Anthropocene and the Dawn of Planetary Civilization. *Anthropology Today* 37(6). 9–12.
Roberts, Neil ([3]2014). *The Holocene. An Environmental History*. Oxford & Malden, Mass.: Blackwell Publishers.
Robin, Libby (2013). Histories for Changing Times: Entering the Anthropocene? *Australian Historical Studies*, 44(3): 329–340.
Robin, Libby, Paul Warde & Sverker Sörlin (eds.) (2013). *The Future of Nature. Documents of Global Change*. New Haven: Yale University Press.
Robin, Libby & Will Steffen (2007). History for the Anthropocene. *History Compass* 5(5). 1694–1719.
Robbins, Joel (2007). Continuity Thinking and the Problem of Christian Culture: Belief, Time, and the Anthropology of Christianity. *Current Anthropology* 48(1). 5–38.
Robbins, P. & S. A. Moore (2013/12). Ecological Anxiety Disorder: Diagnosing the Politics of the Anthropocene. *Cultural Geographies* 20(1). 3–19.
Robbins, Richard A. & Rachel A. Dowty ([7]2019). *Global Problems and the Culture of Capitalism*. New York, N.Y.: Pearson Education.

Robbins, Richard A. & Rachel A. Dowty Beech ([8]2021). *Cultural Anthropology. A Problem-Based Approach*. Thousand Oaks, Cal.: Sage Publishing.
Rojas, David (2013). Welcome to the Anthropocene. On Rio+20 and Environmental Imagination in Climate Change Diplomacy. *Anthropology News*. http://www.anthropology-news.org/index.php/2012/12/14/welcome-to-the-anthropocene-2/
Rockström, Johan (2019). *Eat Good. Das Kochbuch, das die Welt verändert*. Hilsdesheim: Gerstenberg.
Rockström, Johan (2021). "Dieses Jahrzehnt bietet die letzte Gelegenheit, das Ruder herumzureißen". (Interview). Der Spiegel # 20 vom 15.5.2021: 102–106 (Serie „Republik", 21,2).
Rockström, Johan & Mattias Klum (2015). *Big World, Small Planet*. Abundance Within Planetary Boundaries. Stockholm: Bokförlaget Max Ström, and New Haven: Yale University Press.
Rockström, Johan, Will Steffen, Kevin Noon, Åsa Persson, F. Steward Chapin, Eric F. Lambin, Timothy M. Lenton, Marten Scheffer, Carl Folke & Hans Joachim Schellnhuber (2009). A Safe Operating Space for Humanity. Nature 461: 472–475.
Röttger-Rössler, Birgitt (2020). Research across Cultures and Disciplines: Methodological Challenges in an Interdisciplinary and Comparative Research Project on Emotion Socialization. In: Michael Schnegg, & Edward D. Lowe (eds.). *Comparing Cultures. Innovations in Comparative Ethnography*. Cambridge: Cambridge University Press: 180–200.
Röttger-Rössler, Birgitt & Jan Slaby (eds.) (2018). *Affect in Relation. Families, Places, Technologies*. London & New York: Routledge (Routledge Studies in Affective Societies, 1).
Rohmer, Stascha & Georg Toepfer (eds.) (2021). *Anthropozän – Klimawandel – Biodiversität. Transdisziplinäre Perspektiven auf das gewandelte Verhältnis von Mensch und Natur*. Freiburg & München: Verlag Karl Alber.
Rolston, Holmes III ([2]2020a). *A New Environmental Ethics. The Next Millenium for Life on Earth*. New York & Abingdon: Routledge.
Rolston, Holmes III (2020b). The Anthropocene!: Beyond the Natural? In: Stephen M. Gardiner & Allen Thompson (eds.). T*he Oxford Handbook of Environmental Ethics*. New York: Oxford University Press: 62–73.
Rosa, Hartmut (2021). Best Account. Skizze einer systematischen Theorie der modernen Gesellschaft. In: Andreas Reckwitz & Hartmut Rosa: *Spätmoderne in der Krise. Was leistet die Gesellschaftstheorie?* Berlin: Suhrkamp: 151–251.
Roscoe, Paul (2016). Method, Measurement, and Management in IPCC Climate Modeling. *Human Ecology* 44: 655–664.
Rose, Deborah Bird (2013). Slowly: Writing into the Anthropocene. TEXT Special Issue 20: Writing Creates Ecology and Ecology Creates Writing: 1–14.
Rose, Deborah Bird (2015a). The Ecological Humanities. In: Katherine Gibson, Deborah B. Rose & Ruth Fincher (eds.). *Manifesto for Living in the Anthropocene*. Brooklyn, N.Y.: Punctum Books: 1–5.
Rose, Neill L. (2015b). Spheroidal Carbonaceous Fly Ash Particles Provide a Globally Synchronous Stratigraphic Marker for the Anthropocene. *Environmental Sience and Technology* 49: 4155–4162.
Rosling, Hans, with Anna Rosling Rönnlund & Ove Rosling (2018). *Factfulness. Ten Reasons We're Wrong About the World – And Why Things Are Better than You Think*. London: Sceptre.
Rosol, Christoph & Giulia Rispoli (eds.) (2013–2022). Anthropocene Curriculum. Berlin: Max Planck Institute for the History of Science. (https://www.anthropocene-curriculum.org/).
Rosol, Christoph, Georg N. Schäfer, Simon D. Turner, Colin N. Waters, Martin J. Head, Jan Zalasiewicz, Carlina Rossée, Jürgen Renn, Katrin Klingan & Bernd M. Scherer (2023). Evidence and experiment: Curating contexts of Anthropocene geology. *The Anthropocene Review* 10(1). 330–339.
Rossi, Paolo (1984). *The Dark Abyss of Time. The History of the Earth and the History of Nations from Hooke to Vico*. Chicago & London: The University of Chicago Press.
Rothenberg, David (2018). *Tiefenökologie vs. das Anthropozän*. Bonn: Bundeszentrale für politische Bildung.

Rowan, Rory (2014). Notes on Politics after the Anthropocene. In: Elizabeth Johnson & Harlan Morehouse (eds.). After the Anthropocene. Politics and Geographic Inquiry for a New Epoch. *Progress in Human Geography* 38(3). 447–450.

Rubenstein, James M. ([13]2019). *The Cultural Landscape. An Introduction to Human Geography*. Harlow: Perasoon Education.

Ruddiman, William F, Erle C. Ellis, Jed O. Kaplan & Dorian Q. Fuller (2015). Defining the epoch we live in Is a formally designated "Anthropocene" a good idea? *Science* 348(6230): 38–39.

Rudiak-Gould, Peter (2012). Promiscuous Corroboration and Climate Change Translation: A Case Study From the Marshall Islands. *Global Environmental Change* 22: 46–54.

Rudiak-Gould, Peter (2015). The Social Life of Blame in the Anthropocene. *Environment & Society* 6: 48–65.

Rudwick, Martin John Spencer (2005). *Bursting the Limits of Time. The Reconstruction of Geohistory in the Age of Revolution*. Chicago, Ill. & London: The University of Chicago Press.

Rudwick, Martin John Spencer (2014). *Earth´s Deep History. How it was Discovered and Why it Matters*. Chicago, Ill. & London: The University of Chicago Press.

Ruddiman, William F. (2003). The Anthropogenic Greenhouse Era Began Thousands of Years Ago. *Climatic Change* 61, 261–293.

Ruddiman, William F. (2010). *Plows, Plagues, and Petroleum: How Humans Took Control of Climate*. New York: Princeton University Press (zuerst 2005).

Ruddiman, William F. (2013). The Anthropocene. *Annual Review of Earth and Planetary Sciences* 41: 45–68.

Ruddiman, William F. (2016a). The Early Anthropogenic Hypothesis. *Environmental Science* (https://doi.org/10.1093/acrefore/9780199389414.013.192).

Ruddiman, William F. (2016b). The Early Anthropocene Hypothesis: An Update. *Realclimate* (http://www.realclimate.org/index.php/archives/2016/03/the-early-anthropocene-hypothesis-an-update/).

Ruddiman William F. (2018). Three Flaws in Defining a Formal 'Anthropocene'. *Progress in Physical Geography* 42(4). 451–461.

Ruddiman, William F., F. He, S. J. Vavrus & J. E. Kutzbach (2020). The Early Anthropogenic Hypothesis: A Review. *Quaternary Science Reviews*, 240, 106386: 1–14.

Rull, Valentí (2016). The Humanized Earth System (HES). *The Holocene* 26(9). 1513–1516.

Rull, Valentí (2017). The "Anthropocene": Neglects, misconceptions, and possible futures. *EMBO Reports* 18(7): 1056–1060.

Rull, Valentí (2018a). *El antropoceno*. Madrid: CSIC, Los Libros de la Catarata (¿Qué sabemos de?, 90).

Rull, Valentí (2018b). What If the 'Anthropocene' Is Not Formalized as a New Geological Series/Epoch? Quaternary. 1(3) (https://doi.org/10.3390/quat1030024).

Rudolph, Wolfgang (1973). Kultur, Psyche, Weltbild. In: Ders.: *Ethnologie. Zur Standortbestimmung einer Wissenschaft*. Tübingen: Verlag Elly Huth .

Rudolph, Wolfgang & Peter Tschohl (1977). *Systematische Anthropologie*. München: Wilhelm Fink Verlag (Uni-Taschenbücher, 639)

Rutherford, Danilyn (2019). Patchy Anthropocene: Frenzies and Afterlives of Violent Simplifications Wenner-Gren Symposium Supplement. *Current Anthropology* 60, Supplement 20: S183–185.

Sagan, Carl (1994). *Pale Blue Dot. A Vision of the Human Future in Space*. New York, N.Y.: Ballantine Books.

Sahlins, Marshall David & Elman Rogers Service (eds.) (1960). *Evolution and Culture*. Ann Arbor: University of Michigan Press.

Sakkas, Konstantin (2021). War Angst vor dem Anthropozän. Anthropologische, geschichtsphilosophische und ontologische Überlegungen zur Anthropozändebatte. *Internationales Jahrbuch für philosophische Anthropologie* 11(1). 153–169.

Salk, Jonas (1992). Are We Being Good Ancestors? *World Affairs* 1(2). 16–18.

Salvador, Amos (ed.) ([2]1994). *International Stratigraphic Guide (A Guide to Stratigraphic Classification, Terminology and Procedures)*. Boulder, Col.: International Union of Geological Sciences, Geological Society of America.
Samways, Michael (1999). Translocating Fauna to Foreign Lands. Here Comes the Homogenocene. *Journal of Insect Conservation* 3: 65–66.
Sanderson, Eric (2013). *Mannahatta. A Natural History of New York City*. New York: Abrams.
Santana, Carlos (2019). Waiting for the Anthropocene. *British Journal of Philosophical Science* 70: 1073–1096.
Sato, Kohei (2023). *Marx in the Anthropocene. Towards the Idea of Degrowth Communism*. Cambridge etc.: Cambridge University Press.
Sauer, Carl O. (1950). Grassland Climax, Fire, and Man. *Journal of Range Management* 3(1).16–21.
Sayre, Nathan F. (2012). The Politics of the Anthropogenic. *Annual Review of Anthropology* 41:57–70
Scarborough Vernon L. (2018). A framework for facing the past. In: Emily Holt (ed.). *Water and Power in Past Societies*. Albany, NY: State University of New York Press: 297–315.
Scarborough, Vernon L. & Christian Isendahl (2020). Distributed urban network systems in the tropical archaeological record: Toward a model for urban sustainability in the era of climate change. *The Anthropocene Review* 7(3) 208–230.
Schär, Bernhard (2016). *Tropenliebe. Schweizer Naturforscher und niederländischer Imperialismus in Südostasien um 1900*. Frankfurt am Main: Campus (Globalgeschichte, 20).
Schär, Kathrin (2021). *Erdgeschichte(n) und Entwicklungsromane. Geologisches Wissen und Subjektkonstitution in der Poetologie der frühen Moderne. Goethes Wanderjahre und Stifters Nachsommer*. Bielefeld: Transcript.
Schareika, Nikolaus (2004). Lokales Wissen: ethnologische Perspektiven. In: Nikolaus Schareika & Thomas Bierschenk (eds.). *Lokales Wissen - sozialwissenschaftliche Perspektiven*. Berlin & Münster: Lit Verlag: 9–39.
Schareika, Nikolaus (2006). Modelle der interdisziplinären Zusammenarbeit zwischen Ethnologie und Naturwissenschaften. *Sociologus* 56:15–36.
Schatzki, Theodore (2016). Practice Theory as Flat Ontology. In: G. Spaargaren, D. Weenick, & M. Lamers (eds.). *Practice Theory and Research. Exploring the dynamics of social life*. London: Routledge: 28–42.
Scheidler, Fabian (2021). *Der Stoff, aus dem wir sind. Warum wir Natur und Gesellschaft neu denken müssen*. München: Piper Verlag.
Schellnhuber, Hans-Joachim (1999). "Earth System" Analysis and the Second Copernican Revolution. *Nature* 402: C19–C23 (supplement).
Scherer, Bernd M. (2015). Der blinde Fleck der Aufklärung. Zum Verständnis von Natur und Kultur. In: Nina Möllers, Christian Schwägerl & Helmuth Trischler (eds.). *Willkommen im Anthropozän. Unsere Verantwortung für die Zukunft der Erde*. München: Deutsches Museum Verlag, S. 103–107.
Scherer, Bernd M. (2017). Die Monster. In: Jürgen Renn & Bernd Scherer (eds.). *Das Anthropozän. Zum Stand der Dinge*. Berlin: Matthes & Seitz: 241–226.
Scherer, Bernd (2022). *Der Angriff der Zeichen. Denkbilder und Handlungsmuster des Anthropozäns*. Berlin: Matthes & Seitz.
Schiefenhövel, Wulf, Volker Heeschen, Klaus Helfrich, Volker Jacobshagen & Gerd Koch (eds.) (1983). *Mensch, Kultur und Umwelt im zentralen Bergland von West-Neuguinea*. Berlin: Dietrich reimer Verlag.
Schiemann, Gregor ([2]2020). Jenseits der Naturverhältnisse: Natur ohne Menschen. In: Thomas Kirchhoff , Nicole C. Karafyllis, Dirk Evers, Brigitte Falkenburg, Myriam Gerhard, Gerald Hartung, Jürgen Hübner, Kristian Köchy, Ulrich Krohs, Thomas Potthast, Otto Schäfer, Gregor Schiemann, Magnus Schlette, Reinhard Schulz & Frank F. Vogelsang (eds.). *Naturphilosophie. Ein Lehr – und Studienbuch*. Tübingen: Mohr Siebeck: 248–253.
Schill, Caroline, John M. Anderies, Therese Lindahl, Carl Folke, Stephen Polasky, Juan Camilo Cárdenas, Anne-Sophie Crépin, Marco A. Janssen, Jon Norberg & Maja Schlüter (2021). A more

Dynamic Understanding of Human Behaviour for the Anthropocene. *Nature Sustainability*. (https://doi.org/10.1038/s41893-019-0419-7)

Schlaudt, Oliver (2019). Homo Migrans: Schlaglichter aus der Dunkelheit der „Deep History". *Merkur* 848:71–79.

Schlehe, Judith (2021). Nachhaltigkeit als Mobilisierung und Konkretisierung. In: Dies. (eds.). *Nachhaltigkeitsprojekte in Südost- und Ostasien: Landwirtschaft – Markt – Müll*. Freiburg: Institut für Ethnologie Albert-Ludwigs-Universität Freiburg (Freiburger Ethnologische Arbeitspapiere Nr. 43). 12–18.

Schlemm, Annette (2023): *Climate Engineering. Warum wir uns technisch zu Tode siegen, statt die Gesellschaft zu revolutionieren*. Wien & Berlin: Mandelbaum Kritik & Utopie.

Schmelzer, Matthias & Andrea Vetter (^{2}2019). *Degrowth/Postwachstum zur Einführung*. Hamburg Junius Verlag.

Schmitt, Stefan (2021). Der Pate unseres Erdzeitalters. *Die Zeit* Nr. 6 vom 4.2.2021: 32.

Schmidt, Isabell & Andreas Zimmermann (2019). Population Dynamics and Socio-spatial Organization of the Aurignacian: Scalable Quantitative Demographic Data for Western and Central Europe. PLOS ONE 14: e0211562.

Schmidt, Isabell & Andreas Zimmermann (2021). Population dynamics of the Palaeolithic. In: Thomas Litt, Jürgen Richter & Frank Schäbitz (eds.). *The journey of modern humans from Africa to Europe. Culture, environmental interaction and mobility*. Stuttgart: Schweizerbart Science Publishers: 161–174.

Schmidt, Isabell, Birgit Gehlen & Andreas Zimmermann (2021). Population estimates for the Final Palaeolithic (14,000 to 11,600 years cal. BP) of Europe – challenging evidence and methodological limitations. In: Ludovic Mevel, Mara-Julia Weber & Andreas Maier (eds.). *En Mouvement / On the Move / In Bewegung*. Paris: Société préhistorique française (Séances de la Société préhistorique française, 17). 221–237.

Schmidt, Peter R & Stephen A. Mrozowski (eds.) (2013). *The Death of Prehistory*. Oxford etc.: Oxford University Press.

Schmieder, Falko (2014). Urgeschichte der Nachmoderne. Zur Archäologie des Anthropozäns. *Forum Interdisziplinäre Begriffsgeschichte* [E-Journal], 3(2). 43–48.

Schneider, Birgit (2018). *Klimabilder. Eine Genealogie globaler Bildpolitiken von Klima und Klimawandel*. Berlin: Matthes & Seitz.

Schneiderman Jill S. (2016). Naming the Anthropocene. *Philosophia* 5(2).179–201.

Schneiderman, Jill S. (2017). The Anthropocene Controversy. In: Richard Grusin (ed.). *Anthropocene Feminism*. Minneapolis & London: University of Minnesota Press: 169–195 (Center for 21st Cenrtury Studies).

Schneidewind, Uwe und viele Mitarbeiter (2019). *Die Große Transformation. Eine Einführung in die Kunst des gesellschaftlichen Wandels*. Frankfurt am Main: Fischer Taschenbuch Verlag (Forum für Verantwortung)

Schnyder, Peter (eds.) (2020). *Erdgeschichten. Literatur und Geologie im langen 19. Jahrhundert*. Würzburg: Königshausen & Neumann.

Schroer, Markus (2017). Geosoziologie im Zeitalter des Anthropozäns. In: Anna Henkel, Henning Laux & Fabian (eds.). *Raum und Zeit. Soziologische Betrachtungen zur gesellschaftlichen Raumzeit*. Weinheim & Basel: Beltz Juventa: 126–151 (Sonderband der Zeitschrift für theoretische Soziologie, 4).

Schroer, Markus (2019). Geosoziologie: Raum als Territorium. In: Ders.: *Räume der Gesellschaft . Soziologische Studien*. Wiesbaden: Springer VS.

Schroer, Markus (2020). Bündnisse, die Gaia stiftet. Neue Kollektive im Anthropozän. *Zeitschrift für Kultur- und Kollektivwissenschaft* 6(1). 269–300.

Schroer, Markus (2022). *Geosoziologie. Die Erde als Raum des Lebens*. Berlin: Suhrkamp.

Schüttpelz, Erhard (2016). Domestizierung im Vergleich. *Zeitschrift für Medien- und Kulturforschung* 7(2), Schwerpunkt Medien der Natur: 93–109.

Schüttpelz, Erhard (2022). Vor und nach Corona. *Curare. Zeitschrift für Medizinethnologie. Journal of Medical Anthropology* 44 (1–4): 109–113.

Schüttpelz, Erhard, Sina Steglich & Moritz Ingwersen (2022). Allochronie im Anthropozän. *Sciendo*, Sonderheft 7: 107–122.
Schulz, Astrid (2015). Klimaschutz als Weltbürgerbewegung. In: Heike Landschuh, Gerd Michelsen, Udo Ernst Simonis, Jörg Sommer & Ernst U. von Weizsäcker (eds.). *Globale Weltumweltpolotik. Herausforderungen im AnthropozänStutgart: S. Hirzel Verlag* (Jahrbuch Ökologie 2016).126–132.
Schulz, David (2020). *Die Natur der Geschichte. Die Entdeckung der geologischen Tiefenzeit und die Geschichtskonzeptionen zwischen Aufklärung und Moderne.* Berlin & Boston: de Gruyter & Oldenbourg (Ordnungssysteme, Studien zur Ideengeschichte der Neuzeit, 56).
Schwägerl, Christian (2012). *Menschenzeit. Zerstören oder gestalten? Wie wir heute die Welt von morgen erschaffen.* München: Goldmann.
Schwägerl, Christian (2014). *Anthropocene. The Human Era and How It Shapes Our Planet.* Santa Fe & London: Synergetic Press.
Schwägerl, Christian (2024). Keine Epoche für die Menschheit. Spektrum https://www.spektrum.de/news/geologen-lehnen-neues-erdzeitalter-anthropozaen-ueberraschend-ab/2210153).
Schwägerl, Christian & Reinhold Leinfelder (2014). Die menschgemachte Erde. *Zeitschrift für Medien- und Kulturforschung* 5(2). 233–240.
Schwägerl, Christian (2016). *Leben in einer globalen Polis. Das Anthropozän als politische Herausforderung.* Bonn: Bundeszentrale für Politische Bildung (Dossier Anthropozän).
Schwägerl, Christian & Kai Urban (2024). *Ist das Anthropozän jetzt tot?* Riffreporter (https://podcasts.apple.com/de/podcast/ist-das-anthropoz%C3%A4n-jetzt-tot-christian-schw%C3%A4gerl/id1722237973?i=1000650471004).
Schwinger, Elke (eds.) (2019). *Das Anthropozän im Diskurs der Fachdisziplinen.* Marburg: Metropolis-Verlag.
Scoones, Ian (1999). New Ecology and the Social Sciences: What Prospects for a Fruitful Engagement? *Annual Reviews of Anthropology* 28:479–507.
Scott, James C. (1998). *Seeing Like a State. How Certain Schemes to Improve the Human Condition Have Failed.* New Haven, Conn: Yale University Press.
Scott, James C. (2017). *Against the Grain. A Deep History of the Earliest States.* New Haven, Conn: Yale University Press (Yale Agrarian Studies Series).
Scranton, Roy (2015). *Learning to Die in the Anthropocene. Reflections on the End of Civilzation.* San Francisco: City Lights.
Seebens, Hanno, Tim M. Blackburn, Ellie E. Dyer, Piero Genovesi, Philip E. Hulme, Jonathan M. Jeschke, Shyama Pagad, Petr Pyšek, Mark van Kleunen, Marten Winter, Michael Ansong, Margarita Arianoutsou, Sven Bacher, Bernd Blasius, Eckehard G. Brockerhoff, Giuseppe Brundu, César Capinha, Charlotte E. Causton, Laura Celesti-Grapow, Wayne Dawson, Stefan Dullinger, Evan P. Economo, Nicol Fuentes, Benoit Guénard, Heinke Jäger, John Kartesz, Marc Kenis, Ingolf Kühn, Bernd Lenzner, Andrew M. Liebhold, Alexander Mosena, Dietmar Moser, Wolfgang Nentwig, Misako Nishino, David Pearman, Jan Pergl, Wolfgang Rabitsch, Julissa Rojas-Sandoval, Alain Roques, Stephanie Rorke, Silvia Rossinelli, Helen E. Roy, Riccardo Scalera, Stefan Schindler, Kateřina Štajerová, Barbara Tokarska-Guzik, Kevin Walker, Darren F. Ward, Takehiko Yamanaka, and Franz Essl (2018). Global Rise in Emerging Alien Species. Results From Increased Accessibility of New Source Pools. *Proceedings of the National Academy of Sciences of the United States of America* 115: E2264–E2273.
Seithel, Friderike (2000). *Von der Kolonialethnologie zur Advocacy Anthropology. Zur Entwicklung einer kooperativen Forschung und Praxis von Ethnologinnen und indigenen Völkern.* Münster etc.: Lit Verlag (Interethnische Beziehungen und Kulturwandel, 34).
Selcer, Perrin (2021). Anthropocene. In: *Encyclopedia of the History of Science.* https://doi.org/10.34758/zr3n-jj68
Sen, Amartya Kumar (2009). *The Idea of Justice.* London: Penguin and Harvard University Press..
Sepkoski, David (2021). *Catastrophic Thinking. Extinction and the Value of Diversity from Darwin to the Anthropocene.* Chicago & London: The University of Chicago Press (Science.Culture).

Shanster E. V. (1973). Anthropogenic system. In: A. M. Prokhorov (ed.). *Great soviet encyclopedia*. New York: Macmillan & London: Collier Macmillan: 139–144.
Sheldrake, Merlin (2020). *Entangled Life: How Fungi Make Our Worlds, Change Our Minds, and Shape Our Futures*. New York: Random House.
Shellenberger, Michael (2020). *Apocalypse Never. Why Environmental Alarmism Hurts Us All*. New York, N.Y.: Harper Collins.
Shellenberger, Michael & Ted Nordhaus (2011). Evolve: The Case for Modernization as the Road to Salvation. In: Dies. (eds.). *Love Your Monsters. Postenvironmentalism and Anthropocene*. Oakland, Cal.: The Breakthrough Institute: 8–16.
Shoreman-Ouimet, Eleanor & Helen Kopnina (2016). *Culture and Conservation. Beyond Anthropocentrism*. London & New York: Routledge/Earthscan (Routledge Explorations in Environmental Studies). 14–15, 54
Shryock, Daniel & Daniel Lord Smail, with nine co-authors (2011). Deep History. The Architecture of Past and *Present*. Berkeley etc.: University of California Press.
Sieferle, Rolf Peter (1982). *Der unterirdische Wald. Energiekrise und Industrielle Revolution*. Berlin: C. H. Beck.
Sieferle, Rolf Peter (2020). *Rückblick auf die Natur. Eine Geschichte des Menschen und seiner Umwelt*. Berlin: Landt Verlag (Wrtkausgabe, 5; zuerst Luchterhand, 1997).
Sieferle, Rolf Peter, Fridolin Krausmann, Heinz Schandl & Verena Winiwarter (2011). *Das Ende der Fläche. Zum gesellschaftlichen Stoffwechsel der Industrialisierung*. Köln etc.: Böhlau Verlag (Umwelthistorische Forschungen, 2).
Siegler, Martin (2024). Hohlozän (sic!). *Zeitschrift für Medienwissenschaft* 16(30). Was uns ausgeht: 57–59.
Simangan, Dahlia (2019). Situating the Asia Pacific in the Age of the Anthropocene. *Australian Journal of Interational Affairs* 73(6). 564–584.
Simangan, Dahlia (2020a). Where is the Anthropocene? IR in a New Geological Epoch. *International Affairs* 96(1). 211–224.
Simangan, Dahlia (2020b). Where is the Asia Pacific in Mainstream International Relations Scholarship on the Anthropocene? *The Pacific Review*, https://doi.org/10.1080/09512748.2020.1732452).
Simarmata, Hendricus Andy, Anna-Katharina Hornidge & Christoph Antweiler (2020). Assessing Flood-related Vulnerability of Urban Poor: An Empirical Case Study of Kampung Muara Baru, Jakarta. In: Greg Bracken, Paul Rabé, R. Parthasarathy, Neha Sami & Zhang Bing (eds.). *Future Challenges of Cities in Asia*. Amsterdam: Amsterdam University Press (Asian Cities, 11).
Simon, Zoltan Boldizsár (2017). Why the Anthropocene has no history: Facing the unprecedented. *The Anthropocene Review* 4(3). 239 –245.
Simpson, Michael (2020). The Anthropocene as Colonial Discourse. *Environment & Planning D: Society and Space* 38(1). 53–71.
Simmons, Ian Gordon (^{2}1996). *Changing the Face of the Earth. Culture, Environment, History*. Oxfiord & Cambridge, Mass.: Blackwell Publishers.
Skinner, Jonathan (2021). Most Humanistic, Most Scientific: Experiencing Anthropology in the Humanities and Life Sciences. In: Emma Heffernan, Fiona Murphy & Jonathan Skinner (eds.). *Collaborations. Anthropology in a Neoliberal Age*. London: Routledge: 65–84 (Criminal Practice Series).
Sklair, Leslie (2019). The Corporate Capture of Sustainable Development and its Transformation into a 'Good Anthropocene' Historical Bloc. *Civitas* 19 (2). 296–314.
Sklair, Leslie (2021a). Geoethics: A Reality Check from Media Coverage of the Anthropocene. In: Geo-societal Narratives. In: Martin Bohle & Eduardo Marone (eds.). *Geo-Societal Narratives Contextualising Geosciences*. London etc.: Palgrave Macmillan: 127–134.
Sklair, Leslie (ed.) (2021b). *The Anthropocene in Global Media. Neutralizing the Risk*. London & New York: Routledge (Routledge Studies in Environmental Communication and Media).

Sloterdijk, Peter (2017). Das Anthropozän – ein Prozess-Zustand am Rand der Erd-Geschichte? In: Renn, Jürgen & Bernd Scherer (eds.). *Das Anthropozän. Zum Stand der Dinge*. Berlin: Matthes & Seitz: 25–44.

Smil, Vaclav (2013). *Harvesting the Biosphere. What We Have Taken from Nature*. Cambridge, Mass. & London: The MIT Press.

Smil, Vaclav (2019). *Growth. From Microorganisms to Megacities*. Cambridge, Mass. & London: The MIT Press.

Smil, Vaclav (2020). *Numbers Don't Lie. 71 Things You Need to Know About the World*. London: Viking.

Smil, Vaclav (2021). *Grand Transitions. How the Modern World was made*. Oxford: Oxford University Press.

Smil, Vaclav (2022). *How the World Really Works*. London: Viking.

Smith, Bruce D. & Melinda A. Zeder (2013). The Onset of the Anthropocene. *Anthropocene* 4: 8–13. https://doi.org/10.1016/j.ancene.2013.05.001

Smith, Claire (ed.) (^{2}2020). *Encyclopedia of Global Archaeology*. New York: Springer (Springer Reference).

Sodikoff, Genese Marie (ed.) (2012). *The Anthropology of Extinction. Essays on Culture and Species Death*. Bloomington, Ind.: Indiana University Press.

Soentgen, Jens (2018). *Ökologie der Angst*. Berlin: Matthes & Seitz.

Soentgen, Jens (2021a). *Der Pakt mit dem Feuer. Philosophie eines weltverändernden Bundes*. Berlin: Matthes & Seitz.

Soentgen, Jens (2021b). Umweltforschung im Anthropozän. Warum Interdisziplinarität unerlässlich ist. *Merkur* 75(860). 89–95.

Soentgen, Jens (2023). Abschied vom Feuer. *Merkur* 77(887). 81–89.

Sörlin, Sverker (2014). Environmental Turn in the Human Sciences. Will It Become Decisive Enough? *The Institute Letter* (1).12–13. https://www.ias.edu/ideas/2014/sorlin-environment

Sörlin, Sverker (2018). Reform and Responsibility – the Climate of History in Times of Transformaton. *Historisk Tijdskrift* 97(3). 7–23.

Sörlin, Sverker & Graeme Wynn (2016). Fire and Ice in the Academy. The Rise of Integrative Humanities. An Essay. *Literary Review of Canada* 24(6). 14–15. (https://reviewcanada.ca/magazine/2016/07/fire-and-ice-in-the-academy/).

Sommer, Volker (2016). Planet ohne Affen? Zur Zukunft unserer Mitprimaten. In: Wolfgang Haber, Martin Held & Markus Vogt (eds.). *Die Welt im Anthropozän. Erkundungen im Spannungsfeld zwischen Ökologie und Humanität*. München: Oekom: 67–78.

Sommer, Volker (2021). *Unter Mitprimaten. Ansichten eines Affenforschers*. Stuttgart: S. Hirzel Verlag.

SONA - Netzwerk Soziologie der Nachhaltigkeit (eds.). *Soziologie der Nachhaltigkeit*. Bielefeld: Transcript (Soziologie der Nachhaltigkeit, 1).

SONA (Anna Henkel, Björn Wendt, Thomas Barth, Cristina Besio, Katharina Block, Stefan Böschen, Sascha Dickel, Benjamin Görgen, Matthias Groß, Jens Köhrsen, Thomas Pfister, Matthias Schloßberger). (2021) Zur Einleitung: Kernaspekte einer Soziologie der Nachhaltigkeit. In: SONA - Netzwerk Soziologie der Nachhaltigkeit (eds.). *Soziologie der Nachhaltigkeit*. Bielefeld: Transcript (Soziologie der Nachhaltigkeit, 1). 9–31

Sonnemann Ulrich (2011). *Negative Anthropologie. Vorstudien zur Sabotage des Schicksals*. Frankfurt am Main: Syndikat (orig. Reinbek bei Hamburg: Rowohlt, 1969).

Spangenberg, Joachim H. (2011). Sustainability Science: A Review, an Analysis and Some Empirical Lessons. *Environmental Conservation* 38 (3). 275–287.

Spangenberg, Joachim H. (2014). China in the Anthropocene: Culprit, Victim or Last Best Hope for a Global Ecological Civilization? *BioRisk* 9: 1–37.

Spier, Fred (2010). *Big History and the Future of Humanity*. Malden, Mass. & Chichster: Blackwell.

Spivak, Gayatri Chakravorty (1985). The Rani of Sirmur: An Essay in Reading the Archives. *History and Theory* 24(3). 247–272.

Spivak, Gayatri Chakravorty (2003). *Death of a Discipline*. New York: Columbia University Press.

Spivak, Gayatri Chakravorty (2008). *Other Asias*. Malden, Mass. Etc.: Basil Blackwell.
Sprenger, Florian (2019). *Epistemologien des Umgebens. Zur Geschichte, Okologie und Biopolitik künstlicher ,environments'*. Bielefeld: Transcript.
Sprenger, Guido (2021). Can Animism Save the World? Reflections on Personhood and Complexity in the Ecological Crisis. *Sociologus* 71(1): 73–92.
Sprenger, Guido (2022). Counterintuitive Utopias. What modern society can learn from anthropology. *Paideuma* 68: 7–33.
Sprenger, Guido & Kristina Großmann (2018). Introductory Essay. Plural Ecologies in Southeast Asia. *Sojourn: Journal of Social Issues in Southeast Asia* 33(2). ix–xxii.
Springer, Anna-Sophie (2019). *Der Anthropozän-Wortschatz*. Bonn: Bundeszetrale für politische Bildung. https://www.bpb.de/gesellschaft/umwelt/anthropozaen/216925/das-woerterbuch-zum-anthropozaen)..
Springer, Anna-Sophie & Etienne Turpin (eds.) (2015). *Land & Animal & Nonanimal* Berlin: K. Verlag und Haus der Kulturen der Welt.
SQS International Commission on Stratigraphy, Subcommission on Quaternary Stratigraphy (2009). Annual Report 2009. http://quaternary.stratigraphy.org/wp-content/uploads/2018/04/SQSAnnual-report09.doc
Stanley, Stephen (1987). Extinction. New York: Scientific American Books.
Star, Susan Leigh & James Griesemer (1989). Institutional Ecology, "Translations" and Boundary Objects: Amateurs and Professionals in Berkeley´s Museum of Vertebrate Zoology, 1907–31. *Social Studies of Science* 19(3). 387–420.
Steffen, Will (2019). Mid-20th-Century 'Great Acceleration'. In: Jan Zalasiewicz, Colin Waters, Mark Williams & Colin P. Summerhayes (eds.). *The Anthropocene as a Geological Time Unit: A Guide to the Scientific Evidence and Current Debate*. Cambridge etc.: Cambridge University Press: 254–60.
Steffen, Will, Paul J. Crutzen & John R. McNeill (2007). The Anthropocene: Are Humans Now Overwhelming the Great Forces of nature? *A Journal of the Human Environment* 36(8). 614–621.
Steffen, Will, Regina Angelina Sanderson, Peter D. Tyson, Jill Jäger, Pamela A. Matson, Berrien Moore III, Frank Oldfield, Katherine Richardson, Hans-Joachim Schellnhuber, Billie L. Turner, & Robert J. Wasson (2004). *Global Change and the Earth System. A Planet under Pressure*. Berlin: Springer (Global Change – The IGBP Series).
Steffen, Will, Åsa Persson, L. Deutsch, Jan Zalasiewicz, Mark Williams, Katherine Richardson, Carole Crumley, Paul Crutzen, Carl Folke, L. Gordon & M. Molina (2011a). The Anthropocene: From Global Change to Planetary Stewardship. *Ambio A Journal of the Human Environment* 40(7). 739–761.
Steffen, Will, Jacques Grinevald, Paul J. Crutzen & John McNeill (2011b). The Anthropocene: Conceptual and Historical Perspectives. *Philosophical Transactions of the Royal Science Academy* 369: 842–867.
Steffen, Will, Wendy Broadgate, Lisa Deutsch, Owen Gaffney, & Cornelia Ludwig (2015a). The Trajectory of the Anthropocene: The Great Acceleration. *The Anthropocene Review* 2(1). 81–98.
Steffen, Will et al. (2015b). Planetary Boundaries: Guiding Human Development on a Changing Planet. Science 347(6223). 1259855.
Steffen, Will, Johan Rockström, Katherine Richardson, Timothy M. Lenton, Carl Folke, Diana Liverman, Colin P. Summerhayes, Anthony D. Barnosky, Sarah E.N, Michel Crucifix, Jonathan F. Donges, Ingo Fetzer, Steven J. Ladea, Marten Scheffer, Ricarda Winkelmann & Hans Joachim Schellnhuber (2018). Trajectories of the Earth System in the Anthropocene, Supportiug Information: Holocene Variability and Anthropocene Rates of Change. *Proceedings of the National Academy of Sciences* 115(33). 8252–8259.
Steffen, Will, Katherine Richardson, Johan Rockström, Hans-Joachim Schellnhuber, Opha Pauline Dube, Sébastien Dutreuil, Timothy M. Lenton & Jane Lubchenco (2020). The Emergence and Evolution of Earth System Science. *Nature Reviews Earth & Environment* 1: 54–63.

Stegemann, Bernd (2021): *Die Öffentlichkeit und ihre Feinde*. Stuttgart: Klett-Cotta.

Steiner, Christian, Gerhard Rainer, Verena Schröder & Frank Zirkl (2022). *Mehr-als menschliche Geographien. Schlüsselkonzepte, Beziehungen und Methodiken*. Stuttgart: Franz Steiner Verlag. (https://library.oapen.org/handle/20.500.12657/57981).

Steiner, Sherrie M. (2018). *Moral Pressure for Responsible Globalization. Religious Diplomacy in the Age of the Anthropocene*. Boston etc.: Brill Academic (International Studies in Religion and Society, 30).

Steininger, Benjamin & Alexander Klose (2020). *Erdöl. Ein Atlas der Petromoderne*. Berlin: Matthes & Seitz.

Steinmetz, George (photos) & Andrew Revkin (text) (2020). *The Human Planet. Earth at the Dawn of the Anthropocene*. New York: Harry N. Abrams.

Stengers, Isabelle (2008). *Spekulativer Konstruktivismus*. Berlin: Merve (Internationaler Merve Diskurs: Perspektiven der Technokultur).

Stengers, Isabelle (2009). *Aux temps des catastrophes. Résister à la barbarie qui vient*. Paris: La Découverte (Les Empêcheurs de Panser en Rond).

Stengers, Isabelle (2013). *Une autre science est possible! Manifeste pour une ralentissement des sciences*. Paris: La Découverte.

Stensrud, Astrid B. & Thomas Hylland Eriksen (2019). Anthropological Perspectives on Global Economic and Environmental Crises in an Overheated World. In: Dies. (eds.). *Climate, Capitalism, and Communities. An Anthropology of Environmental Overheating*. London: Pluto Press: 1–20.

Stephens, Lucas, Dorion Fuller, Nicole Boivin, Torben Rick, Nicolas Gauthier, Andrea Kay, A. et al. (2019). Archaeological Assessment Reveals Earth's Early Transformation Through Land Use. *Science* 365: 897–902.

Stephens Lucas & Erle C. Ellis (2020). The Deep Anthropocene. A Revolution in Archaeology Has Exposed the Extraordinary Extent of Human Influence over Our Planet's Past and Its Future. *Aeon*, https://aeon.co/essays/revolutionary-archaeology-reveals-the-deepest-possible-anthropocene.

Sterelny, Kim (2021). *The Pleistocene Social Contract. Culture and Cooperation in Human Evolution*. New York: Oxford University Press.

Steward, Julian Haynes (1955). *Theory of Culture Change. The Methodology of Multilinear Evolution*. Urbana, Ill.: University of Illinois Press.

Stichweh, Rudolf (2018). *Sociocultural Evolution and Social Differentiation. The Study of the History of Society and the two Sociologies of Change and Transformation*. Bonn: Ms.

Stiegler, Bernard (2018). *The Neganthropocene*. London: Open Humanities Press.

Stiner, Mary C., Timothy Earle, Daniel Lord Smail & Andrew Shryock (2011). Scale. In: Daniel Shryrock & Daniel Lord Smail, with nine co-authors: *Deep History. The Architecture of Past and Present*. Berkeley etc.: University of California Press: 242–272.

Stirling, Paul (1965). Turkish Village. New York: John Wiley and Sons.

Stoetzer, Bettina (2018). Ruderal Ecologies. Rethinking Nature, Migration, and the Urban Landscape *Cultural Anthropology* 33(2). 295–323.

Stoffle, Richard W., Rebekka Toupal & Nieves Zedeno (2003). Landscape, Nature and Culture: A Diachronic Model of Human-Nature Adaptations. In: Helaine Selin (ed.). *Nature Across Cultures. Views of Nature and the Environment in Non-Western Cultures*. Dordrecht etc.: Kluwer Academic Publishers: 97–114 (Science across Cultures: The history of Non-Western Science, 4).

Stoller, Paul (2014). Welcome to the Anthropocene: Anthropology and the Political Moment. *Huffington Post*, 29.11.2014 (www.huffingtonpost.com/paul-stoller/welcome-to-the-anthropocene_b_6240786.html).

Stoner, Andony & Alexander Melathopoulos (2015). *Freedom in the Anthropocene. Twentieth-Century Helplessness in the Face of Climate Change*. Basingstoke & New York: Palgrave Macmillan (Palgrave Pivot).

Stoppani, Antonio (1873). *Corso di geologia. Vol. II (Geologia stratigrafica).* Milano: G. Bernardoni e G. Brigola Editori.

Strathern, Marilyn (2020). *Relations. An Anthropological Account.* Dubuque, Iowa: Duke University Press.

Strauss, Sarah & Benjamin S. Orlove (eds.) (2003). *Weather, Climate, Culture.* Oxford & New York: Routledge.

Streminger Gerhard (2021). Die *Welt gerät ins Wanken. Das Erdbeben von Lissabon im Jahre 1755 und seine Nachwirkungen auf das europäische Geistesleben. Ein literarischer Essay.* Frankfurt am Main: Alibri Verlag.

Subramanian, Meera (2019). Humans Versus Earth: The Quest to Define the Anthropocene. *Nature* 572: 168–170.

Suess, Eduard (1862). *Der Boden der Stadt Wien.* Wien: W. Braumüller.

Sullivan, Alexis P., Douglas W. Bird & George H. Perry (2017). Human Behaviour as a Long-term Ecological Driver of Non-Human Evolution. *Nature Ecology & Evolution.* (https://doi.org/10.1038/s41559-016-0065).

Summerhayes, Colin P. (2020). *Paleoclimatology. From Snowball Earth to the Anthropocene.* Chichester & New York, NY.: Wiley.

Suzman, James (2020). *Work. A History of How We Spend Our Time.*London etc.: Bloomsbury Publishing.

Svampa, Maristella (2020). *Die Grenzen der Rohstoffausbeutung. Umweltkonflikte und ökoterritoriale Wende in Lateinamerika.* Bielefeld: Bielefeld University Press & Transcript.

Swanson, Heather Anne (2018). Domestication Gone Wild: Pacific Salmon and the Disruption of the Domus. In: Heather Anne Swanson, Marianne Lien & Gro Ween (eds.). *Domestication Gone Wild. Politics and Practices of Multispecies Relations.* Durham, NC: Duke University Press.: 141–158.

Swanson, Heather Anne, Nils Bubandt & Anna L. Tsing (2015). Less Than One But More Than Many. Anthropocene as Science Fiction and Scholarship-in-the-Making. *Environment and Society: Advances in Research* 6: 149–166.

Swyngedouw, Erik (2007). Technonatural Revolutions: The Scalar Politics of Franco's Hydro-Social Dream for Spain, 1939–1975. *Transactions of the Institute of British Geographers, New Series* 32 (1). 9–28.

Swyngedouw, Erik & Henrik Ernstson (2018). Interrupting the Anthropo-ObScene: Immuno-Biopolitics and Depoliticizing Ontologies in the Anthropocene. *Theory, Culture & Society* 35(6). 3–30.

Syvitski, Jaia, Colin N. Waters, John Day, John D. Milliman, Colin Summerhayes, Will Steffen, Jan Zalasiewicz, Alejandro Cearreta, Agnieszka Gałuszka, Irka Hajdas, Martin J. Head, Reinhold Leinfelder, John R. McNeill, Clément Poirier, Neil Rose, William Shotyk, Michael Wagreich & Mark Williams (2020). Extraordinary Human Energy Consumption and Resultant Geological Impacts Beginning Around 1950 CE Initiated the Proposed Anthropocene Epoch. *Communications Earth and Environment* 1: 1–13.

Syvitsky, James P. M. (2012). Anhropocene. An Epoch of Our Making. *Global Change Magazine* 78 http://www.igbp.net/news/features/features/anthropoceneanepochofourmaking.5.1081640c135c7c04eb480001082.html.

Szeman, Imre & Dominic Boyer (eds.) (2017). *Energy Humanities. An Anthology.* Baltimore, Mar: Johns Hopkins University Press.

Szeman, Imre, Jennifer Wenzel & Patricia Yaeger(eds.) (2017). *Fueling Culture. 101 Words for Energy and Environment.* New York: Fordham University Press.

Szerszynski, Bronislaw (2003). Technology, Performance, and Life Itself: Hannah Arendt and the Fate of Nature. *Sociological Review* 51: 203–218.

Szerszynski, Bronislaw (2017). Gods of the Anthropocene: Geo-Spiritual Formations in the Earth's New Epoch. *Theory, Culture and Society* 34(2–3).253–275.

Szerszynski , Bronislaw (2019). How the Earth Remembers and Forgets. In: Adam Bobbette & Amy Donovan (eds.). *Political Ecology. Active Stratigraphies and the Making of Life*. London: Palgrave Macmillan: 219–236.

Taddei, Renzo, Karen Shiratori & Rodrigo C. Bulamah (2022). Decolonizing the Anthropocene. In: Hilary Callan & Simon Coleman (eds.). *The International Encyclopedia of Anthropology*. London: John Wiley & Sons (https://doi.org/10.1002/9781118924396.wbiea2519).

Tainter, Joseph A. (1988). *Collapse of Complex Societies*. Cambridge: Cambridge University Press. (New Studies in Archaeology)

Tang, Shiping (2020). *On Social Evoliution. Phenomenon and Paradigm*. London & New York: Routledge.

Tansley, A. G. (Arthur George) (1923). *Practical Plant Ecology. A Guide for Beginners in Field Studies of Plant Communities*. London: George Allen & Unwin. New York: Dodd, Mead & Company.

Tansley A. G. (Arthur George) (1939). *The British Isles and Their Vegetation*. Cambridge: Cambridge University Press.

Tauger, Mark B. ([2]2021). *Agriculture in World History*. London & New York: Routledge (Themes in World History).

Testot, Laurent (2021). *Die Globalgeschichte des Menschen. Von Faustkeil bis zur Digitalisierung*. Stuttgart: Reclam.

Testot, Laurent & Nathanaël Wallenhorst (2023). *Vortex. Faire face a l´Anthropocene*. Paris: Éditions Payot et Rivages.

The Age of Humans (2021). Living in the Anthropocene (https://www.mag.com/science-nature/age-humans-living-anthropocene-180952866/).

The Economist (2011). Welcome to the Anthropocene. The Geology of the Planet. *The Economist*, (https://www.economist.com/leaders/2011/05/26/welcome-to-the-anthropocene).

Thelen, Tatjana (2021). Entfesselte Verwandtschaft. Rezension zu Strathern 2020: *Soziopolis. Gesellschaft beobachten*. https://soziopolis.de/lesen/buecher/artikel/entfesselte-verwandtschaft/.

The Plantationocene Series. Plantation Worlds, Past and Present (2021). (https://edgeeffects.net/plantationocene-series-plantation-worlds)/.

Thiemeyer, Thomas (2020). Wissenschaftskommunikation in „postfaktischen" Zeiten. *Merkur* 74(854). 71–79.

Thomas, Julia Adeney (2014). History and Biology in the Anthropocene: Problems of Scale, Problems of Value. *American Historical Review* 119: 1587–607.

Thomas, Chris D. (2017a). *Inheritors of the Earth. How Nature is Thriving in an Age of Extinction*. London: Allen Lane.

Thomas, Julia Adeney (2017b). Why Do Only some Places Have History? Japan, the West, and the Geography of the Past. *Journal of World History* 28(2). 187–218.

Thomas, Julia Adeney (2019). Why the'Anthropocene'Is Not'Climate Change'and Why It Matters. *Asia Global Online* (https://www.asiaglobalonline.hku.hk/anthropocene-climate-change/).

Thomas, Julia Adeney (ed.) (2022) *Altered Earth. Getting the Anthropocene Right*. Cambridge etc. Cambridge University Press.

Thomas, Julia Adenay, Prasannan Parthasarathi, Rob Linrothe, Fa-Ti Fan, Kenneth Pomeranz, and Amitav Ghosh (2016). JAS Round tab. on Amitav Ghosh, The Great Derangement: Climate Change and the Unthinkable. *The Journal of Asian Studies* 75 (4). 929–955.

Thomas, Julia Adeney & Jan Zalasiewicz (2020). *Strata and Three Stories*. Munich: Rachel Carson Center (RCC Perspectives: Transformations in Environment and Society, 3).

Thomas, Julia Adeney, Mark Williams & Jan Zalasiewicz (2020). *The Anthropocene. A Multidisciplinary Approach*. Cambridge & Medford, Mass.: Polity Press.

Thomas, William L., Jr. (ed.) (1956). *Man's Role in Changing the Face of the Earth*. 2 Vols. Chicago: University of Chicago Press.

Thompson, Jessica C, David K. Wright, Sarah J. Ivory, Jeong-Heon Choi, Sheila Nightingale Alex Mackay, Flora Schilt, Erik Otárola-Castillo, Julio Mercader, Steven L. Forman, Timothy

Pietsch, Andrew S. Cohen, J. Ramón Arrowsmith, Menno Welling, Jacob Davis, Benjamin Schiery, Potiphar Kaliba, Oris Malijani, Margaret W. Blome, Corey A. O'Driscoll, Susan M. Mentzer, Christopher Miller Seoyoung Heo, Jungyu Choi, Joseph Tembo, Fredrick Mapemba, Davie Simengwa, Elizabeth Gomani-Chindebvu (2021). Early Human Impacts and Ecosystem Reorganization in Southern-Central Africa. *Science Advances* 7 (https://www.science.org/doi/10.1126/sciadv.abf9776).

Thornber, Karen L. (2012). *Ecoambiguity. Environmental Crises and East Asian Literatures*. Ann Arbor: University of Michigan Press.

Thornber, Karen L. (2014). Literature, Asia, and the Anthropocene: Possibilities for Asian Studies and the Environmental Humanities. *The Journal of Asian Studies* 73(4). 989–1000.

Thurner, Stefan, Rudolf Hanel & Peter Klimek (2018). *Introduction to the Theory of Complex Systems*. Oxford: Oxford Universuity Press. 224–312 Evlutionary Processes

Tickell, Crispin (2011). Societal Responses to the Anthropocene. *Philosophical Transactions of the Royal Society A: Mathematical, Physical and Engineering Sciences* 369: 926–932.

Thober, Bejamin (2019). „Natur unter Menschenhand"? Zur Historisierung des Anthropozoikum-Konzepts von Hubert Markl im Hinblick auf die aktuelle Anthropozän-Debatte. *Revue d'Allemagne et des pays de langue allemande* 51(2). 303–319.

Todd, Zoe (2015). Indigenizing the Anthropocene. In Heather Davis & Etienne Turpin (eds.). *Art in the Anthropocene. Encounters Among Aesthetics, Politics, Environments and Epistemologies*, London: Open Humanities Press (Critical Climate Change). 241–254.

Todd, Zoe (2018). The Decolonial Turn 2.0: The Reckoning. Anthro(dendum), 15.6. (https://anthrodendum.org/2018/06/15/the-decolonial-turn-2–0-the-reckoning/).

Toepfer, Georg (2013). *Evolution*. Stuttgart: Philipp Reclam Jun. (Grundwissen Philosophie).

Toepfer, Georg (2018). From Anthropocene to Mediocene? On the Use and Abuse of Stratifying the Earth's Crust by Mapping Time into Space. *Zeitschrift für Medien- und Kulturforschung* 9: 73–84.

Tomasello, Michael (2001). The Cultural Origins of Human Cognition. Cambridge, Mass.: Harvard University Press.

Tomasello, Michael (2019). *Becoming Human. A Theory of Ontogeny*. Cambridge, Mass. & London: The Belknap Press of Harvard University Press.

Toivanen T. K. Lummaa, P. Majava, V. Järvensivu, T. Lähde, Vaden & J. T. Eronen (2017). The many Anthropocenes: A transdisciplinary challenge for the Anthropocene research. *The Anthropocene Review* 4(3). 183–198.

Trautmann, Thomas R., Gillian Feeley-Harnik & John C. Mitani (2011). Deep Kinship. In: Daniel Shryock & Daniel Lord Smail, with nine co-authors: *Deep History. The Architecture of Past and Present*. Berkeley etc.: University of California Press: 160–188.

Traidl-Hoffmann Claudia, Christian Schulz, Martin Herrmann & Babette Simon (eds.) (2021). *Planetary Health. Klima, Umwelt und Gesundheit im Anthropozän*. Berlin etc.: Medizinisch Wissenschaftliche Verlagsgesellschaft und Springer.

Trigger, Bruce G. (1998). *Sociocultural Evolution. Calculation and Contingency*. Oxford: Blackwell Publishers.

Trischler, Helmuth (2013a). Introduction. In: Ders. (ed.). *Anthropocene. Envisioning the Future of the Age of Humans*. Munich: Rachel Carson Center for Environment and Society (RCC Perspectives, 3).

Trischler, Helmuth (ed.) (2013b). *Anthropocene. Envisioning the Future of the Age of Humans*. Munich: Rachel Carson Center for Environment and Society (RCC Perspectives, 3).

Trischler, Helmuth (2016a). Zwischen Geologie und Kultur: Die Debatte um das Anthropozän. In: Anja Bayer & Daniela Seel (eds.). *„all dies hier, Majestät, ist deins": Lyrik im Anthropozän. Anthologie*. Berlin: kookbooks: 269–286.

Trischler, Helmuth (2016b). The Anthropocene. A Challenge for the History of Science, Technology, and the Environment. In: *N.T.M. – Journal of the History of Science, Technology, and Medicine* 24(3). 309–335.

Trischler, Helmuth & Fabienne Will (2017). What Can Historians of Technology Contribute to the Anthropocene Debate? Technosphere, Technocene and the History of Technology. *ICON: Journal of the International Committee for the History of Technology* 23: 1–17.

Trischler, Helmuth & Fabienne Will (2019). Die Provokation des Anthropozäns. In: Martina Heßler & Heike Weber (eds.). *Provokationen der Technikgeschichte. Zum Reflexionszwang historischer Forschung*. Paderborn: Ferdinand Schöningh: 69–106.

Trischler, Helmuth & Fabienne Will (2020). Anthropozän. In: Martina Heßler & Kevin Liggieri (eds.). *Technikanthropologie. Handbuch für Wissenschaft und Studium*. Baden.-Baden: Nomos (Edition Sigma). 236–243.

Trouillot, Michel-Rolphe (1995). *Silencing the Past. Power and the Production of History*. Boston: Beacon.

Tsai, Yen-Ling (2019). Farming Odd Kin in Patchy Anthropocenes. *Current Anthropology*, Supplement 60: S342–S353.

Tsing, Anna Lowenhaupt (1993). *In the Realm of the Diamond Queen. Marginality in an Out-of-the-Way Place*. Princeton, N.J.: Princeton University Press.

Tsing, Anna Lowenhaupt (2000). The Global Situation. *Cultural Anthropology* 15(3). 327–360.

Tsing, Anna Lowenhaupt (2005). *Friction. An Ethnography of Global Connection*. Princeton, N.J.: Princeton University Press.

Tsing, Anna Lowenhaupt (2010a). Arts of Inclusion, or How to Love a Mushroom. *Manoa* 22(2). 191–203.

Tsing, Anna Lowenhaupt (2010b). Worlding the Matsutake Diaspora. Or, Actor-Network Theory Experiment with Holism? In: Ton Otto & Nils Bubandt (eds.). *Experiments with Holism*. Oxford: Wiley–Blackwell: 47–66.

Tsing, Anna Lowenhaupt (2012a). On Non-Scalability. The Living World Is Not Amenable to Precision-Nested Scales. *Common Knowledge* 18(3). 505–524.

Tsing, Anna Lowenhaupt (2012b). Unruly Edges: Mushrooms as Companion Species. *Environmental Humanities* 1: 141–154.

Tsing, Anna Lowenhaupt (2015). *The Mushroom at the End of the World. On the Possibility of Life in Capitalist Ruins*. Princeton & Oxford: Princeton University Press.

Tsing, Anna Lowenhaupt (2019a). Jenseits ökonomischer und ökologischer Standardisierung. In: Friederike Gesing, Michi Knecht, Michael Flitner & Katrin Amelang (eds.). *NaturenKulturen. Denkräume und Werkzeuge für neue politische Ökologien*. Bielefeld: transcript: 53–82.

Tsing, Anna Lowenhaupt (2019b). A Multispecies Ontological Turn? In: Keiichi Ōmura, , Grant Jun Otsuki, Shiho Satsuka & Atsuro Morita (eds.). *The World Multiple. The Quotidian Politics of Knowing and Generating Entangled Worlds*. London & New York: Routledge: 233–248 (Routledge Advances in Sociology).

Tsing, Anna Lowenhaupt (2019c). The political economy of the Great Acceleration, or, how I learned to stop worrying and love the bomb. In: Thomas Hylland Eriksen & Astrid Stensrud (eds.). *Climate, capitalism, and communities: an anthropology of environmental overheating*. London: Pluto Press: 22–40.

Tsing, Anna Lowenhaupt (2020). Die Erde, vom Menschen belagert. In: Marie-Luise Angerer & Naomie Gramlich (eds.). *Feministisches Spekulieren. Genealogien, Narrationen, Zeitlichkeiten*. Berlin: Kulturverlag Kadmos: 111–133.

Tsing, Anna Lowenhaupt, Heather Swanson, Elaine Gan & Nils Bubandt (eds.) (2017). *Arts of Living on a Damaged Planet. Ghosts of the Anthropocene. Monsters of the Anthropocene*. Minneapolis: University of Minnesota Press.

Tsing, Anna Lowenhaupt, Andrew S. Mathews & Nils Bubandt (2019). Patchy Anthropocene: Landscape Structure, Multispecies History, and the Retooling of Anthropology. An Introduction to Supplement 20. *Current Anthropology* 60(Supplement 20). S186–S197.

Tsing, Anna Lowenhaupt, Jennifer Deger, Alder Saxena Keleman & Feifei Zhou (curs. & eds.) (2020). *Feral Atlas. The More-Than-Human Anthropocene*. Stanford, Cal.: Stanford University Press. (https://feralatlas.org/).

Tsing, Anna Lowenhaupt, Jennifer Deger, Alder Keleman Saxena & Feifei Zhou (2024). *Field Guide to the Patchy Anthropocene: The New Nature*. Stanford, Cal.: Stanford University Press.

Turpin, Etienne (ed.) (2013). *Architecture in the Anthropocene. Encounters Among Design, Deep Time, Science and Philosophy*. Ann Arbor: Open Humanities Press.

Turchin, P. , Th. E. Currie, H. Whitehouse, P. François, K. Feeney, D. Mullins, D. Hoyer, Ch. Collins, St. Grohmann, P. Savage, G. Mendel-Gleason, E. Turner, A. Dupeyron, E. Cioni, J. Reddish, J. Levine, G. Jordan, E. Brandl, A. Williams, R. Cesaretti, M. Krueger, A. Ceccarelli, J. Figliulo-Rosswurm, P.-J. Tuan, P. Peregrine, A. Marciniak, J. Preiser-Kapeller, N. Kradin, A. Korotayev, A. Palmisano, D. Baker, J. Bidmead, P. Bol, D. Christian, C. Cook, A. Covey, G. Feinman, Á. D. Júlíusson, A. Kristinsson, J. Miksic, R. Mostern, C. Petrie, P. Rudiak-Gould, B. ter Haar, V. Wallace, V. Mair, L. Xie, J. Baines, E. Bridges, J. Manning, B. Lockhart, A. Bogaard, Ch. Spencer (2018). Quantitative Historical Analysis Uncovers a Single Dimension of Complexity That Structures Global Variation in Human Social Organization. *Proceedings of the National Academy of Sciences* 115: 144–151.

Turner II, B. L. (2023). *The Anthropocene. 101 Questions and Answers for Understandung the Human Impact on the Global Environment*. Newcaste uponTyne: Agenda Publishing.

Turner, Derek D. (2011). *Paleontology. A Philosophical Introduction*. Cambridge etc.: Cambridge University Press (Cambridge Introductions to Philosophy and Biology).

Turner, Fred (2006). *From Counterculture to Cyberculture. Stewart Brand, the Whole Earth Network, and the Rise of Digital Utopianism*. Chicago: University of Chicago Press.

Turner, Jonathan H. (2021). *On Human Nature. The Biology and Sociology of What Made us* Human. London & New York: Routledge.

Tylor, Edward (2005). *Die Anfänge der Cultur. Untersuchungen über die Entwicklung der Mythologie, Philosophie, Religion, Kunst und Sitte.* Stuttgart: Georg Olms (Nachdr. d. Ausg. Leipzig 1873).

Uekötter, Frank (2020). *Im Strudel. Eine Umweltgeschichte der modernen Welt*. Frankfurt am Main: Campus Verlag.

Uekötter, Frank (2021). Wie das Anthropozän in die Welt kam. Zur Ökologie der Modewörter. *Politische Ökologie* 39(167). 18–23.

Uekötter, Frank (ed.) (2018). *Exploring Apocalyptica. Coming to Terms with Envoronmental Alarmism*. Pittsburgh, Pa.: Pittsburg University Press.

Uexküll, Jakob von (^{2}1921). *Umwelt und Innenwelt der Tiere*. Berlin: Springer.

Uhrqvist, Ola & Eva Lövbrand (2014). Rendering Global Change Problematic: The Constitutive Effects of Earth System Research in the IGBP and the IHDP. *Environmental Politics* 23(2). 339–356.

Uller, Tobias & Kevin Laland (2019). *Evolutionary Causation. Biological and Philosophical Reflections*. Cambridge, Mass. & The MIT Press (Vienna Series in Theoretical Biology, 23).

UNEP United Nations Environment Programme (2021). *Making Peace with Nature. A Scientific Blueprint to Tackle the Climate, Biodiversity and Pollution Emergencies*. Nairobi: UNEP https://www.unep.org/resources/making-peace-nature

UNESCO (2018a). Welcome to the Anthropocene". *UNESCO Courier* 2, April-June 2018 (https://www.bic.moe.go.th/images/stories/Courier_Apr-Jun_2018.pdf).

UNESCO (2018b). Lexikon. (https://en.unesco.org/courier/2018-2/lexicon-anthropocene).

Unmüßig, Barbara (2021). Mensch macht Epoche. Die Erzählung vom Anthropozän. *Politische Ökologie* 39(167). 24–29.

Urry, John (2010). Consuming the Planet to Excess. *Theory, Culture & Society* 27(2–3). 191–212.

Vale, Thomas R. (1998). The Myth of Humanized Landscape: An Example from Yosemite National Park: *Natural Areas Journal* 18(3). 231–236.

Van der Leeuw, Sander (2019). *Social Sustainability, Past and Future: Undoing Unintended Consequences for the Earth's Survival*. New York: Cambridge University Press (New Directions in Sustainability and Society).

Van Loyen, Ulrich (2019). Nachwort. In: Déborah Danowski, & Eduardo Viveiros de Castro (2019). *In welcher Welt leben? Ein Versuch über die Angst vor dem Ende*. Berlin: Matthes & Seitz: 158–161.
Van Schaik, Carel (2016). *The Primate Origins of Human Nature*. Hobken, N.J: Wiley-Blackwell (Foundations of Human Biology, 3).
Van Oyen, A. (2018). Material Agency. In: S. L. López Varela (ed.). *The Encyclopedia of Archaeological Sciences*. Oxford: Wiley, pp. 1–5 (https://doi.org/10.1002/9781119188230.saseas0363).
Van Valen, Leigh (1973). A New Evolutionary Law. *Evolutionary Theory* 1: 1–30.
Vann, Elizabeth F. (2013). Culture. In: James G. Carrier & Deborah B. Gewertz (eds.). *The Handbook of Sociocultural Anthropology*. London: Bloomsbury: 30–48.
Vayda, Andrew P. (1983). Progressive Contextualization: Methods for Research in Human Ecology. *Human Ecology* 11:265–281.
Vayda, Andrew P. (2008). Causal Explanation as a Research Goal: A Pragmatic View. In: Bradley B. Walters, Bonnie J. McCay, P. West & S. Lees (eds.). *Against the Grain. The Vayda Tradition in Human Ecology and Ecological Anthropology*. Lanham, Md: AltaMira Press: 317–367.
Vayda, Andrew P. (2013). Causal Explanation for Environmental Anthropologists. In: Helen Kopnina & Eleanor Ouimet (eds.). *Environmental Anthropology. Future Directions*. London, Routledge: 207–224.
Vayda, Andrew P. & Bonnie J. McCay (1975). New Directions in Ecology and Ecological Anthropology. *Annual Review of Anthropology* 4: 293–306.
Vergès, Françoise (2017). Racial Capitalocene. In: Gaye Theresa Johnson & Alex Lubin (eds.). *Futures of Black Radicalism*. Brooklyn, N.Y.: Verso: 72–82.
Vergès, Françoise (2020). *Dekolonialer Feminismus*. Wien: Passagen Verlag (Passagen Thema).
Vertovec, Steven & Robin Cohen (2002). Introduction: Conceiving Cosmopolitanism. In: Dies. (eds.). *Conceiving Cosmopolitanism. Theory. Context, and Practice*. Oxford etc.: Oxford University Press: 1–22.
Vetlesen, Arne Johan (2019). *Cosmologies of the Anthropocene. Panpsychism, Animism and the Limits of Posthumanism*. London: Routledge (Morality, Society and Culture).
Vernadskij, Vladimir L. (1997). *Der Mensch in der Biosphäre. Zur Naturgeschichte der Vernunft*. Frankfurt am Main: Peter Lang. Europäischer Verlag der Wissenschaften.
Vidas, Davor, Jan Zalasiewicz, Colin Summerhayes & Mark Williams (2020). Climate Change and the Anthropocene: Implications for the Development of the Law of the Sea. In: E. Johansen, S. Busch, & I. V. Jakobsen (eds.). The Law of the Sea and Climate Change. Solutions and Constraints. Cambridge: Cambridge University Press: 22–48.
Vince, Gaia (2014). Adventures in the Anthropocene. A Journey to the Heart of the Planet we Made. London, Chatto & Windus.
Visconti, Guido (2014). Anthropocene: Another Academic Invention? *Rendiconti Lincei* 25(3). 381–392.
Vitousek, Peter M., Harold A. Mooney, Jane Lubchenco & Jerry M. Melillo (1997). Human Domination of Earth's Ecosystems. *Science* 277(5325).494–499.
Viveiros de Castro, Eduardo Batalha (2004a). Perspectival Anthropology and the Method of Controlled Equivocation. In: *Tipití: Journal of the Society for the Anthropology of Lowland South America* 2(1). 3–20.
Viveiros de Castro, Eduardo B. (2004b). Exchanging Perspectives: The Transformation of Objects into Subjects in Amerindian Ontologies. *Common Knowledge* 10(3). 463–484.
Viveiros de Castro, Eduardo Batalha (2012). *Cosmological perspectivism in Amazonia and Eösewhere. Four lectures given in the Department of Social Anthropology, Cambridge University, February–March 1998*. Hau (Masterclass Series 1).
Viveiros de Castro, Eduardo B. (2019). On Models and Examples: Engineers and Bricoleurs in the Anthropocene. *Current Anthropology* 60(suppl. 20). S296–S308.
Völker, Oliver (2021). *Langsame Katastrophen. Eine Poetik der Erdgeschichte*. Wallstein Verlag.
Vogt, Markus (2016). Humanökologie – Neuinterpretation eines Paradigmas mit Seitenblick auif die Umweltenzyklika Laudato si´. In: Wolfgang Haber, Martin Held & Markus Vogt (eds.)

(2016). *Die Welt im Anthropozän. Erkundungen im Spannungsfeld zwischen Ökologie und Humanität*. München: Oekom: 93–104.

Von Blanckenburg, Friedhelm (2021). Leben im nahen Untergrund. *Forschung* 2/21: 10–15.

Von Engelhardt, Wolf & Jörg Zimmermann (1982). *Theorie der Geowissenschaft*. Paderborn etc.: Schöningh.

Von Humboldt, Alexander (2014). *Kosmos. Entwurf einer physischen Weltbeschreibung*. Berlin: Eichborn (Die Andere Bibliothek, Folio14) (orig. Stuttgart & Tübingen: J. G. Cotta´scher Verlag, 1845).

Von Redecker, Eva (2020). *Revolution für das Leben. Philosophie der neuen Protestformen*. Frankfurt am Main: S. Fischer Verlag.

Von Storch, Hans & Werner Krauß (2013). *Die Klimafalle. Die gefährliche Nähe von Klimaforschung und Politik*. München: Hanser Verlag.

Vries, Peer H. H. (2001). Are Coal and Colonies Really Crucial? Kenneth Pomeranz and the Great Divergence. *Journal of World History* 12(2, Fall). 407–446.

Wagner, Peter (2021). In den Klimawandel hineingeschlittert? Rezension zu "The Climate of History in a Planetary Age" von Dipesh Chakrabarty. *Soziopolis: Gesellschaft beobachten*. (https://nbn-resolving.org/urn:nbn:de:0168-ssoar-80632-8).

Wagreich, Michael & Erich Draganits (2018). Early Mining and Smelting Lead Anomalies in Geological Archives as Potential Stratigraphic Markers for the Base of an Early Anthropocene. *The Anthropocene Review*, 5(2). 177–201.

Wagreich, Michael & Erich Draganits (2019). Pre-industrial Revolution Start Dates for the Anthropocene. In: Jan Zalasiewicz, Colin Waters, Mark Williams & Colin P. Summerhayes (eds.). *The Anthropocene as a Geological Time Unit: A Guide to the Scientific Evidence and Current Debate*. Cambridge etc.: Cambridge University Press: 246–250.

Wainwright, John (2009). Earth-System Science. In: Noel Castree, David Demeritt, Diana Liverman & Bruce Rhoads (eds.). A *Companion to Environmental Geography*. Oxford: Wiley-Blackwell: 145–67 (Blackwell Companions to Geography).

Walker, Mike J., C. M. Berkelhammer, S. Björck, L. C. Cwynar, D. A. Fisher, A. J. Long, J. J. Lowe, R. M. Newnham, S. O. Rasmussen & H. Weiss (2012). Formal Subdivision of the Holocene Series/Epoch: A Discussion Paper by a Working Group of INTIMATE (Integration of ice-core, marine and terrestrial records) and the Subcommission on Quaternary Stratigraphy (International Commission on Stratigraphy). *Journal of Quaternary Science* 27 (7). 649–659 (https://doi.org/10.1002/jqs.2565).

Walker, Michael J.C., Andrew M. Bauer, Matthew Edgeworth, Erle C. Ellis, Stanley C. Finney, Philip L. Gibbard & Mark Maslin (2024). The Anthropocene is best understood as an ongoing, intensifying, diachronous event. *Boreas* 53(1). 1–3 (https://doi.org/10.1111/bor.12636).

Walker, Mike (= Michael), Phil (Philip) Gibbard & John Lowe (2015). Comment on When did the Anthropocene begin? A mid-twentieth century boundary is stratigraphically optimal" by Jan Zalasiewicz et al. Quaternary International, 383, 196–203. *Quarternary International* 383: 204–207.

Wallace, Alfred Russel (1910). *The World of Life. A Manifestation of Creative Power, Directive Mind and Ultimate Purpose*. London: Chapman & Hall).

Wallace, Alfred Russel (2019). *The Wonderful Century. The Age of New Ideas in Science and Innovation*. Abingdon & New York: Routledge (Routledge Library Editions: Science and Technology in the Nineteenth Century, 10, firstly London: Allen & Unwin, Swan Sonnenschein, 1898).

Wallenhorst, Nathanaël (2019). *L'Anthropocène décodé pour les humains*. Paris: Le Pommier (Essai Pommier!).

Wallenhorst, Nathanaël (2020). *La vérité sur l'anthropocène*. Paris: Éditions Le Pommier/ Humensis.

Wallenhorst, Nathanaël (2023). *A Critical Theory of the Anthropocene*. Berlin: Springer VS (Anthropocene – Humanities and Social Science).

Wallenhorst, Nathanaël & Christoph Wulf (eds.) (2023). *Handbook of the Anthropocene. Humans between Heritage and Future*. Berlin etc.: Springer Nature.

Wallerstein, Immanuel Maurice (1988). *Das moderne Weltsystem. Die Anfänge kapitalistischer Landwirtschaft und die europäische Weltökonomie im 16. Jahrhundert*. Berlin: Syndikat Verlag.

Wallerstein, Immanuel (2004). *World-System Analysis. An Introduction*. Durham & London: Duke University Press.

Wallsten, Björn (2015). *The Urk World. Hibernating Infrastuctures and the Quest for Urban Mining*. Linköping: Linköping University.

Walton, Samantha (2020). Feminism's Critique of the Anthropocene. In J.ennifer Cooke (ed.). *The New Feminist Literary Studies*. Cambridge: Cambridge University Press: 113–128 (Twenty-First-Century Critical Revisions).

Walters, Bradley B., Bonnie J. McCay, Paige West & Susan Lees (eds.) (2008). *Against the Grain: The Vayda Tradition in Human Ecology and Ecological Anthropology*. Lanham, Mar.: AltaMiraPress.

Walters, Bradley B. & Andrew P. Vayda (2018). Event Ecology. In: Hilary Callan (general ed.). The International Encyclopedia of Anthropology. London etc: Wiley-Blackwell: 1–10 (DOI: 9781118924396.wbiea1420).

Wanner, Heinz (^{2}2016). *Klima und Mensch. Eine 12'000-jährige Geschichte*. Bern: Haupt Verlag.

Wapner, Paul (2012). After Nature: Environmental Politics in a Postmodern Age. In: Paul Dauvergne (ed.). *Handbook of Global Environmental Politics*. Cheltenham, UK: Edward Elgar Publishing: 431–442.

Wapner, Paul (2014). The Changing Nature of Nature: Environmental Politics in the Anthropocene. *Global Environmental Polititics* 14(4). 36–54.

Warde, Paul, Libby Robin & Sverker Sörlin (2017). Stratigraphy for the Renaissance: Questions of Expertise for "the Environment" and "the Anthropocene". The *Anthropocene Review* 4(3). 246–258.

Warde, Paul, Libby Robin & Sverker Sörlin (2018). *The Environment. A History of the Idea*. Baltimore, Mar.: Johns Hopkins University Press.

Wark, McKenzie (2017). *Molekulares Rot. Theorie für das Anthropozän*. Berlin: Matthes & Seitz. (orig: "Molecular Red. Theory for the Anthropocene", London & New York, Verso, 2015).

Waters, Colin N. & Jan A. Zalasiewicz (2018). Concrete: The Most Abundant Novel Rock Type of the Anthropocene. In: Dominick A. DellaSala & Michael I. Goldstein (chief. eds.). *Encyclopedia of the Anthropocene*. Dordrecht etc.: Elsevier: 75–85.

Waters, Colin N., Jan A. Zalasiewicz, Mark Williams. Erle Ellis, A. M. Snelling (2014). A *Stratigraphical Basis for the Anthropocene*. London: Geological Society (Special Publications, 395).

Waters, Colin N., Jan Zalasiewicz, Colin Summerhayes, Anthony D. Barnosky, Clément Poirier, Agnieszka Galuszka, Alejandro Cearreta et 17 al. (2016). Review Summary: The Anthropocene is Functionally and Stratigraphically Distinct from the Holocene. *Science* 351(6269). aaad2622.

Waters, Colin N., Jan Zalasiewicz, Colin Summerhayes, Ian J. Fairchild, Neil L. Rose, Neil J. Loader, William Shotyk, Alejandro Cearreta, Martin J. Head, James P.M. Syvitski, Mark Williams, Michael Wagreich, Anthony D. Barnosky, Reinhold Leinfelder, Catherine Jeandel, Agnieszka Gałuszka, Juliana A. Ivar do Sul, Felix Gradstein, Will Steffen, John R. McNeill, Scott Wing, Clément Poirier, Matt Edgeworth (2018). A Global Boundary Stratotype Sections and Points (GSSPs) for the Anthropocene Series: Where and How to Look for a Potential Candidate. Earth-Science Reviews, 178: 379–429.

Waters, Colin et al. (2023a). Executive Summary The Anthropocene Epoch and Crawfordian Age: proposals by the Anthropocene Working Group (https://eartharxiv.org/repository/view/6853/).

Waters, Colin N., Mark Williams, Jan Zalasiewicz, Simon D. Turner, Anthony D. Barnosky, Martuin J. Head, Scott L. Wing, Michael Wagreich, Will Steffen, Colin P. Summerhayes, Andrew B. Cundy, Jens Zinke, Barbara Fiałkiewicz-Kozieł, Reinhold Leinfelder, Peter K. Haff, John R. McNeill, Neil L. Rose, Irka Hajdas, I., Fancine M. G. McCarthy, Andrew Cearreta, Agnieszka Gałuszka, Jan Syvitski, Y. Han An, Z., Ian J. Fairchild, J.A. Ivar do Sul, & Catherine Jeandel (2022). Epochs, events and episodes: Marking the geological impact of humans. *Earth Science Reviews* 234, (104171 https://doi.org/10.1016/j.earscirev.2022.104171).

Waters, Colin N., Martin J. Head, Jan Zalasiewicz Francine M.G. McCarthy, Scott L. Wing, Peter K. Haff, Mark Williams, Anthony D. Barnosky, Barbara Fiałkiewicz-Kozieł, Reinhold

Leinfelder, J. R. McNeill, Neil L. Rose, Will Steffen, Colin P. Summerhayes, Michael Wagreich, A. Zhisheng , Alejandro Cearreta, Andrew B. Cundy, Ian J. Fairchild, Agnieszka Gałuszka, Irka Hajdas, Yongming Han, Juliana A. Ivar do Sul, Catherine Jeandel, Jaia Syvitsku, Simon D. Turner & Jens Zinke: A Response to Merritts et al. (2023b). The Anthropocene is complex. Defining it is not. *Earth-Science Reviews* 238 (https://doi.org/10.1016/j.earscirev.2023.104335).

Waters, H. (2016). Where in the World Is the Anthropocene. *Smithsonian Magazine*. (https://www.smithsonianmag.com/science-nature/age-humans-living-anthropocene-180952866/).

Waterton, Emma ([2]2019). More-than-Representational Landscapes. In: Peter Howard, Ian Thompson, Emma Waterton & Mick Atha (eds.). *The Routledge Companion to Landscape Studies*. London: Routledge: 91–101.

Waterton, Emma & Saul Hayley Saul (2020). Ghosts of the Anthropocene: Spectral Accretions at the Port Arthur Historic Site. *Landscape Research*__: 1–15. https://doi.org/10.1080/01426397.2020.1808957

Watson-Verran, Helen & David Turnbull (1995). Science and Other Indigenous Knowledge Systems. In: Sheila Jasanoff, Gerald E. Markle, James C. Peterson & Trevor Pinch (eds.). *Handbook of Science and Technology Studies*. Oxford: Sage Publications: 115–139.

WBGU, Wissenschaftlicher Beirat der Bundesregierung Globale Umweltveränderungen (2011a). *Welt im Wandel – Gesellschaftsvertrag für eine Große Transformation*. Berlin: WBGU

WBGU, Wissenschaftlicher Beirat der Bundesregierung Globale Umweltveränderungen (2011b). *Welt im Wandel – Gesellschaftsvertrag für eine Große Transformation. Zusammenfassung für Entscheidungsträger*. Berlin: WBGU.

WBGU, Wissenschaftlicher Beirat der Bundesregierung Globale Umweltveränderungen (2019). *Towards our Common Digital Future. Flagship Report*. Berlin: WBGU.

WBGU, Wissenschaftlicher Beirat der Bundesregierung Globale Umweltveränderungen (2016). *Der Umzug der Menschheit. Die transformative Kraft der Städte*. Berlin: WBGU.

WBGU, Wissenschaftlicher Beirat der Bundesregierung Globale Umweltveränderungen (2020). *Landwende im Anthropozän. Von der Konkurrenz zur Integration*. Berlin: WBGU.

WBGU, Wissenschaftlicher Beirat der Bundesregierung Globale Umweltveränderungen (2023). *Gesund leben auf einer gesunden Erde*. Berlin: WBGU.

Weber, Andreas (2019). Enlivenment. Toward a Poetics for the Anthropocene. Cambridge, Mass. & London: The MIT Press (Untimely Meditations, 16) (orig. „Enlivenment. Eine Kultur des Lebens. Versuch einer Poetik für das Anthropozän", Berlin: Matthes & Seitz, Fröhliche Wissenschaft, 79, 2016).

Weber, Elke U. (2017). Breaking cognitive barriers to a sustainable future. *Nature Human Behaviour* 1, 0013 (https://doi.org/10.1038/s41562-016-0013).

Weiss, Philipp, Raffaela Schöbitz (Ill.) (2018). *Am Weltenrand sitzen die Menschen und lachen*. Roman, 5 Bände. Berlin: Suhrkamp.

Weisman, Alan (2008). *The World without us*. London: Picador.

Weller, Robert (2006). *Discovering Nature. Globalization and Environmental Culture in China and Taiwan*. Cambridge: Cambridge University Press.

Welk-Joerger, Nicole (2019). Restoring Eden in the Amish Anthropocene. Environmental Humanities 11 (1). 72–97.

Welsch, Wofgang (2019). Wohin treibt das Anthropozän? In: Ders.: *Wer sind wir?* Wien: New Academic Press: 144–163.

Welz, Gisela (2018). 'Environmental Orientations' and the Anthropology of the Anthropocene. *Anthropological Journal of European Cultures* 27(1). 40–44.

Welz, Gisela (2021). More-Than-Human-Futures: Towards a Relational Anthropology in/of the Anthropocene. *Hamburger Journal für Kulturanthropologie* 13: 36–46.

Wendt, Björn (2021). Utopien, Dystopien und Soziologien der Nachhaltigkeit. Grundrisse eines Forschungsprogramms und Mehrebenenmodells. In: SONA - Netzwerk Soziologie der Nachhaltigkeit (eds.). *Soziologie der Nachhaltigkeit*. Bielefeld: Transcript (Soziologie der Nachhaltigkeit, 1). 155–183.

Wenninger, Andreas, Fabienne Will, Sascha Dickel, Sabine Maasen & Helmuth Trischler (2019). Ein- und Ausschließen: Evidenzpraktiken in der Anthropozändebatte und der Citizen Science. In: Karin Zachmann & Sarah Ehlers (eds.). *Wissen und Begründen. Evidenz als umkämpfte Ressource in der Wissensgesellschaft*. Baden-Baden: Nomos: 31–58.

Werber, Niels (2014). Anthropozän: Eine Megamakroepoche und die Selbstbeschreibung der Gesellschaft. *Zeitschrift für Medien- und Kulturforschung* 5(2): 241–246.

Werbner, Pnina (2008). The Cosmopolitan Encounter: Social Anthropology and the Kindness of Strangers. In: Dies. (ed.). *Anthropology and the New Cosmopolitism. Rooted, Feminist and Vernacular Perspectives* Oxford etc.: Berg Publishers: 47–68 (Association of Social Anthropologists Monographs, 48).

Wergin, Carsten Holger (2018). Collaborations of Biocultural Hope: Community Science Against Industrialisation in Northwest Australia. *Ethnos. Journal of Anthropology* 83 (3). 455–472.

Werner, Oswald (1972). Ethnoscience. *Annual Review of Anthropology* 1: 271–308.

Werner, Oswald & Mark Schoepfle (1987). *Systematic Fieldwork* (2 vols.). Newbury Park: Sage Publications.

West, Page (2016). *Dispossession and the Environment. Rhetoric and Inequality in Papua New Guinea*. Durham & London: Duke University Press.

West, Geoffrey Brian (2017). *Scale. The Universal Laws of Growth, Innovation, Sustainability, and the Pace of Life in Organisms, Cities, Economies, and Companies*. New York, N.Y.: Penguin Press.

White, Frank (^{4}2021). *The Overview Effect: Space Exploration and Human Evolution*. O.O.: Multiverse Publishing.

White, Leslie Alvin (1945). Energy and the Evolution of Culture. *American Anthropologist* 45: 335–356.

White, Leslie Alvin (1949). *The Science of Culture. A Study of Man and Civilization*. New York: Farrar & Strauss, Grove Press.

White Jr., Lynn T. (1967). The Historical Roots of Our Ecological Crisis. *Science* 155(3767). 1203–1207.

White, Peter S. & S. T. A. Pickett (1985). Natural Disturbance and Patch Dynamics: An Introduction. In. S. T. A. Pickett and Peter S. White (eds.). *The Ecology of Natural Disturbance and Patch Dynamics*. San Diego, CA: Academic Press: 3–13.

Whitington, Jerome (2013). Fingerprint, Bellwether, Model Event: Climate Change as Speculative Anthropology. *Anthropological Theory* 13(4).308–328.

Whorf, Benjamin Lee (1956). *Thought and Reality*. Cambridge, Mass.: The MIT Press.

Whyte, Kyle P. (2017). Indigenous climate change studies: Indigenizing futures, decolonizing the Anthropocene. https://kylewhyte.marcom.cal.msu.edu/wp-content/uploads/sites/12/2018/07/IndigenousClimateChangeStudies.pdf

Whyte, Kyle P. (2018). Indigenous Science (Fiction) for the Anthropocene: Ancestral Dystopias and Fantasies of Climate Change Crises. *Environment & Planning E: Nature and Space* 1(1–2). 224–242.

Widdau, Christoph Sebastian (2021). *Einführung in die Umweltethik*. Ditzigen: Philipp Reclam Jun.

Wiertz, Thilo (2015a). *Politische Geographien heterogener Gefüge. Climate Engineering und die Vision globaler Klimakontrolle*. Heidelberg (Dissertation).

Wiertz, Thilo (2015b). Geoengineering. In: Sybille Bauriedl (eds.). *Wörterbuch Klimadebatte*. Bielefeld: Transcript: 87–94.

Wignall, Paul B. (2019). *Extinction. A Very Short Introduction*. Oxford: Oxford University Press (Very Short Introductions).

Wilke, Sabine & Japhet Johnstone (eds.) (2017). *Readings in the Anthropocene. The Environmental Humanities*. New York: Bloomsbury Academic (New Directions in German Studies, 18).

Wilkinson, B. H. (2005). Humans as Geologic Agents: A deep-Time Perspective. *Geology* 33: 161–164.

Will, Fabienne (2021). *Evidenz für das Anthropozän. Wissensbildung und Aushandlungsprozesse an der Schnittstelle von Natur-, Geistes- und Sozialwissenschaften*. Göttingen: Vandenhoeck & Ruprecht (Umwelt und Gesellschaft, 24).

Williams, Mark, Jan Zalasiewicz, Peter K. Haff, Christian Schwägerl, Anthony D. Barnosky & Erle C. Ellis (2015). The Anthropocene Biosphere. *The Anthropocene Review* 1: 1–24.

Williams, Mark, Jan Zalasiewicz (eds.) (2011). The Anthropocene: A New Epoch of Geological Time? *Philosophical Transactions of the Royal Society A: Mathematical, Physical and Engineering Sciences* 369(1938). 85–843.

Williams, Mark, Jan Zalasiewicz, Alan Haywood & Mike Ellis (eds.) (2011). Theme Issue 'The Anthropocene: A New Epoch of Geological Time?' *Philosophical Transactions of the Royal Society A: Mathematical, Physical and Engineering Sciences* 369(1938). 1056–1084.

Williams, Mark, Jan Zalasiewicz, Colin N. Waters, Matt Edgeworth, Carys Bennett, Anthony D. Barnosky, Erle C. Ellis, Michael. A. Ellis, Alejandro Cearreta, Peter K. Haff, Juliana A. Ivar do Sul, Reinhold Leinfelder, John. R. McNeill, J. R. Eric Odada, Naomi Oreskes, Andrew Revkin, Daniel de B. Richter, Will Steffen, Colin Summerhayes, James P. Syvitski, Davor Vidas, Michael Wagreich, Scott L. Wing, Alexander P. Wolfe & An Zhisheng (2016). The Anthropocene: A Conspicuous Stratigraphical Signal of Anthropogenic Changes in Production and Consumption Across the Biosphere. *Earth's Future*, 4: 34–53. https://agupubs.onlinelibrary.wiley.com/doi/full/10.1002/2015EF000339

Williams, Mark, Mark Williams, Matt Edgeworth, Jan Zalasiewicz, Colin N. Waters, Will Steffen, Alexander P. Wolfe, Nic Minter, Alejandro Cearreta, Agnieszka Galuszka, Peter Haff, John McNeill, Andrew Revkin, Daniel Richter, Simon Price & Colin Summerhayes (2020). Underground Metro Systems: A Durable Geological Proxy of Rapid Urban Population Growth and Energy Consumption During the Anthropocene. In: Craig Benjamin, Esther Quaedackers and David Baker (eds.). *The Routledge Companion to Big History. Milton Park & New York: Routledge* (Routledge Companions). 434–455.

Williams, Mark, Francine M. G McCarthy, Alejandro Cearreta, Martin J. Head, Reinhold Leinfelder, Jens Zinke, Anthony D. Barnosky, Kristine L. DeLong (2022). Biological and Paleontological Signatures of the Anthropocene. In: Christoph Rosol & Giulia Rispoli (eds.). *Anthropogenic Markers. Stratigraphy and Context*, Anthropocene Curriculum. Berlin: Max Planck Institute for the History of Science. https://doi.org/10.58049/04x5-fw34.

Williams, Michael (2006). *Deforesting the Earth. From Prehistory to Global Crisis, An Abridgment*. Chicago: The University of Chicago Press.

Williams, Raymond (1980). Ideas of Nature. In: Ders.: *Problems in Materialism and Culture*. London: Verso: 67–85.

Wilson, Edward O. (1984). *Biophilia. The Human Bond with Other Species*. Cambridge, Mass.: Harvard University Press.

Wilson, Edward O. (1998). *Consilience. The Unity of Knowledge*. New York: Alfred A. Knopf.

Wilson, Edward O. (2016). *Half Earth. Our Planet´s Fight for Life*. New York: W. W. Norton.

Wilson, Edward O. (2019). *Genesis. The Deep Origin of Societies*. New York: Norton.

WinklerPrins, Antoinette M. G. (2014). Terra Preta: The Mysterious Soils of the Amazon. In: G. Jock Churchman & Edward R. Landa (eds.). *The Soil Underfoot. Infinite Possibilities for a Finite Resource*. Boca Raton Fla.: CRC Press: 235–246.

Wirth, Sven, Anett Laue, Markus Kurth, Katharina Dornenzweig, Leonie Bossert & Karsten Balgar (eds.) (2016). *Das Handeln der Tiere. Tierliche Agency im Fokus der Human-Animal Studies*. Bielefeld: Transcript.

Wissen, Markus & Ulrich Brand (2022). Emanzipatorische Perspektiven im „Anthropozän". Über die Grenzen des grünen Kapitalismus und die Notwendigkeit einer radikalen Alternative. *PROKLA* # 207, 52(2). 263–281.

Witze, Alexandra (2024). Geologists Reject the Anthropocene as Earth´s New Epoch – after 15 Years of Debate. *Nature* 627: 249–250.

Wobser, Florian (ed.) (2024). *Anthropozän. Interdisziplinäre Perspektiven und philosohische Bildung*. Frankfurt am Main & New York: Campus Verlag.

Wolf, Eric R. (1982). *Europe and the People without History*. Berkeley: University of California Press.

Wolfstone, Irene Friesen (2019). Review of Donna Haraway, Staying With the Trouble: Making Kin in the Chthulucene. *Imaginations* 10(1). 389–391.

Wood, Barry (2020). Big History and the Study of Time. The Underlying Temporalities of Big History. In: Craig Benjamin, Esther Quaedackers and David Baker (eds.). *The Routledge Companion to Big History. Milton Park & New York: Routledge* (Routledge Companions). 37–56.

Worster, D. (2008). Environmentalism Goes Global. *Diplomatic History* 32: 639–641.

Wrigley, Edward Anthony (2016). *The Path to Sustained Growth: England's Transition from an Organic Economy to an Industrial Revolution*. Cambridge: Cambridge University Press.

Wu, Jianguo & Orie L. Loucks (1995). From Balance of Nature to Hierarchical Patch Dynamics: A Paradigm Shift in Ecology. *Quarterly Review of Biology* 70(4).439–466.

Wulf, Christoph (2020a). *Bildung als Wissen vom Menschen im Anthropozän*. Weinheim & Basel: Beltz Juventa.

Wulf, Christoph (2020b). Den Menschen neu denken im Anthropozän. Bestandsaufnahme und Perspektiven. *Paragrana. Internationale Zeitschrift für Historische Anthropologie* 29(1): 13–35.

Wuscher, Patrice, Christophe Jorda, Quentin Borderie, Nathalie Schneider & Laurent Bruxelles (2020). De la formation geologique a la tranchee: trouver et comprendre les sites archeologiques menaces par les travaux d´ amenagement du territoire. *Archimède. Archéologie et Histoire Ancienne* 7: 158–175.

Yan, Yuzhen, Michael L. Bender, Edward J. Brook, Heather M. Clifford, Preston C. Kemeny, Andrei V. Kurbatov, Sean Mackay, Paul A. Mayewski,Jessica Ng, Jeffrey P. Severinghaus & John A. Higgins. (2019). Two-Million-Year-Old Snapshots of Atmospheric Gases from Antarctic Ice. *Nature*, 574: 663–666.

Yoneyama, Shoko (2019). *Animism in Contemporary Japan: Voices for the Anthropocene from post-Fukushima Japan*. London : Routledge (Routledge Contemporary Japan).

Yoneyama, Shoko (2021). Miyazaki Hayao's Animism and the Anthropocene Shoko Yoneyama. *Theory, Culture & Society* 38(7–8). 251–266.

Yusoff, Karen (2013). Geologic Life: Prehistory, Climate, Futures in the Anthropocene. *Environment and Planning D: Society and Space* 31(5). 779–795.

Yusoff, Karen (2015). Geologic Subjects: Nonhuman Origins, Geomorphic Aesthetics, and the Art of Becoming Inhuman. Cultural Geographies 22(3). 383–407.

Yusoff, Karen (2016). Anthropogenesis: Origins, and Endings in the Anthropocene. *Environment and Planning D: Society and Space* 31(5). 779–795.

Yusoff, Karen (2018). *A Billion Black Anthropocenes or None*. Minneapolis, Mn.: University of Minnesota Press (Forerunners: Ideas First).

Zalasiewicz, Jan (2009). *The Earth After Us. What Legacy Will Humans Leave in the Rocks?"*. New York: Oxford University Press.

Zalasiewicz, Jan (2015). Die Einstiegsfrage: Wann hat das Anthropozän begonnen? In: Jürgen Renn & Bernd Scherer (eds.). *Das Anthropozän. Zum Stand der Dinge*. Berlin: Matthes & Seitz,: 160–180.

Zalasiewicz, Jan (2017). Geologie: eine vielschichtige Angelegenheit. *Die Zukunft der Menschheit Spektrum Spezial.* 3.17: 52–60.

Zalasiewicz, Jan (2022). *How to Read a Rock. Our Planet´s Hidden Stories*. Washington, D.C.: Smithsonian Books.

Zalasiewicz, Jan, Martin J. Head, Colin N. Waters, Simon D. Turner, Peter K. Haff , Colin Summerhayes, Mark Williams, Alejandro Cearreta, Michael Wagreich, Ian Fairchild, Neil L. Rose, Yoshiki Saito, Reinhold Leinfelder, Barbara Fiałkiewicz-Kozieł, Zhisheng An, Jaia Syvitski, Agnieszka Gałuszka, Francine M. G. McCarthy, Juliana Ivar do Sul, Anthony Barnosky, Andrew B. Cundy, J. R. McNeill, & Jens Zink (2024a). The Anthropocene within the Geological Time Scale: A response to fundamental questions. *Episodes* 47(1). 65–83.

Zalasiewicz, Jan, Julia Adeney Thomas, Colin N. Waters, Simon Turner & Martin J. Head (2024b). What should theAnthropocene mean? Nature 632: 680–684.

Zalasiewicz, Jan, Scott L. Wing & the Anthropocene Working Group (2024c). What Myths About the Anthropocene Get Wrong. Smithsonian Magazine (https://www.smithsonianmag.com/smithsonian-institution/what-myths-about-the-anthropocene-get-wrong-180984181).

Zalasiewicz, Jan, Mark Williams, et al. (2008a). Are We Now Living in the Anthropocene? *GSA Today* 18 (2). 4–8.

Zalasiewicz, Jan Mark Williams, Alan Smith, Tiffany L. Barry, Angela L. Coe, Paul R. Bown, Patrick Brenchley, David Cantrill, Andrew Gale, Philip Gibbard, F. John Gregory, Mark W. Houndslow, Andrew C. Kerr, Paul Pearson, Robert Knox, John Powell, Colin Waters, John Marshall, Michael Oates, Peter Rawson & Philip Stone (2008b). Are We Now Living in the Anthropocene? *GSA Today* 18 (2): 4–8.

Zalasiewicz, Jan, Colin Waters & Will Steffen (2021a). Remembering the Extraordinary Scientist Paul Crutzen (1933–2021). *Scientific American*. 5. (2.2021 https://www.scientificamerican.com/article/remembering-the-extraordinary-scientist-paul-crutzen-1933-2021/).

Zalasiewicz, Jan, Colin N. Waters, Erle C. Ellis, Martin J. Head, Davor Vidas, Will Steffen, Julia Adeney Thomas, Eva Horn, Colin P. Summerhayes, Reinhold Leinfelder, J. R. McNeill, Agnieszka Gałuszka, Mark Williams, Anthony D. Barnosky, Daniel de B. Richter, Philip L. Gibbard, Jaia Syvitski, Catherine Jeandel, Alejandro Cearreta, Andrew B. Cundy, Ian J. Fairchild, Neil L. Rose, Juliana A. Ivar do Sul, William Shotyk, Simon Turner, Michael Wagreich & Jens Zinke (2021b). The Anthropocene: Comparing Its Meaning in Geology (Chronostratigraphy) with Conceptual Approaches Arising in Other Disciplines. *Earth's Future* 9:1–25.

Zalasiewicz, Jan, Colin N. Waters, Colin Summerhayes & Mark Williams (2018a). The Anthropocene. *Geology Today* 34(5). 177–181.

Zalasiewicz, Jan, Colin Waters, Martin J. Head, Will Steffen, J. P. Syvitski, Davor Vidas, Colin Summerhayes & and Mark Williams (2018b). The Geological and Earth System Reality and the Anthropocene (Comment on Bauer & Ellis 2018). *Current Anthropology* 59(2). 220–223.

Zalasiewicz, Jan, Mark Williams, Colin N. Waters, Anthony D. Barnosky & Peter Haff (2014a). The Technofossil Record of Humans. *Anthropocene Review* 1(1). 34–43.

Zalasiewicz, Jan, Colin N. Waters & Mark Williams (2014b). Human Bioturbation, and the Subterranean Landscape of the Anthropocene. *Anthropocene* 6: 3–9.

Zalasiewicz, Jan, Colin N. Waters, Colin, Mark Williams, A.D. Barnosky, A. Cearreta, Paul Crutzen, Erle Ellis, M.A. Ellis, I.J. Fairchild, Jacques Grinevald & Peter K. Haff (2015). When Did the Anthropocene Begin? A Mid-Twentieth Century Boundary Level is Stratigraphically Optimal. *Quaternary International* 383: 196–203.

Zalasiewicz, Jan, Colin N. Waters, Colin P. Summerhayes, Alexander P. Wolfe, Anthony D. Barnosky, Alejandro Cearreta, Paul Crutzen, Erle Ellis, Ian J. Fairchild, Agnieszka Gałuszka, Peter Haff, Irka Hajdas, Martin J. Head, Juliana A. Ivar do Sul, Catherine Jeandel, Reinhold Leinfelder, John R. McNeill, Cath Neal, Eric Odada, Naomi Oreskes, Will Steffen, James Syvitski, Davor Vidas, Michael Wagreich & Mark Williams (2017a). The Working Group on the Anthropocene: Summary of Evidence and Interim Recommendations. *Anthropocene* 19: 55–60.

Zalasiewicz, Jan, Colin N. Waters, Alexander P. Wolfe, Anthony D. Barnosky, Alejandro Cearreta, Matt Edgeworth, Erle C. Ellis, Ian J. Fairchild, Felix M. Gradstein, Jacques Grinevald, Peter Haff, Martin J. Head, Juliana A. Ivar do Sul, Catherine Jeandel, Reinhold Leinfelder, John R. McNeill, Naomi Oreskes, Clément Poirier, Andrew Revkin, Daniel de B. Richter, Will Steffen, Colin Summerhayes, James P. M. Syvitski, Davor Vidas, Michael Wagreich, Scott Wing & Mark Williams (2017b). Making the Case for a Formal Anthropocene Epoch: An Analysis of Ongoing Critiques. Newsletters on Stratigraphy 50(2). 205–226.

Zalasiewicz, Jan, Colin Waters, Mark Williams & Colin P. Summerhayes (eds.) (2019a). *The Anthropocene as a Geological Time Unit: A Guide to the Scientific Evidence and Current Debate*. Cambridge etc.: Cambridge University Press.

Zalasiewicz, Jan, Colin N. Waters, Martin J. Head, Clément Poirier, Colin P. Summerhayes, Reinhold Leinfelder, Jacques Grinevald, Will Steffen, Jaia Syvitski, Peter Haff, John R. McNeill, Michael Wagreich, Ian J. Fairchild, Daniel D. Richter, Davor Vidas, Mark Williams, Anthony

D. Barnosky, Alejandro Cearreta (2019b). A Formal Anthropocene is Compatible With But Distinct From Its Diachronous Anthropogenic Counterparts: A Response to W.F. Ruddiman's "Three-Flaws in Defining a Formal Anthropocene". *Progress in Physical Geography* 43(3). 319–333.

Zalasiewicz, Jan, Colin Waters, Mark Williams ([2]2020a). The Anthropocene. In: Felix M. Gradstein, James G. Ogg, Mark Schmitz & Gabi Ogg (eds.). A *Geologic Time Scale 2020. Vol. 1.* Dordrecht: Elsevier.

Zalasiewicz, Jan, Colin Waters, Mark Williams ([2]2020b). The Anthropocene. In: Felix M. Gradstein, James G. Ogg, Mark Schmitz & Gabi Ogg (eds.). A *Geologic Time Scale 2020. Vol. 1.* Dordrecht: Elsevier.

Zalasiewicz, Jan, Colin Waters, Simon Turner, Mark Williams & Martin J. Head (2023). Anthropocene Working Group. In: Nathanaël Wallenhorst & Christoph Wulf (eds.). *Handbook of the Anthropocene. Humans between Heritage and Future*. Berlin etc.: Springer: 315–321.

Zalasiewicz, Jan, Mark Williams, Will Steffen & Paul Crutzen (2010). The New World of the Anthropocene. *Environmental Science & Technology* 44(7). 2228–2231.

Zalasiewicz, Jan, Mark Williams, Colin N. Waters, Anthony D. Barnosky, John Palmesino, Ann-Sofi Rönnskog, Matt Edgeworth, Cath Neal, Alejandro Cearreta, Erle C. Ellis, Jacques Grinevald, Peter K. Haff, Juliana A. Ivar do Sul, Catherine Jeandel, Reinhold Leinfelder, John R. McNeill, Eric Odada, Naomi Oreskes, Simon J. Price, Andrew Revkin, Will Steffen, Colin Summerhayes, Davor Vidas, Scott Wing & Alexander P. Wolfe (2017c). Scale and Diversity of the Physical Technosphere: A Geological Perspective. *The Anthropocene Review* 4(1). 9–22.

Zeder, Melinda A. (2018). Why Evolutionary Biology Needs Anthropology: Evaluating Core Assumptions of the Extended Evolutionary Synthesis. *Evolutionary Anthropology: Issues, News and Reviews* 27:267–284.

Zeller, Christian (2020). Eine ökosozialistische Perspektive entwickeln und umsetzen. Vorwort zur deutschen Ausgabe. In: Ian Angus: *Im Angesicht des Anthropozäns. Klima und Gesellschaft in der Krise.* Münster: Unrast Verlag: 9–22.

Ziegler, Suzie Svatek (2019). The Anthropocene in Geography. *Geographical Review* 109(2). 271–280.

Zinkina, Julia, David Christian, Leonid Grinin, Ilya Ilyin, Alexey Andreev, Ivan Aleshkovski, Sergey Shulgin & Andrey Korotayev (2019). A *Big History of Globalization. The Emergence of a Global World System.* Cham: Springer Nature (World-Systems Evolution and Global Futures).

Zong, Raymond (2024). Are We in the 'Anthropocene,' the Human Age? Nope, Scientists Say. *The New York Times*. (https://www.nytimes.com/2024/03/05/climate/anthropocene-epoch-vote-rejected.html).

Zottola, Angela & Claudio de Majo (2022). The Anthropocene: Genesis of a Term and Popularization in the Press. *Text & Talk* 42(4). 453–473.

Index

C. Antweiler, *Anthropology in the Anthropocene*, Anthropocene – Humanities and Social Sciences, https://doi.org/10.1007/978-3-031-74591-1

D

M

N

The manufacturer's authorised representative in the EU is Springer Nature Customer Service Centre GmbH, Europaplatz 3, 69115 Heidelberg, Germany. If you have any concerns regarding our products, please contact ProductSafety@springernature.com

Printed and bound by CPI Group (UK) Ltd, Croydon, CR0 4YY
01/10/2026
02225938-0008